STELLAR COLLAPSE

ASTROPHYSICS AND SPACE SCIENCE LIBRARY

VOLUME 302

STELLAR COLLAPSE

Edited by

CHRIS L. FRYER
Los Alamos National Laboratory,
Los Alamos, NM, U.S.A.
and
University of Arizona
Tucson, AZ, U.S.A.

KLUWER ACADEMIC PUBLISHERS
DORDRECHT / BOSTON / LONDON

A C.I.P. Catalogue record for this book is available from the Library of Congress.

ISBN 1-4020-1992-0 (HB)
ISBN 0-306-48599-0 (e-book)

Published by Kluwer Academic Publishers,
P.O. Box 17, 3300 AA Dordrecht, The Netherlands.

Sold and distributed in North, Central and South America
by Kluwer Academic Publishers,
101 Philip Drive, Norwell, MA 02061, U.S.A.

In all other countries, sold and distributed
by Kluwer Academic Publishers,
P.O. Box 322, 3300 AH Dordrecht, The Netherlands.

Printed on acid-free paper

Printed in the Netherlands.

This book is dedicated to the memory of Hirlo Hicks who taught me the conviction to pursue my dreams.

Contents

List of Figures

List of Tables

Contributing Authors

Shizuka Akiyama is a graduate student at the University of Texas at Austin.

Stephen W. Bruenn is a professor at Florida Atlantic University.

Robert Buras is a graduate student at the Max-Planck-Institut für Astrophysik in Garching.

Adam Burrows is a Professor of the Department of Astronomy, University of Arizona.

Christian Y. Cardall is a joint professor at National Laboratory and the University of Tennessee.

Jinsong Deng is a post-doc at the University of Tokyo.

Chris L. Fryer is a permanent staff scientist at Los Alamos National Laboratory and adjunct professor at the University of Arizona.

Mario Hamuy is a Carnegie Fellow at Carnegie Observatories.

Peter Höflich is a research scientist at the University of Texas at Austin.

Daniel E. Holz is a post-doctoral researcher at the University of Chicago.

Scott A. Hughes is an assistant Professor at the Massachusetts Institute of Technology.

Aimee L. Hungerford is a graduate student jointly at the University of Arizona and Los Alamos National Laboratory.

Koichi Iwamoto is an instructor at Nihon University.

H.-Thomas Janka is a permanent staff scientist at the Max-Planck-Institut für Astrophysik in Garching.

Konstantinos Kifonidis is a post-doctoral research assistant at the Max-Planck-Institut für Astrophysik in Garching.

Alexei Khokhlov is a staff scientist at the Naval Research Laboratory.

Matthias Liebendörfer is a post-doctoral researcher at the Canadian Institute for Theoretical Astrophysics.

Keiichi Maeda is a graduate student at the University of Tokyo.

Christopher D. Matzner is an assistant Professor at the University of Toronto.

Paolo Mazzali is an assistant astronomer at Osservatorio Astronomico di Trieste, Italy.

O.E. Bronson Messer is a post-doctoral research assistant at the University of Tennessee and Oakridge National Laboratory.

Anthony Mezzacappa is a staff scientist at Oak Ridge National Laboratory.

Ken'ichi Nomoto is a professor at the University of Tokyo.

Tomek Plewa is a post-doctoral research assistant at the University of Chicago.

Markus Rampp is a post-doctoral research assistant at the Max-Planck-Institut für Astrophysik in Garching.

Gabe Rockefeller is a graduate student at the University of Arizona.

Todd A. Thompson is a Hubble fellow at the University of California at Berkeley.

Hideyuki Umeda is a post-doc at the University of Tokyo.

Lifan Wang is a post-doctoral researcher at Lawrence Berkeley Laboratory.

Michael S. Warren is a permanent staff scientist at Los Alamos National Laboratory.

J. Craig Wheeler is a Professor at the University of Texas at Austin.

Weiqun Zhang is a graduate student at the University of California in Santa Cruz.

Preface

This book is designed to provide an incoming graduate student in astrophysics much of the background information on stellar collapse. By including the latest results, this book provides the current status of the field.

This is a fast moving field both in theory where computational resources are allowing increasingly sophisticated models and in observations, where new transient objects are being discovered that are opening wide the implications from stellar collapse. Because rapid rate of new discovers, no book on the subject will remain up-to-date for long nor, if we include more than one author, will all of the authors agree on every subject. However, any future work on stellar collapse will build upon the ideas and techniques presented here, and this book provides an ideal starting point to enter into this field.

The first part of this book is devoted to traditional studies of stellar collapse: the search for the mechanism behind core-collapse supernovae, from progenitor stars (Ch. 1) to observed supernovae (Ch. 2), the physics of the neutrino-driven supernova engine (Chs. 3-5) and the theoretical predictions for the nucleosynthetic yields and gamma-rays produced in core-collapse supernovae (Chs. 6-7). Part I poses several problems with the basic spherically symmetric picture of stellar collapse and core-collapse supernovae.

Part II of this book covers the role asymmetries have played in changing our understanding of stellar collapse. Some scientists argue that a new stellar collapse engine is required to explain all supernovae (Chs. 8,9), but most scientists have instead isolated new set of supernova-like outbursts from hypernovae (Ch. 10) to Gamma-ray bursts (Ch. 11-12) which are a small, but very important due to their extreme asymmetries, subset of outcomes from stellar collapse. These asymmetries are caused by asymmetries deep within the core of the collapsing star and can be tested by observations of the gravitational waves emitted in collapse (Ch. 13).

CHRIS L. FRYER

Foreword

This volume is a major compendium of the current research in supernova theory and simuluation, ten major articles by ten major groups that explore the depths of current research. It is truely extraordinary that research in supernova remains so deep and esoteric as the research of these groups testifies. On the other hand it is equally extraordinary that an agreed mechanism has not emerged from $\sim$55 years of research, starting with Zwicky, Hoyle, Fowler and the Burbidges. We all know the conundrum as to why it has taken so long: how can positive kinetic energy, free energy, emerge from a system that is gravitationally bound? Why does it not just collapse and emit its binding energy as neutrinos, or to a black hole with no emission? The fact that a stellar collapse leads to an explosion of the star is a demanding physics challenge, especially to "red blooded males" who love explosions — and even to others. Just as "Nature" abhors a vaccumm, physicists, all those contributing to this volume, abhor not understanding something so "simple" as an explosion. Consequently, I could never, leave it alone.

A Touch of History

The touch of history starts for me with the testing of the US's largest thrmonuclear weapon, 15 megatons at the Bikini Atoll in the Pacific in 1954. To design an experiment to measure the gamma rays and neutrons from such an explosion demanded an intensive education, tutored by Montgomery Johnson, LLNL, Marshal Rosenbluth, and Conrad Longmire, LANL, in hydrodynamics, nuclear physics, radiation energy and pressure, and the competition between diffusion and advection. When one has designed and then analized results from a successful experiment, the confirmation of the physics is awesome. All of these physical processes on which the explosion depends take on a physical reality that is hard to deny. Consequently with the emergence of the space age and this success and understanding, I was asked by Teller and York to consider the effects of such an explosion in space. It soon became apparent to Montgomery and I that the radiation conditions from the surface of a megaton of energy in a cubic meter, $T \sim 5$ kev, would lead to the acceleration of a high opacity surface layer to 100's of MeV per nucleon kinetic energy. (These conditions presaged the concept of the high entropy atmosphere that builds up on a deleptonized neutron star.) Only the magnetic field of the earth or heliosphere would contain such a particle flux. Subsequently the starfish event confirmed just such confinement processes (provided one includes the grad B drift that shifts the conjugate point in latitude.). Wouldn't a supernova do the same thing and produce the cosmic rays of the universe? And so Montgomery and I proceeded to attempt to answer this question and you all know the first attempt to explain a supernova as a "bounce shock", formed when the neutron star elastically rebounds, has failed calculational tests.

Now, as we all recognize, the failure of the bounce shock has led to models depending upon seconds of time, large scale convection driven by the high entopy atmosphere, and thus a much weaker shock. But the story leading up to the first calculations of the bounce shock is more convolved. Again Teller, with Bethe's necessary concurrence, asked me to be the technical representative (knowledge of nuclear weapons) attached to the State Department in Geneva for the negotiations for the test ban treaty in space. The official US, desparately unspoken, position was: what a great idea to get satelites into space to "spy" on the Soviets and Soviets, in turn, thought what a bad idea it would be to allow it to happen. As you know, they did and we did and together we discovered gamma ray bursts rather than each others nuclear explosions. Along the way, we had to justify our sophisticated "spy" satelites rather than an unsophisticated giger counter. So guess who was asked to innocently give a talk on falsly triggering a "giger counter satelite". Well the obvious ploy was to invoke a supernova to do it. After all, those MeV gamma rays would look just like a clandestine nuclear test in space. With such a detection, we would all distrust each other even more, as if that were possible. Well the Soviets, many senior scientists, huddled with their exceedingly capable ambassador, "Scratchy" Serapkin, and after considerable cogitation pronounced with great authority "who knows what supernovae will do." Well they had the authority, because they had far greater knowledge and far greater seniority than myself. Of course a recess was called and I went to the CERN Library for two weeks to soothe my embarrassed feelings and possibly consider what in hell supernovae might do and how. The more I fruitlessly cogitated, the more determined I became not to be, nor the Lab to be, in such an embarrassing postion again. Fortunately Dr. Teller, then the lab director, felt the same way and supernova calculations with Dick White were initiated at the lab with bomb codes.

The first problem was to attempt to calculate the bounce shock starting from the iron-helium thermal decomposition threshold instability. The hardest part was developing numerical hydrodynamics to the point where the properties (speed up) of a strong shock in a density gradient were accurately reproduced. Fortunately, Burgers of South Africa had long before developed an analytical example and very many others contributed the equations of state from relativistic gas to neutron star matter. No matter how we teased the calculations the bounce shock was swallowed by the in-falling matter. Willy Fowler was convinced that thermonuclear burning of the carbon layers would save the day. Not so! Even with half the star's mass, $5M_{odot}$, of detonated carbon! What to do? Fortunately Willy Fowler had insisted that I become somewhat less ignorant about astrophysics. The deal was that I "chat" with his students about explosions while they grill me into learning astrophysics.

We all knew that the normal stellar imploding matter had to transform to neutron star matter by emitting neutrinos, but these would be instantly lost, of would they? Thinking of neutrino fluxes as radiation pressure leading to a ponderamotive force - that was weird. Fortunately Bob Christy could substantiate my early weird thoughts and we were off and running with neutrino driven supernova explosions. Neutrino radiation pressure? That was just another way to make a better explosion. These were not the neutrinos from the deleptonization or neutronization neutrinos, but those from the heat of delayed matter accreting onto the "hard", small (10^6 cm deep), negative gravitational potential ($-c^2/5$), deleptonized neutron star. The deposition approximation depended upon the high temperature of this shocked atmosphere, $\sim 10to15$ Mev and the higher energy neutrino thermal-tail neutrinos, ~ 50 Mev. These neutrinos had high enough energy so that their cross section for deposition was large enough to heat the imploding matter, but small enough to transport the heat to matter to a less deep, higher gravitational potential layer of the star and consequently reverse the accretion. As we all know, after this work was published, some 5 years after its completion but before publication, neutrino neutral current scattering had been discovered leading to trapping of the neutrinos during collapse and consequently a proto-neutron star that had $\times 10$ the radius of the compact neutron star, and therefore a smaller, less negative and less deep gravitational potential. In addition with the smaller mass fluxes the accretion shock was too low in temperature to give rise to the neutrino deposition explosion.

1. The light Problem

One last bit of history: A stellar explosion without emitting light is not a supernova. The problem is that the heat energy of the explosion is nearly all converted to kinetic energy and work against gravity by adiabatic expansion before it can diffuse out of the expanding star and be observed as light. First reaching the understanding of the origin of the supernova light, with Chester McKee, the NI Co Fe tripply forbidden decay, was the most difficult problem and euphoric understanding of my life, some ten years after the start of our supernova calculations, (These were started in 1959, with neutrino driven exposions in 1961, submitted to Reviews of Modern Physics in $\sim$1962, and published 1966 because Chandra took charge, and the ^{56}Ni came in 1969.) I, we , delayed the publication of these supernova calculations by almost two years searching for a solution to the light emission. One can find in Colgate and White, almost in desperation, the only derivation (according to Bethe) of the ensemble fission product decay curve, $dn/dt \propto t^{-1.2}$, but this too (along with Californium from Fowler) did not work. Alone, in a taxi to the Denver airport, the idea emerged. When waiting in line for the plane to Albuquerque, Hubert Reeves was in the adjoining line going to Paris. I yelled across to him as he boarded "Its ^{56}Ni!". There was the briefest pause, and then "Mon Dieu, d'accord", and we are still working on a better understanding of collapse supernova.

2. How I view the Present Problem

It is a weird and unlikely circumstance that a collapse supernova (Type II) should explode. The peculiar mechanism that facilitates this explosion is the formation and preservation of large scale structures in a high entropy atmosphere residing on the surface of a nearly formed neutron star. The high entropy atmosphere is maintained by two sources: the gravitational energy of initial formation of the neutron star, released by diffusion and transport of neutrinos and secondly and possibly dominantly by the gravitational energy released at the suface by additional low entropy matter falling through to the neutron star surface. The preservation of this entropy contrast between up and down flows requires thermal isolation between the low entropy down flows and the high entropy up flows. This entropy contrast allows an efficient Carnot cycle to operate and thus allows the efficient conversion of thermal energy to mechanical, which in turn drives the explosion. The P-V diagram of various up and down going mass elements in the calculations demonstrates the existence of the cycle and its efficiency. Greater thermal isolation should occur in 3-D as opposed to 2-D calculations because of the difference in relative thickness or surface to mass ratio for the same mass flow in 2 and 3-D. This may explain the observed stronger explosion in 3-D calculations.

3. Justification of this View

The spherically symmetric diffusion of heat from a thermonuclearly explosive fuel is extraordinarily stable as the existence of all the various stars attests. In stars the free energy of thermonuclear burn is many orders of magnitude greater than the gravitational binding energy and so even a very small runaway thermonuclear reaction should lead to explosion, but it usually does not happen, novas and SN Ia's being the exception. The lepton degenerate core of a forming neutron star on the other hand, is strongly bound and so there is no explosive free energy available, yet the supernova explodes and ejects nearly the whole star. However, although the free energy of this interior lepton degenerate gas is small compared to the binding energy at the surface of the degenerate gas, it is large compared to the binding energy of the stellar matter at larger radii of the star. The question is how can this free energy be transported to a much larger radius sufficiently rapidly so that the heat of this free energy can not diffuse away more rapidly than the hydrodynamic equilibration time i.e., the explosion time? How also can this free energy be transported without doing excessive work against gravity, i.e., leave "its" mass behind?

The structure of any star is inherently stable because, with the heat generated in the interior, the diffusion of heat from the center is always slow compared to diffusion from the surface because of the very large density gradient. In the case of a collapsed star, after the formation of the neutron star core, the neutrino flux from the continuing deleptonization and the neutrino flux from the subsequent accretion flows lead to a high entropy, neutrino-dominated atmosphere on the surface as opposed to the interior of the neutron star. High entropy in this case means matter whose internal energy is so high that it is not gavitationally bound. (It is confined by the external pressure of accretion.) We think that the properties of convection, truncated at the large scale by diffusion, uniquely solves this problem. The large scale of the Rayleigh - Taylor instability occurs because neutrino diffusion at the neutrino sphere prevents the growth, i.e., truncates the growth of all smaller scale wave lengths than the local scale height. Convection allows the transport of heat without doing work against gravity. A large scale convective element, a plume in a high entropy, relativistic gas of specific heat ratio, $\gamma = 4/3$, allows a plume or large scale mass element to rise and survive without entrainment or mixing for many scale heights of displacement, because the expansion is homologous and faster than Helmholtz mixing. Thus, the plume or mass element can reach a height where a significant change in gravitational potential occurs. In turn, provided the displacement of the mass element remains adiabatic, the Carnot efficiency for converting the high entropy or heat of the plume to useful work remains high. It is this work, we believe, that causes the supernova ejection or explosion. A fraction of the high entropy atmosphere formed adjacent to the neutron star, is transported, adiabatically, by large scale convection to a larger radius without significant loss of heat and thus to a much lower gravitational potential. In a lower gravitational potential it requires less work from the adiabatic expansion to eject the matter. It requires almost a conspiracy of all three pieces of physics, in our view, to cause a collapse supernova to explode.

The beginning physics in this scenario is the formation of a high entropy atmosphere lying on top of a nascent neutron star. This unlikely sounding circumstance is the inevitable result of continuing deleptonization of the contracting neutron star core and as well low entropy and higher density matter falling onto a tightly bound neutron star. Furthermore, the rate of release of this free energy must be great enough to establish a neutrino fire-ball above the neutrinosphere where the energy flux is so high that the radiation energy density in neutrinos becomes opaque to the neutrinos themselves. We will discuss the origin of this atmosphere first while attempting to evaluate the dynamic range of physical processes that might limit its formation. However, a universal understanding of the Carnot cycle in large scale convection in our view is the key to the understanding of the convective transport process. Viewing this process as a wind negates the possibility of counter current flows and therefore of an energy conversion cycle.

The second law of thermodynamics establishes the limiting efficiency for accessing the free energy of two, different temperature reservoirs. The Carnot cycle describes the sequence of deformations necessary to accomplish this efficiency. A necessary part of this cycle to access this free energy is the thermal isolation of the two parts of the cycle of compression from expansion. Similarly the necessary heat flow isolation between the up and down flows of large scale convection determines the efficiency for transporting heat and free energy from the deep gravitational potential. In our view, this isolation is absolutely necessary in order to maintain an efficient Carnot cycle and is therefore the basis for the somewhat greater energy release of the 3-D calculations (Fryer & Warren, 2002) over those performed in 2-D. In 3-D a given low entropy mass flow is cylindrical in shape and therefore thicker against neutrino heat flow transport than the corresponding entropy mass flow in the form of a sheet in 2-D, namely the surface to mass ratio is more favorable in 3-D.

There is still a long way to go in understanding this convective view of collapse supernova explosions from Colgate, Herant, & Benz, 1993, in Herant, Benz, Hix, Fryer, & Colgate 1994, and in Colgate & Fryer, 1994, and now recently the many calculations in this volume. We need to understand in a more lucid fashion the scale of the buoyant elements as determined by the heat diffusion at the neutrinosphere and secondly the expected homologous, shape preserving expansion of the buoyant elements in the $\gamma = 4/3$ high entropy atmosphere. The continuing dedication of all those contributing to this volume, their students, and students' students will achieve it. It has been and still is a great adventure for all of us.

Stirling A. Colgate
MS B227, LANL,
P.O. Box 1663,
Los Alamos, NM 87545

Acknowledgments

The initial set of authors and idea for this book grew out of a special session on stellar collapse at the 200th American Astronomy Society Meeting in Albuquerque. With the basic idea of a graduate level seminar text from the publishers at Kluwer (B. Burton and H. Blom), the authors have designed chapters to give both a review of the field with a taste of the current ideas and problems facing the field. This work was funded under the auspices of the U.S. Dept. of Energy, and supported by its contract W-7405-ENG-36 to Los Alamos National Laboratory, the University of Arizona, and by a DOE SciDAC grant number DE-FC02-01ER41176.

I

CORE-COLLAPSE SUPERNOVAE

Introduction

Chris Fryer

In the late 1930s and early 1940s it was realized that the relatively slowly varying emission from galaxies were disrupted every century by intense sources of light (for a brief time these outbursts could dominate the total light from their host galaxy). But such outbursts were short-lived and they faded over the course of weeks and months and gradually disappeared entirely making it hard to determine the actual source. At the same time, the basic building blocks of matter were becoming better and better understood and the idea of a star made up entirely of neutrons was proposed. Such a neutron star would be very compact. If a massive star collapsed down to such a compact configuration, the potential energy released could easily power a supernova explosion and this was proposed in the late 1930s. But understanding the full details of how this explosion process worked was left for future generations of scientists and a new invention, the computer.

In the 1960s, Colgate and collaborators began the first numerical calculations of explosions from the collapse of a massive star down to a neutron star. These first series of simulations found that when the collapsing stellar core reached nuclear densities, the collapse halted, sending a bounce shock through the star. This shocked stalled, but it was believed that neutrinos leaking out of the hot core (the potential energy of collapse is converted into thermal energy) would revive the shock and drive a supernova explosion. I say "believed" because, at the time, the simplistic models of core-collapse were limited by assuming an efficiency factor for neutrino deposition.

But these early simulations spawned an industry of increasingly detailed models for core-collapse supernovae, causing supernova theorists to push the frontiers of computational astrophysics combining the best numerical methods and detailed physics. The ultimate goal of these simulations has been to model supernovae in their full glory, without efficiency factors or free parameters. And in my mind, these explosive events deserve no less.

C.L. Fryer (ed.), Stellar Collapse, 3–4.

The first part of this book is dedicated to understanding neutrino-driven supernovae and the current status of the field started by Colgate and collaborators back in the 1960s. Chapter 1 discusses the massive stars whose collapse will power the supernova explosion. Chapter 2 reviews the observations. Chapters 3-5 discuss various aspects of the physics necessary to modelling stellar collapse and the neutrino-driven explosion. We end with theoretical predictions for both nucleosynthetic products and gamma-ray lines (Chs. 6-7). No book can be truly complete, and the most evident hole in this book is the lack of discussion of the equation of state used for core-collapse models. There is currently a great deal of work on this subject and I advise the reader to start with papers by Lattimer and collaborators.

The new-to-the-field reader may be concerned that small details in the physics can make an explosion succeed or fail. Indeed, many scientists think that this means that we may be missing critical physics in the explosion mechanism. That may be true, but I would argue that the supernova mechanism must depend upon the details. Why? Because massive stars are not all that different. Yet, in nature,some stars produce neutron stars and strong supernova explosions. Others collapse to form black holes and only weak or no explosions. In nature, supernova explosions seem to depend on the details. And so, as we see in the next 7 chapters, astrophysicists must also worry about those details.

Chapter 1

MASSIVE STAR EVOLUTION

Patrick A. Young and David Arnett
Steward Observatory, the University of Arizona, 933 N. Cherry Avenue, Tucson AZ 85721
payoung@as.arizona.edu, darnett@as.arizona.edu

1. Introduction

Supernovae are of wide interest to the astronomical community because they play a key role in the nucleosynthetic evolution of galaxies, stellar remnant populations, the energetics and phase balance of the interstellar medium (ISM), the interpretations of observations of distant galaxies, models of galaxy formation and evolution, the physics of star formation, and tests of cosmology, among other topics. This book deals largely with the interesting physics and observations associated with the supernova explosion itself and the immediate development of its gaseous and compact remnants. However, in order to contribute to these other fields, the fates of massive stars must be placed in a larger context. It is not sufficient to study the physics of the explosion in isolation; it is also necessary to have a firm grasp on the evolution of supernova progenitors. Quantifying the contribution of massive stars to the ambient radiation field, the dynamics of the ISM, and the chemical evolution of galaxies requires connecting supernovae with the stellar initial mass function (IMF) and understanding all the physics that lies in between.

The life of a star is usually divided up according to the primary energy producing nuclear reaction occurring in the interior. The burning separates, usually, into the following stages: hydrogen, helium, carbon, neon, oxygen, and silicon burning. Table 1 summarizes the properties of each stage. Together these produce all of the abundant elements except for the H and He from the Big Bang. The s-process, r-process, and p-process are important diagnostics (Meyer 1994) as they make many rare nuclei, but involve less than 10^{-6} of the solar abundance pattern. This chapter will treat each stage of evolution and its relevant physics in turn, including the available observational contraints, and will conclude with a brief discussion of the unresolved issues in modelling stellar evolution. We will emphasize how each stage affects the final fate of

C.L. Fryer (ed.), Stellar Collapse, 5–38.

Table 1.1. Bulk Nucleosynthesis Processes

Process	Reaction	Temperature(K)	Cooling	Products
Hydrogen	pp	2×10^7	photons	He
	CNO	3×10^7	photons	He, N, Na
Helium	$3\alpha \rightarrow^{12} \mathrm{C}$	2×10^8	photons	C
	$^{12}\mathrm{C}(\alpha, \gamma)^{16}\mathrm{O}$			O
Carbon	$^{12}\mathrm{C} + {}^{12}\mathrm{C}$	9×10^8	neutrinos	Ne, Na, Mg, Al
Neon	$^{20}\mathrm{Ne}(\gamma, \alpha)^{16}\mathrm{O}$	1.5×10^9	neutrinos	O, Mg, Al
Oxygen	$^{16}\mathrm{O} + {}^{16}\mathrm{O}$	2×10^9	neutrinos	Si, S, Ar, Ca
Silicon	$^{28}\mathrm{Si}(\gamma, \alpha)$	3.5×10^9	neutrinos	Fe

the star and its relevance in a larger context, for example galactic chemical evolution.

The first question upon which progenitor evolution bears is, "What do we mean by massive stars?" In this context we are interested in stars which undergo core collapse and explode as Type II or Type Ib/c supernovae (SNII or SNIb/c). In the past the minimum initial mass for which core collapse occurs was thought to be approximately $10 M_\odot$. In recent years, theoretical and observational considerations have caused this number to creep downward until it hovers now around $8 M_\odot$. In either case, the range of masses spanned is enormous, from 8 to over $100 M_\odot$. The physical conditions range from temperature and densities of 3000 K and 10^{-9} g cm^{-3} at the surface of a red supergiant to $10^{9.5}$ K and 10^9 g cm^{-3} in its core. At all stages of their lives these stars are non-static systems, heavily influenced by hydrodynamic processes; the most massive are in a near-constant state of instability, shedding huge gouts of mass to become Luminous Blue Variables (LBV's) and Wolf-Rayet (WR) stars. The physics of nuclear burning, plasmas, magnetic fields, hydrodynamics, and radiative transfer all couple in determining the behavior and fate of massive stars. The field of stellar evolution has a strong theoretical and logical basis. It is realistic to expect we can do a fairly accurate job with the proper care and attention to detail, but it must be kept in mind that creating the final stellar conditions which lead to a supernova explosion is a complex, and in some areas poorly understood, process rife with exciting research opportunities of its own. Our challenge is not only to reproduce the results observed in nature, but to determine – uniquely – how nature did it.

2. Useful Concepts

In our discussion, several concepts will recur often, so we will give a brief overview before starting in on the lifecycle of a star.

Radiation Entropy

The internal structure, stability, and nucleosynthesis in stars is heavily influenced by the *radiation entropy* of the stellar plasma, $S_\gamma/\Re = 4aT^3/3\rho\Re$ in Boltzmann units. Here T is the temperature, ρ the mass density, a the radiation constant, and $\Re$ the gas constant. Eddington (1926) introduced a related quantity β in his Standard model, $S_\gamma/\Re = 4Y(1-\beta)/\beta$, which appears in the famous quartic equation, $1-\beta = 0.00298\,(m/Y^2)^2\,\beta^4$, where m is the stellar mass in solar units and Y is the number of pressure-contributing particles per baryon (an ideal gas of matter plus black body radiation is assumed). For most stars of interest to us for nucleosynthesis, $S_\gamma/\Re$ lies in the range of 0.3 to 30, while cosmological values range from 10^8 to 10^{10}. Values for stellar quasistatic burning range about unity while explosive values are usually around ten, an interestingly restricted range.

Neutron Excess

The yields of nuclei depend upon the nuclei being produced, but also upon their surviving. Nuclear stability reflects nuclear binding, which exhibits strong systematic properties. These systematics were observed in the Seuss & Urey (1956) abundances by Burbidge et al. (1957 - hereafter BBFH) and Cameron (1957), and still have relevance. Nuclei tend to be more stable if they have equal numbers of neutrons and protons (Z=N), even numbers of neutrons (even N), and even numbers of protons (even Z). However, protons repel by coulomb forces, so that nuclei with larger Z are more tightly bound if they are neutron rich.

Below the iron peak, the abundance of odd-Z elements is sensitive to the relative abundance of neutrons and protons in the environment in which they are formed. The most bound isotopes will have even-N. For example, ^{23}Na has Z=11 and N=12. There is one excess neutron. Similarly, ^{27}Al has Z=13 and N=14, with one excess neutron. If there is a paucity of excess neutrons, Na and Al will be rare.

The *neutron excess* is defined as $\eta = \Sigma_i Y_i (N_i - Z_i)$, the fraction of extra neutrons per baryon. Charge neutrality of the plasma relates this to ratio of electrons to baryons, $Y_e = \Sigma_i Y_i Z_i = (1-\eta)/2$.

Timescales

Several timescales always need to be held in mind when considering problems of stellar evolution. The *Kelvin-Helmholtz* or gravitational contraction timescale is proportional to the gravitational binding energy of the star divided by its luminosity. The *thermal* timescale is related. Both of these depend on the energy leakage rate from the star. This is determined by the opacity of the sellar material. Accurate determination of opacities is indispensible to stellar evolution.

A second set of times involve fluid motion. The *dynamical* timescale is simply the free-fall time, $\tau_{dyn} = R/v_{ff}$, where $v_{ff} = \sqrt{2GM/R}$. This timescale is usually only relevant during core collapse. Much more important is the hydrodynamic timescale, which is simply the sound travel time in the region of interest. During oxygen burning or later stages this time becomes comparable to the evolutionary timescale.

In general, the evolutionary timescale is set by a nuclear burning timescale. This is the time required to process a significant fraction of the fuel in a burning zone and is set by the reaction rates and thermal balance. Nucleosynthesis in stars may be divided into two qualitatively different categories: *quasistatic* or *explosive* burning. All the evolutionary stages involve quasistatic burning, though in the case of Si burning this period is only a few hours. Explosive burning does not affect the evolution of the star, but is often a prime determiner of the final yields. Products of H, He, C, and Ne burning returned to the ISM are largely quasistatic. Based upon β decay rates, BBFH and Cameron(1957) determined that the s-process must be quasistatic and the r-process, with timescales of order seconds, must be explosive. Quasistatic O and Sicore burning produce neutron excesses which are much too large to match the solar system abundances; only explosive burning or partial O shell burning are likely to contribute to the yield.

3. Formation and IMF

In order to quantify the chemical, energetic, and stellar remnant contributions of the massive components of a stellar population to a galactic environment, we must at the very least know how many massive stars there are. The process of star formation produces a characteristic distribution of stellar masses called an initial mass function (IMF), which is usually expressed in one of two forms: dN/dm or $d(\log N)/d(\log m)$. Salpeter described the IMF from 0.3–$10M_\odot$ with a simple power law:

$$\frac{dN}{dm} = am^{-\gamma} \tag{1.1}$$

or

$$\frac{d(\log N)}{d(\log m)} = Am^{-\Gamma} \tag{1.2}$$

where $\Gamma = \gamma - 1 = 1.35$. More complicated forms of the IMF have been proposed, but it is debatable whether these provide any better fit to the available data for high mass stars than a simple power law.

The IMF for low (0.08 – $\sim$ 2.0 $M_\odot$) and intermediate (2.0 – $\sim$ 8.0 $M_\odot$) mass stars has been extensively studied since Salpeter's seminal work in 1955, but high effective temperatures and small number statistics combine to made the determination of a high mass IMF very difficult (please see Massey 1998 for an excellent discussion of the issues involved, and of the high mass IMF in general). It was not until the 1990's that significant progress was made for stars over 10 $M_\odot$.

There are now estimates of the IMF for various environments in the Galaxy ($z \simeq 0.02$), Large Magellanic Cloud (LMC) ($z \simeq 0.008$), and Small Magellanic Cloud (SMC) ($z \simeq 0.002$). Taken together these sample an order of magnitude in metallicity and three orders of magnitude in stellar surface density, from 8.0×10^3 kpc^{-2} in the sparsest OB associations to $> 6.3 \times 10^6$ kpc^{-2} in the prototypical super star cluster R136 in 30 Doradus. Remarkably, the IMF does not seem to vary much within the limitations of the observations over this wide range of conditions. The weighted average of the IMF slopes in the Milky way is $\Gamma = 1.1 \pm 0.1$; that of the LMC and SMC $\Gamma = 1.3 \pm 0.1$. The IMF has been measured from 2.8 to > 120 $M_\odot$ in R136, which contains no less than 39 stars of type O3. The slope was found to be $\Gamma = 1.3$–1.4. These slopes are all consistent with each other to within the observational errors, and even close to the Salpeter slope determined from low and intermediate mass stars in the solar neighborhood (Massey et al.1995a; Massey et al.1995b). This clearly tells us something about the physics of star formation, though exactly what is a very diffucult question.

In addition, the maximum stellar mass found in these associations does not depend upon metallicity, at least for the range observed. There does not appear to be an upper mass cutoff; in an association with a given number of stars, the most massive star observable is consistent with the mass at which we would statistically expect to see of order one star based on the IMF. This means that radiation pressure on grains is not the limiting factor in determining the highest mass star that can form, as has been suggested on theoretical grounds (Massey et al.1995a; Massey et al.1995b). Other possibilities for an upper mass cutoff exist which are consistent with the observations. A sharp downturn in the star formation probability after a turbulent crossing time of the molecular cloud has been suggested by Elmegreen(2000). This requires the maximum stellar mass to be correllated with the Jeans mass of the cloud, leading to the intriguing prospect of supermassive stars in starburst galaxies. Further observations connected to this hypothesis should prove interesting.

Having made this comforting generalization, there are areas in which the high mass IMF does seem to diverge significantly from the values mentioned above. Crude estimates of the relative numbers of massive stars can be made by comparing the numbers of different types of Wolf-Rayet stars (which will be discussed in detail later) in a population. There should be a strong correlation between the ratio of WN and WC stars with metallicity, since WC stars should be harder to make with less metal line opacity to drive mass loss. The majority of observations support this idea. The Local Group starburst Galaxy IC 10, however, has a WC/WN ratio nearly a factor of ten too large for its metallicity. The simplest way to resolve this dilemma is to assume an IMF weighted heavily towards massive stars (Massey et al.1996). This situation does not seem to prevail for "Wolf-Rayet galaxies" in general (Schaerer 1996). More work needs to be done on the IMF of starbursts before firm conclusions can be drawn. On a similar note, the IMF of the first generation of zero metallicity stars (Population III) may be top-heavy or bimodal. The second peak may be around 100 $M_{\odot}$, and maximum masses of several hundred to a thousand solar masses have been suggested (Nakamura & Umemura 2001, for example). Unfortunately, this problem is likely to remain purely theoretical for some time to come until direct observations of the first stars become possible.

Some studies of field stars tentatively reveal an IMF different from clusters. An estimate of the number of high mass field stars can be made by taking a census of stars which are too far from known OB associations for any reasonable birth velocity or kick imparted from the supernova explosion of a binary companion to have carried them in a main sequence lifetime. Such stars do exist, but have an IMF slope of $\Gamma = 4.1 \pm 0.2$ in the LMC and $\Gamma = 3.7 \pm 0.5$ in the SMC (Massey et al.1995a). Elmegreen(1997) has suggested that the uniformity of the IMF in associations is due to the collapse of molucular clouds with a hierarchical or fractal morphology. If there are clouds with a more regular morphology, for example low mass clouds with regular distribution of clumps, the IMF may be biased to a small range of masses. These field stars may form in such non-fractal clouds.

The timescale for the formation of high mass stars is short, of order $\Delta\tau < 1 - 2$ Myr (Massey 1998). It is difficult to determine the degree of coevality of different mass stars in a cluster. In a typical OB association such as Upper Sco, there will be evidence of supernovae, possibly supergiants, early B stars on the turnoff, B, A, and F stars on the main sequence, and lower mass stars still on the pre-MS. Since pre-MS contraction times are longer for lower masses, this suggests that the process of formation begins at roughly the same time for all masses. Claims for the actual age spread can range from 1–2 Myr to > 10

Myr. Exactly where the truth lies probably varies with conditions in the natal molecular cloud. In any case, a large *observed* age spread in a cluster is likely to be dominated by observational errors. For a discussion on the uncertainties involved in estimating the ages of very young stars, see Hartmann(2001).

High mass stars themselves continue accretion even after hydrogen ignition. Radiation pressure from the luminous new star may play a role in stopping the accretion process and determining its final mass. Observations of Galactic starbirth clusters show O and B stars with infrared excesses consistent with the presence of circumstellar envelopes or the accretion disks so characteristic of low mass star formation (Figuerêdo et al. 2002; Conti & Blum 2002). What occurs after the formation of massive stars may depend on the stellar density. The energy input into the ISM by stellar winds from massive stars on the main sequence and in the WR phase can each be comparable to that from a supernova (Abbott 1982). In a cluster with a high density of OB stars, this energy input may serve to disperse the cloud and terminate further star formation. Clusters with a lower density of OB stars show evidence of intermediate mass star formation continuing for several million years (Hillenbrand et al. 1993).

Much work remains to be done in the field of massive star formation. In particular, a theoretical understanding of the star formation process is yet to be perfected. Also, the connection between star formation and the large scale properties and dynamics of the ISM is not sufficiently well understood and is a major area of uncertainty in modelling galaxy evolution. Star formation is patchy and time-variable, and acts upon masses of gas varying by many orders of magnitude. Nevertheless, much has been learned, and the future of the field looks promising indeed.

4. Hydrogen Burning

Let us now assume we've figured out how to construct a realistic stellar population with the right proportion of stars at each mass on the zero age main sequence (ZAMS). We can also begin to model what happens to a star with a given structure when it undergoes core collapse. The question is, which stars from our distribution, if any, end up in that particular final state? How do they get there? It is essential to answer this question if we are to understand the effects of an ensemble of supernovae on their environment. As an example, consider the relative number of neutron stars and black holes in the galaxy. If we are not careful with how we evolve our progenitor stars, it is not difficult to imagine that the real minimum mass star which produces a black hole is 10% smaller than we might predict. For a simple Salpeter IMF that means there are 30% more stars making black holes than we thought. The same principle applies to nucleosynthetic yields, frequency of supernovae and collapsars, ISM energy budgets, and the like. We will concentrate on the issues relevant to making pre-supernova models *for a particular progenitor star* as accurate as possible.

The hydrogen burning main sequence (MS) phase is similar for all stars with convective cores. The energy production of the star is dominated entirely by the CNO cycle. The strong temperature dependence of CNO hydrogen burning ($\epsilon \propto T^{17}$) means that most of the luminosity is generated in a very small region in the core and greatly exceeds that which can be transported by radiative diffusion. Perforce, a convective core must develop. A rough estimate of the mass fraction of the star which is convective can be made with the following fitting formula:

$$q_{cc} = e^{-\alpha} \tag{1.3}$$

where $\alpha = 0.802(3-n)+0.357(k-nr)$, n is an appropriate polytropic index, and k and r are the exponents in the Kramer's opacity expression $\kappa = \kappa_0\rho^r/T^k$. In very high mass stars dominated by radiation pressure (hence with large n) 90% of the star can be convective. This is a rough estimate, and a detailed answer turns out to be non-trivial to calculate for reasons we will discuss shortly.

The size of the convective core has two important consequences for the star. First, it determines how much of the star is processed nucleosynthetically and the final mass of the collapsing core. This will impact the nature of the supernova explosion and its yields. Secondly, it increases the amount of fuel available and hence extends the star's MS lifetime. Also, for very high mass stars $M/L \propto const$, so the lifespan on the main sequence does not continue to decrease with mass as it does for low mass stars. The lifespan approaches a more or less constant value of 2-3 Myr, even for $M > 100\ M_\odot$. The lowest mass supernova progenitors, at $\sim 8\ M_\odot$, probably live about 40 Myr. The helium burning stages typically last about 10% as long, at least for massive stars, since the energy yield of triple alpha burning is about one tenth that of hydrogen burning. Later stages do not contribute much to the total lifetime.

During the MS, the creation of ^{4}He both reduces the number of particles available to provide pressure support and eliminates two electrons per ^{4}He nucleus created, lowering the dominant Thompson scattering opacity. As a result, the central temperature and the luminosity of the star increase. On the HR diagram, stars move towards higher luminosity and lower T_{eff}. The least massive stars with convective cores move almost vertically. As the mass on the ZAMS goes up, more of the increased luminosity goes into mechanical work expanding the star's envelope, and the track in the HR diagram increasingly becomes a horizontal progress towards the red as the star ages. When H is depleted in the core, the entire star contracts, increasing in temperature and briefly moving toward the blue before H shell burning and core He burning are established. Figure 1.1 shows representative evolutionary tracks for stars from 10 to 100 $M_\odot$ at solar metallicity. For lower metallicities, the core must burn at a higher temperature due to the lower abundance of CNO catalysts,

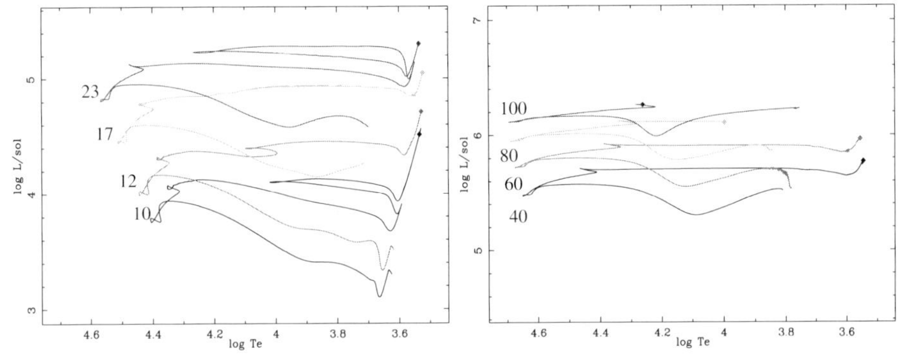

Figure 1.1. Evolutionary tracks for massive stars. The left panel shows tracks for 10, 12, 17, and 23 $M_\odot$ from the pre-MS until near core He exhaustion. The 10 $M_\odot$ star shows a blue loop morphology very similar to intermediate mass stars. Most core He burning takes place on this loop. For the 12 $M_\odot$ star this loop is reduced to a kink on the Hertzsprung gap. At higher masses He burning does not significantly perturb the path of the star across the HR diagram. The 23 $M_\odot$ star undergoes such a loop, but it occurs during He shell burning. The right panel shows tracks for 40, 60, 80, and 100 $M_\odot$. on the same log scale. These stars were evolved until evidence for a significant hydrodynamic instability appeared in the calculations, making further evolution in a 1-D spherically symmetric code inappropriate. (The exact location of this point is very algorithmically dependent.) For the 40 and 60 $M_\odot$ this likely occurs on the Hayashi track. For higher masses it happens while crossing the Hertzsprung gap. In both cases the instability appears to be driven by the H shell, while core He burning is also taking place. All tracks were created with the stellar evolution code TYCHO for solar metallicity (Young et al. 2001; Young et al. 2003).

while the opacities, particularly in the outer parts of the star, decrease. A lower metallicity systematically moves the evolutionary track for a given mass blueward and slightly higher in luminosity. The situation for zero metallicity (Pop III) is more complex and will be dealt with in a separate section.

Hydrogen burning is usually considered uninteresting from a nucleosynthetic point of view, but there are certain features which can be enlightening. The isotopes ^{14}N, ^{17}O, and ^{23}Na are enhanced by roughly a factor of ten. Some ^{13}C is also made, but ^{12}C, ^{15}N, ^{16}O, ^{18}O, ^{19}F, ^{21}Ne, ^{22}Ne, and ^{25}Mg are depleted. Perhaps the most interesting species is ^{26}Al, which decays to ^{26}Mg. The decay from ground state is inhibited by a large angular momentum barrier and has a half life of 7.5×10^5 yr. The ^{26}Al does, however, have a low lying excited state which is not so inhibited and begins to be significantly populated at $T \sim 10^8$ K. Thus at high temperatures ^{26}Al decays much more quickly. However, more ^{26}Al is produced at higher radiation entropies. In low mass stars all the ^{26}Al decays before the end of hydrogen burning, even through the ground state, but in very high mass stars, LBV's, and WR stars, measurements of ^{26}Al can potentially provide a sensitive test of the burning stage, the extent of mass loss, and the history of mixing in the star.

The subject of mixing brings us to a discussion of the remaining open questions for the hydrogen burning main sequence. Traditional calculations treat stellar evolution as a series of snapshots of the star in equilibrium, or at least steady state. The increasing quality of observational data, however, has rendered that approximation inadequate. It is quite clear that compositional mixing extends well beyond the region of the star formally unstable to convection. It is now necessary to consider the hydrodynamics of the radiative region of the star.

Most tests of the mixing are geared toward lower mass stars with surface convection zones, but there are a few possible for massive stars. The first is based upon the abundance of ^{26}Al as discussed above, and has not yet been implemented. The second involves age estimates of clusters determined by examining Li depletion in brown dwarfs. These ages are approximately 50% longer than ages determined with traditional turnoff fitting methods (Stauffer et al. 1998; Stauffer et al. 1999). Extra mixing produces larger cores and hence longer main sequence lifetimes. The youngest cluster upon which this test has been performed is IC 2391, at $\sim$ 50 Myr. This age puts the turnoff firmly in the realm of intermediate mass stars, but the test should in principle be applicable to younger clusters. The final test, which has actually been used on massive stars, is the apsidal motion test on double lined eclipsing binaries. Since stars are not point sources, tidal torques can cause an additional precession of the apsides of the orbit beyond that predicted by general relativity. The magnitude of this precession is a diagnosic of the moment of inertia tensor and thus the density profile of the stars. Apsidal motion tests require larger core sizes than those generated by mixing length theory (Young et al. 2001).

A great deal of progress has been made in tackling this perennial problem. There appear to be three main components which contribute to slow compositional mixing in the radiative region. The first is meridional circulation driven by rotation, in the spirit of Eddington-Sweet mixing (Maeder & Meynet 1989; Schatzman 1999; Pinsonneault et al. 2002). The second and third components arise from the coupling of the convective zones to the radiative regions. The impact of convective plumes on the surface of separation between the convectively stable and unstable regions drives gravity waves at the boundary. This effect occurs both for core/shell convection driven by nuclear burning and surface convection driven by entropy losses at the photosphere. It is a simple consequence of the fluid motions. These gravity waves have two effects. First, they are an effective means of transporting angular momentum within the star, providing a second mechanism whereby rotation influences mixing (Charbonnel & Talon 1999; Talon, Kumar, & Zahn 2002). Secondly, these waves are

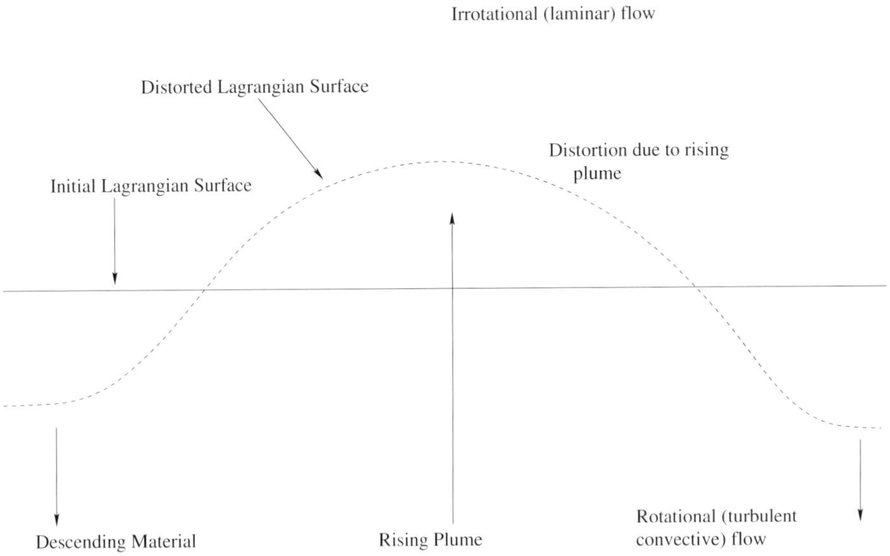

Figure 1.2. Lagrangian (co-moving) fluid surfaces at boundary between convectively stable and unstable regions. The distortion due to a rising plume and a downdraft are shown. The original spherical boundary is distorted as interface material bobs up and down, generating gravity waves. Contrary to the usual assumption in stellar evolution simulations, the convective boundary is neither spherical nor static. Dissipation of the gravity waves actually causes slow circulation in the nominally laminar region outside the convective zone (Young et al. 2003).

well described by a potential flow in the laminar region. The waves are dissipated, primarily by radiative losses. This is an entropy-generating process, which creates vorticity and leads to slow mixing in the radiative zone (Young et al. 2003). Figure 1.2 is a schematic of this process. It emphasizes that the convective boundary is neither spherical nor static, and its coupling with the laminar region must be taken into account. These three processes, when taken together, bid fair to solve most of the outstanding problems with the upper main sequence. Caution should of course be exercised, as there may be additional physics of which we are as yet unaware. The primary effect of the extra mixing in massive stars is to produce larger, higher entropy cores, which will have an impact on nucleosynthesis in subsequent stages of evolution.

The last insight into the mixing in particular would probably have gone unnoticed were it not for multidimensional numerical hydrodynamic simulations. The equations of state for massive stars, and the opacities, which have a profound impact on the energy leakage rate and thus luminosity, color, and structure of stars and their mass loss, have been determined experimentally by High Energy Density (HED) laser experiments (Iglesias & Rogers 1996).

Further advances in stellar evolution are likely to rely heavily upon both laboratory astrophysics and numerical simulations both for physical insight and quantitative calculation.

5. He Burning

The He burning phase truly sets the stage for the final evolution of the star and also turns out to be one of the most complex problems in stellar evolution. It is during this phase that the final core mass and C/O fraction are established, which determine the further evolution of the star all the way to core collapse. Numerous pieces of physics contribute to the final result. From the stellar evolution side, mixing and mass loss are both crucial and not completely understood. During this stage the thermal timscale begins to exceed the evolutionary timescale, further invalidating the normal quasistatic assumption of stellar evolution. Nevertheless, the timescales are still too long for efficient numerical simulation by explicit schemes. On the subatomic scale, the final state is dependent upon the $^{12}\mathrm{C}(\alpha,\gamma)^{16}\mathrm{O}$ reaction , which is the most poorly known major reaction rate in nuclear astrophysics.

In all the stars of interest to this volume, He burning begins quiescently in the core without the complications of a degenerate flash. For the lowest mass SN progenitors, the star has time to cross the entire Hertzsprung gap before core He ignition, and the evolutionary tracks appear similar to intermediate mass stars. More massive stars ignite He progressively earlier until burning begins just after the gravitational contraction of the convective core following H exhaustion. It is not clear whether such stars travel redward enough to come up against the Hayashi track, especially for low metallicities.

The early stages of He burning are dominated by the $^{4}\mathrm{He}(2\alpha,\gamma)^{12}\mathrm{C}$ (triple alpha) reaction. This three body reaction consists of two parts, $^{4}\mathrm{He}(\alpha,)^{8}\mathrm{Be}$ and $^{8}\mathrm{Be}(\alpha,\gamma)^{12}\mathrm{C}$. Hoyle(1954) predicted that the second reaction must be resonant, otherwise the reaction would not be fast enough at stellar energies to produce the observed abundances of carbon and oxygen. Indeed, the appropriate nuclear energy level, predicted by *astronomical* arguments was experimentally confirmed within three years (Cook et al. 1957). This reaction is highly temperature dependent ($\epsilon \propto T^{18.5}$), so He burning cores are always convective. The triple alpha burning is accompanied by other reactions which do not contribute significantly to the energy generation rate but are important nucleosynthetically. At temperatures of $T \sim 1\times10^{8}$ K ^{26}Al and ^{18}O are being destroyed effectively. However, ^{18}O begins to be produced by $^{14}\mathrm{N}(\alpha,\gamma)^{18}\mathrm{F}$ followed by a β^{+} decay to ^{18}O. This has the effect of converting an abundant nucleus with equal number of protons and neutrons to one with a fractional neutron excess of 0.111. This creates a total neutron excess in the material of $\eta \sim 1.5\times10^{-3}$, which is approximately what is needed to account for the solar abundances of weak s-process products. Depletion of He begins to be significant at $T \sim 1.15 \times 10^8$ K. At the same time, $^{18}\mathrm{O}(\alpha,\gamma)^{22}\mathrm{Ne}$, $^{22}\mathrm{Ne}(\alpha,\gamma)^{26}\mathrm{Mg}$, and $^{22}\mathrm{Ne}(\alpha,n)^{25}\mathrm{Mg}$ occur. The last reaction converts the neutron excess of ^{18}O to free neutrons which can participate in the s-process. The neutron flux is dependent upon the rate of $^{22}\mathrm{Ne}(\alpha,n)^{25}\mathrm{Mg}$ relative to He burning, which can be increased by increasing the entropy (i.e. having a larger core).

Because the triple alpha rate is proportional to ρ_α^2, the reaction $^{12}\mathrm{C}(\alpha,\gamma)^{16}\mathrm{O}$ becomes important as He is depleted in the core, even though the ratio of the reaction rates is only some 5%. The exact value of this ratio is not well known due to uncertainties in the $^{12}\mathrm{C}(\alpha,\gamma)^{16}\mathrm{O}$ rate. Direct determination requires measuring cross sections to much better than a nano-barn, an experimentally very difficult proposition. Recent determinations are beginning to converge (Angulo et al. 1999, for example), but quoted uncertainties are still of the order of 40%; in reality they may be higher. This reaction rate is of surpassing importance. Were it much larger, He burning would always proceed to pure ^{16}O. Much smaller, and ^{12}C would always dominate. Yet ^{16}O and ^{12}C are the third and fourth most abundant elements, arguing that the true case lies somewhere between these extremes.

To complicate the situation further, the C/O ratio is strongly influenced by stellar physics. Because of the extra density dependence of triple alpha, destruction of ^{12}C increases at higher entropies. Thus the production of oxygen relative to carbon increases with stellar mass. This also means that mixing during the main sequence can affect the C/O ratio through changes in the size of the He core.

Mixing physics during He burning also affects the ratio. Any admixture of unburned He into the core will increase the destruction of ^{12}C. The convective core should increase in size with time, ingesting He produced by the H burning shell. The nablas in the core very nearly follow the adiabatic gradient. A parcel of unprocessed He rich material from outside the core will be subject to radiative heating and heating from nuclear burning. Simulations of this process (Baleisis, private communication) show that it is quite possible for such a parcel to become positively bouyant over a short distance, limiting its maximum excursion into the convective zone. A von Neumann stability analysis of the coupled advection diffussion equation shows that it is numerically stable, so the problem is tractable, however the equations of stellar structure are very stiff, so an inappropriate change to the gradients in the star by mixing can drive the system to converge at a substantially different solution. If the details of how the ingested material is transported within the convective zone are not handled with great care, a thermal and nuclear instability can result which leads to excessive core growth and too small a C/O ratio. This is akin to the "breathing pulse" problem familiar to modellers of intermediate mass stars (Caloi & Mazzitelli 1993, and references therein). Observations of the ratio of AGB to BHB stars put a limit on the time the stars spends doing core He buring. These limits argue strongly against the existence of breathing pulses in real stars. Most codes fix the problem by arbitrarily decreeing that no new He can be mixed into the core after the He fraction drops below a certain level. A time-dependent treatment may solve the problem by itself, without recourse to *ad hoc* restrictions.

Incomplete He burning in a He shell can also provide a source for ^{12}C, one which is spatially distinct from the core. Nucleosynthesis calculations can get good matches to the measured solar abundances even with this complex interplay by tweaking various adjustable parameters. Plausible choices give plausible results, but we may get the right answer for the wrong reasons.

In the HR diagram He burning and later stars are thought to undergo "blue loops", excursions toward higher T_{eff} and smaller radius. For the lowest mass supernova progenitors this occurs during He core burning, but for stars above 15 $M_\odot$ it happens during He shell burning if at all. (See Figure 1.1 for examples of blue loop morphology.) Observations of external galaxies with the *Hubble Space Telescope* are beginning to overcome the problem of small number statistics for massive stars. Examining the ratio of blue to red supergiants gives an estimate of the relative amounts of time spent at the blue and red ends of the blue loops and an estimate of the duration of core and/or shell He burning. The B/R supergiant ratio provides a sensitive test of convection, hydrodynamic and rotational mixing, and mass loss. Unfortunately, it provides a sensitive test of convection, hydrodynamic and rotational mixing, *and* mass loss, with little power to extricate the effects of any one process. Model tracks often do not reliably fit the B/R ratio or the positions of blue loop stars (Dohm-Palmer & Skillman 2002; Alcaino et al. 2003, for example), betraying the presence of additional physics. In the next section we will more closely examine one aspect of this additional physics as we discuss stars with high rates of mass loss.

6. LBVs and Wolf-Rayets

One characteristic of massive stars is extensive, and at times violent, mass loss. In fact, the highest mass main sequence stars and Wolf-Rayets contribute as much or more kinetic energy input into the ISM as supernovae (Abbott 1982). Aside from its impact on the galactic environment, this mass loss is interesting both as a physical process and because of its profound effect on later stages of evolution. Though all OB stars have respectable mass loss rates, we will consider here the two most spectacularly profligate classes of stars. The Luminous Blue Variables are, as their name implies, very bright, high T_{eff} stars which display irregular variability. These large eruptions which can involve several magnitudes of brightening followed by substantial dimming as dust forms in the expelled material and extincts much of the light from the star.

Some archetypical examples of this class are η Carinae, with its associated large and complex Homunculus Nebula, and P Cygni, famous for spectral line profiles characteristic of outflows. The Wolf-Rayet stars are defined observationally as stars with broad emission lines, primarily of He I and He II, large mass loss rates ($> 10^{-6}\ M_\odot\ \mathrm{yr}^{-1}$), high T_{eff} (25,000 – 100,000 K), and high wind velocities (a few $\times 10^5$ cm s^{-1}). The subclasses of WR stars are based on the dominant emission lines (besides He) of the object. Objects with strong nitrogen emission are classed as WN, and are further divided into stars with hydrogen (WNL) and without (WNE). Stars with He burning products are classified as WC (strong carbon lines) and WO (oxygen) (Vanbeveren & Conti 1980). Wolf-Rayets are effectively the stripped cores of massive stars. Most WRs are burning helium, but some WNL objects may be extremely massive stars on the hydrogen burning main sequence.

We will attempt to place these objects in an evolutionary context and elucidate the physical processes involved. In these stars there appear to be two components to the mass loss. One is a steady wind which arises from several sources, depending on the conditions in the star. This component is well modelled and conforms satisfactorily to observations. The other is characterized by catastrophic eruptions in which several solar masses of material may by ejected in a single event. The underlying physical process in this case is less well understood.

In the O supergiant precursors to these stars, there is a steady line driven wind which is optically thin in the continuum. An important characteristic of such winds is the "bi-stability jump" (Vink et al. 1999). This feature is driven by a rapid change in the ionization equilibrium of the plasma with temperature. At $T_{eff} < 21,000$ K the opacity is dominated by a large number of weak lines. Above 21,000 K most of the opacity is provided by a small number of stronger lines. On the hot side of the jump the mass loss rate drops by a factor of five, and the terminal velocity of the wind increases by a factor of two. Following Lamers & Nugis(2002), we will refer to these as the "line cool" and "line hot" winds. O supergiants appear to be dominated by the line hot winds. As these stars move towards the red with age, the transition to the line cool winds occurs, and may signal the onset of the LBV phase. This transition can occur while the star is still on the main sequence. However, it is possible that some stars become LBVs without ever moving to the cool side of the bi-stability jump.

The steady winds in WR stars have a somewhat different origin. These winds are driven by an optically thick continuum. As mass is lost the radius of the star decreases and the material becomes increasingly hydrogen poor. The temperature at the wind's sonic point correspondingly increases. In order for the continuum driving process to be effective, the opacity must be increasing outward at the sonic point of the wind (Nugis & Lamers 2002). This occurs only at certain temperatures where metals produce bumps in the opacity profile. Two

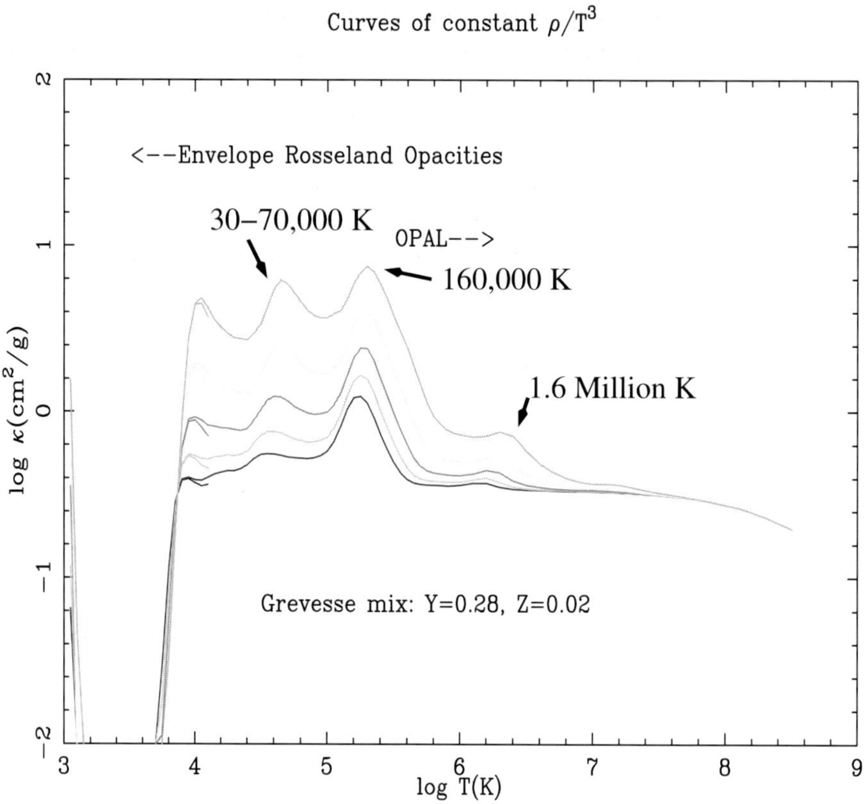

Figure 1.3. OPAL opacities for solar composition Iglesias & Rogers 1996. The opacity peaks of particular interest in LBV's and WR's are indicated. The two lower temperature peaks are implicated in the steady, continuum driven mass loss of WR stars. The peak at 1.6×10^6 K may drive instabilities in the deep envelope which cause large eruptions in LBV's.

important opacity bumps occur at $156,000 \leq T \leq 162,000$ K and $37,000 \leq T \leq 71,000$ K (Iglesias & Rogers 1996). Figure 1.3 shows the OPAL opacity tables for solar composition. The relevant opacity peaks are marked. The second bump overlaps the temperature range for line hot winds, implying that the transition from an LBV to a WNL star may be a gradual process. The transition to the higher temperature continuum driving must occur more abruptly as the star's hydrostatic radius and luminosity decrease and the wind is restructured. These winds can produce both large mass loss rates and high terminal velocities (Lamers & Nugis 2002).

The opacity bumps which cause the continuum driven winds may play an important role in the catastrophic mass loss as well. These bumps cause the star's luminosity to near or exceed the local Eddington luminosity ($L_{\rm Edd}$). Near the surface this drives a strong wind. Deeper in the interior of the star, this may cause a strong hydrodynamic instability which can lead to an eruption of substantial amounts of mass. Figure 1.4 shows a plot of the ratio of radiative to gravitational acceleration (equivalent to $L_{\rm star}/L_{\rm Edd}$) versus mass for a $40\ M_\odot$ star. The 70,000 K and 160,000 K bumps are contained in the final point, which is averaged over the outer envelope. The outermost spike in $L_{star}/L_{\rm Edd}$ corresponds to a third opacity peak at 1.6×10^6 K. L_{star} also reaches a significant fraction of $L_{\rm Edd}$ at the sharp composition gradients bounding the H depleted shell and further depleted core. These $L_{\rm Edd}$ instabilities may initiate strong non-radial pulsational instabilites, especially for radiation pressure dominated equations of state. Preliminary models show the instability occuring in stars during hydrogen shell and He core burning. For stars of $\leq 60\ M_\odot$ the instability appears on the Hayashi track. For higher mass stars the instability appears progressively earlier on the Hertzsprung gap, until it is probably observationally indistinguishable from the main sequence. The mechanism is a candidate for eruptive outbursts in both LBVs and WRs. Instabilities in the deep envelope are especially attractive for LBVs, as the amounts of mass ejected can be truly prodigious. Observations of η Carinae in the thermal infrared reveal a massive cool component to the Homunculus nebula which increases the estimate of the mass ejected in the "Great Eruption" of the $19^{\rm th}$ century from 2 – 3 $M_\odot$ to $> 12\ M_\odot$ (Smith et al. 2003).

Wolf-Rayet stars, being hydrogen depleted massive cores, are believed to be the progenitor objects of SN Ib/c. This means that the initial/final mass relation for these objects is of great importance. Some observational constraints on the parent populations of LBVs and WRs do exist. In the Large Magellanic cloud, early type WRs occur in clusters with turnoff masses as low as $30\ M_\odot$. (Turnoff masses for clusters this young can be somewhat unreliable, but the results of this study do give a good representative feel for the range of masses spanned by these objects.) Late type WN and WC stars have very similar distributions and occur in clusters with turnoff masses of $\geq 45\ M_\odot$. The LBVs in the LMC and the Galaxy appear to come exclusively from stars of $> 85\ M_\odot$. Some Ofpe/WN9 stars of 25 – $35\ M_\odot$ undergo LBV-like outbursts, but are likely more evolved stars which undergo an LBV-like instability during He burning. A striking difference is observed in the SMC. Though the numbers are small, all WN and WC stars appear in clusters with turnoff masses of $\geq 70\ M_\odot$ Massey et al.(2000). This is an unsurprising result, as both the steady and episodic mass loss mechanisms in these objects depend upon metal line opacity. Knowing the initial masses does not necessarily lead to reliable masses for the LBV's and WR's themselves. Unfortunately, theoretical evolutionary calculations have

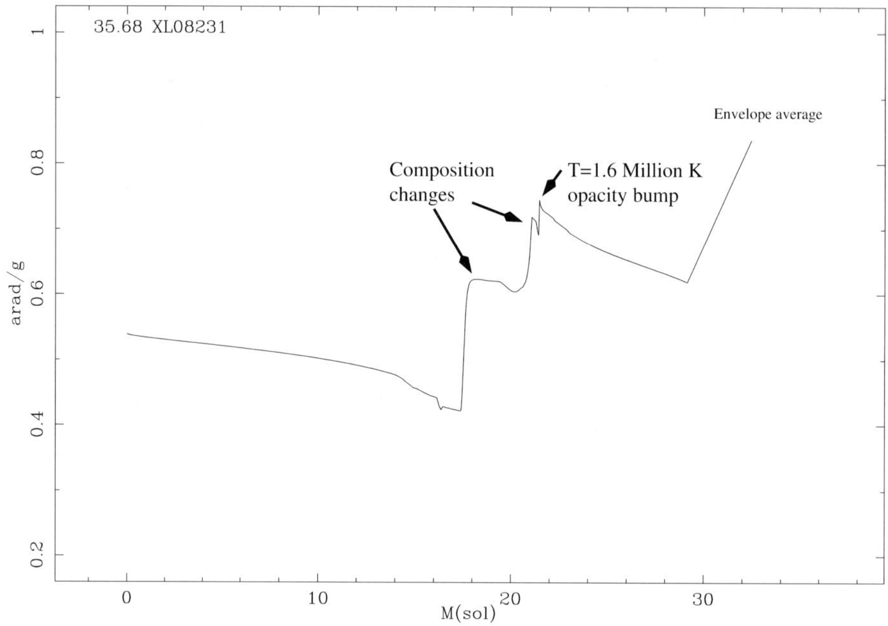

Figure 1.4. Plot of the ratio of radiative to gravitational acceleration ($L_{star}/L_{\rm Edd}$) versus enclosed mass for a 40 $M_\odot$ star. $L_{star}/L_{\rm Edd}$ becomes significant at several points. The outermost corresponds to an opacity peak at 1.6×10^6 K. The interior spikes correspond to sharp composition gradients outside the H burning shell and He burning core.

been limited. The most recent widely used tracks are from Langer et al.(1994), Crowther et al.(1995), and following papers, and Maeder & Meynet(1994). These groups use differing assumptions which lead to differences in final core masses of factors of two. These calculations also do not benefit from more recent advances in our understanding of opacities, mixing, and mass loss. The best we can say at the moment is that the SN Ib/c progenitor cores may range from roughly 5 to 20 $M_\odot$. This area has possibly the most potential for improvement in the near future of any bearing on supernova progenitors.

7. C and Ne Burning

Carbon burning proceeds at $T > 9 \times 10^8$ K by the fusion of $^{12}\mathrm{C} +^{12} \mathrm{C}$, making ^{23}Na, ^{20}Ne, and β unstable ^{23}Mg, as well as free particles (p, α, n). The latter undergo rapid secondary reactions. Neon burning occurs at $T > 1.5\times10^9$ K by photodissociation of $^{20}\mathrm{Ne}(\gamma,\alpha)^{16}\mathrm{O}$ followed often by $^{20}\mathrm{Ne}(\alpha,\gamma)^{24}\mathrm{Mg}$. The α particles also induce (α, n) and (α, p) reactions, and these secondary reactions make the burning of neon resemble that of carbon. The physics of carbon and neon burning have been extensively discussed in Arnett (1996).

Carbon burning is the first burning stage which is cooled by $\nu\bar{\nu}$ emission, not photon diffusion. Photon diffusion is proportional to the divergence of the radiative flux, and thus dependent upon the conditions in the surrounding material. The cooling is fundamentally nonlocal; in fact, the photon cooling rate is set by the second derivative of the star's temperature with repect to space. For neutrino cooling, however, the rate depends upon the local properties of the material, primarily temperature ($s \propto T^k/\rho, k \simeq 11$), and not on its spatial derivatives. This changes the structure of the star.

The rate of nuclear energy generation increases more quickly with temperature than the rate of neutrino cooling for carbon burning and later stages, so nuclear heating can always overcome neutrino cooling in the center of the core (or bottom of a burning shell). Beyond a certain radius cooling dominates. This generates a convective instability. If the convective core is too small, $\langle\epsilon\rangle > \langle s\rangle$, so the core heats, increasing in entropy and growing. Similarly, if the core is too large it cools, decreasing its entropy and shrinking. As fuel is consumed, the heating drops below the rate needed to maintain this thermal balance and the core contracts, heating up and coming back into balance. Hence only a small amount of mass near the center of the star is depleted of fuel. This phenomenon of *core convergence* occurs for all later burning stages and drives the core to near the minimum ignition mass of oxygen (also near the Chandrasekhar mass). We are much indebted to this piece of physics. The transition to neutrino cooling means the size of the carbon burning and later cores is much smaller than the He burning core. As a consequence, much of the oxygen and carbon from triple alpha burning goes unconsumed and is available to make astrophysicists.

It should be noted that the description above is oversimplified. It is important to remember that the flame zone, unlike earlier stages, resides in the convective region. The simple picture assumes that fuel is completely exhausted in a burning region, at which point the burning shell jumps outward to start consuming fresh fuel. Multidimensional simulations (Bazàn & Arnett 1998; Asida & Arnett 2000) indicate that thermal balance is achieved, but unburned fuel can be mixed from regions outside the burning core/shell. The compositional gradients are not as abrupt as normal stellar evolution calculations predict, and there are significant density perturbations and wave physics at the boundary. The process of core convergence still holds, but will be different in detail, as will the position and strength of burning shells and the associated nucleosynthesis.

The timescales of carbon burning range from periods of a few centuries in massive stars to near the Kelvin-Helmholtz timescale for stars of $8 - \sim 15\ M_{\odot}$. Carbon burning lasts much longer in these stars because of (1) lower entropy and hence slower neutrino cooling rates and (2) more ^{12}C relative to ^{16}O from He burning. This may have observational consequences (Smartt et al. 2002) for SNII-P events.

Only fairly recently (Arnett, §9.6 1996) was it realized that the secondary reactions of C and Ne burning provide a timescale for these processes. Explosive C and Ne burning overproduce isotopes of ^{64}Ni, ^{70}Zn, ^{76}Ge, ^{82}Se, and ^{86}Kr. Quasistatic burning is slow enough to allow β decays and avoids this problem. Therefore, the bulk of the yield of C and Ne burning products come from quasistatic burning with little explosive processing.

Carbon ignition at high density ($\rho > 10^9 \mathrm{g\ cm}^{-3}$) is particularly complex because the burning produces effective URCA pairs (^{25}Mg–^{25}Na, ^{23}Na–^{23}Ne, and ^{21}Ne–^{21}F) which, if they are convected, counter the heating from ^{12}C fusion (Arnett, §11.3 1996).

8. O and Si Burning

At the end of Ne burning the most abundant nuclei are ^{16}O, ^{24}Mg, and ^{28}Si. The most common nuclei with a neutron excess are ^{27}Al and ^{29}Si. Oxygen burning proceeds at $T > 2 \times 10^9$ K by the fusion of $^{16}\mathrm{O} +^{16} \mathrm{O}$, and also releases free (p, α, n) which react even more vigorously than for C or Ne burning (Arnett, Ch. 10 1996). After relatively little burning, electron capture and positron decay increase the neutron excess to the point that an explosion would produce not ^{56}Ni, but other nuclei (^{54}Fe, ^{58}Ni, ^{56}Fe, ^{58}Fe, etc.). Supernova light curves show evidence for decay of ^{56}Ni and its daughter ^{56}Co, so we infer that *explosive* oxygen burning is dominant in the ejected material.

Oxygen burning is driven so strongly by neutrino cooling that the entire phase lasts a matter of months. The hydrodynamic timescale approaches the evolutionary timescale, and direct numerical simulation is feasible (Arnett 1994; Bazàn & Arnett 1998; Asida & Arnett 2000). The results are disquieting: the hydrodynamics of the numerical simulations is dramatically different from that assumed in the stellar evolutionary models. Convection in the oxygen shell drives gravity waves which cause slow circulation and mixing; the same basic process is discussed in §3 for hydrogen burning stars. The differences result in much shallower composition gradients in the core and a different structure for the burning shells. There are also density perturbations at the boundary which will influence the transmission of the shock after core collapse. The acoustic flux is also non-trivial, as Mach numbers can exceed several percent, even for weak shell burning (Bazàn & Arnett 1998; Asida & Arnett 2000). Computational strategies must be chosen with care; anelastic codes, for example, are probably not appropriate for oxygen shell burning, but could make earlier stages computationally manageable.

Silicon burning takes place at $T > 3.5 \times 10^9$ K and lasts a few days. Initially, burning follows $^{32}\mathrm{S}(\gamma, \alpha)^{28}\mathrm{Si}$, as oxygen burning produces more ^{32}S than quasiequilibrium allows. Light neutron-rich nuclei such as ^{34}S are depleted early as iron peak species increase. As ^{28}Si is consumed and more mass is pushed toward the iron peak, the faster rates of positron capture and beta decay cause the neutron excess to increase markedly. Near Si depletion η can reach several $\times 10^{-2}$. This neutron excess is much too high to account for the measured solar system abundances, so we infer that this material, too, goes largely into the compact remnant, and ejected material is a product of explosive Si burning.

At any given time, multiple core and/or shell burning episodes are likely to be occuring. The exact ordering of these episodes (or at least models thereof) appears to be sensitive to initial conditions and to the details of mixing. An adequate discussion of the possible variations and contributing factors is far beyond the scope of this chapter; we will attempt to communicate some trends in terms of nucleosynthesis. (Arnett 1996, Chap. 10 discusses the problem in considerable detail; much of this section follows that work and references therein.)

In general, shell burning takes place at higher temperatures than core burning of the same element. Ashes from the innermost shell increase the inert core mass until it reaches a high enough temperature to ignite the next fuel. Burning of one shell usually damps burning in the next shell farther out. This alternating shell buring is similar in principle to the thermal pulses of late AGB stars.

In carbon shell burning the elements from Cu ($Z = 29$) to Kr ($Z = 36$) are made with nearly the solar abundance pattern. The upper limit reflects the neutron magic number at $N = 50$. This abundance pattern will always appear for shell carbon shell burning of Pop I abundance material. This material should be a component of the SN ejecta, and is a prime contributor to the weak s-process. Neon burning brings some lighter neutron rich nuclei underproduced in C burning (^{29}Si, ^{30}Si, ^{31}P, ^{36}S, ^{40}Ar, ^{40}K, and ^{46}Ca) up to solar values.

Quiescent oxygen shell burning does not suffer the anomalously high neutron excess seen for core O burning, and so may form an important component of the ejected material. This is because the oxygen shell burning is incomplete and because the lower densities in the shell inhibit electron captures. Fragile nuclei of high nucleon number ($A > 50$) are destroyed by the high temperatures. Initially ^{16}O and ^{24}Mg are depleted, going to ^{28}Si and ^{32}S. Only near the end of the burning are ^{36}Ar and ^{40}Ca significantly enhanced. The neutron excess moves to heavier nuclei such as ^{34}S. The neutron excess is too high to explain the Si to Ca region, but works well for the iron peak. The shell may be ejected as-is, after incomplete burning, or explosively processed. The oxygen shell, burning either quasistatically or explosively, is a candidate for the site of the p-process Lambert(1992).

In the Si shell, much as in the core, a short period of S burning occurs before Si is depleted as abundances reach new equilibrium values different from oxygen burning. Above $T > 3.5 \times 10^9$ K, canonical Si burning occurs. The neutron excess shifts from ^{34}S to the iron peak, especially ^{54}Fe. A very large neutron excess in the iron peak is generated, driving abundances of species such as ^{55}Mn and ^{54}Fe to factors of several larger than the solar system abundances. Any material significantly depleted in Si will have too large a neutron excess to match observed abundances and is probably not a major component of the ejecta.

It should be noted that the above conclusions are drawn from simplified models and should be taken merely as guidelines. Serious attempts to make accurate SN progenitors should model the shell burning with 3D hydrodynamic codes. Because of the short evolutionary timescales involved and the conscientious application of Moore's Law to computing power, this strategy is becoming realistic. The features of such a model will be discussed in the section on the pre-SN state.

9. Pop III stars

Recent major advances in astronomical observations are threatening to make the behavior of zero metallicity stars more than a theoretical curiousity. Measurements of metal lines in Lyα clouds at high redshift imply substantial chemical enrichment at early times. The *WMAP* satellite has discovered tantalizing hints of the reionization of the universe by the first massive stars in the polarization of the Cosmic Microwave Background. In 2002 a star with [Fe/H] = -5 was found in the Galactic halo. One of the stated mission objectives of the *James Webb Space Telescope* is to detect the light of the first stars in the universe. What can we predict about the properties of these objects?

The problem turns out to be surprisingly tractable. Eliminating heavy elements should not introduce any new stellar physics. Opacity tables and equations of state are available for $z = 0$ compositions. Large nuclear reaction networks are already designed to handle singular matrices gracefully. Numerical issues such as roundoff error may have more chances to crop up, but they are not fundamentally different from those encountered in a normal stellar evolution calculation. A carefully constructed stellar evolution code should be able to model the lifecycle of a zero metallcity star as well as one of solar composition. The same caveats and uncertainties apply as in a normal calculation, of course. Rotation, in particular, is a worry, as it should strongly affect these high entropy stars. Progress in this area will likely require multi-dimensional hydro simulations. In some respects the problem is actually simplified. Since steady mass loss in hot stars is mostly radiatively driven by metal lines, its rate, and thus its effect on the evolution should be much reduced. The question of Pop III star formation is a much more complicated one, but is outside the scope of this chapter. In this section we will concentrate on the differences between Pop

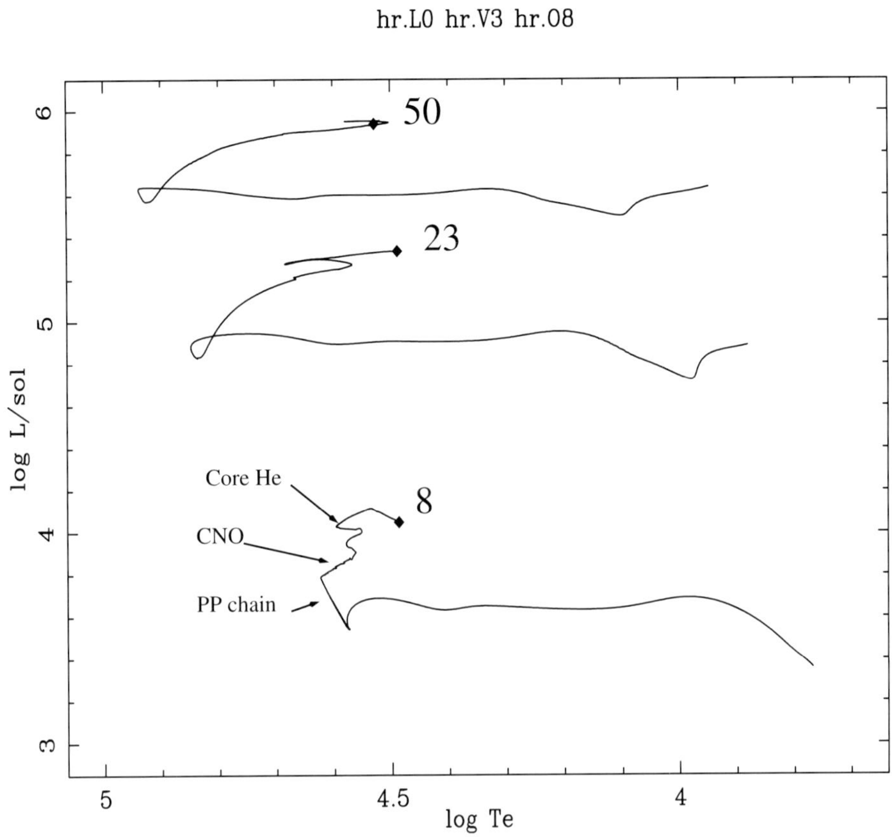

Figure 1.5. Evolutionary tracks up to core He exhaustion for 8, 23, and 50 $M_\odot$ $z = 0$ stars. The dominant nuclear burning processes are labelled for the 8 $M_\odot$ case. For the larger star PP chain and triple alpha burning begin on the "pre-MS" part of the track. CNO burning takes over when the star begins to increase in luminosity and move to the red.

III and Pop I/II evolution described earlier. Figure 1.5 shows tracks for 8, 23, and 50 $M_\odot$ zero metallicity stars up to the point of core He exhaustion.

The greatest differences in the evolution of Pop III stars arise from the lack of CNO catalysts for H burning. The PP chain does not produce enough energy to bring a massive star into thermal balance, so the star continues to move blueward across the HR diagram, contracting and increasing in temperature while PP burning is taking place. As a result of the relatively weak temperature dependence of the PP chain, H is being processed into He in a large fraction of the star, and ^{3}He is only being partially destroyed. A small convective core is established. For most of the mass range of interest, this phase does not last long enough for H to be significantly depleted. Eventually, the temperature exceeds 10^8 K, and triple-alpha burning begins. Once He burning has increased the abundance of C to $\sim 10^{-11}$, CNO burning can begin. The stars come into thermal balance and begin to increase in luminosity and move towards the red. This is what we would recognize as the ZAMS in an evolutionary track of normal composition. The convective core established in the PP phase persists throughout triple alpha and CNO burning. During CNO burning the core temperature remains high enough to maintain a low level of triple-alpha. This has interesting nucleosynthetic consequences, as various proton capture elements are produced in the core, albeit at a low level since the abundance of He burning products is still small. Upon H exhaustion, core He burning begins immediately and takes place entirely on the blue side of the HR diagram. The ignition of C occurs on what we would normally identify as the Hertzsprung gap. By this point the core evolution should proceed much as it does in higher metallicity stars, since the material has been processed through complete He burning. The nucleosynthetic details will differ because of the abundance of proton capture elements, but the energetics should be similar. The hydrogen burning shell will burn at a higher temperature, due to the low abundance of CNO. Evolutionary calculations have rarely been performed beyond the point of C ignition, so the extent to which nuclear burning products are dredged up and ejected prior to the supernova explosion is uncertain: Heger & Woosley(2002); Chieffi & Limongi(2002). Since the timescales of the later burning stages are not large compared to a convective turnover time in the envelope, it is unlikely that much processed material is brought to the surface if dredge-up occurs at all.

The situation for the lowest mass potential SN progenitors ($8 - 10\ M_\odot$) is somewhat different. Though central temperature and density continue to increase during the PP phase, they do so slowly. On the HR diagram the star follows a track reminiscent of a low mass star, moving upwards in luminosity and slightly towards the blue. The temperature remains below 10^8 K, so triple-alpha burning takes place at a very low rate. More than half of the central hydrogen is depleted by PP chain burning before the He burning products are abundant enough for CNO burning to begin in earnest. The small convective core present at the beginning of hydrogen burning disappears entirely for a time. When CNO burning begins the evolutionary track takes a turn towards the red, as for a higher metallicity intermediate mass star. Full scale He core burning takes place entirely on the blue side of the HR diagram, so no Red Giant Branch exists for this population. During shell burning there is likely to be one or more gredge-up phases, enriching the entire star in heavy elements. As a result, later stages are similar but not identical to corresponding stages in higher metallicity stars. Models from different groups differ in detail. A more complete discussion can be found in Siess, Livio, & Lattanzio(2002) and Chieffi et al.(2001) and references therein.

Population III stars have one possible characteristic which makes them unique. Because of their low opacities, it is hypothesized that zero metallicity stars have an extremely high mass cutoff, as high as $10^3\ M_\odot$. The physics of stars with such high temperatures and radiation entropies opens an alternative fate to core collapse. These stars may undergo an e^-e^+ pair instability during oxygen burning which leads to explosive disruption of the star. This mechanism and its consequences are discussed in the next section.

10. Supermassive Stars and Pair Instability

The potentially top-heavy IMF of Pop III stars brings us to the interesting subject of supermassive stars. Much of the following discussion follows Bond, Arnett, & Carr(1984) and Arnett (1996). Bond et al. (1984) define Very Massive Objects (VMOs) as those stars unstable during oxygen burning due to e^-e^+ pair production. Depending upon composition, these stars range in mass from $\sim 10^2\ M_\odot$ to the mass limit for the general relativistic instability. Zel'dovich & Novikov(1971) estimate this cutoff, above which the term Supermassive Objects (SMOs) is used, to be $\sim 6 \times 10^4\ M_\odot$ without nuclear burning. With nuclear energy generation from hydrogen burning this increases to $\sim 5 \times 10^5\ M_\odot$ for Population I and II and $\sim 2 \times 10^5\ M_\odot$ for Pop III . Fricke (1973); Fricke (1974) find smaller values, and predict that Pop III SMOs will always become black holes. During helium burning, the cutoff for the general relativistic instability drops to $\sim 4 \times 10^4\ M_\odot$, and more massive cores will already be collapsing before oxygen burning. Such objects should end up as black holes.

Much of relevance to the early evolution of such stars is discussed in §6, since VMOsVery Massive Objects (VMOs) are the likely progenitors of LBVs and the most massive Wolf-Rayets. Therefore we proceed directly to oxygen burning and the pair instability. Evolution after He burning is complete is so rapid that mass loss becomes irrelevant. Carbon and neon burning, too, are only minor perurbations to the final evolution, since little ^{12}C remains after high entropy He burning, and Ne burning has a small energy output. These species will have a much more significant impact on the nucleosynthetic yields.

At sufficiently high temperatures ($T > 10^9$ K if $\rho = 10^3$g cm^{-3}), photons will be sufficiently energetic to produce e^-e^+ pairs when scattering off the coulomb field of nuclei ($\gamma + Z \rightarrow Z + e^- + e^+$). A significant fraction of the energy of the photon gas goes into increasing the number of free particles without much increasing the temperature of the material. This corresponds thermodynamically to a large increase in the specific heat of the plasma. For high entropies (i.e. very massive stars) this can drive $\Gamma_1 < 4/3$, making the core of star unstable (Clayton 1968). (Using the adiabatic gammas is a simplification, of course; pair annihilation and neutrino cooling makes the star significantly non-adiabatic. This does not change the argument qualitatively.) The star will undergo a runaway contraction, which can drive explosive core nucleosynthesis.

The impact of pair creation increases with increasing entropy. For oxygen cores with masses at the lowest end of the instability range (very approximately 40 – 50 $M_\odot$) the contraction phase is stopped by an oxygen flash which does not disrupt the star. The star settles into nuclear-driven pulsations. The equation of state is very soft, so small perturbations can result in large amplitude motions. Thus the oscillations of stars at this end of the mass range must be carefully modelled in order to determine whether the star eventually explodes or evolves on to core collapse.

For larger stars (approximately 50 – 150 $M_\odot$) the collapse phase extends to higher temperatures, and the star burns oxygen explosively and is completely disrupted, leaving no remnant. The ejected oxygen mass appears to be roughly constant at several tens of solar masses. Masses of Si–Ca species and ^{56}Ni start out negligible, but reach several tens of solar masses for initial oxygen cores of $> 100 M_\odot$. These supernovae should be spectacular. Arnett (1996) calculates an explosion energy of 10^{53} erg for a 120 $M_\odot$ core. The bolometric lightcurve calculated for a $100 M_\odot$ core had a peak at $\log(L/L_\odot) \simeq 10.6$ which lasted 100 days. These events, while exceedingly rare in the recent universe, should be observable out to high redshift if we are lucky enough to have one occur.

In larger cores, the collapse proceeds to sufficiently high temperatures that nuclei are dissociated into alphas and alphas are dissociated into nucleons. These processes act like pair creation to drive the instability even more strongly. At this point (roughly for oxygen cores of $> 150 M_\odot$), there is no energy source which can stop the collapse, and the star presumably becomes a black hole without exploding.

11. Pre-supernova State

Mathematically, explosions are exponentiating solutions, which are sensitive to initial conditions. What, then, are the properties in the immediately pre-SN star which influence the explosion most?

The most global factor affecting the explosion is the entropy structure and mass of the core. This begins to be determined all the way back in hydrogen burning, when the mass of the nuclear burning processed core is first set. Care is required in modelling the mixing in early stages of the evolution to get this right at the end. This also influences the neutron excess and distribution of nuclear buring products.

These last two factors achieve their final distribution during late stages of shell burning. At the same time, the final important properties of the star are being set. The details of the explosion will depend upon the angular momentum and velocity distribution of the stellar material and any density anisotropies. All

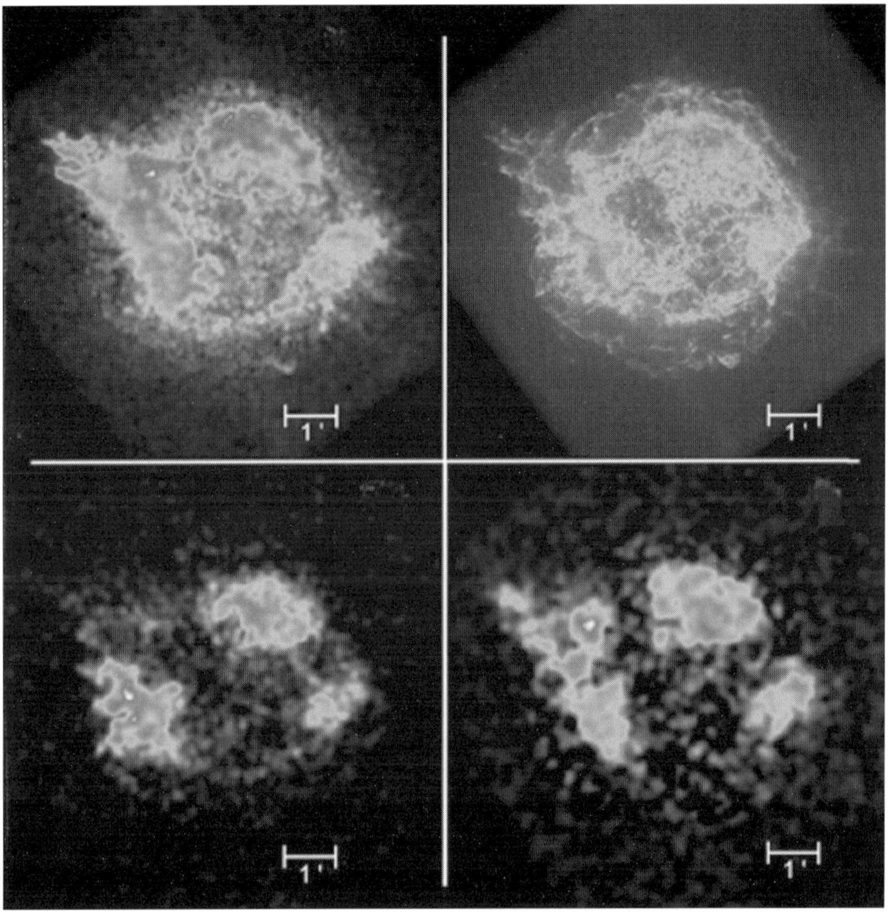

Figure 1.6. This Chandra x-ray image of Cassiopea A shows inhomogeneity in the products of explosive nucleosynthesis (NASA/GSFC/U.Hwang et al.). The upper left image is broadband, and in clockwise order are images of emission from silicon, iron, and calcium.

four of these quantities depend upon the hydrodynamics of shell burning, which simulations have shown to be quite complex.

Figure 6 shows the inhomogeneity in the products of explosive nucleosynthesis in the supernova remanant Cassiopea A. Silicon is produced by explosive or quasistatic oxygen burning. Calcium is most likely produced by explosive oxygen burning. Iron can be produced by oxygen burning that continues into silicon burning. All have different shapes in the image. This could be due to different degrees of explosive burning, with large burning making Ca, little burning leaving the Si from quasistatic burning intact, and extreme burning making Fe. It could also be due to inhomogeneity in the progenitor.

Optical data (Fesen 2001), infrared data (Douvion, Lagange, & Cesarcky 1999) from ISOCAM, x-ray data (Hughes et al.2000) from Chandra and (Willingale et al. 2003) from XMM-Newton, and gamma-ray lines (Vink et al. 2001) are giving a much improved picture of Cas A. In addition to Si, Ca and Fe, the optical data include O and Ar (incomplete oxygen burning) and N and H (pre-explosion mass loss?). The infrared include a silicate layer, a S and Ar layer, and a Ne layer. The gamma-lines give ^{44}Ti. All show evidence for inhomogeneity in composition and kinematics. The best choice for the progenitor seems to be fairly massive, possibly a 10 to 30 solar mass Wolf-Rayet star, but details of the identification are sketchy.

If highly evolved stars no longer maintain the symmetry of spherical shells, what will be the consequences for nucleosynthesis yields? Suppose the star to be made not of spherical shells, but of a patchwork of blobs. Neutrino cooling would lower the entropy of a blob causing it to sink, moving past its neighbors. This would continue until the temperature rose enough to ignite a fuel. This would increase the entropy, making the blob buoyant, and causing it to rise. We might imagine a cycle of repeated rising and falling, with some fuel consumed each time. The temperature at which falling was reversed would depend upon the composition of the blob, so that on average there *would* be a layering according to composition, but it would be broken by fluctuations. Some burning flashes would be more vigorous, giving larger excursions, spreading ashes in ways not seen in one dimensional simulations. Could a few vigorous oxygen flashes make ^{56}Ni and carry it out far enough to explain the early appearance of ^{56}Co gamma-ray lines in SN1987A? What would be the effect of a supernova shock on the inhomogeneous oxygen-rich layer? Are nuclear-energized pulsations driven by the accelerating contraction of the core during silicon burning? How does rotation couple to the burning, and does this qualitatively change the nature of the inital core collapse model? The range of possiblities is larger than yet considered. Fortunately, the late shell burning stages are amenable to direct numerical simulation in multiple dimensions, so the situation is likely to improve.

12. Modelling Issues

We have a firm grasp of how stars evolve and what basic physics are involved. Stars are very unlikely to surprise us in a fundamental way. Because of this, there is a common perception among astronomers outside the field that stellar evolution is a "solved problem." In many cases, however, we still do not have the quantitative understanding needed in other fields dependent upon knowledge of the behavior of stars. This situation is rapidly changing with improved observations and increased computing power driving our theoretical understanding. Many of the unresolved issues have been discussed in the text; we will briefly summarize them here.

In order to follow the evolution of a star over millions or billions of years, certain extreme simplifications need to be made in the calculations. Most of the shortcominings of stellar evolution arise from two of these assumptions. The first is that stellar evolution can be accurately treated as a series of snapshots of the star in equilibrium or steady state configurations. The second is the assumption of spherical symmetry, allowing the use of one-dimensional stellar evolution codes. Both of these assumptions eliminate the possibility of hydrodynamic or magnetohydrodynamic effects on the structure and evolution of a star.

Our most pervasive problem lies in the realm of mixing, which affects every star from its pre-MS evolution all the way through the process of core collapse. During the longer evolutionary stages convection is usually calculated using either mixing length or full spectrum of turbulence theory. Mixing length theory would appear to commit every sin we can think of in this context, being an ensemble average of one-D motions of isolated blobs, but ironically it usually gives quite satisfactory answers for the convective zone itself. Except for the outer envelopes, the nablas are extremely close to adiabatic, and therefore the convection depends weakly on the theory used.

Problems arise at the convective boundary and in the radiative regions. The boundary is non-spherical and non-static and of necessity couples the convective and laminar regions. (See Figure 1.2 for a schematic of this interaction.) The laminar region itself has hydrodynamic behavior. It appears now that three main processes contribute to mixing outside the convective core: meridional circulation driven by rotation, angular momentum transport in the laminar regions by gravity waves driven by convective plumes, and slow circulation driven by non-isentropic dissipation of the gravity waves. This is an excellent example of the power of integrating hydrodynamics and stellar evolution. These behaviors would probably not have been diagnosed or understood without multi-D numerical simulations. Now that the processes have been identified and studied, analytic approximations of each have been incorporated with substantial success in stellar evolution codes. The numerical simulations are invaluable for identifying new processes and spot-checking the results from simplified one-dimesnional (1-D) evolution codes.

More difficulties are a consequence of the steady state assumption. Most evolution codes assume instantaneous mixing in convective zones. During hydrogen burning this is not much of a problem. During He core and shell burning, when unprocessed material is ingested into the convective zone, the consequences can be much more dire. The result can be "breathing pulses", thermal instabilities which cause excessive core growth. Most codes impose arbitrary limits on the ingestion of fresh helium into the burning region. A time-dependent treatment may obviate the problem. Much interesting nucleosynthesis goes on during shell burning, and depends sensitively on the abundances of trace elements, the temperature and entropy at which they are processed, and the free neutron flux. In light of this, it would be surprising if instantaneous mixing did not give us the wrong result.

A more obvious example of both the problems and their solutions is that of catastrophic mass loss in massive stars. There is ample observational evidence of eruptive mass loss in LBVs and Wolf-Rayets, but a 1-D stellar evolution code is totally unequipped to handle it. A well crafted code will, however, identify conditions under which the star should become unstable. The observed durations of eruptions are relatively short, on the order of days to weeks. This kind of timescale is beginning to be within the means of 3-D stellar structure codes with realistic physics, for example the Djehuty project at Lawrence Livermore National Labs. When a potential instability is identified in an evolution code, the strongly hydrodynamic phase can be calculated with a more appropriate tool and returned to the evolution code when the star has settled down to a longer lived quiescent state. While not perfect, the final star calculated by this process should be much more realistic than one produced by a 1D code with a steady mass loss rate alone. The series of multi-D WR or LBV models should also be more realistic and more relevant to actual observations of such objects.

Simulations show late stages of core and shell burning are strongly hydrodynamic, with density perturbation of several percent, Mach numbers of $> 10^{-2}$, and strong interactions with surrounding laminar regions. During these stages the nuclear timescale can be even shorter than the hydrodynamic timescale. Clearly, a spherically symmetric, quasistatic model is inappropriate. The inhomgeneities generated during this stage will imprint themselves directly upon the supernova explosion. Fortunately, the problem can be calculated explicitly and in multi-D. The remainder of the star has adjustment times much longer than the evolutionary time and can largely be ignored (Bazàn & Arnett 1998; Asida & Arnett 2000).

These considerations suggest a novel strategy for producing SN models which are both more realistic and useful to the larger astronomical community. Traditional stellar evolution codes are adequate for large parts of a star's life. These can be improved substantially by incorporating results from numerical hydro simulations of snapshots of these evoltuionary stages. Situations that are strongly hydrodynamic (i.e. eruptive mass loss, O buring) are likely to be of short duration. These stages can potentially be calculated explicitly in multi-D from a stellar evolution model, with the final product returned to the 1-D code for continued evolution. The end result should be a SN progenitor with a much stronger relationship to a *particular* initial star, allowing supernova modelling to become a better tool for studying entire astronomical systems, as well as the explosions in isolation.

Table 1.2. Stellar Evolution with Detailed Yields

Reference	Masses/$M_\odot$	Metallicity	Networks	Comment
Rauscher, Heger Hoffman,& Woosley (2002)	15,19,20, 21, 25	solar	19 700-2200	NON-SMOKER rates
Heger & Woosley (2002)	65-130, steps of $\Delta M = 5$	solar	19 304,477(exp)	He cores
Thielemann, Nomoto & Hashimoto (1996)	13, 15, 20, 25	solar	299	
Maeder & Meynet et al.	0.8-120	Z= 0.001-0.08	19 rotation	mass loss
Chiefi, Limongi, & Straniero (1998)	25	solar	149	pre-MS to collapse
Limongi, Straniero, & Chieffi (1998)	13-25	solar	149	pre-MS to collapse
Chiefi & Limongi (2002)	15,20,25, 35,50,80	Z=0	179	pre-MS to collapse
Limongi, & Chieffi (1998)	13-35	solar	267	pre-MS to collapse
The, El Eid, Meyer (2000)	15,20,25,30	solar	659	Ne ignition
Aubert, Prantzos, Baraffe (1996)	20	solar	96	Ne-exhaustion Göttigen code

13. Current Work

The current state of the art for supernova nucleosynthesis simulations is illustrated by the work of several groups, in which extensive networks have been coupled to the stellar evolutionary equations. Several are summarized below in Table 1.2; the list is quite incomplete, but fairly representative. Woosley, Heger, & Weaver (2002) have recently produced an extensive review. In addition, the intermediate mass stars make an important contribution (see above).

References

Abbott, D. C. 1982, ApJ, 263, 723

Alcaino, G., Alvarado, F., Borissova, J., & Kurtev, R. 2003, A&A, 400, 917

Angulo, C. et al. 1999, Nucl. Phys. A656, 3

Arnett, D. 1994, ApJ, 427, 932

Arnett, D. 1996, *Supernovae and Nucleosynthesis* (Princeton: Princeton Univ. Press)

Asida, S. M. & Arnett, D. 2000, ApJ, 545, 435

Aubert,O., Prantzos, N., & Baraffe, I., 1996, A&A, 312, 845

Bazàn, G. & Arnett, D. 1998, ApJ 494, 316
Blum, R. D., Conti, P. S., & Damineli, A. 2000, AJ, 119, 1860
Blum, R. D., Damineli, A., & Conti, P. S. 2001, AJ, 121, 3149
Bond, J. R., Arnett, W. D., & Carr, B. J. 1984, ApJ, 290, 825
Burbidge, E. M., Burbidge, G., Fowler, W. A., & Hoyle, F. 1957, Rev. Mod. Phys., 29, 547
Caloi, V. & Mazzitelli, I. 1993, A&A, 271, 139
Cameron, A. G. W., 1957, Chalk River Report CRL-41
Charbonnel, C. & Talon, S. 1999, A&A 351, 635
Chieffi, A., Dominguez, I., Limongi, M., & Straniero, O. 2001, ApJ, 554, 1159
Chieffi, A., & Limongi, M. 2002, ApJ, 577, 281
Clayton, D. D. 1968, *Stellar Evoltuion and Nucleosynthesis* (New York: McGraw-Hill)
Conti, P. S. & Blum, R. D. 2002, ApJ, 564, 827
Cook, C. W., Fowler, W. A., Lauritsen, C. C., & Lauritsen, T. 1957, Phys. Rev., 107, 508
Crowther, P. A., Hillier, D. J., & Smith, L. J. 1995, A&A, 293, 172
Dohm-Palmer, R. & Skillman, E. D. 2002, AJ, 123, 1433
Douvion, T., Lagange, P. O., & Cesarcky, C. J. 1999, A&A, 532, L111
Eddington, A. S., 1926, *The Internal Constitution of the Stars*, (New York: Dover), reprinted 1959
Elmegreen, B. G. 1997, ApJ, 486, 944
Elmegreen, B. G. 2000, ApJ, 539, 342
Fesen, R. A., 2001, ApJS, 133, 161
Figuerêdo, E., Blum, R. D., Damineli, A., & Conti, P. S. 2002, AJ, 124, 2739
Fricke, K. J. 1973, ApJ, 183, 941
Fricke, K. J. 1974, ApJ, 189, 535
Hartmann, L. 2001, AJ, 121, 1030
Heger, A., & Woosley, S. E, 2002, ApJ, 567, 532
Hillenbrand, L. A., Massey, P., Strom, S. E., & Merrill, K. M. 1993, AJ, 106, 1906
Hoyle, F. 1954, ApJS, 1, 121
Hughes, J. P., Rakowski, C. E., Burrows, D. N., & Slane, P. O. 2000, ApJ, 528, 109
Iglesias, C. A. & Rogers, F. J. 1996, ApJ, 464, 943
Kumar, P. & Quataert, E. J. 1997, ApJ, 475, L143
Lambert, D. L. 1992, A. and Ap. Rev.
Lamers, H. J. G. L. M. & Nugis, T. 2002, A&A, 395, L1
Langer, N., Hamann, W.-R., Lennon, M., Najarro, F., Pauldrach, A. W. A., & Puls, J. 1994, A&A, 290, 819
Limongi, M., Straniero, O., & Chieffi, A., 2000, ApJS, 129, 625
Limongi, M., & Chieffi, A., 2003, arXiv:astro-ph/0304185

Maeder, A. & Meynet, G. 1989, A&A, 84, L89
Maeder, A. & Meynet, G. 1994, A&A, 287, 803
Maeder, A., & Meynet, G., 2000a, A&A, 361, 101
Maeder, A., & Meynet, G., 2000b, A&A, 361, 159
Maeder, A., & Meynet, G., 2001, A&A, 373, 555
Massey, P. 1998, in *The Stellar Initial Mass Function*, ed. G. Gilmore & D. Howell, San Francisco: Astronomical Society of the Pacific, 17
Massey, P., Bianchi, L., Hutchings, J. B., & Stecher, T. P. 1996, ApJ, 469, 629
Massey, P., Johnson, K. E., & DeGioia-Eastwood, K. 1995a, ApJ, 454, 151
Massey, P., Lang, C. C., DeGioia-Eastwood, K., & Garmany, C. D. 1995b, ApJ 438, 188
Massey, P., Waterhouse, E., & DeGioia-Eastwood, K. 2000, AJ, 119, 2214
Meyer, B. 1994, ARA&A, 32, 153
Meynet, G., & Maeder, A., 2002, A&A, 390, 561
Meynet, G., & Maeder, A., 2003, arXiv:astro-ph/0105051
Nakamura, F. & Umemura, M. 2001, ApJ, 548, 19
Nugin, T. & Lamers, H. J. G. L. M. 2002, A&A, 389, 162
Pinsonneault, M. H., Steigman, G., Walker, T. P., & Narayanan, V. K. 2002, ApJ, 574, 398
Rauscher, T., Heger, A., Woosley, S. E., & Hoffman, R. D., 2002, ApJ, 576, 323
Salpeter E. E. 1955, ApJ, 121, 161
Schaerer, D. 1996, ApJ, 467, L17
Schatzman, E. 1999, Ap&SS, 265, 97
Siess, L., Livio, M., & Latanzio, J. 2002, ApJ, 570, 329
Seuss, H. E. & Urey, H. C. 1956, Rev. Mod. Phys, 28, 53
Smartt, S. J., Gilmore, G. E., Tout, C. A., & Hodgkin, S. T. 2002, ApJ, 565, 1087
Smith, N., Gehrz, R. D., Hinz, P. M., Hoffmann, W. F. Hora, J. L., Mamajek, E. E., & Meyer, M. R. 2003, AJ, 125, 1458
Stauffer, J. R., Schultz, G., & Kirkpatrick, J. D. 1998, ApJ, 499, L199
Stauffer, J. R., Barrado y Navascués, D., Bouvier, J., Morrison, H. L., Harding, P., Luhman, K. L., Stanke, T., McCaughrean, M., Terndrup, D. M., Allen, L., & Assouad, P. 1999, ApJ, 527, 219
Talon, S., Kumar, P., & Zahn, J-P. 2002, ApJ, 574, L175
The, L.-S., El Eid, M., & Meyer, B. S., 2000, ApJ, 533, 998
Thielemann, F.-K., & Arnett, D. 1985, ApJ, 295,604
Thielemann, F.-K., Nomoto, K., & Hashimoto, M., 1996, ApJ, 460, 408
Vanbeveren, D. & Conti, P. S. 1980, A&A, 88, 230
Vink, J. S., de Koter, A., & Lamers, H. J. G. L. M. 1999, A&A, 350, 181
Vink, F., Laming, J. M., Kaastra, J. S., Bleeker, J. A. M., Bloemen, H., & Oberlack, U. 2001, ApJ, 560, L79

Willingale, R., Bleeker, J. A. M., van der Heyden, K. J., & Kaastra, J. S. 2003, A&A, 398, 1021
Woosley, S. E., Heger, A., & Weaver, T. A. 2002, Rev. Mod. Phys., 74, 1015
Young, P. A., Knierman, K. A., Rigby, J. R., & Arnett, D. 2003, ApJ, in press
Young, P. A., Mamajek, E. E., Arnett, D., & Liebert, J. 2001, ApJ, 556, 230
Zel'dovich, Ya. B., & Novikov, I. D. 1971, *Relativistic Astrophysics*, (Chicago: University of Chicago Press)

Chapter 2

REVIEW ON THE OBSERVED AND PHYSICAL PROPERTIES OF CORE COLLAPSE SUPERNOVAE

Mario Hamuy
Carnegie Observatories
mhamuy@ociw.edu

Abstract Core-collapse supernovae prove to comprise the most common general class of exploding star in the Universe and they come in a great variety of flavors. The wide range of luminosities, expansion velocities, and chemical abundances displayed by these objects is evidence for large variations in explosion energy and in the properties of their progenitors. This paper summarizes observed and physical properties of all types of core collapse supernovae. Despite the great diversity displayed by these objects, several regularities emerge which suggest that 1) there is a continuum in the properties of these objects, 2) the mass of the envelope is one of the driving parameters of the explosion, or it is correlated with some other property of the core, with the latter determining the outcome of the explosion, and 3) the physics of the core and explosion mechanism of all core collapse supernovae are not be fundamentally different, regardless of the external appearance of the supernova. Far above in energy scale and ^{56}Ni production lies SN 1998bw, the only supernova firmly associated with a gamma-ray burst.

Keywords: supernovae,nucleosynthesis,abundances

Introduction

Supernovae (SNe) owe their name to astronomers Baade and Zwicky who, in the 1930's, realized that these objects were much more luminous and rarer than common novae (Baade 1938; Zwicky 1938). Their high luminosities (comparable to that of their host galaxies) and broad spectral lines led them to conclude that SNe were very energetic explosions. They went a step further and hypothesized that a SN resulted from the "transformation of an ordinary

C.L. Fryer (ed.), Stellar Collapse, 39–64.

star into a collapsed neutron star", a remarkable idea for its time which lies at the heart of modern models for SNe that result from the gravitational collapse of the cores of massive stars.

"Core-collapse SNe" (CCSNe, hereafter) prove to comprise the most common general class of exploding star in the Universe (Cappellaro et al. 1999), each releasing $\sim 10^{51}$ ergs of mechanical energy and enriching the interstellar medium with several solar masses of new chemical elements. Their astrophysical importance is no longer limited to the central role they play in the chemical evolution of the Universe and in the shaping of the galaxies themselves, but now extends to the possibility that a fraction of them might be the source of the enigmatic gamma-ray bursts (GRBs) .

CCSNe come in a great variety of flavors. In this paper, I summarize the photometric and spectroscopic properties of all types of CCSNe and their physical parameters. Despite the great diversity displayed by these objects several regularities emerge which provide valuable clues and a better insight on their explosion mechanism, a matter that still remains quite controversial (see Janka et al. 2003, Ch. 3, Cardall 2003, and Burrows 2000 for recent reviews).

1. Supernova Classification

Observers classify SNe according to the presence or absence of certain elements in their atmospheres based on spectroscopic observations. By the time Baade and Zwicky introduced the SN class, spectra of these objects had already been obtained. Minkowski (1941) published the first paper on this subject where he introduced two main SN spectroscopic types: the Type II class comprises SNe with prominent hydrogen lines, whereas the Type I class is defined by the absence of hydrogen in their spectra. This classification scheme has evolved as more spectra have become available (see Filippenko 1997 for a detailed review). Five distinct SN types can be distinguished from spectra obtained near maximum light (Fig. 2.1):

- classical Type II: These objects have prominent Balmer lines exhibiting P-Cygni profiles.
- Type IIdw : The members of this class have strong hydrogen lines in emission. They can be distinguished from the classical Type II SNe by the lack of absorption in their Balmer lines. Chugai (1997a) introduced this designation to reflect the fact that these SNe undergo significant interaction with a "dense wind" produced by the SN progenitor prior to explosion.
- Type Ia: They are characterized by a strong absorption attributed to Si II $\lambda\lambda$ 6347,6371.
- Type Ib: These objects are distinguished by spectra with no evident Balmer lines, weak or absent Si II $\lambda\lambda$ 6347,6371, and strong He I $\lambda\lambda$ 4471, 5876, 6678,

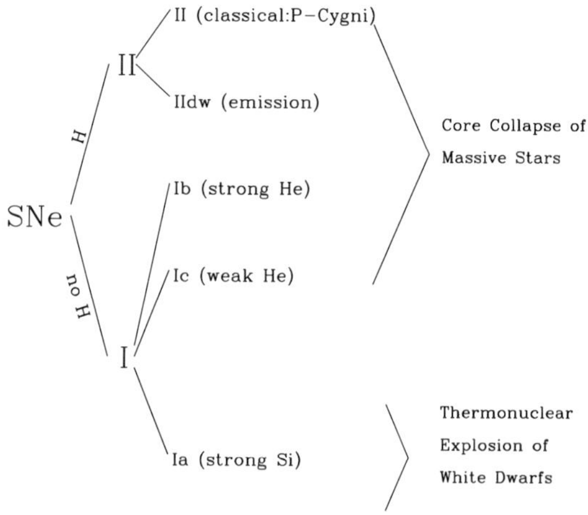

Figure 2.1. Supernova Classification

and 7065 lines. Bertola (1964) reported the first observations of this class of SNe but the "Ib" designation was introduced later by Elias et al. (1985).

• Type Ic: The members of this class are characterized by weak or absent hydrogen and helium lines, and no evident Si II $\lambda\lambda$ 6347,6371. They show Ca II H&K in absorption, the Ca II near-IR triplet with a P-Cygni profile, and O I λ 7774 in absorption. The "'Ic" class was introduced by Wheeler & Harkness (1986).

During the past years a new class of SN seems to be emerging, which is characterized by a smooth and featureless spectrum at early epochs. The current interpretation is that these objects have the usual lines observed in SNe Ic but with an extreme Doppler broadening caused by unusually high expansion velocities. There are three members of this class (SN 1997ef, SN 1998bw, and SN 2002ap), which are often called "hypernovae" or SNe Id. One of them (SN 1998bw) proved to be a remarkable event because it was found at the same temporal and spatial location as GRB980425 (Galama et al. 1998). In the rest of this paper I will refer to these objects as Type Ic hypernovae to reflect the observational fact that their expansion velocities are unusually high.

SNe II, Ib, and Ic occur near star forming regions and have never been observed in elliptical galaxies, which leads to the idea that their progenitors are massive stars born with more than $\sim$8 $M_\odot$ that undergo core collapse, leaving a neutron star or black hole as a remnant and launching an explosion of their envelopes. Type Ia SNe, on the other hand, are observed in all types of galaxies. Given their lack of hydrogen, it is thought that they arise from white dwarfs that explode as they approach the Chandrasekhar mass ($\sim 1.4\ M_\odot$) after a period of mass accretion from a binary companion, leaving no compact remnant behind them.

Theorists give less importance to the external appearance of SNe (spectra) but to their hearts (the origin of the explosion), and distinguish two fundamentally different SN types regardless of their spectroscopic appearance: core collapse and thermonuclear SNe. In this sense SNe Ib and SNe Ic are thought to be physically much closer to SNe II than to SNe Ia, even though SNe Ia, Ib, and Ic all share the same prefix (owing to the lack of hydrogen in their atmospheres).

2. The Properties of Core Collapse Supernovae

Stars born with $\sim$8-10 $M_\odot$ can reach temperatures and densities sufficiently high to produce O/Ne/Mg cores. More massive stars can proceed even further and end up with Fe nuclei. When nuclear burning ceases in these stars the core becomes unstable and gravitational collapse follows, leading to the formation of a neutron star or black hole. It is thought that this is the place where a supernova is launched, but the mechanism by which the core implosion triggers the explosion of the stellar mantle still remains a difficult theoretical problem (see Janka et al. 2003, for example). The standard paradigm is that most of the gravitational energy released during the collapse is carried away by neutrinos ($\sim 10^{53}$ ergs) and that a small fraction ($\sim 1\%$) of this energy is deposited in the bottom of the star's envelope. This produces a shock wave that propagates through the interior of the star and emerges on the surface a few hours later. This is a generic model for CCSNe, whose optical luminosities comprise only a small fraction of the energy released by the gravitational collapse, and whose optical spectra vary solely by the ability of their progenitors to retain their H-rich and He-rich envelopes prior to explosion and/or by the density of the medium in which they explode. In what follows I proceed to discuss the properties of all spectroscopic types of CCSNe.

Classical Type II Supernovae

These SNe are thought to have massive progenitors with extended H-rich envelopes that undergo little interaction with the circumstellar medium (CSM). They are characterized by optical spectra dominated by Balmer lines exhibiting P-Cygni profiles. Many of these SNe have been aimed with radio telescopes, yet only a handful of nearby events such as SN 1987A and SN 1999em have been detected at these frequencies (Weiler et al. 2002). The low radio luminosity of these objects implies that they explode in low-density environments. Based on

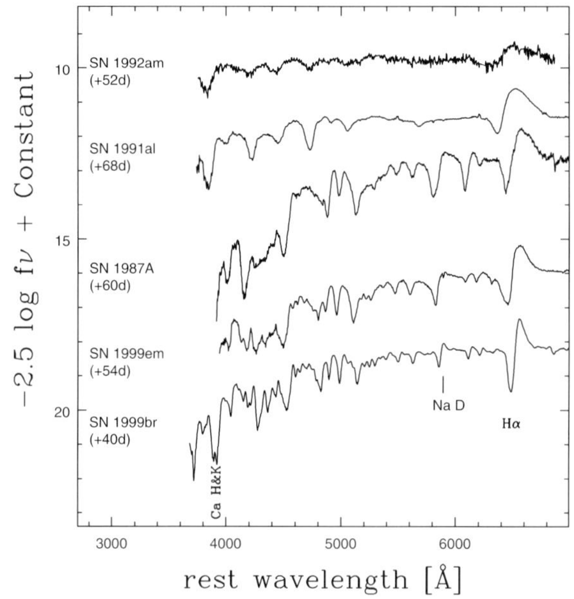

Figure 2.2. Optical spectra of classical Type II SNe obtained ~60 days after explosion, sorted by the width of their spectral lines. The spectrum of SN 1987A is from Phillips et al. (1988), that of SN 1999em is from Leonard et al. (2002a), and the rest are from Hamuy (2001).

the LOTOSS (Li et al. 2000) discoveries reported to the IAU Circulars in 2001, I estimate that classical SNe II constitute ~45% of all CCSNe.

Figure 2.2 displays optical spectra of a sample of classical SNe II taken ~60 days after explosion. The most prominent feature of these objects is Hα exhibiting a P-Cygni profile, which is characteristic of an expanding atmosphere (Kirshner & Kwan 1974; Jeffery & Branch 1990). Besides the Balmer series these spectra show strong Ca H&K, Na I D, and Fe lines. Clearly this sample of SNe reveals a great degree of individuality. In particular it is possible to observe a wide range in the line widths, from the "narrow-line" SN 1999br to the "broad-line" SN 1992am, which suggests a significant range in expansion velocities.

A sample of absolute V lightcurves of classical SNe II is shown in Fig. 2.3. SN 1999em is among the best observed objects of this class which is characterized by a long plateau (~110 days) of nearly constant luminosity. The plateau is followed by a sudden drop in luminosity and a linear tail with a slope of ~0.01 mag day^{-1}. The plateau corresponds to the optically thick phase in which the SN has a well-defined photosphere. Since the opacity is dominated by e^{-} scattering, the photosphere lies at the shell where H recombines. As the

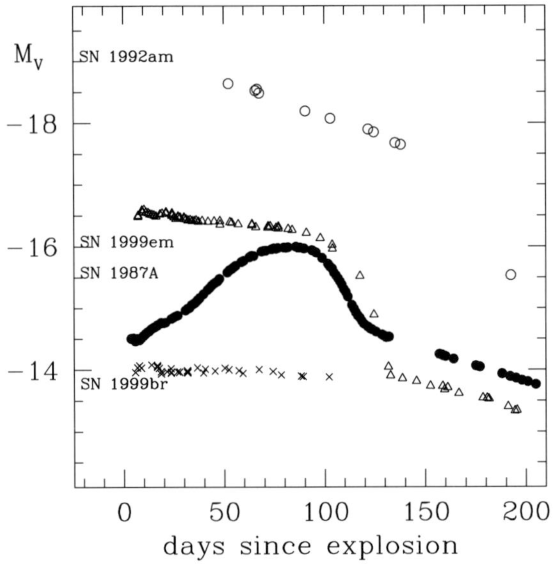

Figure 2.3. Absolute V lightcurves of classical Type II SN. The observations are from Hamuy & Suntzeff (1990) for SN 1987A; from Hamuy (2001) for SN 1992am and SN 1999br; and from Suntzeff et al. (2003) for SN 1999em.

ejecta expands a recombination wave recedes through the envelope. The end of the plateau corresponds to the time when the photosphere reaches the He-rich envelope. Owing to much lower opacities the photosphere recedes faster and the luminosity drops promptly until the SN becomes transparent. Up to this point the lightcurve is powered primarily by the internal energy of the SN previously deposited by the shock wave that ensued from core collapse. During the nebular phase, on the other hand, the lightcurve is powered by the radioactive decay of $^{56}Co \rightarrow {}^{56}Fe$, at a rate corresponding to the e-folding time of the ^{56}Co decay (111.26 days). ^{56}Co is the daughter of ^{56}Ni (with a half-life of 6.1 days), so the luminosity of the tail is determined by the amount of ^{56}Ni freshly synthesized in the explosion. Plateau SNe (SNe II-P) such as SN 1999em comprise the vast majority of spectroscopically classical SNe II. This photometric class was first identified by Barbon et al. (1979). Two more examples are shown in Fig. 2.3: SN 1999br and SN 1992am. It is evident that there is a great range ($\sim$5 mag) in luminosity within this group.

SN 1987A is the most well-observed SN to date and its lightcurve (Fig. 2.3) proves clearly different than that of SNe II-P. It is characterized by a steady rise during three months. After maximum light the SN displayed a fast decline phase

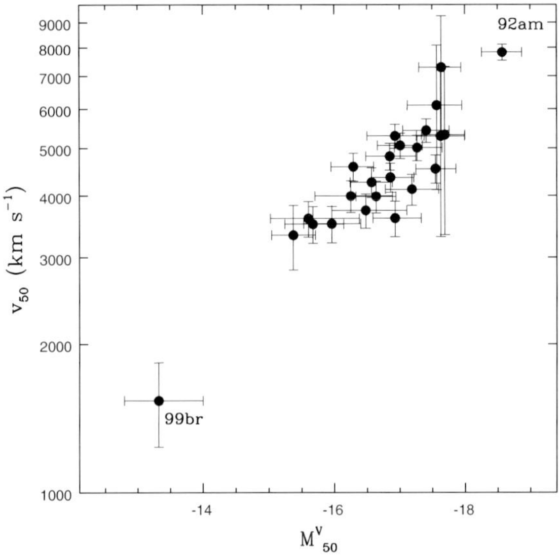

Figure 2.4. Expansion velocities from Fe II λ5169 versus absolute *V* magnitude, both measured in the middle of the plateau (day 50) of 24 Type II SNe II-P.

of ~20 days, followed by a linear tail at the rate expected for $^{56}Co \rightarrow {}^{56}Fe$. Its peculiar shape has been attributed to the relatively small radius of its blue supergiant progenitor Sk 202-69 (Woosley et al. 1987). Unlike SNe II-P which explode as red supergiants (Arnett 1996), most of the shock deposited energy in SN 1987A went into adiabatic expansion, thus leading to a dimmer plateau and to a lightcurve promptly powered by $^{56}Ni \rightarrow {}^{56}Co \rightarrow {}^{56}Fe$ (Blinnikov et al. 2000). The lightcurve shape of SN 1987A reflects the combination of an ever decreasing deposition rate with an ever increasing escape probability for the photons from the SN interior as the ejecta gets thinner. SN 1998A (Woodings et al. 1998) and SN 2000cb (Hamuy 2001) are two other clear examples of events with SN 1987A-like lightcurves and, hence, with blue supergiant progenitors. The relatively compact progenitors of these SNe have been attributed to low metallicities and to mass loss to a binary companion prior to explosion.

Figures 2.2 and 2.3 suggest that SNe with brighter plateaus have higher ejecta velocities, and viceversa. Figure 2.4 shows expansion velocities versus plateau luminosities for the 24 SNe II-P having sufficient photometric and spectroscopic data (see Hamuy 2003 for details). Despite the great diversity displayed by SNe II-P, these objects show a tight luminosity-velocity correlation. This result suggests that, while the explosion energy increases so do the kinetic and internal energies. In fact, as shown in Fig. 2 of Hamuy (2003), the luminosity-velocity correlation is also present in the theoretical models of Litvinova & Nadezhin (1985) (LN85).

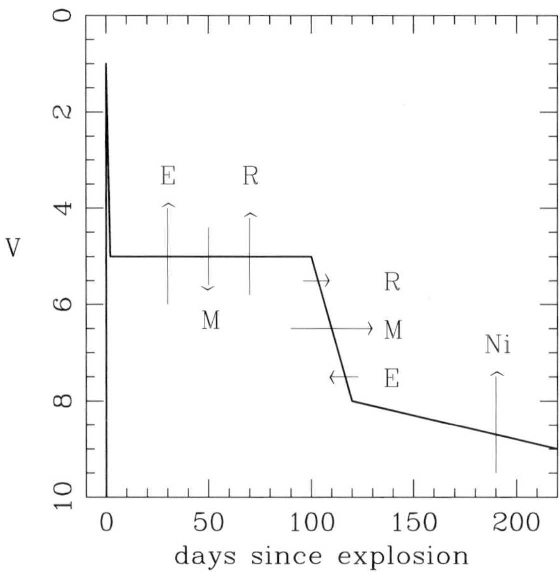

Figure 2.5. Effect of explosion energy (E), ejected mass (M), initial radius (R), and nickel mass (Ni) on the lightcurve of a SN II-P.

Using the hydrodynamic models of LN85 it is possible to derive physical parameters such as energy (E), ejected mass (M), and initial Radius (R) for SNe II-P. In such models the lightcurve is shaped by these parameters and their effect on the plateau phase is illustrated in Fig. 2.5. While the plateau luminosity is particularly sensitive to E, its duration depends largely on M. The models of LN85 yield specific calibrations for three observables: the plateau luminosity, its duration, and the photospheric velocity, which can be used to solve for E, M, and R. One problem with the LN85 models is that they do not cover a wide range in energy and mass so the results derived from this calibration often involve extrapolating their formulas. Although it is necessary to expand the parameter space explored by LN85 before we can firmly believe this method, in its current form it can still provide useful insights on the nature of these objects. As mentioned above, the luminosity of the radioactive tail can be used to find the mass of ^{56}Ni (M_{Ni}) synthesized in the explosion, assuming that all the γ-rays from $^{56}\mathrm{Co} \rightarrow {}^{56}\mathrm{Fe}$ are fully thermalized in the interior. Fig. 2.6 shows the 13 SNe II-P for which there is sufficient data to perform such analysis.

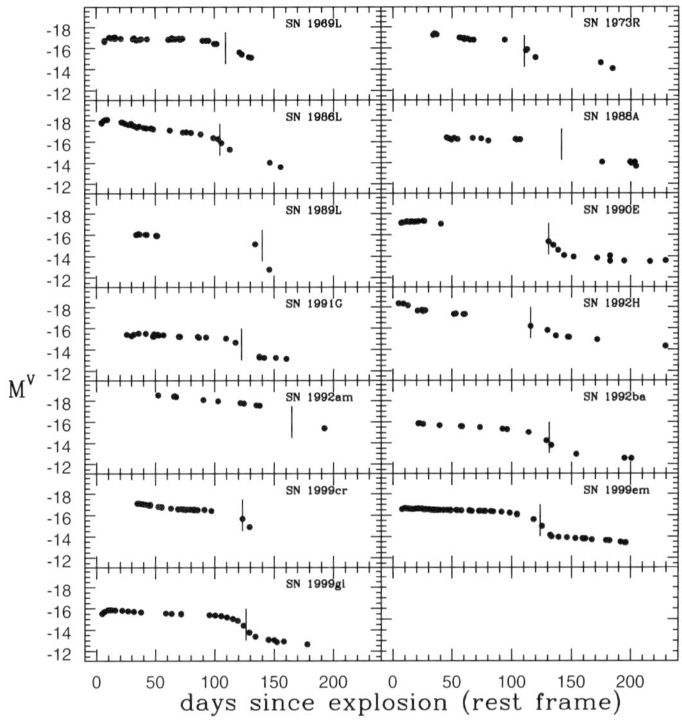

Figure 2.6. Extinction corrected absolute V-band lightcurves of 13 plateau SNe II. The vertical bars indicate the end of the plateau phase for each SN. The observations are from Ciatti et al. (1971) for SN 1969L; Ciatti & Rosino (1977) for SN 1973R; Phillips & Kirhakos (2003) for SN 1986L; Ruiz-Lapuente et al. (1990), Benetti et al. (1991), Turatto et al. (1993b) for SN 1988A; Schmidt (2003) for SN 1989L; Schmidt et al. (1993) and Benetti et al. (1994) for SN 1990E; Blanton et al. (1995) for SN 1991G; Tsvetkov (1994) and Clocchiatti et al. (1996a) for SN 1992H; Hamuy (2001) for SN 1992am, SN 1992ba, and SN 1999cr; Suntzeff et al. (2003) and Leonard et al. (2002a) for SN 1999em; and Leonard et al. (2002b) for SN 1999gi.

Table 2.1 summarizes the resulting parameters and those independently derived for SN 1987A (Arnett 1996), SN 1997D and SN 1999br (Zampieri et al. 2003), which reveals the following,

Table 2.1. Physical Parameters for Classical Type II Supernovae

SN	*Energy* ($\times 10^{51}$ *ergs*)	*Ejected Mass* ($M_\odot$)	*Initial Radius* ($R_\odot$)	*Nickel Mass* ($M_\odot$)
1969L	$2.3^{+0.7}_{-0.6}$	28^{+11}_{-8}	204^{+150}_{-88}	$0.082^{+0.034}_{-0.026}$
1973R	$2.7^{+1.2}_{-0.9}$	31^{+16}_{-12}	197^{+128}_{-78}	$0.084^{+0.044}_{-0.030}$
1986L	$1.3^{+0.5}_{-0.3}$	17^{+7}_{-5}	417^{+304}_{-193}	$0.034^{+0.018}_{-0.011}$
1988A	$2.2^{+1.7}_{-1.2}$	50^{+46}_{-30}	138^{+80}_{-42}	$0.062^{+0.029}_{-0.020}$
1989L	$1.2^{+0.6}_{-0.5}$	41^{+22}_{-15}	136^{+118}_{-65}	$0.015^{+0.008}_{-0.005}$
1990E	$3.4^{+1.3}_{-1.0}$	48^{+22}_{-15}	162^{+148}_{-78}	$0.062^{+0.031}_{-0.022}$
1991G	$1.3^{+0.9}_{-0.6}$	41^{+19}_{-16}	70^{+73}_{-31}	$0.022^{+0.008}_{-0.006}$
1992H	$3.1^{+1.3}_{-1.0}$	32^{+16}_{-11}	261^{+177}_{-103}	$0.129^{+0.053}_{-0.037}$
1992am	$5.5^{+3.0}_{-2.1}$	56^{+40}_{-24}	586^{+341}_{-212}	$0.256^{+0.099}_{-0.070}$
1992ba	$1.3^{+0.5}_{-0.4}$	42^{+17}_{-13}	96^{+100}_{-45}	$0.019^{+0.009}_{-0.007}$
1999cr	$1.9^{+0.8}_{-0.6}$	32^{+14}_{-12}	224^{+136}_{-81}	$0.090^{+0.034}_{-0.027}$
1999em	$1.2^{+0.6}_{-0.3}$	27^{+14}_{-8}	249^{+243}_{-150}	$0.042^{+0.027}_{-0.019}$
1999gi	$1.5^{+0.7}_{-0.5}$	43^{+24}_{-14}	81^{+110}_{-51}	$0.018^{+0.013}_{-0.009}$
		SNe from Other Sources		
1987A[a]	1.7	15	42.8	0.075
1997D[b]	0.9	17	128.6	0.006
1999br[b]	0.6	14	114.3	$0.0016^{+0.0011}_{-0.0008}$

[a] From Arnett (1996).
[b] From Zampieri et al. (2003).

• There is a wide range in explosion energies, from 0.6 to 5.5 foes (1 foe = 10^{51} ergs) among classical SNe II.
• The ejected masses encompass a broad range between 14 and 56 $M_\odot$. Note that, while stars born with more than 8 $M_\odot$ can in principle undergo core collapse, they do not show up as classical SNe II. Perhaps they become white dwarfs (Heger et al. 2003) or they undergo significant mass loss before explosion and are observed as other SN spectroscopic types (Chugai 1997a). Note also that stars as massive as 50 $M_\odot$, which are expected to have strong stellar winds (Heger et al. 2003), seem able to retain a significant fraction of their H envelope and explode as SNe II if their metallicities are low enough. Although these results prove interesting, it must be mentioned that the derived masses are quite uncertain because the LN85 calibration only extends up to 16 $M_\odot$. Of some concern is the sharp contrast found between the ejected masses derived from the LN85 calibration for SN 1999gi and SN 1999em and the values independently obtained by Smartt et al. (2001, 2002). Based on upper limits of the luminosities of the progenitors of these two nearby SNe from pre-discovery images and stellar evolutionary tracks, they derived upper mass limits of 9^{+3}_{-2} and

12^{+1}_{-1} $M_\odot$ for SN 1999gi and SN 1999em, which prove at odds with the values of 43^{+24}_{-14} and 27^{+14}_{-8} $M_\odot$ obtained for these objects from the LN85 calibration. Part of the discrepancy might arise from the LN85 calibration, but it could be due also to the distances adopted by Smartt et al. for the SN host galaxies (in fact with the new Cepheid distance to SN 1999em the upper mass limit rises to $\sim 20 M_\odot$; Leonard 2003), or uncertain stellar evolutionary models for massive stars.

• Except for SN 1987A, within the uncertainties the initial radii correspond to those measured for red supergiants (van Belle et al. 1999), which lends support to the view that the progenitors of SNe II-P have extended atmospheres at the time of explosion (Arnett 1996).

• The Ni masses produced by these SNe vary by a factor of $\sim$100, from 0.0016 to 0.26 $M_\odot$.

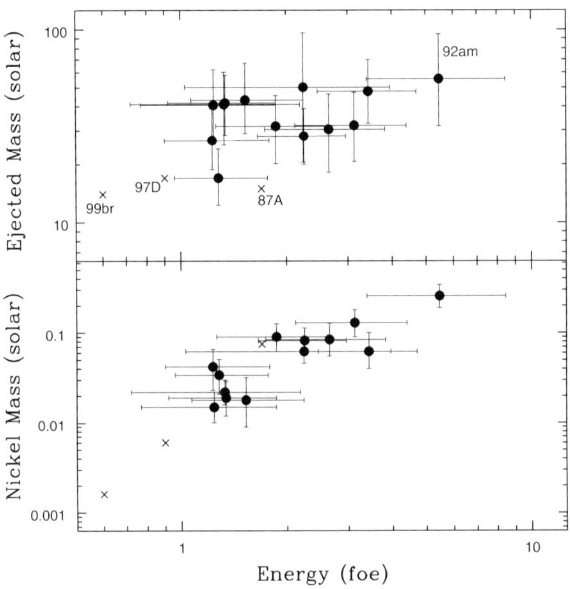

Figure 2.7. Ejected mass and nickel mass of classical SNe II, as a function of explosion energy. Filled circles correspond to the 13 SNe II-P for which the LN85 models could be applied, and the three crosses correspond to SN 1987A, SN 1997D, and SN 1999br, which have been modeled in detail by Arnett (1996) and Zampieri et al. (2003).

Fig. 2.7 shows M and M_{Ni} as a function of E for these 16 SNe II. Despite the large error bars, this figure reveals a couple of correlations. First (top panel), the explosion energy appears to be correlated with the ejected mass, in the sense that more energetic SNe eject greater masses. This suggests that the outcome of the core collapse is somehow determined by the mass of the envelope, or that the mass of the envelope is correlated with some property of the core (e.g. mass), with the latter determining the outcome. Second (bottom panel), SNe with greater energies produce more nickel, a result previously suggested by Blanton et al. (1995). This could mean that greater temperatures and more nuclear burning are reached in such SNe, and/or that less mass falls back onto the neutron star/black hole in more energetic explosions.

Type IIdw Supernovae

A distinct class of SNe II can been identified which, unlike classical SNe II, are believed to be strongly interacting with a "dense wind" produced by the SN progenitor prior to explosion. These SNe have strong radio emission caused by the interaction with the CSM (Chevalier 1998). Models of the radio observations imply high mass-loss rates $\sim 10^{-4}\ M_{\odot}\ yr^{-1}$ for their progenitors (Weiler et al. 2002). SNe IIdw comprise $\sim$30% of all CCSNe and $\sim$40% of all SNe II.

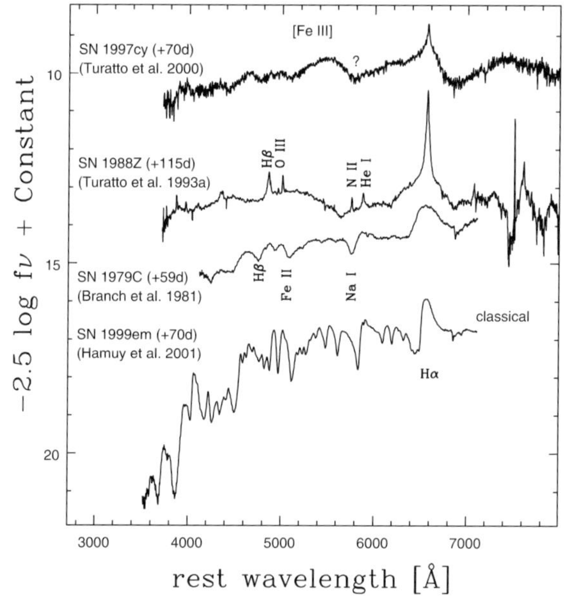

Figure 2.8. Optical spectra of SNe IIdw obtained $\sim$3 months after explosion, compared to the classical Type II SN 1999em.

Fig. 2.8 shows optical spectra for a sample of SNe IIdw, compared to the classical SN 1999em. Evidently these SNe show a strong degree of individuality, but they are unified by the *lack of absorption* in the Balmer lines. Their spectra are dominated by strong Hα broad emission (SN 1979C) , sometimes with a superposed narrow (FWHM$\sim$200 km s^{-1}) emission (SN 1988Z and SN 1997cy). One of most well-observed and recent additions to the IIdw events is SN 1998S (Leonard et al. 2000). When the narrow component is present the SN is classified as IIn (standing for "narrow"; Schlegel 1990). Occasionally a narrow P-Cygni profile can be observed, such as in SN 1994aj (Benetti et al. 1998) and SN 1996L (Benetti et al. 1999), in which case the SN is typed as IId (the "d" stands for "double" profile).

The interaction of the SN envelope and the CSM is very difficult to model. In the models of Chugai (1997a) this collision produces an outer shock wave that propagates in the unshocked CSM, and an inner shock wave that propagates inward through the SN envelope. In between the two shocks is located a cool and dense shell that produces broad emission lines by excitation from the X-rays produced in the inner and outer shocks. Broad emissions are also thought to arise in the undisturbed SN ejecta excited by the X-rays. Chugai (1997b) claims that the absence of the absorption component of the Hα profile observed in classical SNe II is a consequence of the excitation mechanism: while the atmospheres of the classical SNe II are excited by internal energy of the explosion and by radioactivity, SNe IIdw are predominantly excited by the shocks. In these models the narrow emission component in SNe IIdw is thought to originate in the undisturbed photoionized CSM. The narrow P-Cygni profile observed in SNe IId is attributed to the recombined unshocked CSM.

The lightcurves of some SNe IIdw are compared to that of the classical SN 1999em in Fig. 2.9. The strong degree of individuality seen in their spectra is also reflected in their lightcurves. SN 1979C belongs to the photometric class of "linear" SNe introduced by Barbon et al. (1979). SN 1979C displayed a post-maximum decline phase at a rate of 0.04 mag day^{-1} for $\sim$100 days, followed by a slower decline tail at $\sim$0.01 mag day^{-1}. SN 1980K is another clear example of a linear event (Barbon et al. 1982a). SN 1988Z and SN 1997cy, on the other hand, showed very different behaviors. They both faded slowly, at 0.01 mag day^{-1}. This rate is not very different than that of the classical plateau SN 1999em, but the main difference is that this slow evolution in SN 1988Z and SN 1997cy extended for several hundred days.

Despite the great photometric diversity among SNe IIdw, these objects share the property of being generally more luminous than the classical SNe II. This can be appreciated in Fig. 2.10, which compares the distribution of peak magnitudes of classical and Type IIdw SNe. The high luminosities of Type IIdw SNe constitute another optical signature of the collision of SN envelopes with dense winds, which efficiently converts kinetic energy into light. Often, this mechanism predominates over the usual SN radiation mechanism, thus leading to a powerful optical display (Chugai 1997a).

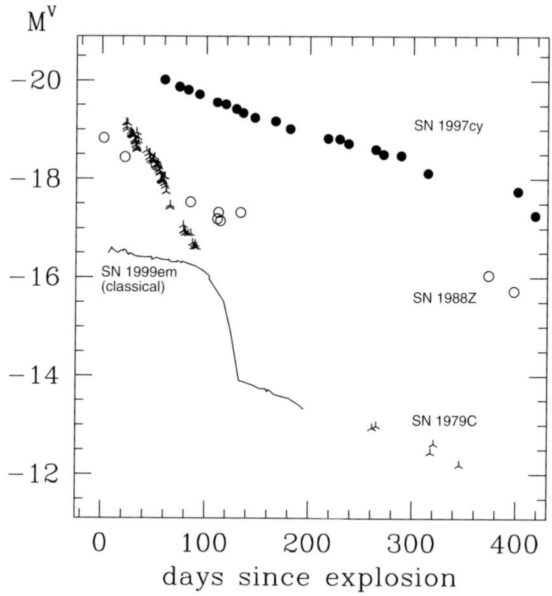

Figure 2.9. Absolute V lightcurves of three SNe IIdw, compared to the classical Type II SN 1999em (solid line; Suntzeff et al. 2003). The observations are from Balinskaya et al. (1980), De Vaucouleurs et al. (1981), and Barbon et al. (1982b) for SN 1979C ; Turatto et al. (1993a) for SN 1988Z; and Germany et al. (2000) for SN 1997cy.

Table 2.2. Physical Parameters for Type IIdw Supernovae

SN	*Energy* ($\times 10^{51}$ *ergs*)	*Ejected Mass* ($M_\odot$)	*Initial Radius* ($R_\odot$)	*Nickel Mass* ($M_\odot$)	*Mass − loss Rate* ($M_\odot\ yr^{-1}$)
1979C[a]	1-2	6	6000	<0.1	1×10^{-4}
1988Z[b]	1	<1	...	...	1-7$\times10^{-4}$
1997cy[c]	30	5	...	...	4.0$\times10^{-4}$

[a] From Blinnikov & Bartunov (1993) and Weiler et al. (2002).
[b] From Chugai & Danziger (1994) and Weiler et al. (2002).
[c] From Turatto et al. (2000) and Nomoto et al. (2000).

Table 2.2 summarizes physical parameters for three SNe IIdw derived from models involving circumstellar interaction. In general these SNe eject less mass than the classical SNe II, between 1-6 $M_\odot$. The models require large mass-loss rates before explosion. Two of them exploded with normal energies compared to classical SNe II, but one case (SN 1997cy) was much more energetic than any other SN II. Chugai (1997a) proposes that 1) the progenitors of SNe IIdw

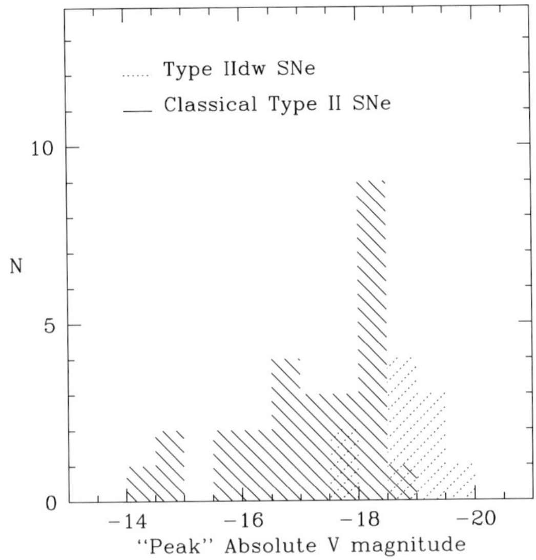

Figure 2.10. Distribution of absolute peak magnitudes of classical and Type IIdw SNe.

are stars born with 8-10 $M_\odot$ that undergo significant mass loss owing to helium shell flashes during the asymptotic red supergiant branch stage, and 2) the large observed diversity among SNe IIdw is due to large variations in the wind density near the SN progenitor. This hypothesis is consistent with the conclusions reached above from the LN85 models, that the progenitors of the classical SNe II are born with more than 14 $M_\odot$.

Type Ib and Ic Supernovae

The distinguishing feature of Type Ib and Ic SNe is the lack of conspicuous hydrogen spectral lines. Their progenitors are thought to be massive stars that lose most of their H-rich (and perhaps their He-rich) envelopes via strong winds or transfer to a binary companion via Roche overflow. Evidence for this hypothesis has been recently revealed by the Type Ic SN 1999cq, which showed intermediate-width helium emission lines caused by the interaction of the SN ejecta with an almost pure helium wind lost by the progenitor (Matheson et al. 2000). Type Ib and Ic SNe are as radio loud as SNe IIdw, yet the mass-loss rates of their progenitors are lower, between 10^{-5}-10^{-6} $M_\odot\ yr^{-1}$ (Weiler et al. 2002). Approximately 25% of all CCSNe fall in the SNe Ib and SNe Ic category.

Fig. 2.11 shows near-maximum optical spectra of various members of the Type I core collapse family. SN 1984L is one of the prototypes of the Type Ib class, which is characterized by strong He I $\lambda\lambda$4471, 5876, 6678, and 7065 lines. These objects are not seen very often and comprise only $\sim$1% of all CCSNe. SN 1987M was the first well established member of the Ic family and SN 1994I proves to be one of the best observed objects of this class. The strongest lines are the Ca II H&K and O I λ7774 absorptions, and the P-Cygni profile of the Ca II IR triplet.

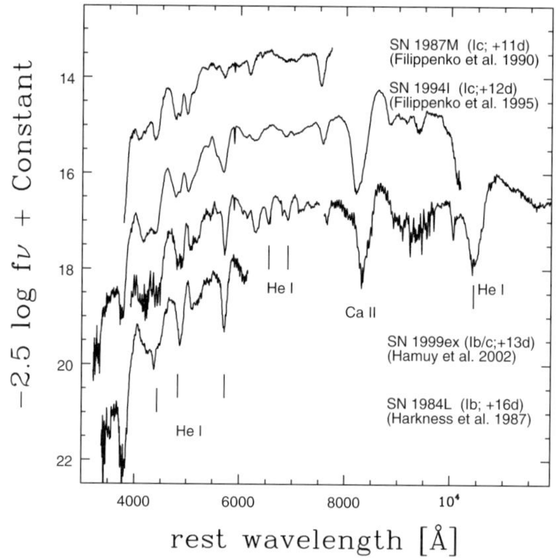

Figure 2.11. Optical spectra of Type Ib and Type Ic SNe, taken $\sim$14 days after maximum-light.

Significant effort has been put over recent years into determining the presence of helium in the spectra of SNe Ic in order to better understand the nature of these objects. Recently Matheson et al. (2001) compiled and analyzed a large collection of spectra of SNe Ib and Ic obtained at Lick Observatory. This study showed no compelling evidence for He in absorption in the spectra of SNe Ic and no gradual transition from the Ib to the Ic class, which supported the idea that these objects are produced by different progenitors. By contrast, a closer look at the SN 1994I spectrum suggests deeper troughs at the wavelengths of the He I lines, especially at 4471, 4921, and 5876 Ang (Filippenko et al. 1995; Clocchiatti et al. 1996b). The spectrum of SN 1999ex shown in Fig. 2.11 reveals unambiguous evidence for He I absorptions of moderate strength in the optical region, thus suggesting the existence of an intermediate Ib/c case and a link between the Ib and Ic classes.

The presence of hydrogen in SNe Ib has been recently analyzed by Branch et al. (2002) by comparison to synthetic spectra. Their conclusion is that the 6300 Ang absorption feature seen in the Type Ib SN 1954A, SN 1999di and SN 2000H is Hα, and that hydrogen appears to be present in SNe Ib in general. A similar analysis led Branch (2002) to conclude that Hα is also present in the spectrum of the intermediate Type Ib/c SN 1999ex. The presence of hydrogen in Type Ic SNe, on the other hand, is still questionable.

The spectral models of Branch et al. (2002) show that small optical depths are required to fit the H and He lines and that a mild reduction in the optical depths would make SNe Ib look like SNe Ic. This supports the concept that SNe Ib and SNe Ic are not fundamentally different. Their spectroscopic appearance may not even reflect the physical presence of H or He in their atmospheres but only the way these elements are mixed in the envelopes.

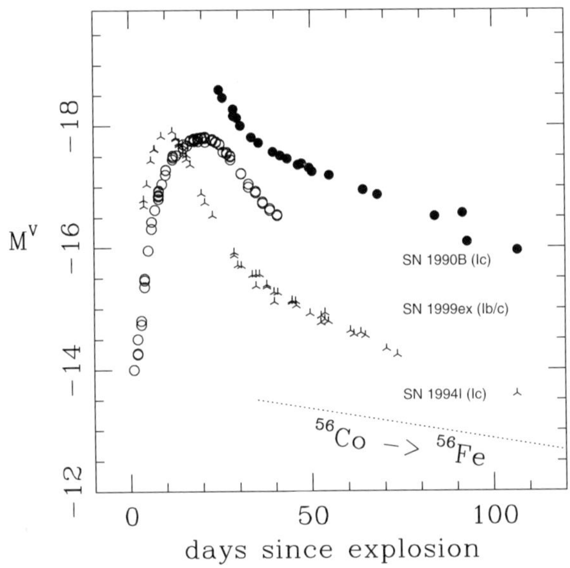

Figure 2.12. Absolute V lightcurves of three Type I CCSNe. The observations are from Clocchiatti et al. (2001) for SN 1990B; Richmond et al. (1996) for SN 1994I; and Stritzinger et al. (2002) for SN 1999ex. The dotted line shows the lightcurve expected from ^{56}Co $\rightarrow$ ^{56}Fe.

The lightcurves of some of the best observed Type I CCSNe are shown in Fig. 2.12. Evidently the lightcurves do not follow a single template, yet they share the same common features: a rise during 2-3 weeks, a fast dimming during $\sim$30 days, followed by a slow decline phase at a rate significantly greater than that expected for the radioactive decay of ^{56}Co $\rightarrow$ ^{56}Fe (shown with a dotted line).

Unlike classical SNe II-P, the early-time lightcurves of SNe I are not powered by shock-deposited energy. This is attributed to the fact that these objects have much more compact progenitors, so that the energy deposited in the envelope by the central collapse is largely spent in adiabatic expansion, which leads to a dimmer and brief plateau (Woosley et al. 1987; Shigeyama et al. 1990). Hence, from the very beginning their lightcurves are powered by radioactive heating. While the peak is determined by the amount of ^{56}Ni synthesized in the explosion, the width depends on the ability of the photons to diffuse out from the SN interior. The width of the lightcurve is determined by the diffusion time which increases with the envelope mass and decreases with expansion velocity. The early-time lightcurve, therefore, provides useful constraints on the ^{56}Ni mass, envelope mass, and kinetic energy (Arnett 1996). The great range in peak luminosities and lightcurve widths displayed by SNe I suggests a wide range in mass, velocity, and ^{56}Ni. The late-time decline rate reveals that a fraction of the γ-rays from ^{56}Co $\rightarrow$ ^{56}Fe escape from the SN ejecta without being thermalized and, therefore, can be used to quantify the degree of ^{56}Ni mixing in the SN interior. In the next section I summarize the physical parameters for these objects.

Type Ic Hypernovae

In the past five years three SNe (SN 1997ef, SN 1998bw, and SN 2002ap) have been found to display very peculiar spectra compared to the standard types described above. Since they show an overall similarity to each other, they seem to define a new spectroscopic class of SNe.

Their maximum-light spectra are shown in Figure 2.13 along with the normal Type Ic SN 1994I. Their main characteristic is that they are extremely smooth and featureless. They do not show obvious signatures of hydrogen or helium. The few spectral features displayed by these SNe are very broad and hard to identify without spectral modeling. The comparison with SN 1994I suggests that these weird SNe are Type Ic events but with higher than normal expansion velocities. This would lead to significant Doppler broadening and blending of the spectral lines and, hence, to featureless spectra. Judging from the line widths the velocities increase from SN 2002ap, to SN 1997ef, and SN 1998bw. Their unusual expansion velocities suggest that these objects are hyper-energetic so they are collectively called "hypernovae". Some people prefer to call them Type Id events.

The absolute V lightcurves of the three hypernovae are shown in Fig. 2.14 along with the normal Type Ic SN 1994I (solid line). They all share the generic shape of normal SNe Ib and SNe Ic and, like these objects, the hypernovae show

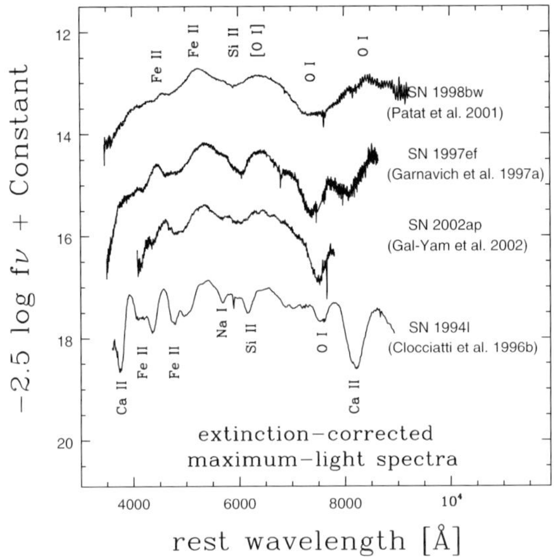

Figure 2.13. Maximum-light optical spectra of the three Type Ic hypernovae, compared to the normal Type Ic SN 1994I.

great dispersion in lightcurve widths and luminosities. While both SN 1997ef and SN 2002ap showed normal luminosities, the former had a broad lightcurve and the latter evolved much faster. SN 1998bw showed a slow luminosity evolution and was clearly overluminous compared to the other Type Ic events, which suggests a large ^{56}Ni production.

SN 1998bw was not only remarkable for its great expansion velocities and luminosity, but also because it exploded at nearly the same location and time as GRB980425 (Galama et al. 1998). No GRB s were detected at the position of the other two hypernovae , on the other hand. Since GRBs could arise in relativistic jets (MacFadyen et al. 2001), it could well be that all hypernovae produce GRBs and that only observers within a range of viewing angles from the jet axis can detect them.

Nomoto and collaborators have modeled SNe Ib as helium stars that lose their hydrogen envelopes prior to explosion by mass transfer to a binary companion, and SNe Ic as C/O bare cores that lose their helium envelopes in a second stage of mass transfer. In all cases they assume that these SNe are spherically symmetric. Given the high levels of polarization displayed by some of these objects (e.g. Wang et al. 2001, Leonard et al. 2002c), this might not be an accurate approximation. Keeping in mind this caveat we list in Table 2.3 their

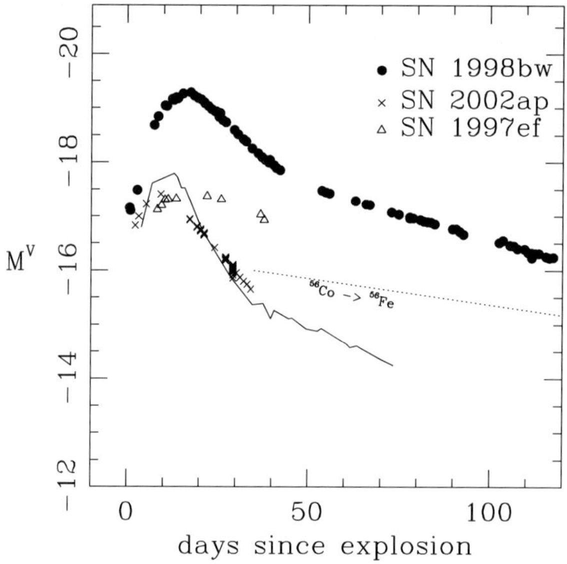

Figure 2.14. Absolute V lightcurves of the three Type Ic hypernovae, compared to the normal Type Ic SN 1994I (solid line). The observations are from Garnavich et al. (1997a) and Garnavich et al. (1997b) for SN 1997ef; Galama et al. (1998), McKenzie & Schaefer (1999), Sollerman et al. (2000), and Patat et al. (2001) for SN 1998bw; and Gal-Yam et al. (2002) for SN 2002ap.

Table 2.3. Physical Parameters for Type Ib and Type Ic Supernovae

SN	*Type*	*Energy* ($\times 10^{51}$ *ergs*)	*Ejected Mass* ($M_\odot$)	*Initial Radius* ($R_\odot$)	*Nickel Mass* ($M_\odot$)	*Reference*
1983I	Ic	1.0	2.1	3.7	0.15	Shigeyama et al. 1990
1983N	Ib	1.0	2.7	3.0	0.15	Shigeyama et al. 1990
1984L	Ib	1.0	4.4	1.9	0.15	Shigeyama et al. 1990
1994I	Ic	1.0	0.9	...	0.07	Nomoto et al. 2000
2002ap	Ic	7.0	3.75	...	0.07	Mazzali et al. 2002
1997ef	Ic	8.0	7.6	...	0.15	Nomoto et al. 2000
1998bw	Ic	60.0	10.0	...	0.50	Nomoto et al. 2000

results and the physical parameters they derive for the three hypernovae and four normal Type Ib and Ic events.

Fig. 2.15 shows how the ejected masses and ^{56}Ni yields vary with explosion energy for the seven Type Ib and Type Ic SNe (crosses) listed in Table 2.3, along with the 16 classical SNe II (filled circles) shown in Fig. 2.7. The top panel reveals that SNe Ib and SNe Ic appear to follow the same pattern shown by classical SNe II, namely, that more energetic SNe eject greater masses. The main difference between both subtypes, of course, is the vertical offset caused by the strong mass loss suffered by SNe Ib and SNe Ic prior to explosion. This suggests that the mass of the envelope is one of the driving parameters of the explosion for CCSNe, or that it is correlated with some other property of the core, with the latter determining the outcome of the collapse.

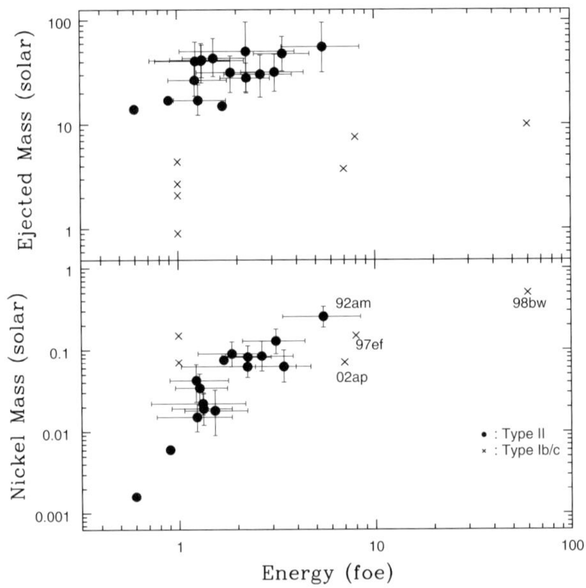

Figure 2.15. Ejected mass and nickel yield of CCSNe, as a function of explosion energy. Filled circles are the same 16 classical SNe II shown in Fig. 2.7, and crosses correspond to the seven SNe Ib and SNe Ic listed in Table 2.3.

These plots permit one to appreciate that the distinguishing characteristic of the three hypernovae is the great explosion energy, between 7 and 60 foes, far above the normal Ib and Ic events which are characterized by energies $\sim$1 foe (note, however, that this conclusion is based on spherically symmetric models; Höflich et al. (1999) showed that the energy of SN 1998bw could be as low as $\sim$2 foes when an ellipsoidal geometry for the ejecta is assumed). When the whole sample of CCSNe is considered there is a continous distribution of energies below 8 foes. Within this regime it appears that some Type II events such as SN 1992am reach explosion energies comparable to that of the hypernovae SN 1997ef and SN 2002ap. *This demonstrates that, in terms of energy, the definition of hypernova is not clear cut.* Whether the energy distribution is continuous above 8 foes remains to be seen when more data become available.

This will permit us to understand if SN 1998bw belongs to a separate class of object or if it just lies at the extreme of the family of CCSNe. At the moment it is fair to say that there is only one firm SN/GRB association and, within the SN context, this object was clearly exceptional regarding its explosion energy.

The bottom panel reveals that SN 1998bw was not only remarkable in explosion energy (60 foes), but also in a high nickel yield (0.5 $M_\odot$). Despite their greater than normal energies, none of the other two Type Ic hypernovae (SN 1997ef and SN 2002ap) produced unusually higher nickel masses compared to the other CCSNe. In fact the Type II SN 1992am ejected even more ^{56}Ni (0.26 $M_\odot$) than these two hypernovae. This suggests that there is continuum in ^{56}Ni yields below 0.3 $M_\odot$ for all CCSNe. More data are needed to ascertain whether this continuity extends up to SN 1998bw.

Leaving aside SN 1998bw, it proves interesting that all other CCSNe share the same location in the ^{56}Ni/energy plane. Since both of these parameters carry imprints from the physical conditions very close to the center of the explosion, this suggests that the physics of the core and explosion mechanisms of all CCSNe are not be fundamentally different.

3. Summary and Discussion

• The general properties of the different CCSNe mentioned above are summarized in Table 2.4.

Table 2.4. General Properties of Core Collapse Supernovae

Spectroscopic Type	*Relative Frequency (Observed)*[a]	*Photometric Type*	*Mass − loss Rate* (10^{-6} $M_\odot$ yr^{-1})	*Ejected Mass* ($M_\odot$)
II	45%	plateau,87A-like	low	14-56
IIdw	30%	linear,slow	10-700	<6
Ib+Ic	25%	bell-like	1-26	1-10

[a] Based on the Lick Observatory Supernova Search (Li et al. 2000) discoveries reported to the IAU Circulars in 2001. No attempt has been made here to correct the observed frequencies from selection biases. The intrinsic SN rates must await the analysis of the observation history of the Lick Observatory Supernova Search galaxy sample.

• There are clear examples of intermediate spectroscopic cases in the CCSN family. With intermediate He I line strengths, SN 1999ex provides a clear link between the Ib and Ic classes (Hamuy et al. 2002). The metamorphosis of SN 1993J from a Type II SN into a Ib event (Filippenko et al. 1993) provides evidence for an intermediate IIb case. This is strong evidence for a continous spectroscopic sequence (II-IIb-Ib-Ib/c-Ic) among CCSNe, which reflects the ability of the SN progenitors to retain their H-rich and He-rich envelopes prior to explosion. Another factor that defines the appearance of a SN is the density of the medium in which they explode, which is determined by the history of mass-loss of the SN progenitor.

• There is a wide range in explosion energies (0.6-60 foes), ejected masses, and ^{56}Ni yields (0.0016-0.5 $M_{\odot}$) among CCSNe, even within the same spectroscopic type. Classical SNe II show a trend in the sense that SNe with greater envelope masses produce more energetic explosions (Fig. 2.15). Type Ib and Ic appear to follow the same pattern. This suggests that the mass of the envelope is one of the driving parameters of the explosion, or that it is correlated with some other property of the core, with the latter determining the outcome of the collapse.

• Type II, Ib, and Ic SNe share the same location in the ^{56}Ni/energy plane. Since both of these parameters carry imprints from the physical conditions very close to the center of the explosion, this suggests that the physics of the core and explosion mechanisms of all CCSNe are not be fundamentally different, regardless of the external appearance of these objects.

• At the moment it is fair to say that there is only one firm SN/GRB association (SN 1998bw), and this object was clearly exceptional regarding energy and nickel production within the SN context.

Acknowledgments

I am very grateful to Dave Branch, Inma Domínguez, Doug Leonard, and Schuyler van Dyk for reading a draft of this review, which led to useful discussions and a significant improvement of the paper. Support for this work was provided by NASA through Hubble Fellowship grant HST-HF-01139.01-A awarded by the Space Telescope Science Institute, which is operated by the Association of Universities for Research in Astronomy, Inc., for NASA, under contract NAS 5-26555.

References

Arnett, D. (1996), Supernovae and Nucleosynthesis, an investigation of the history of matter, from the Big Bang to the present, (New Jersey: Princeton Univ. Press)

Baade, W. (1938), ApJ, 88, 285

Balinskaya, I. S., Bychkov, K. V., & Neizvestny, S. I. (1980), A&A, 85, L19

Barbon, R., Ciatti, F., & Rosino, L. (1979), A&A, 72, 287

Barbon, R., Ciatti, F., & Rosino, L. (1982a), A&A, 116, 35

Barbon, R., Ciatti, F., Rosino, L., Ortolani, S., & Rafanelli, P. (1982b), A&A, 116, 43

Benetti, S., Cappellaro, E., & Turatto, M. (1991), A&A, 247, 410

Benetti, S., Cappellaro, E., Turatto, M., Della Valle, M., Mazzali, P. A., & Gouiffes, C. (1994), A&A, 285, 147
Benetti, S., Cappellaro, E., Danziger, I. J., Turatto, M., Patat, F., & Della Valle, M. (1998), MNRAS, 294, 448
Benetti, S., Turatto, M., Cappellaro, E., Danziger, I. J., & Mazzali, P. A. (1999), MNRAS, 305 811
Bertola, F. (1964), Ann. Ap. 27, 319
Blanton, E. L., Schmidt, B. P., Kirshner, R. P., Ford, C. H., Chromey, F. R., & Herbst, W. (1995), AJ, 110, 2868
Blinnikov, S. I., & Bartunov, O. S. (1993), A&A, 273, 106
Blinnikov, S., Lundqvist, P., Bartunov, O., Nomoto, K., & Iwamoto, K. (2000), ApJ, 532, 1132
Branch, D., Falk, S. W., Marshall, L. M., Rybski, P., Uomoto, A. K., & Wills, B. J. (1981), ApJ, 244, 780
Branch, D. et al. (2002), ApJ, 566, 1005
Branch, D. (2002), in Proceedings of "A Massive Star Odyssey, from Main Sequence to Supernova" IAU Symposium No. 212, eds. K. A. van der Hucht, A. Herrero, & C. Esteban, in press (astro-ph/0207197)
Burrows, A. (2000), Nature, 403, 727
Cappellaro, E., Evans, R., & Turatto, M. (1999), A&A, 351, 459
Cardall, C. Y. (2003), to Proceedings of the "4th International Workshop on the Identification of Dark Matter", (World Scientific), in press (astro-ph/0212438)
Chevalier, R. A. (1998), ApJ, 499, 810
Chugai, N. N., & Danziger, I. J. (1994), MNRAS, 268, 173
Chugai, N. N. (1997a), ARep, 41, 672
Chugai, N. N. (1997b), Ap&SS, 252, 225
Ciatti, F., Rosino, L., & Bertola, F. (1971), MmSAI, 42, 163
Ciatti, F., & Rosino, L. (1977), A&A, 56, 59
Clocchiatti, A., et al. (1996a), AJ, 111, 1286
Clocchiatti, A., Wheeler, J. C., Brotherton, M. S., Cochran, A. L., Wills, D., Barker, E. S., & Turatto, M. (1996b), ApJ, 462, 462
Clocchiatti, A., et al. (2001), ApJ, 553, 886
De Vaucouleurs, G., De Vaucouleurs, A., Buta, R., Ables, H. D., & Hewitt, A. V. (1981), PASP, 93, 36
Elias, J. H., Matthews, K., Neugebauer, G., & Persson, S. E. (1985), ApJ, 296, 379
Filippenko, A. V., Porter, A. C., & Sargent, W. L. W. (1990), AJ, 100, 1575
Filippenko, A. V., Matheson, T., & Ho, L. C. (1993), ApJ, 415, L103
Filippenko, A. V. et al. (1995), ApJ, 450, L11
Filippenko, A. V. (1997), ARA&A, 35, 309
Galama, T. J., et al. (1998), Nature, 395, 670
Gal-Yam, A., Ofek, E. O., & Shemmer, O. (2002), MNRAS, 332, L73

Garnavich, P., Jha, S., Kirshner, R., Challis, P., Balam, D., Brown, W., & Briceno, C. (1997a), IAUC 6786
Garnavich, P., Jha, S., Kirshner, R., Challis, P., Balam, D., Berlind, P., Thorstensen, J., & Macri, L. (1997b), IAUC 6798
Germany, L. M., Reiss, D. J., Sadler, E. M., Schmidt, B. P., & Stubbs, C. W. (2000), ApJ, 533, 320
Hamuy, M., & Suntzeff, N. B. (1990), AJ, 99, 1146
Hamuy, M., et al. (2001), ApJ, 558, 615
Hamuy, M. (2001), Ph.D Thesis, The University of Arizona
Hamuy, M. et al. (2002), AJ, 124, 417
Hamuy, M. (2003), ApJ, 582, in press (astro-ph/0209174)
Harkness, R. P. et al. (1987), ApJ, 317, 355
Heger, A., Fryer, C. L., Woosley, S. E., Langer, N., & Hartmann, D. H. (2003), ApJ, submitted (astro-ph/0212469)
Höflich, P., Wheeler, J. C., & Wang, L. (1999), ApJ, 521, 179
Janka, H.-T., Buras, R., Kifonidis, K., Rampp, M., & Plewa, T., (2003), this volume (astro-ph/0212314)
Jeffery, D. J., & Branch, D. (1990), in Jerusalem Winter School for Theoretical Physics: Supernovae, Vol. 6, ed. J. C. Wheeler, T. Piran, & S. Weinberg (Singapore: World Scientific), 149
Kirshner, R. P., & Kwan, J. (1974), ApJ, 193, 27
Leonard, D. C., Filippenko, A. V., Barth, A. J., & Matheson, T. (2000), ApJ, 536, 239
Leonard, D. C., et al. (2002a), PASP, 114, 35
Leonard, D. C., et al. (2002b), AJ, 124, 2490
Leonard, D. C., Filippenko, A. V., Chornock, R., & Foley, R. J. (2002c), PASP, 114, 1333
Leonard, D. C. (2003), private communication
Li, W. D. et al. (2000), in Cosmic Explosions, ed. S. S. Holt & W. W. Zhang (New York: AIP), 103
Litvinova, I. Y., & Nadezhin, D. K. (1985), SvAL, 11, 145
MacFadyen, A. I., Woosley, S. E., & Heger, A. (2001), ApJ, 550, 410
Matheson, T., Filippenko, A. V., Chornock, R., Leonard, D. C., & Li, W. (2000), AJ, 119, 2303
Matheson, T., Filippenko, A. V., Li, W., Leonard, D. C., & Schields, J. C. (2001), AJ, 121, 1648
Mazzali, P. A., et al. (2002), ApJ, 572, L61
McKenzie, E. H., & Schaefer, B. E. (1999), PASP, 111, 964
Minkowski, R. (1941), PASP, 53, 224
Nomoto, K. et al. (2000), in Gamma-ray Bursts, 5th Huntsville Symposium, AIP Conf. Ser., Vol 526, ed. R. Marc Kippen, R. S. Mallozzi, & G. J. Fishman (New York: Melville), 622

Patat, F., et al. (2001), ApJ, 555, 900
Phillips, M. M., Heathcote, S. R., Hamuy, M., & Navarrete, M. (1988), AJ, 95, 1087
Phillips, M. M., & Kirhakos, S. (2003), private communication
Richmond, M. W., et al. (1996), AJ, 111, 327
Ruiz-Lapuente, P., Kidger, M., López, R., & Canal, R. (1990), AJ, 100, 782
Schlegel, E. M. (1990), MNRAS, 244, 269
Schmidt, B. P., et al. (1993), AJ, 105, 2236
Schmidt, B. P. (2003), private communication
Shigeyama, T., Nomoto, K., Tsujimoto, T., & Hashimoto, M. (1990), ApJ, 361, L23
Smartt, S. J., Gilmore, G. F., Trentham, N., Tout, C. A., & Frayn, C. M. (2001), ApJ, 556, L29
Smartt, S. J., Gilmore, G. F., Tout, C. A., & Hodgkin, S. T. (2002), ApJ, 565, 1089
Sollerman, J., Kozma, C., Fransson, C., Leibundgut, B., Lundqvist, P., Ryde, F., & Woudt, P. (2000), ApJ, 537, L127
Stritzinger et al. (2002), AJ, 124, 2100
Suntzeff, N. B., et al. (2003), private communication
Tsvetkov, D. Y. (1994), AstL, 20, 374
Turatto, M., Cappellaro, E., Danziger, I. J., Benetti, S., Gouiffes, C., & Della Valle, M. (1993a), MNRAS, 262, 128
Turatto, M., Cappellaro, E., Benetti, S., & Danziger, I. J. (1993b), MNRAS, 265, 471
Turatto, M., et al. (2000), ApJ, 534, L57
van Belle, G.T., et al. (1999), AJ, 117, 521
Wang, L., Howell, D. A., Höflich, P., & Wheeler, J. C. (2001), ApJ, 550, 1030
Weiler, K. W., Panagia, N., Montes, M. J., Sramek, R. A., & Van Dyk, S. D., (2002), ARA&A, 40, 387
Wheeler, J. C., & Harkness, R. P. (1986), in Galaxy distances and deviations from universal expansion, ed. B. F. Madore & R. B. Tully (Dordrecht, Reidel), 45
Woodings, S. J., Williams, A. J., Martin, R., Burman, R. R., & Blair, D. G. (1998), MNRAS, 301, L5
Woosley, S. E., Pinto, P. A., Martin, P. G., & Weaver, T. A. (1987), ApJ, 318, 664
Zampieri, L., Pastorello, A., Turatto, M., Cappellaro, E., Benetti, S., Altavilla, G., Mazzali, P., & Hamuy, M. (2003), MNRAS, in press, (astro-ph/021017)
Zwicky, F. (1938), PASP, 50, 215

Chapter 3

EXPLOSION MECHANISMS OF MASSIVE STARS

A Critical Review of Possibilities and Perspectives

H.-Thomas Janka,
Robert Buras, Konstantinos Kifonidis, Markus Rampp
Max-Planck-Institut für Astrophysik,
Karl-Schwarzschild-Str. 1, D-85741 Garching, Germany
thj@mpa-garching.mpg.de

and Tomek Plewa
Dept. of Astron. and Astrophysics and Center for Astrophysical Thermonuclear Flashes,
The University of Chicago, Chicago, IL 60637, U.S.A.
and Nicolaus Copernicus Astronomical Center, Bartycka 18, 00716 Warsaw, Poland

Abstract One of the central problems in supernova theory is the question how massive stars explode. Understanding the physical processes that drive the explosion is crucial for linking the stellar progenitors to the final remnants and for predicting observable properties like explosion energies, neutron star and black hole masses, nucleosynthetic yields, explosion anisotropies, and pulsar kicks. In this article we review different suggestions for the explosion mechanism and discuss the constraints that can or cannot be deduced from observations. The prompt hydrodynamical bounce-shock mechanism has turned out not to work for typical stellar iron cores and empirical values of the compressibility of bulk nuclear matter. Magnetohydrodynamical models on the other hand contain a number of imponderabilities and are still far behind the level of refinement that has been achieved in nonmagnetic simulations. In view of these facts the neutrino-driven mechanism must still be considered as the standard paradigm to explain the explosion of ordinary supernovae, although its viability has yet to be demonstrated convincingly. Since spherically symmetric models do not yield explosions, the hope rests on the helpful effects of convection inside the nascent neutron star, which could boost the neutrino luminosity, and convective overturn in the neutrino-heated region behind the stalled shock, which increases the efficiency of neutrino-energy transfer in this layer. Here we present the first two-dimensional simulations of these processes which have been performed with a Boltzmann solver for the neutrino transport and a state-of-the-art description of neutrino-matter interactions.

C.L. Fryer (ed.), Stellar Collapse, 65–97.

Although our most complete models fail to explode, convection brings them encouragingly close to a success. An explosion could be obtained by just a minor modification of the neutrino transport, in which case the exploding model fulfills important requirements from observations. We discuss necessary improvements on the route to finally successful models.

Keywords: Supernovae, Neutrinos, Radiation-Hydrodynamics

1. Introduction

Supernova explosions of massive stars are powered by the gravitational binding energy that is released when the initial stellar core collapses to a compact remnant and its radius shrinks from typically a few thousand kilometers to little more than ten kilometers. For solar-metallicity progenitors with main-sequence masses of less than about 20–25 $M_\odot$ the compact leftover will be neutron star. In case of more massive stars a black hole will be formed, most likely by the fallback — on a timescale of seconds to hours — of matter that does not become unbound in the stellar explosion. But also the direct collapse of the stellar core to a black hole on a multiple of the dynamical timescale is possible (Woosley, Heger, & Weaver 2002; Heger et al. 2002).

Since the collapse proceeds essentially adiabatically the total energy of the stellar core is conserved during the implosion. The gravitational energy is temporarily stored as internal energy, mainly of degenerate electrons and electron neutrinos. If rotation plays a significant role in the progenitor core, a major fraction of the potential energy may also be converted to rotational energy of the nascent neutron star (or black hole).

The disruption of the massive star in a supernova explosion now means that some fraction of the energy in these reservoirs has to be transferred from the compact central object to the outer stellar layers. What are the physical mechanisms to mediate this energy transfer to the ejecta? And on what timescale do they work? Proposals in the literature include the hydrodynamical bounce-shock, neutrinos, or magnetic fields. The former would initiate the explosion on a dynamical timescale, whereas the latter two can establish the energy transfer only on the secular timescales of neutrino diffusion or magnetic field amplification, respectively.

Unfortunately, observations have so far not been able to yield direct insight into the processes in the stellar center at the onset of the explosion. The hope is that a future Galactic supernova will change this situation by allowing the measurements of a large number of neutrinos and possibly of a gravitational wave signal in great detail. The few neutrino events discovered in connection with Supernova 1987A were a clear signal of stellar core collapse and neutron star formation, but they were not sufficient to reveal the secrets of the explosion. Up to now we have to exploit the less direct information that is provided by the supernova light, by the characteristic properties of supernovae and diffuse and compact supernova remnants, and by the nucleosynthesis of heavy elements which takes place in the immediate vicinity of the newly formed neutron star.

Section 2 will discuss the constraints for the explosion mechanism that are associated with such observations. In Sect. 3 we shall briefly review the different suggestions that have been brought forward to explain the explosions of massive stars and will critically evaluate our knowledge on grounds of theoretical considerations. In Sect. 4 we shall summarize the status of detailed hydrodynamical supernova simulations and their implications for our understanding of the delayed explosion mechanism by neutrino-energy deposition behind the supernova shock. In Sect. 5 we shall present the first results of a new generation of multi-dimensional supernova models which employ a Boltzmann solver for the neutrino transport and a state-of-the-art description of neutrino-matter interactions. Section 6 will conclude with an appraisal of the results and an outlook on open ends.

2. Observational Facts

Progress in our understanding of the processes that lead to the explosion of massive stars is mainly based on elaborate numerical modeling, supplemented by theoretical analysis and constrained by a growing data base of observed properties of supernovae. The latter may carry imprints from the physical conditions very close to the center of the explosion. Observable features at very large radii, however, can be linked to the actual energy source of the explosion only indirectly through a variety of intermediate steps and processes. Any interpretation with respect to the mechansim that initiates the explosion therefore requires caution.

A viable model for the explosion mechanism of massive stars should ultimately be able to explain the observed explosion energies, nucleosynthetic yields (in particular of radioactive isotopes like ^{56}Ni, which are created near the mass cut), and the masses of the compact remnants (neutron stars or black holes) and their connection with the progenitor mass.

Recent evaluations of photometric and spectroscopic data for samples of well-observed Type-II plateau supernovae reveal a wide continuum of kinetic energies and ejected nickel masses. Faint, low-energy cases seem to be nickel-poor whereas bright, high-energy explosions tend to be nickel-rich and associated with more massive progenitors (Hamuy 2002). This direct correlation between stellar and explosion properties, however, is not apparent in an independent analysis by Nadyozhin (2002) who speculates that more than one stellar parameter (rotation or magnetic fields besides the progenitor and core mass) might determine the explosion physics. A large range of nickel masses and explosion energies was also found for Type Ib/c supernovae (Hamuy 2002).

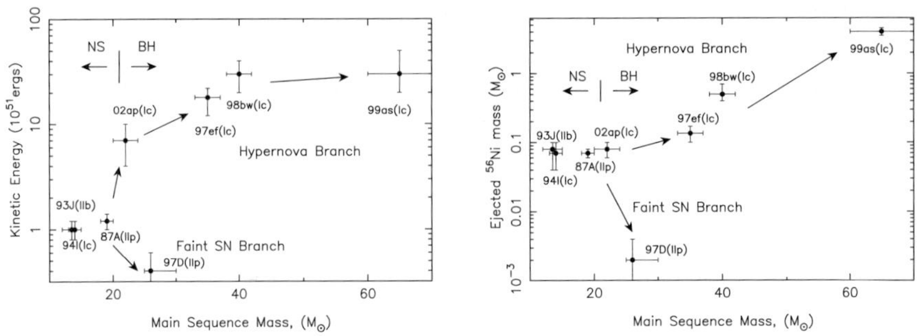

Figure 3.1. Explosion energy (left) and ejected ^{56}Ni mass as functions of the main sequence mass of the progenitor stars for several supernovae and hypernovae [from Nomoto et al. (2002)].

Interpreting results obtained by the fitting of optical lightcurves and spectra, Nomoto et al. (2002) proposed that explosions of stars with main sequence masses above 20–25 $M_\odot$ split up to a branch of extraordinarily bright and energetic events ("hypernovae") at the one extreme and a branch of faint, low-energy or even "failed" supernovae at the other. Stars with such large masses might collapse to black holes rather than neutron stars. The power of the explosion could depend on the amount of angular momentum in the collapsing core, which in turn can be sensitive to a number of effects such as stellar winds and mass loss, metallicity, magnetic fields, binarity or spiraling-in of a companion star in a binary system.

Another constraint for supernova theory comes from measured neutron star and black hole masses. A direct mass determination may not be possible for compact remnants inside supernova explosions, but the combined information about progenitor (ejecta) mass, kinetic energy, and nickel mass of a supernova (as deduced from spectroscopic and photometric data) may be used to determine the mass cut in a progenitor model by reproducing the known parameters with numerical simulations. Masses of binary pulsars (Thorsett & Chakrabarty 1999; Stairs et al. 2002) and X-ray binaries (Bailyn et al. 1998) also add to our knowledge about compact supernova remnants. In this case, however, binary effects make it even more difficult to establish a link to progenitor properties and therefore conclusions on the supernova mechanism should be drawn with great caution.

Complete and incomplete silicon and oxygen burning behind the supernova shock depend on the strength of the shock and the structure of the progenitor. The abundances of corresponding nucleosynthetic products, if observed in an individual supernova or supernova remnant, can therefore yield information about the exploding star and the explosion energy. Their spatial distribution can also reveal anisotropies and mixing phenomena during the explosion as in case of, e.g., Supernova 1987A or the Vela supernova remnant. Another serious constraint for the explosion mechanism and the conditions near the mass cut comes from a comparison of integral yields of supernova nucleosynthesis with Galactic abundances. The ejected mass of closed-neutron shell isotopes with $N = 50$ of Sr, Y, and Zr, for example, is limited to less than about $10^{-4}\,M_\odot$ per event [Hoffman et al. (1996)]. These stable nuclei are easily created through the capturing of α-particles in the neutron-rich matter around the mass cut and thus experience conditions in the close vicinity of the neutron star. The crucial parameters are entropy and neutron excess. Both are determined by the exposure to high neutrino fluxes, because neutrinos exchange energy and lepton number with the stellar medium.

Anisotropic processes and large-scale mixing between the deep interior and the hydrogen layer had to be invoked in case of Supernova 1987A to explain the shape of the lightcurve, the unexpectedly early appearance of X-ray and γ-ray emission, and Doppler features of spectral lines (for a review, see Nomoto et al. 1994). Fifteen years after the explosion, the expanding debris exhibits an axially symmetric deformation (Wang et al. 2002). Supernova 1987A therefore seems to possess an intrinsic, global asymmetry. The same conclusion was drawn for other core-collapse supernovae (Type-II as well as Ib/c) based on the fact that their light is linearly polarized at a level around 1% with a tendency to increase at later phases when greater depths are observed (Wang et al. 2001; Leonhard et al. 2001). This has been interpreted as evidence that the inner portions of the explosion, and hence the mechanism itself, are strongly non-spherical (Höflich, Wheeler, & Wang 1999; Wheeler et al. 2000), possibly associated with a "jet-induced" explosion (Wang et al. 2002; Khokhlov et al. 1999). This is a very interesting and potentially relevant conjecture. It does, however, not necessarily constrain the nature of the physical process that mediates the energy transfer from the collapsed core of the star to the ejecta and thereby creates the asphericity.

The high space velocities of young pulsars with a mean around $400\,\mathrm{km\,s^{-1}}$ (Lyne & Lorimer 1994; Cordes & Chernoff 1998) have also been used to argue for an intrinsically asymmetric explosion mechanism. It is very likely that the neutron stars obtained their recoil at the onset of the explosion or shortly afterwards (for a review, see Lai 1991; Lai, Chernoff, & Cordes 2001). But it is not clear whether the mechanism and the anisotropies of the explosion are directly connected with the pulsar kicks. Also a possible "jet" and perhaps a "counter jet" in the gaseous filaments of the Cas A supernova remnant may neither be linked to the process that started the supernova explosion nor to the runaway velocity of the compact remnant. The latter could have been accelerated (maybe by anisotropic neutrino emission) *after* the explosion was launched. The jet-like features — if at all caused by processes at the supernova center and not by local inhomogeneities at the forward shock — may have

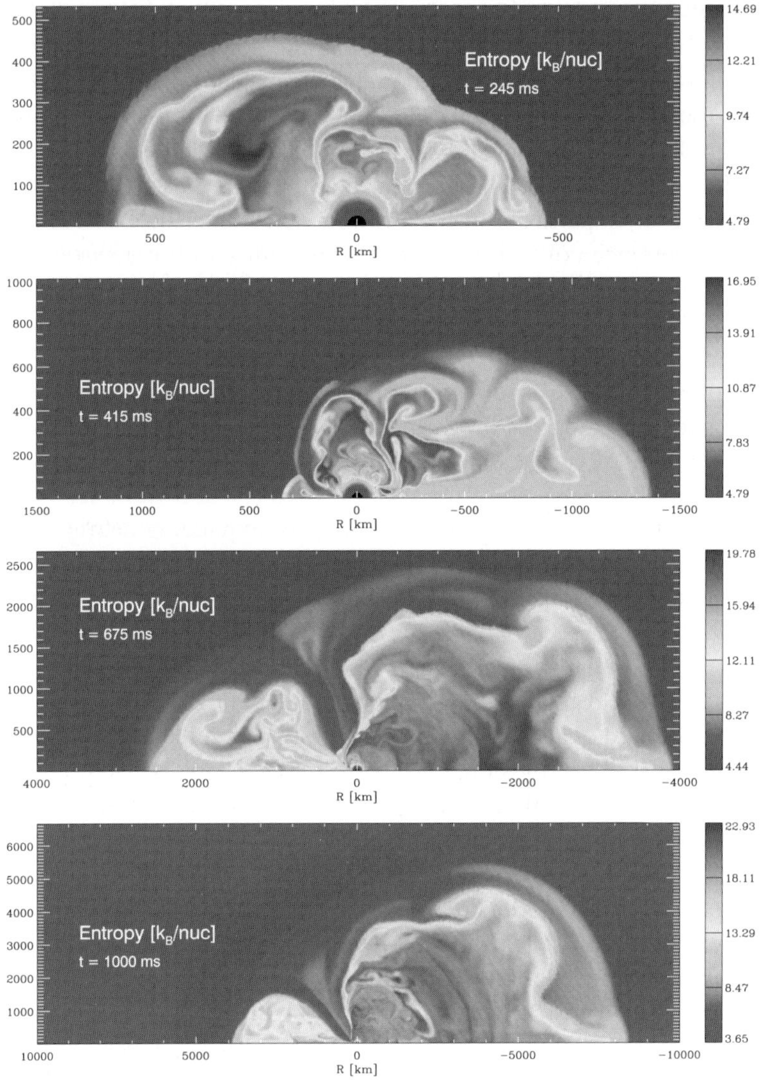

Figure 3.2. Explosion that is driven by neutrino-energy deposition in combination with convective overturn in the region behind the supernova shock. The anisotropy of the neutrino- and shock-heated ejecta is growing in time and becomes very large due to an increasing contribution of the $m = 0$, $l = 1$ mode in the convective pattern. The snapshots (from top to bottom) show the entropy distribution (values between about 4 and 23 k_B per nucleon) at post-bounce times $t_\mathrm{pb} = 245$ ms, 415 ms, 675 ms, and 1000 ms. Note that the radial scales of the figures differ. The neutron star is at the origin of the axially symmetric (2D) grid and plays the role of an isotropic neutrino "light bulb" (Plewa et al., in preparation).

been produced even later, for example by bipolar outflow from a neutron star or a black hole that has accreted from a disk of fallback matter. Interestingly, inspecting the images given by Gotthelf et al. (2001), the dislocation of the compact source relative to the geometrical center of the remnant appears not to coincide with the direction of the possible "jets" (compare also Thorstensen, Fesen, & van den Bergh 2001).

Rotation plus magnetic fields were proposed as the "most obvious" way to break the spherical symmetry and to explain the global asphericity of core-collapse supernovae (Wheeler, Meier, & Wilson 2002; Akiyama et al. 2002; Wheeler 2002). It was argued that current numerical calculations may be missing a major ingredient necessary to yield explosions. A proper treatment of rotation *and* magnetic fields may be necessary to fully understand when and how stellar collapse leads to an explosion. Of course, this might be true. But a confirmation or rejection will require computer models with ultimately the full physics.

It must be stressed, however, that current observations do not necessitate such conclusions and hydrodynamical simulations suggest other possible explanations. Strong convection in the neutrino-heating region behind the supernova shock can account for huge anisotopies of the inner supernova ejecta, even without invoking rotation. If the explosion occurs quickly, much power remains on smaller scales until the expansion sets in and the convective pattern gets frozen in. If, in contrast, the shock radius grows only very slowly and the explosion is delayed for several 100 ms after bounce, the convective flow can merge to increasingly larger structures. In two-dimensional (2D) hydrodynamic calculations including cooling and heating by neutrinos between the neutron star and the shock (with parameter choices for a central, *isotropic* neutrino "light bulb" which enabled explosions), we (Plewa et al., in preparation) found situations where the convective pattern revealed a contribution of the $l = 1, m = 0$ mode that was growing with time and was even dominant at about one second after bounce (Fig. 3.2). Herant (1995) already speculated about such a possibility. Certainly three-dimensional (3D) calculations of the full sphere (and without the coordinate singularity on the axis of the spherical grid) are indispensable to convincingly demonstrate the existence of this phenomenon[1].

3. Theoretical Possibilities

The primary energy source for powering supernovae of massive stars is the gravitational binding energy of the newly formed proto-neutron star or proto-black hole (energy from nuclear reactions contributes at a minor level). To initiate and drive the explosion, energy has to be transferred to the stellar matter above the mass cut to be finally converted to kinetic energy of the ejecta. This could be achieved by hydrodynamical shocks, neutrinos, or magnetic fields as mediators. Collimated outflows or jets of relativistic plasma might also play a role for very special circumstances. Depending on the mediator, the conditions for efficient energy transfer, the corresponding timescale, and the tapped energy reservoir are different. A lot of work has been spent in the past 40 years to identify viable supernova mechanisms and to study the involved physics. The space here is not sufficient to review all important papers. We shall therefore only focus on some key contributions without attempting completeness.

Hydrodynamical (Prompt) Mechanism

When the homologously collapsing inner part of the stellar core is abruptly stopped at core bounce by the stiffening of the nuclear equation of state (EoS), a hydrodynamical shock is formed at the boundary to the supersonically infalling outer part (Colgate & White 1966). In this case the initial energy of the shock is equal to the kinetic energy of the inner core "at the last moment of good homology" (Mönchmeyer 1993):

$$E_{\rm sh,i} \approx 6\times 10^{51}\,\frac{M_{\rm ic}}{\rm M_\odot}\left(\frac{v_{\rm max}}{4\times 10^9\,{\rm cm\,s^{-1}}}\right)^2\ {\rm erg}\,, \tag{3.1}$$

when $M_{\rm ic}$ is the mass of the homologously collapsing part of the stellar core and $v_{\rm max}$ its maximum infall velocity. There is general agreement now, supported by a large number of numerical calculations (e.g., Bruenn 1985; Bruenn 1989a; Bruenn 1989b; Myra et al. 1987; Myra & Bludman 1989; Swesty, Lattimer, & Myra 1994), that this energy is not sufficient for the shock to reach the surface of the iron core: The energy losses from behind the shock by photodisintegration of nuclei to α-particles and free nucleons ($\sim 1.5\times 10^{51}$ erg per $0.1\,{\rm M_\odot}$) as well as neutrino escape after shock breakout through the neutrinosphere (some 10^{51} erg) are so severe that the shock stalls within milliseconds. Only in case of extremely small iron cores ($M \lesssim 1.1\,{\rm M_\odot}$;Baron & Cooperstein 1990) or an extraordinarily soft nuclear EoS (Baron, Cooperstein, & Kahana 1985; Baron et al. 1987) is the prompt hydrodynamical shock able to disrupt the star. Nevertheless, the hope that the prompt hydrodynamical bounce-shock mechanism might work for some stars in a mass window around $10\,{\rm M_\odot}$ has not been given up completely (Hillebrandt, Wolff, & Nomoto 1984; Sumiyoshi et al. 2001).

Magnetohydrodynamical (MHD) Mechanism

Soon after the discovery of pulsars the role of rotation and magnetic fields in the supernova explosion was scrutinized (Ostriker & Gunn 1971; Bisnovatyi-Kogan 1971). Because of flux conservation a seed field in the stellar core can grow significantly during collapse, although for realistic initial fields not to a strength that magnetic pressure could be dynamically important. But winding of field lines in the case of differential rotation, which is natural after the collapse of a spinning core, can further amplify the toroidal field component. If the magnetic pressure becomes comparable to the thermal pressure, magnetohydrodynamical forces can drive an explosion (Müller & Hillebrandt 1979), and might accelerate axial jets (LeBlanc & Wilson 1970; Wheeler, Meier, & Wilson 2002). Also buoyancy instabilities of highly magnetized matter could produce mass motions (Meier et al. 1976). In this case rotational energy of the spinning proto-neutron star is converted to magnetic energy, a part of which might then end up as kinetic energy of the ejecta. Hence,

$$E_{\rm ej} \lesssim 5\times 10^{52}\,\frac{M_{\rm ns}}{1.5\,{\rm M}_\odot}\left(\frac{R_{\rm ns}}{10\,{\rm km}}\right)^2\left(\frac{T_{\rm rot}}{1\,{\rm ms}}\right)^{-2}\ {\rm erg}\,, \qquad (3.2)$$

when $T_{\rm rot}$ is the spin period of the forming neutron star with mass $M_{\rm ns}$. The field, however, grows linearly with time and therefore the timescale to reach sufficiently strong fields is long compared to the accretion timescale of the collapsed core, unless unrealistically large initial values are assumed (Meier et al. 1976; Müller & Hillebrandt 1979). Recently it was suggested that the magnetorotational instability (MRI), which was discussed in the context of accretion disks by Balbus & Hawley (1998), might also be in action in the supernova core (Akiyama et al. 2002). Because of an exponential growth, this would reduce the field amplification timescale drastically and would allow for much smaller initial seed fields, too. It remains to be shown, however, whether the field can grow to a dynamically effective strength in a self-consistent model for reasonable assumptions about the rotation of stellar cores and despite of diverse effects that can lead to field saturation or hinder field growth. Moreover, it is unclear whether axial jets can be initiated by the MRI at the conditions in collapsing stellar cores. Implications for supernova nucleosynthesis are unexplored, but might be problematic if MHD explosions are frequent (Meier et al. 1976). The amplification and transport of magnetic fields in convective regions inside the nascent neutron star (Thompson & Duncan 1993) and in the neutrino-heating layer behind the supernova shock (Thompson & Murray 2001) also deserve further attention with respect to the generation of pulsar fields.

Neutrino-Heating (Delayed) Mechanism

While neutrinos drain energy from the shock-heated matter immediately after bounce, the situation changes somewhat later when the density and temperature behind the shock have dropped. High-energy neutrinos, emitted abundantly from the hot accretion layer that covers the nascent neutron star, are absorbed with a small, but finite probability (typically $\sim 10\%$) in the postshock gas. If this energy deposition is large enough, the stalled supernova shock can be revived. Since this can happen on a timescale of hundred or more milliseconds after bounce, the explosion is called "delayed" in contrast to the prompt bounce-shock mechanism (Wilson 1985; Bethe & Wilson 1985). The idea that the energy transfer to the ejecta is mediated by neutrinos goes back to (Colgate & White 1966). During core collapse and subsequent contraction, the gravitational binding energy of the forming neutron star is first stored as internal energy (degeneracy energy of leptons and thermal energy of nucleons) and only slowly carried away by neutrinos on the diffusion timescale of several seconds. The kinetic energy required for the supernova explosion is of order 10^{51} erg and hence tiny compared to the huge reservoir of binding energy,

$$E_{\rm kin} \ll \frac{3}{5}\frac{GM_{\rm ns}^2}{R_{\rm ns}} \sim \frac{1}{10}M_{\rm ns}c^2 \sim 3\times 10^{53}\ {\rm erg}\,, \tag{3.3}$$

where the second expression results from the first by using $R_{\rm ns} \sim 3R_{\rm s} \equiv 6GM_{\rm ns}/c^2$. But just a smaller fraction (about 10%) of this energy is actually radiated in neutrinos during the first hundreds of milliseconds until the explosion starts.

Neutrino heating behind the shock is essentially *unavoidable*. The temperature in this region drops roughly like $T \propto r^{-1}$ (close to isentropic conditions), which means that the energy loss rate of the stellar plasma by neutrinos produced in $e^{\pm}$ captures on protons and neutrons falls like $Q_\nu^- \propto T^6 \propto r^{-6}$. In contrast, the energy deposition by ν_e-absorption on neutrons and $\bar\nu_e$-absorption on protons decreases much more slowly. The heating rate varies roughly proportional to the neutrino energy density (which scales with the summed luminosity of ν_e plus $\bar\nu_e$, L_ν, divided by the square of the inverse distance r from the neutrinosphere) and the mean squared energy $\langle\epsilon_\nu^2\rangle$ (because of the dependence of the absorption cross section on the neutrino energy ϵ_ν), i.e., $Q_\nu^+ \propto L_\nu\,\langle\epsilon_\nu^2\rangle\,/r^2$. This means that there must be a radial position, the so-called "gain radius" (Bethe & Wilson 1985), exterior of which neutrino heating dominates neutrino cooling. The heating rate for matter that is disintegrated to free neutrons and protons can be estimated as

$$Q_\nu^+ \sim 550\ \frac{L_{\nu,53}\langle\epsilon_{\nu,15}^2\rangle}{r_7^2}\ \frac{\rm MeV}{\rm s} \tag{3.4}$$

per nucleon for a typical luminosity of ν_e plus $\bar\nu_e$ of 10^{53} erg/s, 15 MeV for the rms neutrino energy, and r measured in units of 10^7 cm. The explosion energy should scale as

$$E_{\rm exp} \sim \frac{\Delta M}{m_{\rm u}}\ Q_\nu^+\Delta t_{\rm h}\ , \tag{3.5}$$

where ΔM is the neutrino-heated mass, $m_{\rm u}$ the nucleon mass, and $\Delta t_{\rm h}$ the timescale of neutrino-energy deposition. The latter can be assumed to be roughly the time until nucleons have absorbed an energy comparable to their binding energy in the gravitational potential of the neutron star, i.e.,

$$\Delta t_{\rm h} \sim \frac{G M_{\rm ns} m_{\rm u}}{Q_\nu^+ \, r} \sim 40 \ {\rm ms} \tag{3.6}$$

for $M_{\rm ns} \approx 1.5\,{\rm M}_\odot$. These arguments suggest that the gradual (in contrast to impulsive) energy deposition by neutrinos might not be responsible for a major fraction of the net explosion energy of a supernova. Instead, neutrino heating helps to compensate the gravitational binding and thus triggers the outward expansion of the supernova ejecta. The energy increases further by ongoing energy transfer of neutrinos to the accelerating matter, and finally reaches positive values due to the energy release in recombinations of free nucleons to nuclei. This recombination energy accounts for a dominant part of the supernova energy. Numerical simulations and analytic considerations (Janka 2001) suggest a mass ΔM between a few $10^{-2}\,{\rm M}_\odot$ and about $10^{-1}\,{\rm M}_\odot$. Correspondingly, the explosion energy could range between some 10^{50} erg and several 10^{51} erg, when about 8 MeV are set free by each nucleon that recombines. These numbers naturally match the energies of observed normal supernovae (cf. Sect. 2).

Jet-Powered Explosions

Stars with main sequence masses beyond 20–25 $\rm M_\odot$ seem to be associated with much more powerful explosions with energies up to several 10^{52} erg (Sect. 2). Such energies are probably out of reach for the neutrino-heating mechanism as described above. The idea has been coined (Popham, Woosley, & Fryer 1999; MacFadyen & Woosley 1999) that the core of such stars collapses to a black hole, which then continues to accrete the infalling matter of the progenitor star ("collapsar" or "failed supernova"; Woosley 1993). Provided that the stellar core has retained enough angular momentum during its pre-collapse evolution (or was spun up by a merger with a binary companion), the accretion will proceed at very high rates through a disk or torus on the timescale of viscous angular momentum transport, which is much longer than the dynamical timescale. Therefore the efficiency of energy release can in principle be much higher than in case of spherical accretion. In terms of the rest-mass energy of the accreted matter one can get $E_{\rm acc} = \xi M_{\rm acc} c^2$ with maximum values between $\xi = 0.057$ for a non-rotating black hole and $\xi = 0.42$ for an extreme Kerr hole that accretes from a corotating thin disk (Shapiro & Teukolsky 1983). Since the mass accretion rates are hypercritical and the densities in the disk correspondingly high, photons cannot escape and the energy is set free in form of neutrinos rather than electromagnetic radiation (Popham, Woosley, & Fryer 1999). The efficiency may be reduced compared to the quoted maximum values if the density and temperature are too low for rapid neutrino production or the disk is very dense and thus becomes intransparent to neutrinos. In these cases the energy is advected into the black hole by the accretion flow instead of being carried away by neutrinos (Di Matteo, Perna, & Narayan 2002).

Alternatively, magnetohydrodynamical processes may drive outflows from the disk (e.g., Meier, Koide, & Uchida 2001; Daigne & Mochkovitch 2002; Drenkhahn 2002; Drenkhahn & Spruit 2002). It is also possible that the rotational energy of the spinning black hole is extracted via the Blandford and Znajek mechanism (Blandford & Znajek 1977) through magnetic fields that are anchored in the black hole and couple it to the surrounding disk.

By the annihilation of neutrinos and antineutrinos to $e^{\pm}$-pairs (Woosley 1993) or the mentioned MHD processes, polar jets could be launched to propagate through the star along the rotation axis (MacFadyen & Woosley 1999; Aloy et al. 2000; Zhang, Woosley, & MacFadyen 2002). In case the jets remain collimated and highly relativistic until they break out of the stellar surface (which requires that the star does not possess an extended hydrogen envelope; Matzner 2002) gamma-ray bursts (GRBs)

may be triggered. Nonrelativistic jets that expand laterally and sweep up the surrounding progenitor gas, or energetic winds from the accretion torus might be the driving force behind enormously powerful stellar explosions (MacFadyen, Woosley, & Heger 2001) which are observed as so-called "hypernovae" and GRB-supernovae (Iwamoto et al. 1998). For an overview and more information, see Woosley, Zhang, & Heger (2002).

The different mechanisms for massive star explosions as described above require different properties of the progenitor stars and different physical conditions in the collapsed stellar cores. The latter are incompletely known, but some of the requirements are more likely fulfilled than others, some combinations of necessary conditions may be more common and more typical, while others might be realized only in rare cases and for very special, exceptional circumstances. It is possible, if not probable, that the diversity of the observed explosions of massive stars and the large number of associated phenomena mean that more than one mechanism is at work. It may also be, however, that the majority of ordinary supernovae can be ascribed to the same central process which might then depend extremely sensitively on yet to be determined parameters.

4. Do Neutrino-Driven Explosions Work?

The neutrino-driven mechanism (Wilson 1985; Bethe & Wilson 1985) involves a minimum of controversial assumptions and uncertain degrees of freedom in the physics of collapsing stars. It relies on the importance of neutrinos and their energetic dominance in the supernova core. After the detection of neutrinos in connection with Supernova 1987A and the overall confirmation of theoretical expectations for the neutrino emission, this is not a speculation any more but an established fact. Of course, this does not mean that such a minimal input is sufficient to understand the cause of supernova explosions and to explain all observable properties of supernovae. But at least it can be taken as a good reason to investigate how far one can advance with a minimum of imponderabilities. Detailed numerical simulations are needed to answer the crucial question in this context: Is neutrino-energy deposition efficient enough to drive the explosion?

Spherically symmetric simulations with the current input physics (neutrino interactions and the equation of state of dense matter) do not yield explosions by the neutrino-heating mechanism. There is no controversy about that. All computations are in agreement, independent of Newtonian or relativistic gravity and independent of the neutrino transport being treated in an approximate way by flux-limited diffusion methods (e.g., Myra et al. 1987; Myra & Bludman 1989; Bruenn 1993; Bruenn, De Nisco, & Mezzacappa 2001) or very elaborately by solving the frequency- and angle-dependent Boltzmann transport equation (Rampp & Janka 2000; Mezzacappa et al. 2001; Liebendörfer et al. 2001; Liebendörfer et al. 2002).

Whether neutrinos succeed in reviving the stalled shock depends on the efficiency of the energy transfer to the postshock layer, which in turn increases with the neutrino luminosity and the hardness of the neutrino spectrum. Wilson and collaborators (Wilson & Mayle 1988; Wilson & Mayle 1993; Mayle, Tavani, & Wilson 1993; Totani et al. 1998) have obtained explosions in one-dimensional (1D) simulations for more than ten years now. In these models it is, however, *assumed* that neutron-finger convection in the hot neutron star boosts the neutrino luminosities. Moreover, a special equation of state with a high abundance of pions in the nuclear matter was used (Mayle, Tavani, & Wilson 1993), which again leads to higher neutrino fluxes from the neutron star and thus to enhanced energy-deposition behind the shock. Both assumptions are not generally accepted.

Two-dimensional (Herant, Benz, & Colgate 1992; Herant et al. 1994; Shimizu, Yamada, & Sato 1994; Burrows, Hayes, & Fryxell 1995; Janka & Müller 1996; Mezzacappa et al. 1998; Shimizu et al. 2001) and 3-dimensional simulations (Shimizu, Yamada, & Sato 1993; Fryer & Warren 2002) as well as theoretical considerations (Bethe 1990) have shown that the neutrino-heating layer is unstable to convective overturn. The associated effects support shock revival and can lead to explosions even in cases where spherical models fail. In the multi-dimensional situation accretion and expansion can occur simultaneously. Downflows of cooler, low-entropy matter that has fallen through the shock, coexist with rising bubbles of high-entropy, neutrino-heated gas (Fig. 3.2). On the one hand, the downflows carry cool material close to the gain radius where it absorbs energy readily from the intense neutrino fluxes. On the other hand, heated matter rises in bubbles and can expand and cool quickly, which reduces the energy loss by the reemission of neutrinos. This also increases the postshock pressure and hence pushes the shock farther out. The gain layer grows and thus more gas can accumulate in the neutrino-heating region. The gas also stays longer in the gain layer, in contrast to one-dimensional models where the matter behind the accretion shock has negative velocity and is quickly advected down to the cooling layer. When the gas arrives there, neutrino emission sets in and extracts again the energy which had been absorbed from neutrino heating shortly before. Due to the combination of all these effects postshock convection enhances the efficiency of the neutrino-heating mechanism. Therefore the multi-dimensional situation is *generically different* from the spherically symmetric case.

Nevertheless, the existence of convective overturn in the neutrino-heating layer does not guarantee explosions (Janka & Müller 1996; Mezzacappa et al. 1998). For insufficient neutrino heating the threshold to an explosion will not be overcome. Since neutrinos play a crucial role, an accurate description of the neutrino physics — transport and neutrino-matter interactions — is indispensable to obtain conclusive results about the viability of the neutrino-driven mechanism. All previously published multi-dimensional explosion models, however, have employed some crude approximations or simplifications in the treatment of neutrinos.

5. A New Generation of Multi-Dimensional Supernova Simulations

In order to take a next step of improvement in supernova modelling, we have coupled a new moment-closure Boltzmann solver for the neutrino transport to the PROMETHEUS hydrodynamics code. The combined program is called MuDBaTH (**Mu**lti-**D**imensional **B**oltzm**a**nn **T**ransport and **H**ydrodynamics) and allows for spherically symmetric as well as multi-dimensional simulations (Rampp & Janka 2002). Below we present some results of 1D and our first 2D supernova simulations with this new code.

Technical Aspects and Input Physics

For the integration of the equations of hydrodynamics we employ the Newtonian finite-volume code PROMETHEUS (Fryxell, Müller, & Arnett 1989), which was supplemented by additional problem specific features (Keil 1997). PROMETHEUS is a direct Eulerian, time-explicit implementation of the Piecewise Parabolic Method (PPM) of Colella & Woodward (1994). As a second-order Godunov scheme employing a Riemann solver it is particularly well suited for following discontinuities in the fluid flow like shocks or boundaries between layers of different chemical composition. A notable advantage in the present context is its capability of tackling multi-dimensional problems with high computational efficiency and numerical accuracy. Our code makes use of the "Consistent Multifluid Advection (CMA)" method (Plewa & Müller 1999) for ensuring accurate advection of different chemical components in the fluid, and switches from the original PPM method to the more diffusive HLLE solver of Einfeldt (1988) in the vicinity of strong shocks to avoid spurious oscillations (the so-called "odd-even decoupling" phenomenon) when such shocks are aligned with one of the coordinate lines in multidimensional simulations (Quirk 1994; Kifonidis 2000; Plewa & Müller 2001).

The Boltzmann solver scheme is described in much detail elsewhere (Rampp & Janka 2002). The integro-differential character of the Boltzmann equation is tamed by applying a variable Eddington factor closure to the neutrino energy and momentum equations (and the simultaneously integrated first and second order moment equations for neutrino number). For this purpose the variable Eddington factor is determined from the solution of the Boltzmann equation, and the system of Boltzmann equation and its moment equations is iterated until convergence is achieved. Employing this scheme in multi-dimensional simulations in spherical coordinates, we solve the radius- (and energy-) dependent moment equations on the different angular bins of the numerical grid but calculate the variable Eddington factor only once on an angularly averaged stellar background. This approximation is good only for situations without significant global deformations. Since the iteration of the Boltzmann equation has to be done only once per time step, appreciable amounts of computer time can be saved (Rampp & Janka 2002). We point out here that it turned out to be necessary to go an important step beyond this simple "ray-by-ray" approach. Physical constraints, namely the conservation of lepton number and entropy within adiabatically moving fluid elements, and numerical requirements, i.e., the stability of regions which should not develop convection according to a mechanical stability analysis, make it necessary to take into account the coupling of neighbouring rays at least by lateral advection terms and neutrino pressure gradients (Buras et al., in preparation).

General relativistic effects are treated only approximately in our code (Rampp & Janka 2002). The current version contains a modification of the gravitational potential by including correction terms due to pressure and energy of the stellar medium and neutrinos, which are deduced from a comparison of the Newtonian and relativistic equations of motion. The neutrino transport contains gravitational redshift and time dilation, but ignores the distinction between coordinate radius and proper radius. This simplification is necessary for coupling the transport code to our basically Newtonian hydrodynamics. Although a fully relativistic treatment would be preferable, tests showed that these approximations seem to work satisfactorily well (Liebendörfer et al., in preparation), at least as long as there are only moderate ($\sim$10–20%) deviations of the metric coefficients from unity and the infall velocities do not reach more than 10–20% of the speed of light in decisive phases of the evolution.

As for the neutrino-matter interactions, we discriminate between two different sets of input physics. On the one hand we have calculated models with conventional ("standard") neutrino opacities, i.e., a description of the neutrino interactions which follows closely the one used by Bruenn and Mezzacappa and collaborators (Bruenn 1985; Mezzacappa & Bruenn 1993a; Mezzacappa & Bruenn 1993b). It assumes nucleons to be uncorrelated, infinitely massive scattering targets for neutrinos. In these reference runs we have usually also added neutrino pair creation and annihilation by nucleon-nucleon bremsstrahlung (Hannestad & Raffelt 1998). Details of our implementation of these neutrino processes can be found elsewhere (Rampp & Janka 2002).

A second set of models was computed with an improved description of neutrino-matter interactions. Besides including nucleon thermal motions and recoil, which means a detailed treatment of the reaction kinematics and allows for an accurate evaluation of nucleon phase-space blocking effects, we take into account nucleon-nucleon correlations (following Burrows & Sawyer 1998; Burrows & Sawyer 1999), the reduction of the nucleon effective mass, and the possible quenching of the axial-vector coupling in nuclear matter (Carter & Prakash 2002). In addition, we have implemented weak-magnetism corrections as described by Horowitz (2002). The sample of neutrino processes was enlarged by also including scatterings of muon and tau neutrinos and antineutrinos off electron neutrinos and antineutrinos and pair annihilation reactions between neutrinos of different flavors [i.e., $\nu_{\mu,\tau} + \bar{\nu}_{\mu,\tau} \longleftrightarrow \nu_e + \bar{\nu}_e$; Buras et al. (2002)].

Our current supernova models are calculated with the nuclear equation of state of Lattimer & Swesty (1991) using an incompressibility modulus of $K = 180\,\mathrm{MeV}$ (other values of the incompressibility of bulk nuclear matter in the Lattimer & Swesty EoS cause only minor differences, see Thompson, Burrows, & Pinto 2002). For the density regime below 6×10^7 g/cm^3 we switch to an equation of state that considers electrons, positrons and photons, and nucleons and nuclei with an approximative treatment for composition changes due to nuclear burning and shifts of nuclear statistical equilibrium (Rampp & Janka 2002).

Models and Results

We have performed a number of core-collapse simulations in spherical symmetry and then followed the post-bounce evolution in one and two dimensions. All described calculations were started from a $15\,\mathrm{M}_\odot$ progenitor star, Model s15s7b2, provided to us by S. Woosley. Adopting the naming, we label our models by s15N for Newtonian runs and s15G for runs with approximate treat-

Table 3.1. Input physics for our set of computed models. See Sect. 5 for details.

Model	*Dim.*	*Gravity*	*ν Reactions*	*Transport*	*Wedge*[a]	*Zones*
s15Nso"70161d.a	1D	Newtonian	standard	Case A		
s15Nso"70161d.b	1D	Newtonian	standard	Case B		
s15Nso"70162d.a	2D	Newtonian	standard	Case A	$\pm 27^\circ$	20
s15Gso"70161d.b	1D	approx. GR	standard	Case B		
s15Gso"70161d.b*	1D	approx. GR	standard[b]	Case B		
s15Gio"70161d.a	1D	approx. GR	improved[c]	Case A		
s15Gio"70162d.a	2D	approx. GR	improved	Case A	$\pm 43.2^\circ$	32
s15Gio"70161d.b	1D	approx. GR	improved	Case B		
s15Gio"70162d.b	2D	approx. GR	improved	Case B	$\pm 43.2^\circ$	32

[a] Angular wedge of the spherical coordinate grid around the equatorial plane.
[b] Calculation without neutrino-pair creation by nucleon-nucleon bremsstrahlung.
[c] Calculation without the neutrino-antineutrino processes of Buras et al. (2002).

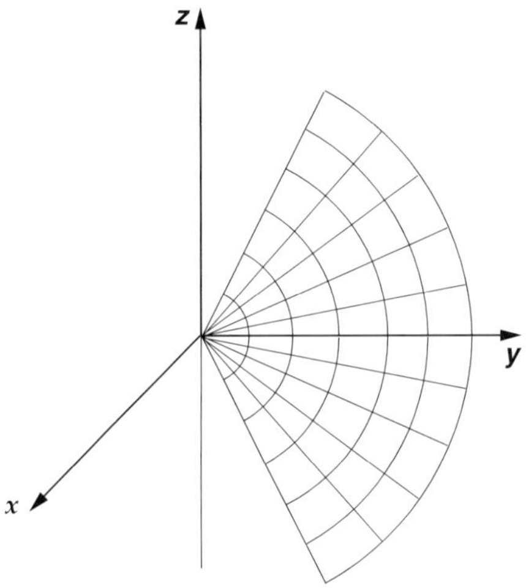

Figure 3.3. Computational grid for the 2D simulations. The wedge is placed around the equatorial plane of the spherical coordinate grid. Azimuthal symmetry around the z-axis is assumed, and periodic conditions are used at the lateral grid boundaries.

ment of general relativity , followed by letters "so" when "standard neutrino opacities" were used and by "io" in case of our state-of-the-art improvement of the description of neutrino-matter interactions. The model names have a suffix that discriminates between 1D (""70161d") and 2D simulations (""70162d"). For the 2D models we used a spherical coordinate grid with 20 or 32 equidistant zones within an angular wedge from -27° to $+27^\circ$ or from -43.2° to $+43.2^\circ$, respectively, around the equatorial plane and assumed periodic conditions at the boundaries (Fig. 3.3).

We have varied yet another aspect in our simulations. Some of the models were computed with a version of the transport code where the velocity depen-

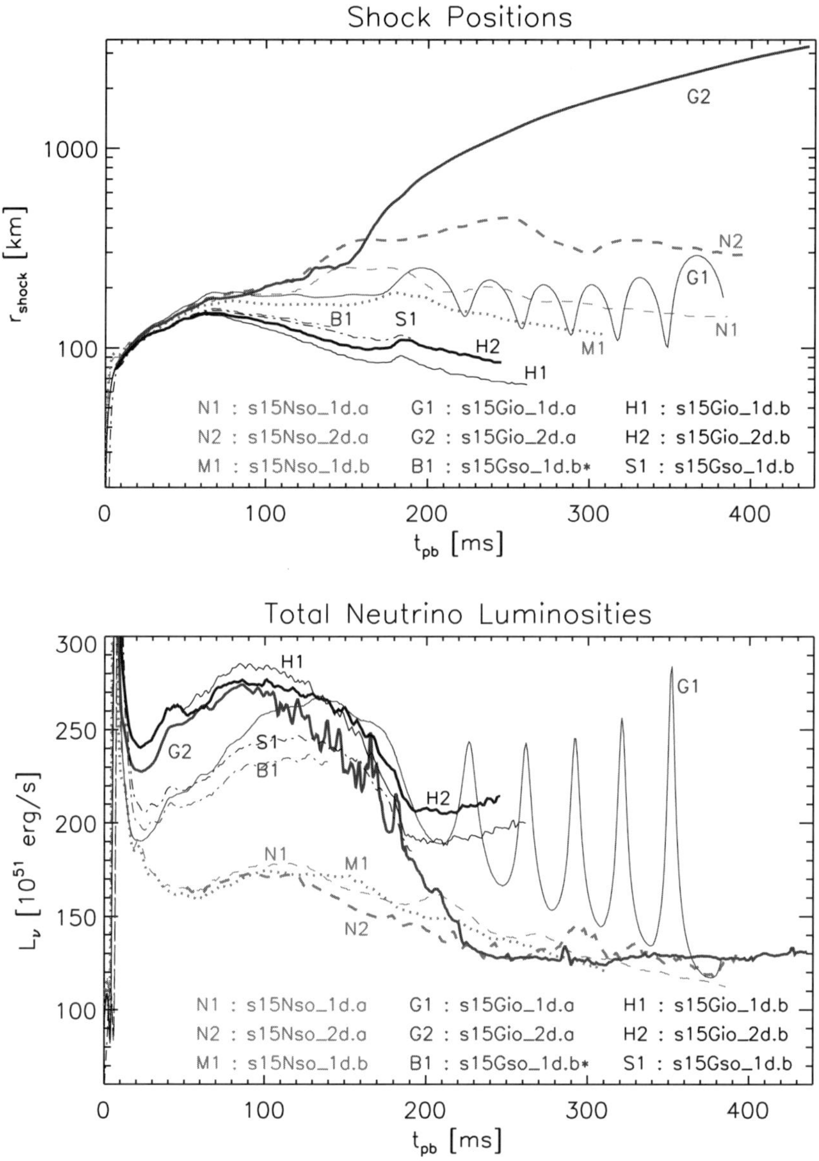

Figure 3.4. Shock trajectories (top) and total neutrino luminosities (sum of luminosities of ν_e, $\bar{\nu}_e$, and muon and tau neutrinos and antineutrinos; bottom) for all models. Bold lines correspond to two-dimensional simulations. Model H2 (s15Gio"70162d.b) is the most complete 2D simulation. It was computed with the improved set of neutrino opacities, relativistic effects, and all velocity-dependent terms retained in the neutrino transport equations.

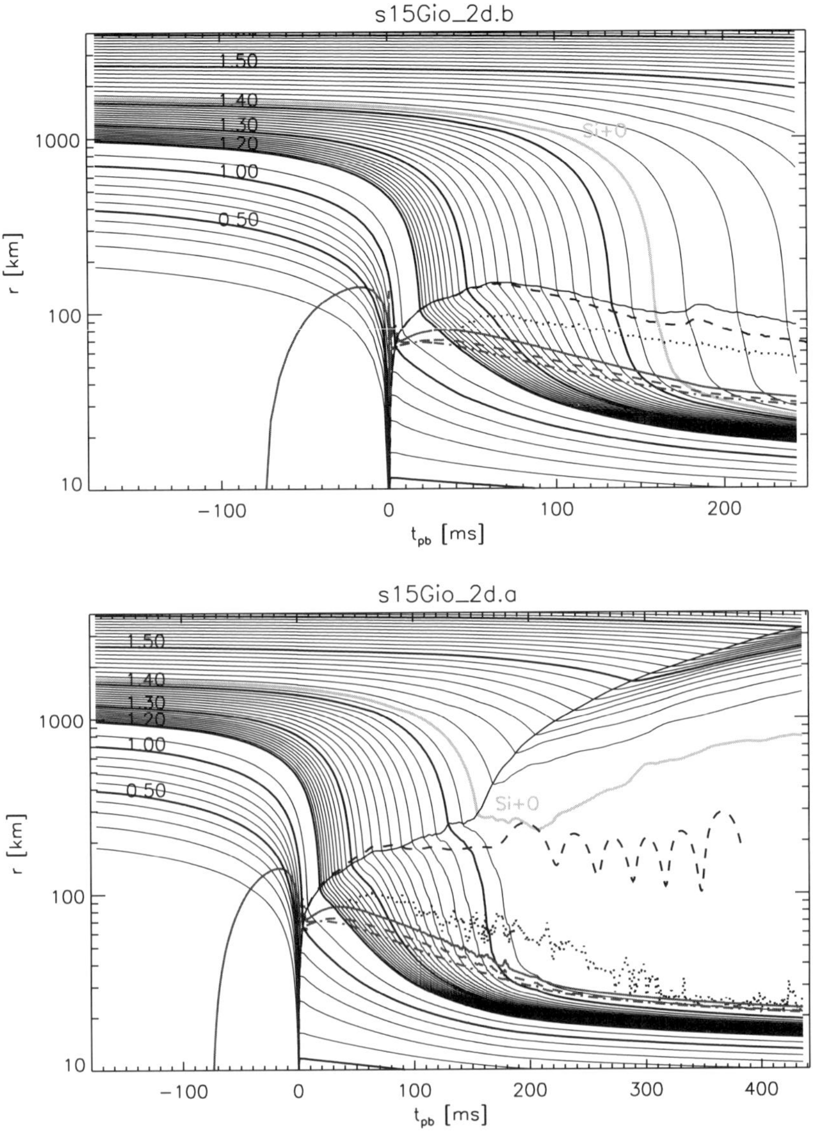

Figure 3.5. Trajectories of mass shells (time being normalized to bounce) for the non-exploding (top) and the exploding 2D model. In the latter case one can see the shock starting a rapid expansion at about 150 ms after bounce. The dashed lines indicate the shock positions in the corresponding 1D simulations, where no explosions were obtained. The angle-averaged gain radius is given by the dotted line, and the neutrinospheres of ν_e, $\bar{\nu}_e$ and heavy-lepton neutrinos (lower solid, dashed, and dash-dotted lines, respectively) are also marked.

dent (Doppler shift and aberration) terms in the neutrino momentum equation (and the corresponding terms in the Boltzmann equation for the antisymmetric average of the specific intensity; see Rampp & Janka 2002) were omitted. These terms are formally of order v/c and should be small for low velocities. This simplification of the neutrino transport, our so-called "Case A", was used in the models which have names ending with the letter "a". Models with neutrino transport including all velocity dependent terms also in the neutrino momentum equation ("Case B") can be identified by the ending "b" of their names. Table 3.1 provides an overview of the computed models.

The shock trajectories and corresponding total neutrino luminosities of all models are displayed in Fig. 3.4. No explosions, neither with the standard nor with the improved description of the neutrino opacities, were obtained with the most complete implementation of the transport equations, Case B (see lines labeled with M1, H1, H2, S1, B1, corresponding to, respectively, Models s15Nso"70161d.b, s15Gio"70161d.b, s15Gio"70162d.b, s15Gso"70161d.b and s15Gso"70161d.b*). The shock trajectories of this sample of models form a cluster that is clearly separated from the models computed with the transport version of Case A, which generally show a larger shock radius and therefore more optimistic conditions for explosions.

The simplification of the neutrino transport in Case A has a remarkable consequence which is obvious from a comparison of the 2D runs of Models s15Gio"70162d.a and s15Gio"70162d.b, which both were performed with our approximation of relativistic effects and the state-of-the-art improvement of neutrino-matter interactions (cf. Sect. 5). While Model s15Gio"70162d.b fails to explode, the stalled shock in Model s15Gio"70162d.a is successfully revived by neutrino heating because very strong convection can develop in the gain region. The time evolution of both models is displayed by the trajectories of mass shells in Fig. 3.5.

The reason for this dramatic difference can be understood in the following way. Some of the velocity dependent terms (those in which derivatives with respect to the neutrino energy do not show up) in the neutrino momentum equation have a simple formal interpretation: In regions with mass infall (*negative* velocity) they effectively act like a *reduction* of the neutrino-medium interaction on the right hand side of this equation. The corresponding changes can be 10% or more for neutrino energies in the peak of the spectrum, depending also on time, radius, and the size of the postshock velocities. As a consequence of the omission of these terms in Model s15Gio"70162d.a, neutrinos are impeded in their streaming. Therefore the comoving-frame neutrino (energy) density is increased (see Fig. 3.6). This is associated with slightly lower neutrino losses in the cooling layer around the neutrinosphere and a significantly enhanced neutrino heating between gain radius and shock.

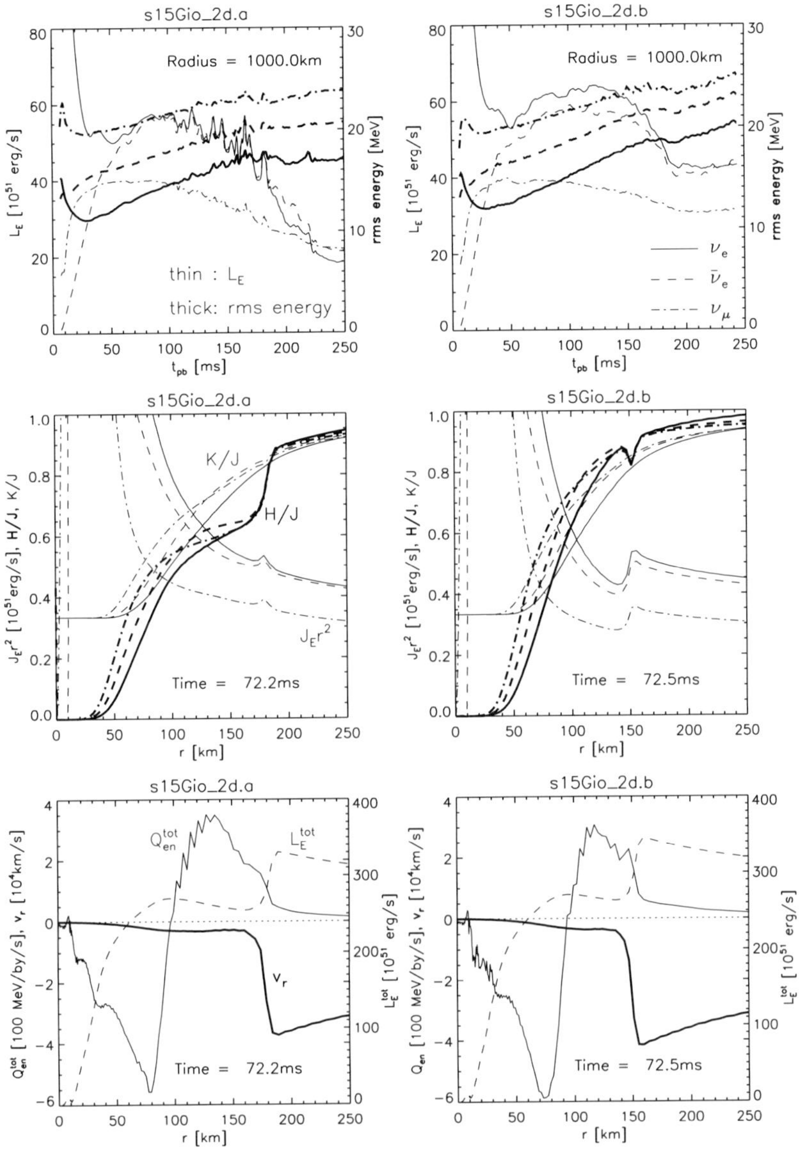

Figure 3.6. Comparison of the exploding 2D model (Model s15Gio"70162d.a; left) and the non-exploding 2D model (Model s15Gio"70162d.b; right). The top panels show the (comoving-frame) neutrino luminosities and rms energies as functions of time after bounce. The middle panels display the neutrino energy moment J, multiplied by r^2 and the flux factor H/J and Eddington factor K/J for both models at post-bounce time $t = 72.5$ ms, which corresponds roughly to the moment of maximum shock expansion in Model s15Gio"70162d.b (cf. Fig. 3.5). The lower panels give the total neutrino luminosities and neutrino heating rates in the comoving frame of the stellar fluid at the same time.

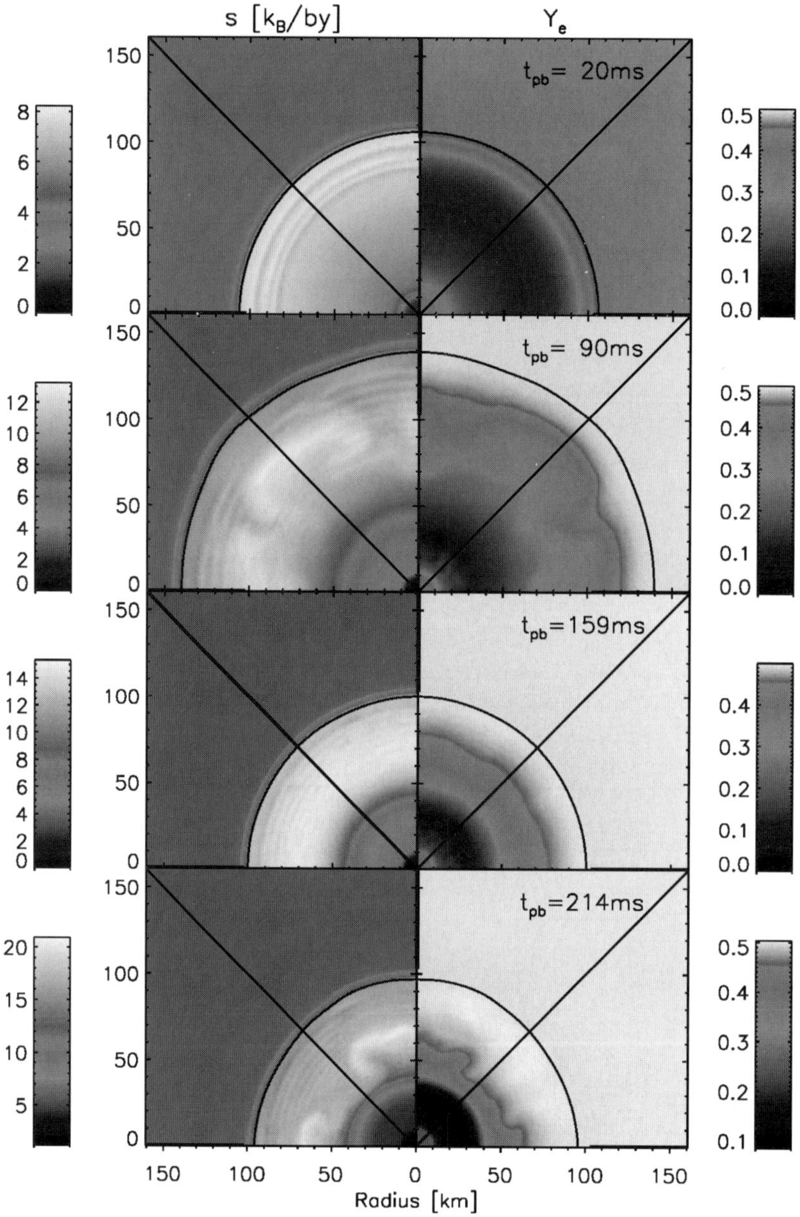

Figure 3.7. Convection in the neutrino-heating region for the non-exploding 2D model (Model s15Gio"70162d.b) at the post-bounce times indicated in the plots. The figures show the entropy distribution (left) and the electron fraction (proton-to-baryon ratio). A wedge of $\pm 43.2^{\circ}$ around the equatorial plane (marked by the diagonal solid lines) of the spherical coordinate grid was used for the computation.

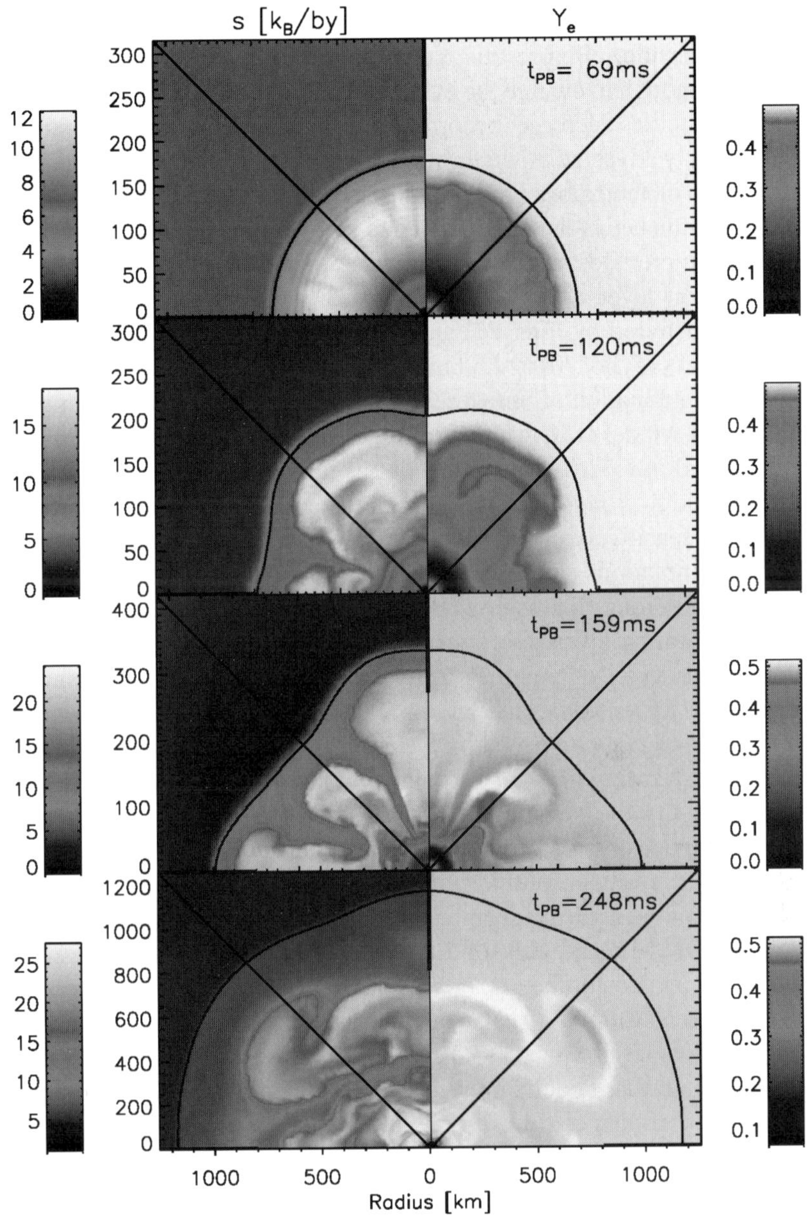

Figure 3.8. Same as Fig. 3.7, but for the exploding 2D model, Model s15Gio"70162d.a. Note the different radial scales of the four snapshots.

Although the differences are moderate (10–30%, depending on the quantity) the accumulating effects during the first 80 ms after bounce clearly damp the shock expansion and finally lead to a dramatic shock recession after the initial phase of expansion. Before this happens postshock convection has not become strong enough to change the evolution. With the onset of contraction, the postshock velocities decrease (become more negative) quickly, neutrino-heated matter is rapidly advected down through the gain radius and loses its energy by the reemission of neutrinos. The larger negative values of the postshock velocity enhance the influence of the velocity-dependent terms and reduce the heating efficiency even more. Because of this disastrous feedback, the gain region shrinks to a very narrow layer, a fact which suppresses the convective activity lateron. This is demonstrated by Figs. 3.7 and 3.8, which show that the convection is weak in Model s15Gio"70162d.b but very strong in Model s15Gio"70162d.a.

Due to a combination of unfavorable effects and a continuously amplifying negative trend, Model s15Gio"70162d.b remains below the explosion threshold while Model s15Gio"70162d.a is just above that critical limit. In the vicinity of the threshold the long-time evolution of the collapsing stellar core therefore depends very sensitively on "smaller details" of the neutrino transport.

The two simulations also demonstrate that a sufficiently large shock radius for a sufficiently long time is crucial for the growth of convective overturn in the neutrino-heating layer. Only when convection behind the shock becomes strong it can be decisive for getting an explosion. This latter fact is known from previous multi-dimensional simulations (Janka & Müller 1996; Mezzacappa et al. 1998) and is again confirmed by a comparison of Model s15Gio"70162d.a with the corresponding one-dimensional simulation, Model s15Gio"70161d.a. It is interesting that the Newtonian 2D run, Model s15Nso"70162d.a (with standard opacities) does marginally not explode, whereas Model s15Gio"70162d.a with relativistic corrections (and state-of-the-art neutrino reactions) succeeds. The influence of the neutrino opacities can be directly seen by comparing Models s15Gso"70161d.b and s15Gio"70161d.b. It is not dramatic, but sufficient to justify the inclusion of the improvements described in Sect. 5. Neutrino-pair creation by bremsstrahlung makes a minor difference during the considered phases of the evolution (Model s15Gso"70161d.b* vs. Model s15Gso"70161d.b).

We also found that Ledoux convection sets in *below* the neutrinosphere already shortly after bounce (Fig. 3.9). It is persistent until the end of our simulations and slowly digs farther into the star as already found in the simulations of Keil, Janka, & Müller (1996). But the convective activity is so deep inside the neutron star that its effects on the ν_e and $\bar{\nu}_e$ luminosities and on the supernova dynamics are insignificant.

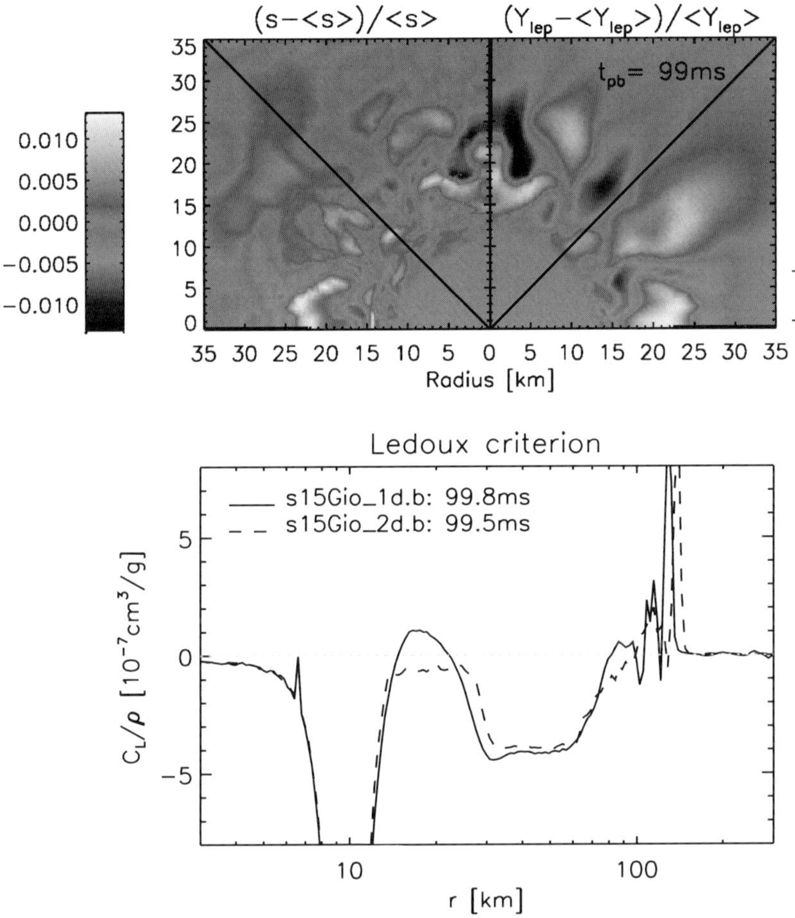

Figure 3.9. Convection below the neutrinosphere at about 100 ms after core bounce in Model s15Gio"70162d.b. The upper figure shows the relative fluctuations of entropy (left) and lepton fraction (right), normalized to the mean values at a given radius. The lower figure displays the Ledoux criterion for the 1D Model s15Gio"70161d.b (without convection; solid line) and the 2D Model s15Gio"70162d.b (dashed line). Positive values indicate convectively unstable conditions, once in an inner region (located at about 15–30 km) below the neutrinosphere and another time in the neutrino-heating layer behind the shock.

6. Conclusions and Outlook

We believe that our 2D models with a Boltzmann solver for the neutrino transport and a state-of-the-art description of neutrino-matter interactions have considerably reduced the uncertainties associated with the treatment of the neutrino physics in previous multi-dimensional simulations. With the most complete implementation of the transport physics we could not obtain explosions.

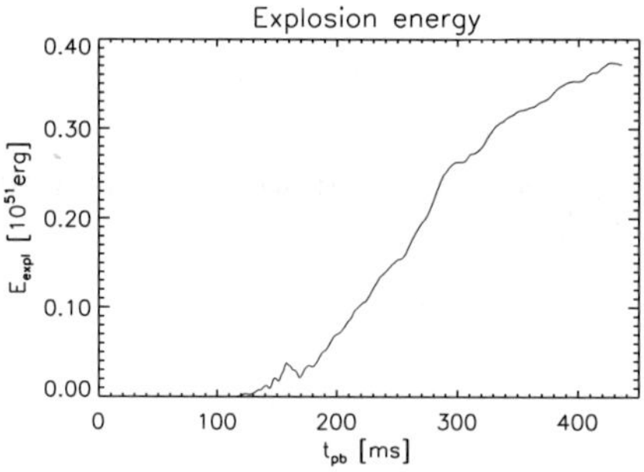

Figure 3.10. Explosion energy as function of time for Model s15Gio"70162d.a.

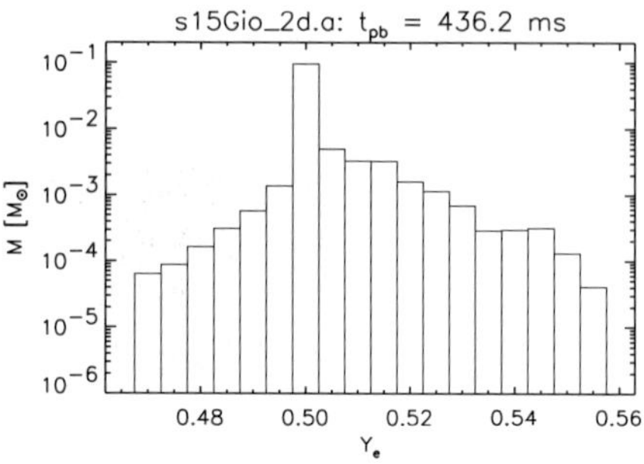

Figure 3.11. Ejecta mass vs. Y_e for Model s15Gio"70162d.a at 436 ms after bounce. The total mass of ^{56}Ni is about 0.1 $M_\odot$.

This result suggests that the neutrino-driven mechanism fails with the employed input physics, at least in case of the considered 15 $M_\odot$ star. We do not think that the approximate treatment of general relativistic effects is likely to jeopardize this conclusion. A comparison with fully relativistic one-dimensional calculations (Liebendörfer et al., in preparation) is very encouraging. Because of the remarkable similarity of the shock trajectories of different progenitors in spherical symmetry (Liebendörfer et al. 2002), it is probable that our negative conclusion is also valid for other pre-collapse configurations with a similar structure. Significant star-to-star variations of the progenitor properties with a non-monotonic dependence on the stellar mass (Woosley, Heger, & Weaver 2002), however, suggest that multi-dimensional core-collapse simulations of a larger sample of stars are needed before one can make final, more generally valid statements.

Another concern may be the omission of lateral neutrino fluxes in our multi-dimensional transport approximation. We describe the neutrino losses from the hot spots that appear at locations where the low-entropy downflows plunge into the neutrinospheric layer, by just a radial flux. In reality the neutrinos produced in this semi-transparent matter would be radiated away more isotropically, a fact which could lead to a larger reabsorption probability in the surrounding medium. The corresponding increase of the heating efficiency might not be very large, but could still be relevant because of the close proximity of our models to an explosion. The supernova problem is highly nonlinear and surprises may lurk behind every corner.

It would therefore be premature to conclude that the neutrino-driven mechanism fails and that not even postshock convection can alter this unquestioned outcome of all current spherical models. Besides studying other progenitors with multi-dimensional simulations, one should also investigate the effects of rotation during the neutrino-heating phase and long-time post-bounce evolution of a supernova. Previous simulations found weaker explosions in case of rotating models (Fryer & Heger 2000). But these calculations were performed with a much simplified treatment of the neutrinos and showed powerful, nearly prompt explosions already in the absence of rotation. It cannot be excluded that even a moderate amount of rotation might have significant consequences when added to nonrotating models which explode only with a long time delay after core bounce or even fail to do so. Naturally, three-dimensional simulations are ultimately desirable and could reveal generic differences compared to 2D, in particular in combination with rotation. The recent results of (Fryer & Warren 2002), which suggest a large degree of similarity between 2D and 3D results, are groundbreaking but again must be interpreted with caution. The serious approximations of the neutrino physics and the rapid development of the explosion in these models do not allow too far-reaching conclusions.

Another major uncertainty of current supernova models is associated with the incompletely known physics in the nuclear and supranuclear medium. Not only does the high-density EoS affect the core bounce, shock formation, and prompt shock propagation, it also determines the internal properties of the hot neutron star and thus its neutrino emission, both indirectly and directly. Hadronic degrees of freedom in addition to nucleons in the supranuclear matter, e.g., hyperons, pions or kaons, can reduce the stiffness of the EoS at very high densities. The possible effects on the supernova explosion are essentially unexplored. A softening of the EoS can, for example, lead to a faster contraction and heating of the nascent neutron star. This in turn also determines the density and temperature profile in the vicinity of the neutrinosphere where most of the neutrino flux is produced during the crucial phase of delayed shock revival.

Our successfully exploding 2D model, Model s15Gio"70162d.a, at least demonstrates that simulations which include the effects of postshock convection are rather close to an explosion. Therefore modest changes of the neutrino emission and transport (in our case the omission of the velocity-dependent terms in the neutrino-momentum equation) seem to be already sufficient to push them beyond the critical threshold. The properties of the explosion in this case are very encouraging and may support one's belief in the basic viability of the delayed explosion mechanism. At 440 ms after bounce the shock has arrived at a radius of more than 3000 km and is expanding with about $10000\,\mathrm{km\,s^{-1}}$ (Fig. 3.4). The explosion of this model does not become very energetic (Fig. 3.10). The energy is just $\sim 4 \times 10^{50}$ erg at that time and grows only very slowly. This may not be a serious problem if one recalls the large spread of energies of observed supernovae (Supernova 1999br, for example, is estimated to have an ejecta mass of $14\,M_\odot$ and an explosion energy of about 6×10^{50} erg; Hamuy 2002). Moreover, the explosion energy is a very steep function of the physical conditions in models which are above the explosion threshold (Janka & Müller 1995). A significant increase of the energy may therefore not require large changes.

Since the explosion starts rather late (at $\sim 150\,\text{ms}$ post bounce), the proto-neutron star has accreted enough matter to have attained an initial baryonic mass of $1.4\,\text{M}_\odot$. Therefore our simulation does not exhibit the problem of previous successful multi-dimensional calculations which produced neutron stars with masses on the lower side of plausible values ($\sim 1.1\,\text{M}_\odot$). Also another problem of published explosion models (e.g., Herant et al. 1994; Burrows, Hayes, & Fryxell 1995; Janka & Müller 1996; Fryer 1999) has disappeared: The ejecta mass with $Y_e \lesssim 0.47$ is less than $\sim 10^{-4}\,\text{M}_\odot$ in our calculation (Fig. 3.11), thus fulfilling a constraint pointed out by Hoffman et al. (1996) for supernovae if they should not overproduce the $N = 50$ (closed neutron shell) nuclei, in particular ^{88}Sr, ^{89}Y and ^{90}Zr, relative to the Galactic abundances. Of course, final statements about explosion energy, ejecta composition, and the neutron star mass (which may change due to later fallback, especially when the explosion energy remains low) require to follow the evolution for a longer time.

Final Words

There is still a long way to go until we will have arrived at a "standard model" for massive star explosions. Current numerical simulations are not advanced enough to convincingly and self-consistently demonstrate the viability of any of the explosion scenarios described in Sect. 3. Despite of respectable progress in treating the neutrino transport, general relativity, and multi-dimensional aspects in the most detailed numerical calculations, the models are far from being complete. Neglecting potentially important physics like, e.g., magnetic fields, one cannot be sure that one is searching for the actual cause of massive star explosions at the right place. Conversely, the existing simulations also do not bring us in a position to reject the idea that neutrinos deliver the required energy in the ejecta.

With the lack of direct observational information and being confronted with the fact that our knowledge about the physics in collapsing stellar cores is incomplete, there is no alternative to systematically explore the space of possibilities by hydrodynamical simulations. Since "the output of complex numerical models is as inscrutable as nature herself" (Emanuel 1991), it seems advisable to first develop reliable and accurate supernova models with a minimum of questionable assumptions and then to move on step by step to expanding them towards larger complexity and more completeness.

For the time being, one's favorate choice of the supernova mechanism is more a matter of taste and belief than based on undisputable facts and logical reasoning. The "ifs" and "buts" in current models are still too many. One should be cautious not to put too much money on any single possibility right now.

Acknowledgments

We are indebted to K. Takahashi for providing us routines to calculate the improved neutrino-nucleon interactions, and to C. Horowitz for correction formulae for the weak magnetism. We also thank M. Liebendörfer for making output data of his simulations available to us for comparisons. The Institute for Nuclear Theory at the University of Washington is acknowledged for its hospitality and the Department of Energy for support during a visit of the Summer Program on Neutron Stars, during which most of the work leading to Fig. 3.2 was done. HTJ, RB and MR are grateful for support by the Sonderforschungsbereich 375 on "Astroparticle Physics" of the Deutsche Forschungsgemeinschaft. TP was supported in part by the US Department of Energy under Grant No. B341495 to the Center of Astrophysical Thermonuclear Flashes at the University of Chicago, and in part by the grant 2.P03D.014.19 from the Polish Committee for Scientific Research. He performed his simulations on the CRAY SV1-1A at the Interdisciplinary Centre for Computational Modelling in Warsaw. The 2D simulations with Boltzmann neutrino transport were only possible because a node of the new IBM "Regatta" supercomputer was dedicated to this project by the Rechenzentrum Garching. Computations were also done on the NEC SX-5/3C of the Rechenzentrum Garching, and on the CRAY T90 and CRAY SV1ex of the John von Neumann Institute for Computing (NIC) in Jülich.

Notes

1. Another interesting possibility was pointed out by Blondin, Mezzacappa, & DeMarino (2002). Their 2D (and 3D; Mezzacappa et al. 2002) calculations revealed hydrodynamical instabilities in the accretion flow behind the stalled shock, which instigate large-scale modes even in the absence of neutrino heating. Whether these calculations are of relevance for the supernova problem has be to seen.

References

Aloy, M.A., Müller, E., Ibáñez, J.M.ª, Martí, J.M.ª, & MacFadyen, A.I. 2000, *Ap. J. (Letters)*, 531, L119

Akiyama, S., Wheeler, J.C., Meier, D.L., & Lichtenstadt, I. 2002, *Ap. J.*, in press (`astro-ph/0208128`)

Bailyn, C.D., Jain R.K., Coppi, P., & Orosz, J.A. 1998, *Ap. J.*, 499, 367

Balbus, S.A. & Hawley, J.F. 1998, Reviews of Modern Physics, 70, 1

Baron, E., Bethe, H.A., Brown, G.E., Cooperstein, J., & Kahana, S. 1987, Phys. Rev. Lett., 59, 736

Baron, E. & Cooperstein, J. 1990, *Ap. J.*, 353, 597

Baron, E., Cooperstein, J., & Kahana, S. 1985, Phys. Rev. Lett., 55, 126

Bethe, H.A. 1990, Rev. Mod. Phys., 62, 801
Bethe, H.A. & Wilson, J.R. 1985, *Ap. J.*, 295, 14 (1985)
Bisnovatyi-Kogan, G.S. 1971, Soviet Astronomy AJ, 14, 652
Blandford, R.D. & Znajek, R.L. 1977, MNRAS, 179, 433
Blondin, J.M., Mezzacappa, A., & DeMarino, C. 2002, *Ap. J.*, in press (astro-ph/-0210634)
Bruenn, S.W. 1985, *Ap. J. Suppl.*, 58, 771
Bruenn, S.W. 1989a, *Ap. J.*, 340, 955
Bruenn, S.W. 1989b, *Ap. J.*, 341, 385
Bruenn, S.W. 1993, in *Nuclear Physics in the Universe*, ed. M.W. Guidry & M.R. Strayer (Bristol: IOP) 31
Bruenn, S.W., De Nisco, K.R., & Mezzacappa, A. 2001, *Ap. J.*, 560, 326
Buras, R., Janka, H.-Th., Keil, M.-Th., Raffelt, G., & Rampp, M. 2002, *Ap. J.*, submitted (astro-ph/0205006)
Burrows, A. & Sawyer, R.F. 1998, Phys. Rev. C., 58, 554
Burrows, A. & Sawyer, R.F. 1999, Phys. Rev. C., 59, 510
Burrows, A., Hayes, J., & Fryxell, B.A.. 1995, *Ap. J.*, 450, 830
Carter, G.W. & Prakash, M. 2002, Physics Letters B, 525, 249
Colella, P. & Woodward, P.R. 1984, J. Computational Physics, 54, 174
Colgate, S.A. & White, R.H. 1966, *Ap. J.*, 143, 626
Cordes, J.M. & Chernoff, D.F. 1998, *Ap. J.*, 505, 315
Daigne, F. & Mochkovitch, R. 2002, *Astr. Ap.*, 388, 189
Di Matteo, T., Perna, R., & Narayan, R. 2002, *Ap. J.*, 579, 706
Drenkhahn, G. 2002, *Astr. Ap.*, 387, 714
Drenkhahn, G. & Spruit, H.C. 2002, *Astr. Ap.*, 391, 1141
Emanuel, K.A. 1991, Ann. Rev. Fluid Mech., 23, 179
Fryer, C.L. 1999, *Ap. J.*, 522, 413
Fryer, C.L. & Heger, A. 2000, *Ap. J.*, 541, 1033
Fryer, C.L. & Warren, M.S. 2002, *Ap. J. (Letters)*, 574, L65
Fryxell, B.A., Müller, E., & Arnett, W.D. 1989, (Preprint MPA-449, Garching: Max-Planck-Institut für Astrophysik)
Gotthelf, E.V., Koralesky, B., Rudnick, L., Jones, T.W., Hwang, U., & Petre, R. 2001, *Ap. J. (Letters)*, 552, L39
Hamuy, M. 2002, *Ap. J.*, in press (astro-ph/0209174)
Hannestad, S. & Raffelt, G. 1998, *Ap. J.*, 507, 339
Heger, A., Woosley, S.E., Fryer, C.L., & Langer, N. 2002, in: *From Twilight to Highlight — The Physics of Supernovae*, ed. W. Hillebrandt & B. Leibundgut (Springer Series "ESO Astrophysics Symposia", Berlin: Springer) (astro-ph/0211062)
Einfeldt, B. 1988, SIAM Jour. Numer. Anal., 25, 294
Herant, M. 1995, Physics Rep., 256, 117
Herant, M., Benz, W., & Colgate, S.A. 1992, *Ap. J.*, 395, 642

Herant, M., Benz, W., Hix, W.R., Fryer, C.L., & Colgate, S.A. 1994, *Ap. J.*, 435, 339
Hillebrandt, W., Wolff, R.G., & Nomoto, K. 1984, *Astr. Ap.*, 133, 175
Höflich, P., Wheeler, J.C., & Wang, L. 1999, *Ap. J.*, 521, 179
Hoffman, R.D., Woosley, S.E., Fuller, G.M., & Meyer, B.S. 1996, *Ap. J.*, 460, 478
Horowitz, C.J. 2002, Phys. Rev. D, 65, 043001-1
Iwamoto, K., *et al.* 1998, Nature, 395, 672
Janka, H.-Th. 2001, *Astr. Ap.*, 368, 527
Janka, H.-Th. & Müller, E. 1995, *Ap. J. (Letters)*, 448, L109
Janka, H.-Th. & Müller, E. 1996, *Astr. Ap.*, 306, 167
Janka, H.-Th., Buras, R., & Rampp, M. 2002, in *Proceedings of the 7th Int. Symposium on Nuclei in the Cosmos*, Nuclear Physics A, in press
Keil, W. 1997, PhD Thesis, Technische Universität München
Keil, W., Janka, H.-Th., & Müller, E. 1996, *Ap. J. (Letters)*, 473, L111
Kifonidis, K. 2002, PhD Thesis, Technische Universität München
LeBlanc, J.M. & Wilson, J.R. 1970, *Ap. J.*, 161, 541
Khokhlov, A.M., Höflich, P., Oran, E.S., Wheeler, J.C., Wang, L., & Chtchelkanova, A.Yu. 1999, *Ap. J. (Letters)*, 524, L107
Lai, D. 2001, in *Physics of Neutron Star Interiors*, ed. D. Blaschke, N.K. Glendenning, & A.D. Sedrakian, (Lecture Notes in Physics, 578, Berlin: Springer) 424
Lai, D., Chernoff, D.F., & Cordes, J.M. 2001, *Ap. J.*, 549, 1111
Lattimer, J.M. & Swesty, F.D. 1991, Nucl. Phys. A, 535, 331
Leonhard, D.C., Filippenko, A.V., Ardila, D.R., & Brotherton, M.S. 2001, *Ap. J.*, 553, 86
Liebendörfer, M., Messer, O.E.B., Mezzacappa, A., Hix, W.R., Thielemann, F.-K., & Langanke, K. 2002, in *Proc. 11th Workshop on Nuclear Astrophysics*, ed. W. Hillebrandt & E. Müller (Report MPA/P13, Garching: Max-Planck-Institut für Astrophysik) 126 (astro-ph/0203260)
Liebendörfer, M., Mezzacappa, A., Thielemann, F., Messer, O.E.B., Hix, W.R., & Bruenn, S.W. 2001, Phys. Rev. D., 63, 3004
Lyne, A.G., & Lorimer, D.R. 1994, Nature, 369, 127
MacFadyen, A.I. & Woosley, S.E. 1999, *Ap. J.*, 524, 262
MacFadyen, A.I., Woosley, S.E., & Heger, A. 2001, *Ap. J.*, 550, 410
Matzner, C.D. 2002, MNRAS, submitted (astro-ph/0203085)
Mayle, R.W., Tavani, M., & Wilson, J.R. 1993, *Ap. J.*, 418, 398
Meier, D.L., Koide, S., & Uchida, Y. 2001, Science, 291, 84
Meier, D.L., Epstein, R.I., Arnett, W.D., & Schramm, D.N. 1976, *Ap. J.*, 204, 869

Mezzacappa, A., *et al.* 2002, in: *From Twilight to Highlight — The Physics of Supernovae*, ed. W. Hillebrandt & B. Leibundgut (Springer Series "ESO Astrophysics Symposia", Berlin: Springer)

Mezzacappa, A. & Bruenn, S.W. 1993, *Ap. J.*, 405, 637

Mezzacappa, A. & Bruenn, S.W. 1993, *Ap. J.*, 410, 740

Mezzacappa, A., Liebendörfer, M., Messer, O.E.B., Hix, W.R., Thielemann, F.-K., & Burenn, S.W. 2001, Phys. Rev. Lett., 86, 1935

Mezzacappa, A., Calder, A.C., Bruenn, S.W., Blondin, J.M., Guidry, M.W., Strayer, M.R., & Umar, A.S. 1998, *Ap. J.*, 495, 911

Mönchmeyer, R. 1993, PhD Thesis, Technische Universität München

Müller, E. & Hillebrandt, W. 1979, *Astr. Ap.*, 80, 147

Myra, E.S. & Bludman, S.A. 1989, *Ap. J.*, 340, 384

Myra, E.S., Bludman, S.A., Hoffman, Y., Lichenstadt, I., Sack, N., & van Riper, K.A. 1987, *Ap. J.*, 318, 744

Nadyozhin, D.K. 2002, *Astr. Ap.*, submitted (Preprint MPA 1458, Garching: Max-Planck-Institut fur Astrophysik)

Nomoto, K., Shigeyama, T., Kumagai, S., Yamaoka, H., Suzuki, T. 1994, in *Supernovae, Les Houches Session LIV*, ed. S.A. Bludman, R. Mochkovitch, & J. Zinn-Justin (Amsterdam: Elsevier/North-Holland) 489

Nomoto, K., Maeda, K., Umeda, H., Ohkubo, T., Deng, J., & Mazzali, P. 2002, in *A Massive Star Odyssey, from Main Sequence to Supernova, Proc. IAU Symposium 212*, ed. K.A. van der Hucht, A. Herrero, & C. Esteban (San Francisco: ASP) in press (`astro-ph/0209064`)

Ostriker, J.P. & Gunn, J.E. 1971, *Ap. J. (Letters)*, 164, L95

Shapiro, S.L. & Teukolsky, S.A. 1983, *Black Holes, White Dwarfs, and Neutron Stars* (New York: Wiley)

Plewa, T. & Müller, E. 1999, *Astr. Ap.*, 342, 179

Plewa, T. & Müller, E. 2001, Computer Physics Communications, 138, 101

Popham, R., Woosley, S.E., & Fryer, C.L. 1999, *Ap. J.*, 518, 356

Quirk, J.J. 1994, Int. J. Num. Meth. Fluids, 18, 555

Rampp, M. & Janka, H.-Th. 2000, *Ap. J. (Letters)*, 539, L33

Rampp, M. & Janka, H.-Th. 2002, *Astr. Ap.*, 396, 361

Shimizu, T., Yamada, S., & Sato, K. 1993, Publ. Astron. Soc. Japan, 45, L53

Shimizu, T., Yamada, S., & Sato, K. 1994, *Ap. J. (Letters)*, 432, L119

Shimizu, T.M., Ebisuzaki, T., Sato, K., & Yamada, S. 2001, *Ap. J.*, 552, 756

Stairs, I.H., Thorsett, S.E., Taylor, J.H., & Wolszczan, A. 2002, *Ap. J.*, in press (`astro-ph/0208357`)

Sumiyoshi, K., Terasawa, M., Mathews, G.J., Kajino, T., Yamada, S., & Suzuki, H. 2001, *Ap. J.*, 562, 880

Swesty, F.D., Lattimer, J.M., & Myra, E.S. 1994, *Ap. J.*, 425, 195

Thompson, C. & Duncan, R.C. 1993, *Ap. J.*, 408, 194

Thompson, C. & Murray, N. 2001, *Ap. J.*, 560, 339

Thompson, T.A., Burrows, A., & Pinto, P.A. 2002, *Ap. J.*, submitted (astro-ph/0211194)
Thorsett, S.E. & Chakrabarty, D. 1999, *Ap. J.*, 512, 288
Thorstensen, J.R., Fesen, R.A., & van den Bergh, S. 2001, *Ap. J.*, 122, 297
Totani, T., Sato, K., Dalhed, H.E., & Wilson, J.R. 1998, *Ap. J.*, 496, 216
Wang, L., Howell, D.A., Höflich, P., & Wheller, J.C. 2001, *Ap. J.*, 550, 1030
Wang, L., Wheeler, J.C., Höflich, P., Khokhlov, A., *et al.* 2002, *Ap. J.*, in press (astro-ph/0205337)
Wheeler, J.C. 2002, AAPT/AJP Resource Letter, American J. of Physics, in press (astro-ph/0209514)
Wheeler, J.C., Meier, D.L., & Wilson, J.R. 2002, *Ap. J.*, 568, 807
Wheeler, J.C., Yi, I., Höflich, P., & Wang, L. 2000, *Ap. J.*, 537, 810
Wilson, J.R. 1985, in *Numerical Astrophysics*, ed. J.M. Centrella, J.M. LeBlanc, R.L. Bowers, & J.A. Wheeler (Boston: Jones and Bartlett) 422
Wilson, J.R. & Mayle, R. 1988, Phys. Rep., 163, 63
Wilson, J.R. & Mayle, R. 1993, Phys. Rep., 227, 97
Woosley, S.E. 1993, *Ap. J.*, 405, 273
Woosley, S.E., Heger, A., & Weaver, T.A. 2002, Reviews of Modern Physics, in press
Woosley, S.E., Zhang, W., & Heger, A. 2002, in: *From Twilight to Highlight — The Physics of Supernovae*, ed. W. Hillebrandt & B. Leibundgut (Springer Series "ESO Astrophysics Symposia", Berlin: Springer) (astro-ph/0211063)
Zhang, W., Woosley, S.E., & MacFadyen, A.I. 2002, *Ap. J.*, submitted (astro-ph/0207436)

Chapter 4

NEUTRINO TRANSPORT IN CORE COLLAPSE SUPERNOVAE

Anthony Mezzacappa
Physics Division, Oak Ridge National Laboratory, Oak Ridge, TN 37831-6354

Matthias Liebendörfer
Physics Division, Oak Ridge National Laboratory, Oak Ridge, TN 37831-6354
Department of Physics and Astronomy, University of Tennessee, Knoxville, TN 37996 *

Christian Y. Cardall
Physics Division, Oak Ridge National Laboratory, Oak Ridge, TN 37831-6374
Department of Physics and Astronomy, University of Tennessee, Knoxville, TN 37996
Joint Institute for Heavy Ion Research, Oak Ridge, TN 37831-6374

O.E. Bronson Messer
Physics Division, Oak Ridge National Laboratory, Oak Ridge, TN 37831-6354
Department of Physics and Astronomy, University of Tennessee, Knoxville, TN 37996
Joint Institute for Heavy Ion Research, Oak Ridge, TN 37831-6374

and Stephen W. Bruenn
Department of Physics, Florida Atlantic University, Boca Raton, FL 33431

Abstract Neutrino production, transport, and interaction is arguably the single-most important component of a core collapse supernova model. Neutrinos are believed to be responsible for powering these supernovae, in part or entirely, and their production and transport set the stage for the radiation magnetohydrodynamics of stellar core collapse and bounce, which provides the initial conditions for

*Current Affiliation: CITA, University of Toronto M5S 3H8

C.L. Fryer (ed.), Stellar Collapse, 99–131.

the post–stellar-core-bounce dynamics. Neutrino transport is governed by multidimensional, phase-space, integro-partial differential kinetic equations. The solution of these equations dominates the computational challenge in simulating this supernova class. We present the neutrino transport and neutrino radiation hydrodynamics equations involved, and their finite differencing, and briefly discuss their solution. We use the spherically symmetric (spatially one-dimensional) case to illustrate the equations and the issues involved, but give the general formalism for the spatially multidimensional case as well. We conclude by briefly discussing the implications of the now experimentally measured nonzero neutrino masses.

Keywords: Supernovae, Neutrinos, Transport, Kinetic Theory

1. The Core Collapse Paradigm and the Role of Neutrino Transport

Stars more massive than $\sim$10 $M_\odot$ evolve to an onion-like configuration, with an iron core surrounded by successive layers of silicon, oxygen, carbon, helium, and finally hydrogen. In addition to iron group nuclei, the core is composed of electrons, positrons, photons, and a small fraction of protons and neutrons. The pressure in the core, which supports it against the inward pull of gravity, is dominated at this stage by the electrons, and the balance between the electron pressure and gravity is only marginally stable. As a result of electron capture on the free protons and nuclei in the core and as a result of dissociation, whereby nuclei in the core break up under the extreme densities and temperatures, electron and thermal pressure support are reduced, and the core becomes unstable and collapses on itself.

Beginning with central densities $\sim 10^{9-10}$ g/cm^3, the collapse proceeds through nuclear matter densities ($\sim 1-3\times 10^{14}$ g/cm^3). At this point the core undergoes a phase transition from a two-phase system of nucleons and nuclei to bulk nuclear matter. As a result, the pressure at the center of the core increases dramatically, initiating a core rebound.

The velocity of infalling matter in the core increases linearly with radius, characteristic of the "homologous" collapse expected of a fluid whose pressure is dominated by relativistic , degenerate electron pressure. With increasing radius, the infall velocity exceeds the local sound speed; i.e., the infall becomes supersonic. Any information about the rebounding innermost part of the core is signaled via pressure waves that propagate out from the center at the speed of sound to other parts of the core. When these waves reach the point at which the infall is supersonic—i.e., the "sonic point"—they are swept in as fast as they attempt to propagate outward. The net result: No information about the rebounding inner core reaches the infalling outer core, which in turn sets up a density, pressure, and velocity discontinuity in the flow; i.e., a shock wave. This shock wave will ultimately be responsible for propagating outward through the star, disrupting the star in a core collapse supernova.

If the shock were to propagate outward without stalling, we would have what has become known as a "prompt shock mechanism." Most of the models completed to date suggest that this does not occur. Because the shock loses energy in dissociating the iron nuclei that pass through it as it propagates outward and as the material in the core infalls through it, and because the shock loses additional energy in the form of electron neutrinos that escape the core after they are formed when the electrons and newly liberated protons (via dissociation) combine through electron capture, the shock stalls in the iron core, and it is thought to be revived by what has become known as the "delayed shock mechanism," originally proposed by Wilson, 1985 and Bethe and Wilson, 1985. The size of the inner core piston (in mass or radius) that drives the initial shock wave and, consequently, the energy imparted to the shock at bounce, and the location of the initial shock in the iron core and, consequently, the amount of iron it must dissociate on its way out of the core, are dictated by the amount of electron capture that occurs during core collapse. This, in turn, is dictated by the production and transport of the electron neutrinos produced during this epoch of the stellar core's evolution. Already, in this, the first epoch of a supernova, neutrino transport plays a central role.

At the time the shock stalls, the core configuration is composed of a central radiating object called the proto-neutron star. The proto-neutron star has a relatively cold inner part, composed of unshocked bulk nuclear matter, together with a hot "mantle" of nuclear matter that has been shocked but not expelled. As time elapses, the hot, shocked mantle of the proto-neutron star radiates neutrinos of all three flavors, along with their antineutrinos; i.e., electron, muon, and tau neutrinos and antineutrinos. These neutrinos are radiated from the "neutrinosphere." The neutrinosphere is defined in a manner similar to the photosphere of the Sun. In practice, there is a neutrinosphere for every neutrino energy and flavor owing to the fact that the neutrino cross sections are energy and flavor dependent. But for the purposes of discussion here, it is sufficient to refer to one neutrinosphere.

The neutrinos are radiated from the cooling proto-neutron star at a staggering rate of 10^{45-46} Watts. One can think of this central object as a luminous neutrino "lightbulb," which will ultimately be responsible for heating the material behind the shock, imparting energy to it, and in turn playing a role in reviving the shock, which will cause it to propagate outward again and ultimately disrupt the star. The neutrino heating is mediated primarily by electron neutrino and antineutrino absorption on neutrons and protons, respectively. $\sim 10^{53}$ erg of energy is released in neutrinos from the proto-neutron star. The $\sim 10^{51}$ erg of energy associated with the explosion is only a small fraction of this energy.

Thus, in this, the second epoch of a supernova, the neutrinos continue to play a central role.

Also potentially important are modes of convection and other fluid instabilities that may occur in the proto-neutron star and/or directly beneath the stalled shock that may (a) aid the explosion hydrodynamically and/or (b) serve to boost the neutrino heating and/or neutrino heating efficiency. And, of course, stellar rotation and magnetic fields will most likely contribute to the explosion in a significant way. Nonetheless, this additional physics is intimately linked with the neutrino transport in the core (e.g., see Mezzacappa et al., 1998a).

For additional background on core collapse supernovae, the reader is referred to: Brown et al., 1982; Shapiro and Teukolsky, 1983; Bethe and Brown, 1985; Burrows, 1987; Bethe, 1990; Mueller, 1991. For a complete discussion of the state of the art in spherically symmetric supernova models, the reader is referred to: Rampp and Janka, 2000; Mezzacappa et al., 2001; Liebendörfer et al., 2001; Thompson et al., 2002. For a complete discussion of the state of the art in multidimensional supernova models, the reader is referred to Symbalisty, 1984; Wilson andMayle, 1993; Herant et al., 1994; Burrows et al., 1995; Janka and Müller, 1996; Mezzacappa et al., 1998b; MacFadyen and Woosley, 1999; Fryer and Heger, 2000; Fryer and Warren, 2002; Janka, 2003.

2. Neutrino Transport Preliminaries

Neutrinos propagate through the proto-neutron star and interact with the protons, neutrons, and electrons in this central object via absorption, scattering, and other interaction channels. Because the cross sections for neutrino interactions are neutrino-energy dependent, with typically reduced cross sections for reduced energies, neutrinos of lower energies have longer mean free paths. The neutrinosphere is located at a point in the core at which the neutrino mean free paths become comparable to the size of the proto-neutron star. For mean free paths that are much less, which occurs deep within the core below the neutrinosphere, the neutrinos interact with the core many times before escaping. Consequently, these neutrinos diffuse out of the core, and their transport is well described by diffusion theory. On the other hand, for mean free paths that are much larger than the size of the proto-neutron star, which occurs well outside it, the neutrinos do not interact with the core material, and they stream out of the core unimpeded. Their transport at this point is well described by radial free streaming. At the neutrinospheres, the neutrinos are not transported by diffusion, nor are they radially free streaming. Their transport is significantly more complex and is well described only by solutions of the full Boltzmann neutrino kinetic equations. An additional complexity is introduced here by the fact that the neutrinosphere is neutrino energy and flavor dependent. In reality, there are multiple neutrinospheres. The neutrino interaction cross sections and, hence, the mean free paths are energy and flavor dependent.

A solution to the Boltzmann equation describes the time evolution of the neutrino distribution function which, at each instant of time and each location in space, gives the distribution of neutrinos in direction cosine and energy. Thus, the Boltzmann equation is a phase space equation—i.e., an equation in the multidimensional space of all possible spatial locations, direction cosines, and energies. Therefore, even a "one-dimensional" supernova simulation, in which spherical symmetry is assumed, is in reality a three-dimensional simulation.

The Boltzmann equation for neutrinos in core collapse supernovae for spherically symmetric flows will be given in the next section. For our purposes here, it is sufficient to write it in operator form as

$$\frac{\partial f}{\partial t} = O[f] \tag{4.1}$$

where f is the distribution function, which is a function of space and neutrino direction cosine and energy, and where O is an operator that is the sum of "transport" and "collision" operators. A solution of the Boltzmann equation for the neutrino distribution function then gives an infinite hierarchy of "moments" of the distribution function, defined by

$$\begin{aligned}
\psi^0 &= \int d\mu \mu^0 f \\
\psi^1 &= \int d\mu \mu^1 f \\
\psi^2 &= \int d\mu \mu^2 f \\
&\cdot \\
&\cdot \\
&\cdot \\
\psi^n &= \int d\mu \mu^n f
\end{aligned} \tag{4.2}$$

where we have assumed, as an example, spherical symmetry, in which the neutrino distribution function at any point in space is a function of a single direction cosine, μ, and energy, E.

A solution of the Boltzmann equation (4.1) is equivalent to solving an infinite hierarchy of equations for the "moments" of the distribution function, which can be written as

$$\int d\mu \mu^0 \frac{\partial f}{\partial t} = \int d\mu \mu^0 O[f]$$

$$\int d\mu\mu^1 \frac{\partial f}{\partial t} = \int d\mu\mu^1 O[f]$$
$$\int d\mu\mu^2 \frac{\partial f}{\partial t} = \int d\mu\mu^2 O[f]$$
$$\vdots$$
$$\int d\mu\mu^n \frac{\partial f}{\partial t} = \int d\mu\mu^n O[f] \tag{4.3}$$

It is instructive to note that for each component of the stellar core—i.e., the photons, electrons, positrons, protons, neutrons, and nuclei—one can write down a kinetic equation for the component distribution function. Each kinetic equation, in turn, *induces* an infinite series of moment equations, as in equation (4.3). Under certain conditions and assumptions, these series close and give rise to the familiar hydrodynamics equations for the component fluids. (For more on this, see Balescu, 1975.)

Because solving the Boltzmann equation is computationally intensive, even in one-dimensional simulations that assume spherical symmetry, historically a number of approximations have been implemented, ranging from very simple to rather sophisticated. A brief history will be given in the next section. For our purposes later, we note that, in equation (4.3), ψ^0 (ψ^1) is the neutrino energy density (flux) for the neutrino energy E.

3. A Brief History

A solution to the supernova problem has eluded research efforts for nearly four decades, beginning with the earliest simulations of Colgate and White, 1966. Until recently, dictated mainly by available technology, the transport of neutrinos was simulated with a variety of increasingly sophisticated approximations, from simple "leakage" schemes (e.g., see Van Riper and Lattimer, 1981; Baron et al., 1985), to "two-fluid" approaches (e.g., see Hillebrandt et al., 1984; Cooperstein et al., 1986), and finally "multigroup flux-limited diffusion" (e.g., see Arnett, 1977; Bowers and Wilson, 1982; Bruenn, 1985; Myra et al., 1987).

Multigroup flux-limited diffusion (MGFLD) "closes" the neutrino radiation hydrodynamics hierarchy of equations (4.3) at the level of the zeroth moment (the neutrino energy density per group) by imposing a relationship between the flux per group (the first moment) and the gradient of the zeroth moment. For example,

$$\psi^1 = -\frac{c\Lambda}{3}\frac{\partial \psi^0}{\partial r} + ..., \tag{4.4}$$

$$\Lambda = \frac{1}{1/\lambda + |\partial\psi^0/\partial r|/3\psi^0}, \tag{4.5}$$

where λ is the neutrino mean free path (Bruenn, 1985). (Other forms for the flux-limiter Λ can be found in Bowers and Wilson, 1982; Levermore and Pomraning, 1981; Myra et al., 1987.) The term "multigroup" means that the transport is carried out for each neutrino energy separately. Because of the energy dependence of the neutrino interactions and, consequently, mean free paths, neutrinos of different energies may behave very differently. The term "flux limited" means that the flux is kept from exceeding the maximum flux, corresponding to neutrinos streaming out radially at the speed of light. If diffusion theory were used to describe the transport in this instance, the neutrino flux would be superluminal. This happens because diffusion theory assumes that the neutrinos always propagate a distance given by their mean free path, even if this distance exceeds the distance that could be traversed by the neutrinos in a time Δt, moving at the speed of light (Mihalas and Mihalas, 1975).

Whereas the limits $\lambda \to 0$ and $\lambda \to \infty$ produce the correct diffusion and free streaming fluxes, it is in the critical intermediate regime where the MGFLD approximation is, generally speaking, of unknown accuracy, necessitating, at the very least, comparisons with Boltzmann transport solutions to gauge the efficacy of using MGFLD for neutrino transport in stellar cores. The extreme sensitivity of supernova models to differences in the computed neutrino luminosities, spectra, and distributions in direction cosine (e.g., see Burrows and Goshy, 1993; Janka and Müller, 1996; Mezzacappa et al., 1998b) provides an additional compelling reason to gauge the MGFLD approximation and to consider Boltzmann transport models.

Moreover, in detailed one-dimensional simulations that have implemented elaborate MGFLD neutrino transport (e.g., see Bruenn, 1993; Wilson and Mayle, 1993; Swesty and Lattimer, 1994; Bruenn et al., 2001), explosions were not obtained unless the neutrino heating was boosted by additional phenomena, such as neutron fingers (Wilson and Mayle, 1993). These results presented at least two possibilities: (1) Failures to produce explosions in the absence of additional phenomena, such as neutron fingers, had resulted from neutrino transport approximations. (2) Additional phenomena, such as convection, neutron fingers, rotation, and magnetic field effects, are, in fact, essential.

With present day supercomputers, it has become possible to simulate Boltzmann neutrino transport in core collapse supernovae. For spherically symmetric models, this has been accomplished (Mezzacappa et al., 2001; Liebendörfer et al., 2001) both for the (a) Newtonian gravity, $O(v/c)$ and (b) fully general relativistic cases. These simulations have addressed the two possibilities presented above. With state of the art input nuclear and weak interaction physics, explosions are not obtained in spherical symmetry even with the use of Boltzmann neutrino transport. Thus, we are now left with two other possibilities: (1) Significant changes in the nuclear and weak interaction physics may change the outcomes of spherically symmetric models in a qualitative way. (2) Additional physical phenomena not admitted by spherical symmetry (fluid instabilities, rotation, magnetic field effects) are necessary ingredients in the explosion recipe.

In addition to Boltzmann transport, recent simulations of stellar core collapse and postbounce evolution have been performed with a multigroup variable Eddington factor (MGVEF) scheme (Burrows et al., 2000; Rampp and Janka, 2002). In most practical implementations of this scheme, an approximation to the Boltzmann equation is solved in order to compute the Eddington factors (ratio of moments) needed to close the hierarchy of multigroup radiation moment equations, equation (4.3), although a full Boltzmann solution can be performed together with a solution of the moment equations. In a MGVEF approach, the hierarchy is closed at the level of the first moment, the multigroup radiation momentum density, whereas in a MGFLD approach only the zeroth moment is evolved. A detailed comparison of the results from Boltzmann and MGVEF simulations show excellent agreement in the spherically symmetric case (Liebendoerfer et al., 2003). Notably, in neither the Boltzmann nor the MGVEF cases was an explosion obtained.

Whether or not multidimensional effects such as convection play a significant role in the explosion mechanism will not displace neutrino transport from its central role in core collapse supernova dynamics. Neutrinos either power or significantly contribute to powering these supernovae, and, in either case, neutrino transport is central to the dynamics (stellar core density, entropy, and velocity profiles) of collapse and bounce, determining the postbounce environment within which convection and magnetic fields, for example, will operate. Moreover, it is not known *a priori* how sensitive the supernova mechanism will remain to the neutrino luminosities, spectra, and angular distributions when multidimensional models with all potentially relevant phenomena included are performed, particularly models with magnetic fields admitting MHD phenomena. Some of the sensitivity studies conducted thus far did not assume spherical symmetry (Janka andMüller, 1996). In light of these results, we must anticipate that multidimensional models will remain sensitive to the value of the neutrino quantities and how accurately they are computed. In addition, we do expect the limitations of MGFLD to be accentuated in multidimensional implementations. For these reasons, performing accurate (Boltzmann) neutrino transport in multidimensional models is necessary.

As one might imagine, the neutrino transport approximations implemented in past multidimensional simulations have, by necessity, been less sophisticated than those implemented in detailed one-dimensional simulations (e.g., see Herant et al., 1994; Burrows et al., 1995; Janka and Müller, 1996; Mezzacappa et al.,

1998b; Fryer and Heger, 2000; Fryer and Warren, 2002). One notable exception is the recent implementation of MGVEF along radial rays in two-dimensional models (Buras et al., 2003). Once again, an explosion was not obtained in these models despite the inclusion of convection and rotation. (N.B. With the exception of Wilson's spherically symmetric models that invoke neutron fingers, no simulation performed with multifrequency neutrino transport has yielded an explosion. Note also that the existence of neutron fingers is a matter of debate. For example, see Bruenn and Dineva, 1996.) Nonetheless, multidimensional models with multidimensional, multifrequency neutrino transport must be performed before final conclusions can be drawn (from models that do not include magnetic fields). And the latter models must also ultimately include magnetic fields.

This article outlines the partial differential equations solved and the finite differencing and numerical algorithms used to obtain these solutions for spherically symmetric stellar core collapse with Boltzmann neutrino transport. The general formalism for two- and three-dimensional flows is also presented, but the greater simplicity of the spherically symmetric case is best suited to convey the essential problem and key issues.

4. Neutrino Radiation Hydrodynamics Equations

Hydrodynamics with Radiation Coupling

The Newtonian-gravity, $O(v/c)$, Lagrangian hydrodynamics equations are (Mihalas and Mihalas, 1975; in what follows, r is the Eulerian radial coordinate, and m is the Lagrangian mass coordinate):

$$\frac{\partial r}{\partial m} = \frac{1}{4\pi r^2 \rho}, \tag{4.6}$$

$$\begin{aligned}\frac{\partial \epsilon}{\partial t} &= \frac{-P}{\rho}\frac{\partial \ln\rho}{\partial t} \\ &- \sum_{\nu_e,\bar{\nu}_e} \frac{2\pi c\, C_0}{h^3 c^3}\int dE E^3 d\mu \left(\frac{j}{\rho} - \tilde{\chi} F\right) \\ &+ \sum_{\nu_e,\bar{\nu}_e,\nu_{\mu,\tau},\bar{\nu}_{\mu,\tau}} \frac{2\pi C_0}{h^6 c^6}\int dE E^3 d\mu F \int dE' E'^2 d\mu' \tilde{R}^{\rm out}_{\rm NES}\left(\frac{1}{\rho} - F\right) \\ &- \sum_{\nu_e,\bar{\nu}_e,\nu_{\mu,\tau},\bar{\nu}_{\mu,\tau}} \frac{2\pi C_0}{h^6 c^6}\int dE E^3 d\mu \left(\frac{1}{\rho} - F\right) \int dE' E'^2 d\mu' \tilde{R}^{\rm in}_{\rm NES} F \\ &+ \sum_{\nu_e,\nu_{\mu,\tau}} \frac{2\pi C_0}{h^6 c^6}\int dE E^3 d\mu F \int dE' E'^2 d\mu' \tilde{R}^{\rm abs}_{\rm PAIR}(\bar{F})\end{aligned} \tag{4.7}$$

$$- \sum_{\nu_e,\nu_{\mu,\tau}} \frac{2\pi C_0}{h^6c^6} \int dEE^3 d\mu(\frac{1}{\rho} - F) \int dE'E'^2 d\mu' \tilde{R}^{\rm em}_{\rm PAIR}(\frac{1}{\rho} - \bar{F}),$$

$$\begin{aligned}
\frac{\partial v}{\partial t} &= -\frac{Gm}{r^2} - 4\pi r^2 \frac{\partial P}{\partial m} \\
&+ \sum_{\nu_e,\bar{\nu}_e} \frac{2\pi C_0}{h^3c^3} \int dEE^3 d\mu\mu\, \tilde{\chi} F \\
&+ \sum_{\nu_e,\bar{\nu}_e,\nu_{\mu,\tau},\bar{\nu}_{\mu,\tau}} \frac{2\pi C_0}{h^6c^7} \int dEE^5 d\mu\mu F \int d\mu' R_{\rm IS} \\
&- \sum_{\nu_e,\bar{\nu}_e,\nu_{\mu,\tau},\bar{\nu}_{\mu,\tau}} \frac{2\pi C_0}{h^6c^7} \int dEE^5 d\mu\mu \int d\mu' R_{\rm IS} F \\
&+ \sum_{\nu_e,\bar{\nu}_e,\nu_{\mu,\tau},\bar{\nu}_{\mu,\tau}} \frac{2\pi C_0}{h^6c^7} \int dEE^3 d\mu\mu F \int dE'E'^2 d\mu' \tilde{R}^{\rm out}_{\rm NES}(\frac{1}{\rho} - F) \\
&- \sum_{\nu_e,\bar{\nu}_e,\nu_{\mu,\tau},\bar{\nu}_{\mu,\tau}} \frac{2\pi C_0}{h^6c^7} \int dEE^3 d\mu\mu(\frac{1}{\rho} - F) \int dE'E'^2 d\mu' \tilde{R}^{\rm in}_{\rm NES} F \\
&+ \sum_{\nu_e,\nu_{\mu,\tau}} \frac{2\pi C_0}{h^6c^7} \int dEE^3 d\mu\mu F \int dE'E'^2 d\mu' \tilde{R}^{\rm abs}_{\rm PAIR} \bar{F} \\
&- \sum_{\nu_e,\nu_{\mu,\tau}} \frac{2\pi C_0}{h^6c^7} \int dEE^3 d\mu\mu(\frac{1}{\rho} - F) \int dE'E'^2 d\mu' \tilde{R}^{\rm em}_{\rm PAIR}(\frac{1}{\rho} - \bar{F}).
\end{aligned} \tag{4.8}$$

In equation (4.7) and (4.8),

$$\frac{\partial}{\partial t} = \left(\frac{\partial}{\partial t}\right)_m, \tag{4.9}$$

$$\frac{\partial}{\partial m} = \left(\frac{\partial}{\partial m}\right)_t. \tag{4.10}$$

The quantities ρ, ϵ, v, and P are the fluid rest-mass density, specific internal energy (internal energy per gram), velocity, and pressure, respectively. The flavor sums indicate that the equations contain one such term for each of the specified neutrino flavors (electron, muon, and tau).

To parameterize the neutrino momentum, we use spherical momentum coordinates: $p = E/c$, θ, and ϕ. p is the magnitude of the neutrino momentum, and E is the neutrino energy. In equations (4.7) and (4.8), $\mu \equiv \cos\theta$, which is the neutrino direction cosine relative to the outward-pointing radial vector. All of these quantities are measured by our comoving observers. In equations (4.7) and (4.8), $C_0 = 1.602 \times 10^{-6}$ erg MeV^{-1}. (Our neutrino energies are given in MeV, and the product hc is expressed in MeV cm.)

The third term in the energy equation is the energy exchange between the neutrinos and the fluid via emission and absorption of neutrinos by nucleons and nuclei. The integrand contains the neutrino emissivity, $j(E)$, the neutrino opacity, $\chi(E)$, and the specific neutrino distribution function, $F \equiv f/\rho$. $\bar{F}$ is the specific distribution function for the corresponding antineutrino. The neutrino opacity, $\tilde{\chi}$, corrected for stimulated absorption, is defined by:

$$j - \tilde{\chi} f \equiv j - (j + \chi) f = (1 - f) j - \chi f. \tag{4.11}$$

The term $1 - f$ in equation (4.11) describes Fermionic blocking of states.

The next two terms in the energy equation give the energy exchange between the neutrinos and the fluid as a result of neutrino–electron scattering. The scattering kernels, $\tilde{R}^{\rm in/out}_{\rm NES}(\mu, \mu', E, E') \equiv \rho R^{\rm in/out}_{\rm NES}(\mu, \mu', E, E')$, are the in- and outscattering kernels, respectively, describing the scattering of an electron neutrino into or out of (μ, E). μ and E, are the incoming direction cosine and energy, and μ' and E' are the outgoing direction cosine and energy.

$\tilde{R}^{\rm em/abs}_{\rm PAIR}(\mu, \mu', E, E') \equiv \rho R^{\rm em/abs}_{\rm PAIR}(\mu, \mu', E, E')$ are the pair emission and absorption kernels, respectively, describing the production of a neutrino/anti-neutrino pair of energy E/E' and direction cosine μ/μ'.

The last seven terms in the velocity equation (4.8) are the neutrino stress terms associated with the neutrino–matter interactions described above. In this equation, $R_{\rm IS}(\mu, \mu', E)$ is the kernel for "isoenergetic" scattering of electron neutrinos on neutrons, protons, and nuclei.

Neutrino Transport

The neutrino distribution functions are evolved using the $O(v/c)$ Boltzmann equation (Castor, 1972; Bruenn, 1985; Mezzacappa and Bruenn, 1993b). If we include emission, absorption, isoenergetic scattering of neutrinos by nucleons and nuclei, neutrino–electron scattering, and pair emission and absorption, the Boltzmann equation is

$$\begin{aligned} \frac{1}{c}\frac{\partial F}{\partial t} &+ 4\pi\mu \frac{\partial(r^2 \rho F)}{\partial m} \\ &+ \frac{1}{r}\frac{\partial[(1-\mu^2)F]}{\partial \mu} \\ &+ \frac{1}{c}\left(\frac{\partial \ln\rho}{\partial t} + \frac{3v}{r}\right)\frac{\partial[\mu(1-\mu^2)F]}{\partial \mu} \\ &+ \frac{1}{c}\left[\mu^2\left(\frac{\partial \ln\rho}{\partial t} + \frac{3v}{r}\right) - \frac{v}{r}\right]\frac{1}{E^2}\frac{\partial(E^3 F)}{\partial E} \end{aligned}$$

$$
\begin{aligned}
&= \frac{j}{\rho} - \tilde{\chi} F \\
&+ \frac{1}{c}\frac{1}{h^3c^3}E^2 \int d\mu' R_{\mathrm{IS}} F \\
&- \frac{1}{c}\frac{1}{h^3c^3}E^2 F \int d\mu' R_{\mathrm{IS}} \\
&+ \frac{1}{h^3c^4}\left(\frac{1}{\rho} - F\right) \int dE' E'^2 d\mu' \tilde{R}^{\mathrm{in}}_{\mathrm{NES}} F \\
&- \frac{1}{h^3c^4} F \int dE' E'^2 d\mu' \tilde{R}^{\mathrm{out}}_{\mathrm{NES}} \left(\frac{1}{\rho} - F\right) \\
&+ \frac{1}{h^3c^4}\left(\frac{1}{\rho} - F\right) \int dE' E'^2 d\mu' \tilde{R}^{\mathrm{em}}_{\mathrm{PAIR}} \left(\frac{1}{\rho} - \bar{F}\right) \\
&- \frac{1}{h^3c^4} F \int dE' E'^2 d\mu' \tilde{R}^{\mathrm{abs}}_{\mathrm{PAIR}} \bar{F}.
\end{aligned}
\tag{4.12}
$$

The Boltzmann equation is solved for each neutrino flavor independently. The electron neutrinos and antineutrinos are solved for simultaneously, as are the μ/τ neutrinos and antineutrinos, because each neutrino flavor is coupled to its corresponding antineutrino through the pair emission and absorption channel.

The mass derivative term on the left-hand side of the Boltzmann equation describes the propagation of neutrinos with respect to the Lagrangian mass coordinate, m. Outwardly propagating neutrinos have $\mu > 0$, whereas inwardly propagating neutrinos have $\mu < 0$. The first μ-derivative term describes the rate of change of the neutrino propagation direction with respect to the outward radial direction as the neutrino propagates inward or outward in mass. The second μ-derivative term describes the aberration in the neutrino propagation direction measured by an observer who is instantaneously comoving with the fluid. Because the fluid is accelerating, two neighboring comoving observers will measure different direction cosines. The energy-derivative term describes the shift in the neutrino energy measured by comoving observers. This is a Doppler shift resulting from the change in the velocity, with time and/or radius, of an accelerated fluid; two neighboring comoving observers will measure different frequencies.

On the right-hand side of equation (4.12), the first two terms describe the change in the neutrino distribution function resulting from the absorption and emission of neutrinos by nucleons and nuclei. The next two terms describe the isoenergetic inscattering and outscattering, respectively, of neutrinos by nucleons and nuclei. The fourth and fifth terms describe non-isoenergetic neutrino–electron scattering, and the last two terms describe pair emission and absorption. As in equations 4.7 and 4.8, $\bar{F}$ is the corresponding antineutrino distribution function for each neutrino species evolved with equation (4.12).

The neutrino interactions included in equation (4.12) are not complete, but as far as the functional form of the emissivities and opacities is concerned, they are. For example, scattering of neutrinos on nucleons is not isoenergetic when nucleon recoil and correlations are taken into account, but the functional form of the scattering kernel in this case would then simply take on the form of the non-isoenergetic neutrino–electron scattering kernel. For a complete discussion of the state of the art in neutrino emissivities and opacities, the reader is referred to Burrows (2003).

Lepton Conservation

The hydrodynamics equations governing the evolution of the fluid's density, specific internal energy and velocity are expressions of conservation laws for the fluid mass, energy, and momentum. When coupled to the radiation field, conservation of energy extends to the radiation field and, in addition, lepton number must be conserved. (This assumes, as we do here, that neutrinos are massless and do not undergo flavor transformations. Neutrinos indeed have mass and mix, and lepton number in reality is not strictly conserved, but if we assume massless neutrinos, lepton number conservation is required for consistency and for quality control of the simulation.) Therefore, in addition to the hydrodynamics equations and the neutrino transport equations, we also must solve, in a coupled fashion, the fluid electron fraction equation:

$$\begin{aligned} \frac{\partial Y_{\mathrm{e}}}{\partial t} &= -\frac{2\pi m_{\mathrm{B}} c}{h^3 c^3} \int dE E^2 d\mu \left(\frac{j_{\nu_e}}{\rho} - \tilde{\chi}_{\nu_e} F_{\nu_e} \right) \\ &- \int dE E^2 d\mu \left(\frac{j_{\bar{\nu}_e}}{\rho} - \tilde{\chi}_{\bar{\nu}_e} F_{\bar{\nu}_e} \right). \end{aligned} \tag{4.13}$$

The Y_{e} evolved by equation (4.13) is actually the difference of the electron fraction minus the positron fraction: $Y_{\mathrm{e}} = Y_{\mathrm{e}+} - Y_{\mathrm{e}-}$. To see that this equation describes lepton number conservation, it is sufficient to point out that lepton number is exchanged between electrons and electron neutrinos via electron capture and its inverse process and between positrons and electron antineutrinos via positron capture and its inverse. The right-hand side of equation (4.13) gives the change in the electron fraction resulting from these processes.

5. Finite Differencing of the Neutrino Transport Equations

Time Derivative of the Neutrino Distribution Function

The finite differencing of the time derivative of the neutrino distribution function is straightforward. It is implemented in the following way:

$$\frac{\partial F}{\partial t} = \frac{F_{i',j',k'} - \overline{F}_{i',j',k'}}{dt}. \tag{4.14}$$

The quantities with an overbar refer to the previous value of the variables. All other quantities refer to the updated values after the time step $dt = t - \bar{t}$. In everything that follows, we adopt the convention that $i^{'} \equiv i+1/2$, $j^{'} \equiv j+1/2$, and $k^{'} \equiv k + 1/2$.

Collision Term

The finite difference representation of the collision term in equation (4.12) is

$$\begin{aligned}
&\frac{j^{n+1}_{i',k'}}{\rho^{n+1}_{i'}} - \tilde{\chi}^{n+1}_{i',k'} F_{i',j',k'} \\
+\ &\frac{1}{ch^3c^3} E^2_{k'} \sum_{l=1}^{jmax} w_{l'} \, (R_{\rm IS})^{n+1}_{i',j',l',k'} F_{i',l',k'} \\
-\ &\frac{1}{ch^3c^3} E^2_{k'} F_{i',j',k'} \sum_{l=1}^{jmax} w_{l'} \, (R_{\rm IS})^{n+1}_{i',j',l',k'} \\
+\ &\frac{1}{ch^3c^3} (1/\rho^{n+1}_{i'} - F_{i',j',k'}) \sum_{m=1}^{kmax} \Delta E_{m'} E^2_{m'} \sum_{l=1}^{jmax} w_{l'} \\
\times\ &(\tilde{R}^{\rm in}_{\rm NES})^{n+1}_{i',j',l',k',m'} F_{i',l',m'} \\
-\ &\frac{1}{ch^3c^3} F_{i',j',k'} \sum_{m=1}^{kmax} \Delta E_{m'} E^2_{m'} \sum_{l=1}^{jmax} w_{l'} \\
\times\ &(\tilde{R}^{\rm out}_{\rm NES})^{n+1}_{i',j',l',k',m'} (1/\rho^{n+1}_{i'} - F_{i',l',m'}) \\
+\ &\frac{1}{ch^3c^3} (1/\rho^{n+1}_{i'} - F_{i',j',k'}) \sum_{m=1}^{kmax} \Delta E_{m'} E^2_{m'} \sum_{l=1}^{jmax} w_{l'} \\
\times\ &(\tilde{R}^{\rm em}_{\rm PAIR})^{n+1}_{i',j',l',k',m'} (1/\rho^{n+1}_{i'} - \bar{F}_{i',l',m'}) \\
-\ &\frac{1}{ch^3c^3} F_{i',j',k'} \sum_{m=1}^{kmax} \Delta E_{m'} E^2_{m'} \sum_{l=1}^{jmax} w_{l'} \\
\times\ &(\tilde{R}^{\rm abs}_{\rm PAIR})^{n+1}_{i',j',l',k',m'} \bar{F}_{i',l',m'}
\end{aligned} \tag{4.15}$$

It is important to note that (a) the collision term in our approach is differenced implicitly with respect to time (i.e., defined at the advanced values of the quantities that are being updated) and (b) the scattering kernels are not approximated by a truncated Legendre expansion in $\cos\theta$. Thus, modulo discretization errors, our approach yields a solution of the true Boltzmann equation. Moreover, for certain neutrino interactions, the Legendre expansion of the scattering kernels does not converge; i.e., the kernels can be infinite, although integrably finite (e.g., see Mezzacappa and Bruenn, 1993a). Thus, expanding the kernels in this way is not generally valid.

Neutrino Propagation in Mass

The term $\mu\partial(r^2\rho F)/\partial m$ in equation (4.12) accounts for the neutrinos that are propagating into and out of a spherical mass shell. In the free streaming limit, the advected neutrino number in a time step (as measured by a comoving observer) is large with respect to the neutrino number in the zone. Upwind differencing of the advection term is appropriate to limit destabilizing errors in the advection fluxes. For discrete direction cosines $\mu_{j'}$, the direction of the neutrino "wind" is simply given by the sign of $\mu_{j'}$. (A neutrino with $\mu = 0$ propagates tangential to the mass shell). In diffusive conditions, however, the neutrino flux may be orders of magnitude smaller than the prevailing nearly isotropic neutrino density in the mass shell. In this situation, an asymmetric differencing can easily lead to an overestimate of the first angular moment because of improper cancellations among the contributions of the isotropic component of the neutrino field. Based on transport coefficients $\beta_{i,k'}$, we therefore interpolate between upwind differencing in free streaming regimes ($\beta_{i,k'} = 1$) and centered differencing ($\beta_{i,k'} = 1/2$) in diffusive regimes. Moreover, the equilibrium constraint (73) in Mezzacappa and Bruenn, 1993b only holds for $\beta_{i,k'} = 1/2$. We therefore use transport coefficients defined according to

$$\beta_{i,k'} = \begin{cases} 1/2 & \text{if} \quad 2dr_i > \lambda_{i,k'}, \\ \left(2dr_i/\lambda_{i,k'} + 1\right)^{-1} & \text{otherwise.} \end{cases} \tag{4.16}$$

That is, the transport coefficients at zone edges depend on the ratio of the distance dr_i between zone centers and the neutrino angle and energy dependent mean free path $\lambda_{i,k'}$.

We then discretize the propagation term as:

$$\mu\frac{\partial r^2\rho F}{\partial m} = \frac{\mu_{j'}}{dm_{i'}}\left[4\pi r_{i+1}^2\rho_{i+1}F_{i+1,j',k'} - 4\pi r_i^2\rho_i F_{i,j',k'}\right] \tag{4.17}$$

with

$$\rho_i F_{i,j',k'} = \beta_{i,k'}\rho_{i'-1}F_{i'-1,j',k'} + \left(1-\beta_{i,k'}\right)\rho_{i'}F_{i',j',k'} \tag{4.18}$$

for outward propagating neutrinos ($\mu_{j'} > 0$), and

$$\rho_i F_{i,j',k'} = \left(1-\beta_{i,k'}\right)\rho_{i'-1}F_{i'-1,j',k'} + \beta_{i,k'}\rho_{i'}F_{i',j',k'} \tag{4.19}$$

for inward propagating neutrinos $(\mu_{j'} < 0)$.

Change in the Neutrino Direction Cosine from Spatial Propagation

For any neutrino, the neutrino direction cosine increases toward unity as it propagates through space. Hence, neutrino propagation causes angular advection relative to a fixed grid of neutrino direction cosines $\mu_{j'}$. This contribution to the evolution of the neutrino distribution function is described by the term $\partial[(1-\mu^2)F]/r\partial\mu$ in equation (4.12). The finite difference representation is chosen to maintain equilibrium between the radiation field and stationary matter under conditions in which such an equilibrium would exist (Lewis and Miller, 1984; Mezzacappa and Bruenn, 1993a). We set

$$\frac{\partial[(1-\mu^2)F]}{r\partial\mu} = \frac{3\left[r_{i+1}^2 - r_i^2\right]}{2\left[r_{i+1}^3 - r_i^3\right]}\frac{1}{w_{j'}}\left(\zeta_{j+1}F_{i',j+1,k'} - \zeta_j F_{i',j,k'}\right). \tag{4.20}$$

The differencing of the coefficients, $\zeta = 1 - \mu^2$, is defined by

$$\zeta_{j+1} - \zeta_j = -2\mu_{j'}w_{j'}. \tag{4.21}$$

The angular integration of the term $\partial[(1-\mu^2)F]/r\partial\mu$ produces the zeroth and second angular moments of the neutrino distribution function. Its finite difference representation is therefore not as sensitive to cancellations in the diffusive limit as the differencing of the spatial advection term. Upwind differencing can be justified. The angular "wind" always points towards $\mu = 1$. However, for reasons of completeness and consistency, we also use centered differencing in the diffusive regime. With angular transport coefficients $\gamma_{i',k'} \equiv \beta_{i',k'}$, we interpolate the values of the neutrino distribution function to angular zone edges by

$$F_{i',j,k'} = \gamma_{i',k'}F_{i',j'-1,k'} + \left(1 - \gamma_{i',k'}\right)F_{i',j',k'}. \tag{4.22}$$

Conservation of Lepton Number and Energy

Moving beyond general arguments for the necessity of accurate neutrino transport, quantitative consideration of the energetics shows how severe the requirements are on one aspect of accuracy: energy conservation. Virtually all of the gravitational potential energy ($\sim 10^{53}$erg) released during collapse is eventually converted into intense neutrino fluxes lasting several seconds. Observed ejecta kinetic energies $\sim 10^{51}$erg imply that only about 1% of the neutrinos' energy need be transferred to the (soon-to-be ejected) matter surrounding the nascent neutron star. Because it is difficult to argue with any rigor about the physical impact of any energy lost or gained due to numerical error, the total energy should be conserved to a precision corresponding to the phenomena of interest in the problem. Hence, a simulation's result for explosion energy accurate to, say, 10% would require total energy conservation to an accuracy of about 0.1%. Allowing for systematic error accrual, the total energy would have to be conserved at a level of $0.1\%/N$ per time step, where $N \sim 10^5$ is the total number of time steps in the simulation.

The integration of the Boltzmann equation over momentum space, spanned by the neutrino direction cosine and energy, and mass gives the conservation laws for neutrino number and energy. We define J^N and H^N to represent the zeroth and first μ number moments of the distribution function:

$$\begin{aligned} J^N &= \int_{-1}^{1}\int_0^{\infty} FE^2 dE d\mu, \\ H^N &= \int_{-1}^{1}\int_0^{\infty} FE^2 dE \mu d\mu. \end{aligned}$$

Integration of equation (4.12) over μ and E with E^2 as the measure of integration gives the following evolution equation for J^N:

$$\frac{\partial J^N}{\partial t} + \frac{\partial}{\partial m}\left[4\pi r^2 \rho H^N\right] - \int \frac{j}{\rho} E^2 dE d\mu + \int \chi F E^2 dE d\mu = 0. \quad (4.23)$$

The aberration and frequency shift terms in equation (4.12) do not contribute because $(1-\mu^2)$ vanishes at $\mu = \pm 1$ and $E^3 F$ is zero for $E = 0$ and $E = \infty$. One more integration over rest mass m from the center of the star to its surface gives the evolution equation for the total neutrino number.

Slightly less straightforward is the derivation of total energy conservation. Define the energy moments

$$\begin{aligned} J^E &= \int FE^3 dE d\mu \\ H^E &= \int FE^3 dE \mu d\mu \\ K^E &= \int FE^3 dE \mu^2 d\mu \\ Q^E &= \int FE^3 dE \mu^3 d\mu, \end{aligned}$$

By taking the zeroth and first angular moments of the energy moment ($\int E^3 dE$ $\{\partial F/\partial t = O[F]\}$) of the Boltzmann equation, the latter weighted by the fluid velocity, v,—i.e., $\int E^3 dE d\mu\{\partial F/\partial t = O[F]\}$ and $(v \int E^3 dE d\mu\{\partial F/\partial t = O[F]\}$— one finds

$$\frac{\partial J^E}{\partial t} + \frac{\partial}{\partial m}\left[4\pi r^2 \rho H^E\right] - \left(\frac{\partial \ln\rho}{\partial t} + \frac{2v}{r}\right) K^E + \frac{v}{r}\left(J^E - K^E\right)$$

$$- \int \frac{j}{\rho} E^3 dE d\mu + \int \chi F E^3 dE d\mu = 0, \tag{4.24}$$

$$\begin{aligned} v\frac{\partial H^E}{\partial t} &+ \frac{\partial}{\partial m}\left[4\pi r^2 v\rho K^E\right] - 4\pi r^2 \rho \frac{dv}{dm} K^E - \frac{v}{r}\left(J^E - K^E\right) \\ &- v\left(\frac{\partial \ln\rho}{\partial t} + \frac{2v}{r}\right) H^E + v\int \chi F E^3 dE \mu d\mu = 0. \end{aligned} \tag{4.25}$$

Combining, we obtain the lab frame energy conservation equation:

$$\begin{aligned} 0 &= \frac{\partial}{\partial t}\left(J^E + vH^E\right) + \frac{\partial}{\partial m}\left[4\pi r^2 \rho\left(vK^E + H^E\right)\right] \\ &- \int \frac{j}{\rho} E^3 dE d\mu + \int \chi F E^3 dE d\mu + v\int \chi F E^3 dE \mu d\mu \\ &- \frac{1}{4\pi r^2 \rho}\frac{\partial}{\partial t}\left(4\pi r^2 \rho v\right) H^E. \end{aligned} \tag{4.26}$$

Note that $J^E + vH^E$ is the lab frame neutrino energy density as expressed in terms of the comoving frame moments J^E and H^E. Integration of equation (4.26) over enclosed mass leads to an equation for total neutrino energy evolution. The first term on the RHS is the time evolution of the conserved energy. The second term corrects for surface effects, namely the work at the surface against radiation pressure, ρK^E, and the surface luminosity, $4\pi r^2 \rho H^E$. The second line represents the energy exchange between radiation and matter by emission, absorption, and radiation stress. If one aims to check energy conservation to machine precision, it is important to be aware that taking the $O(v/c)$ limit in the comoving frame does not lead to the identical physical description as taking the $O(v/c)$ limit in the lab frame. For that reason, we keep higher order terms in the third line of equation (4.26) for compatibility with the physical description of the $O(v/c)$ Boltzmann equation (4.12) in the comoving frame. These higher-order terms can become nonnegligible.

We observe that the expressions $(\partial \ln\rho/\partial \mathrm{t} + 2\mathrm{v}/\mathrm{r})\mathrm{K}^\mathrm{E}$ and $K^E 4\pi r^2 \rho \partial v/\partial m$ in equations (4.24) and (4.25) cancel based on the hydrodynamic continuity equation

$$\frac{\partial \ln\rho}{\partial t} + \frac{2v}{r} = -4\pi r^2 \rho \frac{\partial v}{\partial m}. \tag{4.27}$$

Similarly, the terms involving $(J^E - K^E)v/r$ cancel. As these cancelling terms are of $O(v/c)$ and proportional to the neutrino density, they grow during a core collapse simulation and become quite large at and after bounce. Hence, it is important for a satisfactory global energy conservation to guarantee their cancellation in the finite difference representation as well. Identifying the origin of the terms $(\partial \ln\rho/\partial \mathrm{t} + 2\mathrm{v}/\mathrm{r})\mathrm{K}^\mathrm{E}$ and $K^E 4\pi r^2 \rho \partial v/\partial m$, we find that $(\partial \ln\rho/\partial t + 2v/r)K^E$ originates from the zeroth moment of the observer correction

$$\left(\mu^2\left(\frac{\partial \ln\rho}{\partial t}+\frac{2v}{r}\right)-\left(1-\mu^2\right)\frac{v}{r}\right)\frac{1}{E^2}\frac{\partial}{\partial E}\left[E^3 F\right], \tag{4.28}$$

in the Boltzmann equation (4.12), and $K^E 4\pi r^2\rho\partial v/\partial m$ originates from the first moment of the space propagation term, $\mu\partial\left[4\pi r^2\rho F\right]/\partial m$, in the same equation. The terms $\left(J^E - K^E\right)v/r$ stem from the zeroth moment of the same observer correction (4.28) and the first moment of the angular propagation term $\partial\left[\left(1-\mu^2\right)F\right]/(r\partial\mu)$ in the Boltzmann equation (4.12). The requirement of global energy conservation in the lab frame therefore imposes interdependencies on the finite differencing of the space propagation term, the angular propagation term, and the terms describing the observer corrections Liebendörfer, 2000).

In particular, to determine the "matched" finite differencing for

$$A \equiv \mu^2\left(\frac{\partial \ln\rho}{\partial t}+\frac{2v}{r}\right) \tag{4.29}$$

$$B \equiv (1-\mu^2)\frac{v}{r} \tag{4.30}$$

we begin by multiplying the discrete representation of the left-hand side of the Boltzmann equation (4.12) (*sans* the terms describing the observer corrections) by $1+\mu\bar{v}_{i+1}$ (in what follows, unless otherwise specified the indices are i', j', and k'):

$$\begin{aligned}
& (1+\mu\bar{v}_{i+1})E\frac{F-\bar{F}}{cdt} \\
+ & (1+\mu\bar{v}_{i+1})E\frac{4\pi\mu}{dm}[\bar{r}_{i+1}^2\bar{\rho}_{i+1}F_{i+1}-\bar{r}_i^2\bar{\rho}_i F_i] \\
+ & (1+\mu\bar{v_{i+1}})E\frac{3(\bar{r}_{i+1}^2-\bar{r}_i^2)}{2(\bar{r}_{i+1}^3-\bar{r}_3^2)}\frac{1}{w}[\zeta_{j+1}F_{j+1}-\zeta_j F_j] \\
= & \frac{(1+\mu v_{i+1})EF-(1+\mu\bar{v}_{i+1})E\bar{F}}{cdt} \\
- & \frac{\mu v_{i+1}EF-\mu\bar{v}_{i+1}EF}{cdt} \\
+ & \frac{4\pi\mu}{dm}[(1+\mu\bar{v}_{i+1})E\bar{r}_{i+1}^2\bar{\rho}_{i+1}F_{i+1}-(1+\mu\bar{v}_i)E\bar{r}_i^2\bar{\rho}_i F_i] \\
- & \frac{4\pi\mu^2}{dm}[\bar{v}_{i+1}\bar{r}_i^2\bar{\rho}_i EF_i-\bar{v}_i\bar{r}_i^2\bar{\rho}_i EF_i] \\
+ & \frac{3(\bar{r}_{i+1}^2-\bar{r}_i^2)}{2(\bar{r}_{i+1}^3-\bar{r}_3^2)}\frac{1}{w}[\zeta_{j+1}EF_{j+1}-\zeta_j EF_j]
\end{aligned} \tag{4.31}$$

$$
\begin{aligned}
&+ \quad \frac{3(\bar{r}_{i+1}^2 - \bar{r}_i^2)}{2(\bar{r}_{i+1}^3 - \bar{r}_3^2)} \bar{v}_{i+1} \frac{1}{w} [\mu\zeta_{j+1} EF_{j+1} - \mu\zeta_j EF_j] \\
&= \quad \frac{(1+\mu v_{i+1})EF - (1+\mu\bar{v}_{i+1})E\bar{F}}{cdt} \\
&- \quad EF \frac{\mu v_{i+1} - \mu\bar{v}_{i+1}}{cdt} \\
&+ \quad \frac{3(\bar{r}_{i+1}^2 - \bar{r}_i^2)}{2(\bar{r}_{i+1}^3 - \bar{r}_3^2)} \frac{1}{w} [\zeta_{j+1} EF_{j+1} - \zeta_j EF_j] \\
&+ \quad \frac{4\pi\mu}{dm} [(1+\mu\bar{v}_{i+1})E\bar{r}_{i+1}^2 \bar{\rho}_{i+1} F_{i+1} - (1+\mu\bar{v}_i)E\bar{r}_i^2 \bar{\rho}_i F_i] \\
&- \quad \frac{4\pi\mu^2}{dm} \bar{r}_i^2 \bar{\rho}_i EF_i [\bar{v}_{i+1} - \bar{v}_i] \\
&+ \quad \frac{3(\bar{r}_{i+1}^2 - \bar{r}_i^2)}{2(\bar{r}_{i+1}^3 - \bar{r}_3^2)} \bar{v}_{i+1} \frac{1}{w} [\mu\zeta_{j+1} EF_{j+1} - \mu\zeta_j EF_j].
\end{aligned}
$$

As noted, the total energy equation is obtained when summing equations (4.24) and (4.25) and then integrating over m (the integration in μ and E has already taken place). In this sequence of integrations, the term involving A in equation (4.24) cancels with the term $-4\pi r^2 \rho K^E dv/dm$ in equation (4.25). We must ensure this happens in the discrete limit. Identifying the appropriate velocity gradient term in equation (4.31) and focusing on the appropriate integration (in this case, over m), we require that (below, the term involving A comes from the zeroth moment of first term in the observer correction (4.28) *after an integration by parts*; the term involving the velocity gradient is the next to last term in equation (4.31), corresponding to the first moment of the spatial propagation term in the Boltzmann equation (4.12)):

$$
\begin{aligned}
& \quad \sum_{i=1,imax-1} A_{i'} F_{i'} dm_{i'} \\
&- \quad \sum_{i=1,imax-1} 4\pi\mu^2 \bar{r}_i^2 \bar{\rho}_i F_i (\bar{v}_{i+1} - \bar{v}_i) \\
&= \quad \sum_{i=1,imax-1} A_{i'} F_{i'} dm_{i'} \\
&- \quad \sum_{i=1,imax-1, j\le jmax/2} 4\pi\mu^2 \bar{r}_i^2 (\beta_i \bar{\rho}_{i'} F_{i'} + (1-\beta_i)\bar{\rho}_{i'-1} F_{i'-1}) \\
&- \quad \sum_{i=1,imax-1, j\ge jmax/2+1} 4\pi\mu^2 \bar{r}_i^2 (\beta_i \bar{\rho}_{i'-1} F_{i'-1} + (1-\beta_i)\bar{\rho}_{i'} F_{i'}) \\
&= \quad \sum_{i=1,imax-1} A_{i'} F_{i'} dm_{i'}
\end{aligned}
$$

$$
\begin{aligned}
&- \sum_{i=1,imax-1,j\leq jmax/2} 4\pi\mu^2\bar{r}_i^2(\bar{v}_{i+1}-\bar{v}_i)\beta_i\bar{\rho}_{i'}F_{i'} \\
&- \sum_{i=1,imax-2,j\leq jmax/2} 4\pi\mu^2\bar{r}_{i+1}^2(\bar{v}_{i+2}-\bar{v}_{i+1})(1-\beta_{i+1})\bar{\rho}_{i'}F_{i'} \\
&- \sum_{i=1,imax-1,j\geq jmax/2+1} 4\pi\mu^2\bar{r}_i^2(\bar{v}_{i+1}-\bar{v}_i)(1-\beta_i)\bar{\rho}_{i'}F_{i'} \\
&- \sum_{i=1,imax-2,j\geq jmax/2+1} 4\pi\mu^2\bar{r}_{i+1}^2(\bar{v}_{i+2}-\bar{v}_{i+1})\beta_{i+1})\bar{\rho}_{i'}F_{i'} \\
&= 0,
\end{aligned} \tag{4.32}
$$

which gives (Liebendörfer, 2000)

$$
A_{i'} = 4\pi\mu^2\frac{\bar{\rho}_{i'}}{dm_{i'}}[\bar{r}_i^2(\bar{v}_{i+1}-\bar{v}_i)\beta_i + \bar{r}_{i+1}^2(\bar{v}_{i+2}-\bar{v}_{i+1})(1-\beta_{i+1})] \tag{4.33}
$$

for $j \leq jmax/2$ and

$$
A_{i'} = 4\pi\mu^2\frac{\bar{\rho}_{i'}}{dm_{i'}}[\bar{r}_i^2(\bar{v}_{i+1}-\bar{v}_i)(1-\beta_i) + \bar{r}_{i+1}^2(\bar{v}_{i+2}-\bar{v}_{i+1})\beta_{i+1}] \tag{4.34}
$$

for $j \geq jmax/2+1$.

Similarly, defining B' according to

$$
B \equiv \frac{3}{2}\frac{\bar{r}_{i+1}^2-\bar{r}_i^2}{\bar{r}_{i+1}^3-\bar{r}_i^3}\bar{v}_{i+1}B', \tag{4.35}
$$

and again focusing on the appropriate integration (in this case, over μ), we require that (below, the term involving B' comes from the zeroth moment of the second term in the observer correction (4.28), *after an integration by parts*; the second term is the last term in equation (4.31), corresponding to the first moment of the angular propagation term):

$$
\begin{aligned}
0 &= \sum_{j=1,jmax} B'F_{j'}w_{j'} \\
&+ \sum_{j=1,jmax} \frac{2}{w_{j'}}[\mu_{j'}\alpha_{j+1}F_{j+1} - \mu_{j'}\alpha_j F_j]w_{j'} \\
&= \sum_{j=1,jmax} B'F_{j'}w_{j'} \\
&+ \sum_{j=1,jmax} 2[\mu_{j'}\alpha_{j+1}(\gamma F_{j'} + (1-\gamma)F_{j'+1})
\end{aligned}
$$

$$
\begin{aligned}
- & \quad \mu_{j'}\alpha_j(\gamma F_{j'-1} + (1-\gamma)F_{j'})] \\
= & \sum_{j=1,jmax} B^{'} F_{j'} w_{j'} \\
+ & \sum_{j=1,jmax} 2[\mu_{j'}\alpha_{j+1}\gamma - \mu_{j'}\alpha_j(1-\gamma)]F_{j'} \\
+ & \sum_{j=2,jmax} 2\mu_{j'-1}\alpha_j(1-\gamma)F_{j'} \\
+ & \sum_{j=1,jmax-1} (-2)\mu_{j'+1}\alpha_{j+1}\gamma F_{j'} \\
= & \sum_{j=1,jmax} B^{'} F_{j'} w_{j'} \\
+ & \sum_{j=1,jmax} 2\gamma\alpha_{j+1}(\mu_{j'} - \mu_{j'+1})F_{j'} \\
+ & \sum_{j=1,jmax} 2(1-\gamma)\alpha_j(\mu_{j'-1} - \mu_{j'})F_{j'},
\end{aligned}
\tag{4.36}
$$

which gives (Liebendörfer, 2000)

$$
\begin{aligned}
& B_{j'} \\
= & \frac{3}{2}\frac{\bar{r}_{i+1}^2 - \bar{r}_i^2}{\bar{r}_{i+1}^3 - \bar{r}_i^3}\bar{v}_{i+1}[2\gamma\alpha_{j+1}\frac{\mu_{j'+1} - \mu_{j'}}{w_{j'}} \\
+ & 2(1-\gamma)\alpha_j\frac{\mu_{j'} - \mu_{j'-1}}{w_{j'}}].
\end{aligned}
\tag{4.37}
$$

Frequency Shift from Observer Motion

For comoving observers, we must account for the observed shift in the frequency of a neutrino measured by two neighboring observers moving with different velocities. Motivated by the definitions (4.29) and (4.30), the observer correction, equation (4.28), can be written as

$$
0 = E^3\left(\frac{\partial F}{\partial t}\right)_E + \left(\mu^2 A - B\right)E\frac{\partial}{\partial E}\left[E^3 F\right]. \tag{4.38}
$$

It has been found in Bruenn (1985) and Mezzacappa and Bruenn (1993a) that equation (4.38) has an analytical solution. One may write (Mezzacappa and Bruenn, 1993a) the prefactor in equation (4.38) as the time derivative of the quantity $R_f = r^{(3\mu^2-1)}\rho^{(\mu^2)}$:

$$
\frac{\partial \ln R_f}{\partial t} = \mu^2 A - B.
$$

If we transform from the "Eulerian" variable, $x = E$, to the "Lagrangian" variable, $y = E/R_f$, equation (4.38) becomes

$$\begin{aligned} 0 &= \left(\frac{\partial}{\partial t}\left[E^3F\right]\right)_E + \frac{\partial R_f}{R_f^2\partial t}E \times R_f\frac{\partial}{\partial E}\left[E^3F\right] \\ &= \left(\frac{\partial}{\partial t}\left[E^3F\right]\right)_E - \left(\frac{\partial\left[E/R_f\right]}{\partial t}\right)_E \frac{\partial\left[E^3F\right]}{\partial\left[E/R_f\right]} = \left(\frac{\partial}{\partial t}\left[E^3F\right]\right)_{E/R_f}. \end{aligned}$$

For a small section of energy phase space $E^2\Delta E = \left(E_2^3 - E_1^3\right)/3$, this relationship leads to

$$\left(\frac{\partial}{\partial t}\left[E^2F\Delta E\right]\right)_{E/R_f} = 0. \tag{4.39}$$

The validity of equation (4.39) for arbitrary neutrino distributions F leads to the following interpretation: The neutrinos initially residing in the energy interval $E^2\Delta E$ move, according to the observer correction, along constant E/R_f in the phase space of a comoving observer. This allows us to determine the evolution of any neutrino property. For example, the neutrino specific energy in this phase space interval, $d\epsilon = E^3F\Delta E$, evolves according to

$$\left(\frac{\partial}{\partial t}\left[E^3F\Delta E\right]\right)_{E/R_f} = E^2F\Delta E\left(\frac{\partial E}{\partial t}\right)_{E/R_f} = \frac{\partial \mathrm{ln R_f}}{\partial t}d\epsilon. \tag{4.40}$$

A finite difference representation of equations (4.39) and (4.40) has been given in Mezzacappa and Bruenn, 1993a. Consider a neutrino energy group k', with neighboring groups $k' + dk$, $dk = \pm 1$. Equation (4.39) tells us that the number of neutrinos before the comoving observer correction, $F_{i',j',k'}E_{k'}^2dE_{k'}$, is equal to the number of neutrinos after the correction. The distribution after the correction has a diminished number of neutrinos $F_{i',j',k'}E_{k'}^2dE_{k'} - n_{i',j',k'}^-$ in group k' and an additional number of neutrinos $n_{i',j',k'+dk}^+$ in the neighboring group $k' + dk$,

$$F_{i',j',k'}E_{k'}^2dE_{k'} - \left[\left(F_{i',j',k'}E_{k'}^2dE_{k'} - n_{i',j',k'}^-\right) + n_{i',j',k'+dk}^+\right] = 0. \tag{4.41}$$

Equation (4.40) defines a similar correction for the specific neutrino energy in group k'

$$\begin{aligned} F_{i',j',k'}E_{k'}^3dE_{k'} &- \left[\left(F_{i',j',k'}E_{k'}^3dE_{k'} - E_{k'}n_{i',j',k'}^-\right) + E_{k'+dk}n_{i',j',k'+dk}^+\right] \\ &= -\left(\mu_{j'}^2A_{i',k'} - B_{i',j',k'}\right)F_{i',j',k'}E_{k'}^3dE_{k'}dt. \end{aligned} \tag{4.42}$$

where $A_{i',k'}$ and $B_{i',j',k'}$ are given by equations (4.33), (4.34), and (4.37). Equations (4.41) and (4.42) uniquely define the solution

$$\begin{aligned} n_{i',j',k'}^- &= \left(\mu_{j'}^2A_{i',k'} - B_{i',j',k'}\right)\frac{dE_{k'}}{E_{k'+dk} - E_{k'}}E_{k'}^3F_{i',j',k'}dt \\ n_{i',j',k'}^+ &= n_{i',j',k'-dk}^-, \end{aligned} \tag{4.43}$$

which leads, by the update $F_{i',j',k'} = \overline{F}_{i',j',k'} + \left(n^+_{i',j',k'} - n^-_{i',j',k'}\right) / \left(E^2_{k'} dE_{k'}\right)$, to the following finite difference representation of the frequency shift term in the Boltzmann equation (4.12):

$$\begin{aligned} & \frac{1}{E^2_{k'} dE_{k'}} \\ \times & \left[\left(\mu^2_{j'} A_{i',k'-dk} - B_{i',j',k'}\right) \frac{dE_{k'-dk}}{E_{k'} - E_{k'-dk}} E^3_{k'-dk} F_{i',j',k'-dk} \right. \\ - & \left. \left(\mu^2_{j'} A_{i',k'} - B_{i',j',k'}\right) \frac{dE_{k'}}{E_{k'+dk} - E_{k'}} E^3_{k'} F_{i',j',k'} \right] . \end{aligned} \tag{4.44}$$

Angular Aberration from Observer Motion

We are left with the task of finding a finite difference representation for the angular aberration in equation (4.12). Because the global energy depends on $(1 + u_{i'}\mu_{j'}) F_{i',j',k'}$, any correction to the neutrino propagation direction also affects energy conservation. Therefore, it is desirable to construct a numerical implementation of angular aberration that conserves the specific neutrino luminosity $\mu_{j'} F_{i',j',k'} w_{j'}$ (Liebendörfer, 2000), much like our numerical implementation of the frequency shift preserved the specific neutrino energy.

With $\zeta = 1 - \mu^2$, the angular aberration term can be rewritten as

$$(\frac{\partial F}{\partial t})_\mu = (A + B/\zeta) \frac{\partial}{\partial \mu} [\zeta \mu F] . \tag{4.45}$$

As before, we search for an analytic solution. For the quantity $R_a = r^3 \rho$, we find

$$\frac{d\mathrm{lnR_a}}{dt} = A + B/\zeta.$$

It is now possible to rewrite equation (4.45) in terms of the "Lagrangian" variable $y = \zeta^{-1/2}\mu/R_a$ instead of the "Eulerian" variable $x = \mu$. After a multiplication by $\zeta\mu$, equation (4.45) becomes:

$$\begin{aligned} 0 &= \zeta\mu \left[\left(\frac{\partial F}{\partial t}\right)_\mu + \alpha (A + B/\zeta) \frac{\partial}{\partial \mu} [\zeta \mu F] \right] \\ &= \left(\frac{\partial}{\partial t} [\zeta \mu F]\right)_\mu + \zeta^{-1/2} \mu \frac{\partial R_a}{R_a^2 \partial t} \times \zeta^{3/2} R_a \frac{\partial}{\partial \mu} [\zeta \mu F] \\ &= \left(\frac{\partial}{\partial t} [\zeta \mu F]\right)_\mu \\ & - \left(\frac{\partial \left[\zeta^{-1/2}\mu/R_a\right]}{\partial t}\right)_\mu \frac{\partial [\zeta \mu F]}{\partial \left[\zeta^{-1/2}\mu/R_a\right]} = \left(\frac{\partial}{\partial t} [\zeta \mu F]\right)_{\zeta^{-1/2}\mu/R_a} . \end{aligned}$$

And once again, the interpretation is clear. The neutrinos initially residing in the interval $(1-3\mu^2)\,\Delta\mu = \zeta_2\mu_2 - \zeta_1\mu_1$ are shifted by angular aberration along constant $\mu/\left(\sqrt{\zeta}R_a\right)$ in the phase space of a comoving observer:

$$\left(\frac{\partial}{\partial t}\left[\left(1-3\mu^2\right)F\Delta\mu\right]\right)_{\zeta^{-1/2}\mu/R_a} = 0. \tag{4.46}$$

We may now evaluate the change in other neutrino properties; for example, the specific luminosity, $d\ell = (1-3\mu^2)\,\mu F\Delta\mu$,

$$\begin{aligned}\left(\frac{\partial}{\partial t}\left[\left(1-3\mu^2\right)\mu F\Delta\mu\right]\right)_{\zeta^{-1/2}\mu/R_a} &= \left(1-3\mu^2\right)F\Delta\mu\left(\frac{\partial\mu}{\partial t}\right)_{\zeta^{-1/2}/R_a} \\ &= \zeta\frac{\partial \mathrm{lnR_a}}{\partial t}d\ell.\end{aligned} \tag{4.47}$$

We identify the bin size $\left(1-3\mu_{j'}^2\right)\Delta\mu_{j'} = w_{j'}$ with our Gaussian quadrature weights. Equation (4.46) leads to the condition for number conservation:

$$F_{i',j',k'}w_{j'} - \left[\left(F_{i',j',k'}w_{j'} - n^-_{i',j',k'}\right) + n^+_{i',j'+dj,k'}\right] = 0$$

and equation (4.47) leads to a prescription for the numerical evolution of the specific luminosity:

$$\begin{aligned}F_{i',j',k'}\mu_{j'}w_{j'} &- \left[\left(F_{i',j',k'}\mu_{j'}w_{j'} - \mu_{j'}n^-_{i',j',k'}\right) + \mu_{j'+dj}n^+_{i',j'+dj,k'}\right] \\ &= -\left(\zeta_{j'}A_{i',k'} + B_{i',j',k'}\right)F_{i',j',k'}\mu_{j'}w_{j'}dt.\end{aligned}$$

where $dj = \pm 1$. The change in the neutrino distribution from angular aberration is then:

$$F_{i',j',k'} = \overline{F}_{i',j',k'} + \left(n^+_{i',j',k'} - n^-_{i',j',k'}\right)/w_{j'} \tag{4.48}$$

with

$$\begin{aligned}n^-_{i',j',k'} &= \left(A_{i',k'} + B_{i',j',k'}/\zeta_{j'}\right)\frac{w_{j'}}{\mu_{j'+dj}-\mu_{j'}}\zeta_{j'}\mu_{j'}F_{i',j',k'}dt \\ n^+_{i',j',k'} &= n^-_{i',j'-dj,k'}.\end{aligned}$$

This leads to the following finite difference representation of the angular aberration term in the Boltzmann equation (4.12):

$$\begin{aligned}\frac{1}{w_{j'}}&\left[\left(A_{i',k'} + B_{i',j'-dj,k'}/\zeta_{j'-dj}\right)\frac{w_{j'-dj}}{\mu_{j'}-\mu_{j'-dj}}\zeta_{j'-dj}\mu_{j'-dj}F_{i',j'-dj,k'}\right. \\ -&\left.\left(A_{i',k'} + B_{i',j',k'}/\zeta_{j'}\right)\frac{w_{j'}}{\mu_{j'+dj}-\mu_{j'}}\zeta_{j'}\mu_{j'}F_{i',j',k'}\right].\end{aligned} \tag{4.49}$$

We apply the aberration corrections with $dj = +1$ for $\mu \leq 0$ and $dj = -1$ for $\mu > 0$. This is not upwind differencing and, therefore, runs the risk of producing negative neutrino distribution functions. However, there are three reasons to accept this shortcoming: (i) The angular aberration correction is generally small, with the exception of aberration in the vicinity of strong shocks with large velocity gradients. (ii) As long as the angle-integrated neutrino density is positive, transiently negative contributions from backward directions on the Gaussian angular grid ($\mu < 0$) allow the computer representation of strongly forward peaked neutrino fluxes that would not be representable with positive neutrino distribution functions on a grid with limited angles $\mu_{\max} < 1$. (iii) The chosen direction of dj guarantees that no neutrinos are shifted off the grid. This is a prerequisite for number and energy conservation.

6. The General Case

As the previous section illustrates, one of the greatest challenges encountered when developing the numerical methods to obtain a physically reliable solution of the neutrino transport equations in core collapse supernovae is to achieve simultaneous lepton number and energy conservation or *balance*, the former for massless neutrinos and Newtonian gravity, where lepton number is conserved and where the concept of a conserved energy can be defined in multiple spatial dimensions, the latter for the general relativistic case in two and three spatial dimensions. The first step in meeting this challenge is to derive the conservative lepton number and energy conservation/balance equations, which at a more fundamental level means finding the corresponding evolved variables that naturally lend themselevs to a conservative decription of the neutrino radiation field. As we will see, both the conservation/balance equations and the fundamental variables can be extracted from the underlying kinetic theory at a point in the theory where the fundamental physical laws are manifestly conservative (Cardall and Mezzacappa, 2002).

The distribution function f represents the density of neutrinos in phase space. The phase space for neutrinos of definite mass m, M_m, is filled with trajectories $\left(x^{\mu}(\lambda), p^{\hat{i}}(\lambda)\right)$, or "states". As a collection of neutrinos evolves, the number of neutrinos in each state changes due to collisions. If one considers a 6-dimensional hypersurface Σ in M_m, the ensemble-averaged number $N[\Sigma]$ of occupied states crossing Σ is

$$N[\Sigma] = \int_{\Sigma} f\omega, \tag{4.50}$$

where ω is the surface element.

An equation governing the evolution of the distribution function is obtained by considering a closed 6-dimensional hypersurface ∂D bounding a region D in M_m. The net number of occupied states emerging from D is, from equation

(4.50) and the generalized Stokes' theorem,

$$N[\partial D] = \int_{\partial D} f\omega = \int_D d(f\omega). \tag{4.51}$$

This equation relates a volume integral to a surface integral and is key to obtaining conservative formulations of kinetic theory. In fact, in rushing headlong in deriving the general relativistic Boltzmann equation

$$p^{\hat{\mu}}\left(\Lambda^{\bar{\mu}}{}_{\hat{\mu}}e^{\mu}{}_{\bar{\mu}}\frac{\partial f}{\partial x^{\mu}} - \Gamma^{\hat{\nu}}{}_{\hat{\rho}\hat{\mu}}p^{\hat{\rho}}\frac{\partial f}{\partial p^{\hat{\nu}}}\right) = C[f]. \tag{4.52}$$

from equation (4.51), equation (4.51) is easily overlooked. Here $\Lambda^{\bar{\mu}}_{\hat{\mu}}$ is the Lorentz boost between the orthonormal lab and (fluid) comoving frames and $e^{\mu}_{\bar{\mu}}$ is the tetrad transformation between the coordinate basis and the orthonormal lab frame. The integrand on the right-hand side of equation (4.51), $d(f\omega)$, is in conservative form. Manipulating it just enough to bring it into the form $d(f\omega) = F[f]\Omega$, where Ω is the volume element in phase space, yields a conservative formulation of neutrino number kinetics, and manipulating the exterior derivative $d(v_\mu p^\mu f\omega)$, for an arbitrary timelike vector v_μ, yields a conservative formulation of neutrino four-momentum kinetics. Defining the *specific neutrino number flux vector*

$$\mathcal{N}^{\mu} \equiv \mathcal{L}^{\mu}{}_{\hat{\mu}}p^{\hat{\mu}}f \tag{4.53}$$

and the *specific neutrino stress-energy tensor*

$$\mathcal{T}^{\mu\nu} \equiv \mathcal{L}^{\mu}{}_{\hat{\mu}}\mathcal{L}^{\nu}{}_{\hat{\nu}}p^{\hat{\mu}}p^{\hat{\nu}}f, \tag{4.54}$$

where $\mathcal{L}^{\hat{\mu}}{}_{\mu} = e^{\mu}_{\bar{\mu}}\Lambda^{\bar{\mu}}_{\hat{\mu}}$ is the transformation to the comoving frame, the conservative kinetic equations for neutrino number and four-momentum are:

$$\begin{aligned}
&\frac{1}{\sqrt{-g}}\frac{\partial}{\partial x^{\mu}}(\sqrt{-g}\mathcal{N}^{\mu}) - \\
&E(\mathbf{p})\left|\det\left[\frac{\partial \mathbf{p}}{\partial \mathbf{u}}\right]\right|^{-1}\frac{\partial}{\partial u^{\hat{i}}}\left(\frac{1}{E(\mathbf{p})}\left|\det\left[\frac{\partial \mathbf{p}}{\partial \mathbf{u}}\right]\right|\Gamma^{\hat{j}}{}_{\hat{\mu}\hat{\nu}}p^{\hat{\nu}}\frac{\partial u^{\hat{i}}}{\partial p^{\hat{j}}}\mathcal{L}^{\hat{\mu}}{}_{\mu}\mathcal{N}^{\mu}\right) \\
&= C[f],
\end{aligned} \tag{4.55}$$

$$\begin{aligned}
&\frac{1}{\sqrt{-g}}\frac{\partial}{\partial x^{\nu}}(\sqrt{-g}\mathcal{T}^{\mu\nu}) + \Gamma^{\mu}{}_{\nu\rho}\mathcal{T}^{\nu\rho} - \\
&E(\mathbf{p})\left|\det\left[\frac{\partial \mathbf{p}}{\partial \mathbf{u}}\right]\right|^{-1}\frac{\partial}{\partial u^{\hat{i}}}\left(\frac{1}{E(\mathbf{p})}\left|\det\left[\frac{\partial \mathbf{p}}{\partial \mathbf{u}}\right]\right|\Gamma^{\hat{j}}{}_{\hat{\nu}\hat{\rho}}p^{\hat{\rho}}\frac{\partial u^{\hat{i}}}{\partial p^{\hat{j}}}\mathcal{L}^{\hat{\nu}}{}_{\nu}\mathcal{T}^{\mu\nu}\right) \\
&= F^{\mu}{}_{\nu}\mathcal{N}^{\nu} + \mathcal{L}^{\mu}{}_{\hat{\mu}}p^{\hat{\mu}}C[f].
\end{aligned} \tag{4.56}$$

where $\sqrt{-g}$ is the determinant of the spacetime metric. The adjective "specific" often denotes a quantity measured per unit mass, but in this context we use it to denote the neutrino flux and stress-energy in a given invariant momentum space volume element. While the distribution function f obeys the Boltzmann equation (4.52), the specific neutrino number flux and stress-energy satisfy the conservative equations (4.55) and (4.56). In stating that equations (4.55) and (4.56) constitute conservative formulations of neutrino kinetics, we mean that the connection to the corresponding balance equations is transparent. We can use elementary calculus to form the familiar invariant momentum space volume element

$$\frac{d^3p}{E(\mathbf{p})} = \frac{1}{E(\mathbf{p})}\left|\det\left[\frac{\partial \mathbf{p}}{\partial \mathbf{u}}\right]\right| d^3u, \tag{4.57}$$

where a transformation from orthonormal momentum components $\{p^{\hat{i}}\}$ to some other set of coordinates (e.g., momentum space spherical coordinates $\{u^{\hat{i}}\} = \{|\mathbf{p}|, \vartheta, \varphi\}$) has been performed. Multiplying equations (4.55) and (4.56) by equation (4.57) and integrating, the terms with momentum space derivatives are obviously transformed into vanishing surface terms. The results are the balance equations

$$\frac{1}{\sqrt{-g}}\frac{\partial}{\partial x^\mu}(\sqrt{-g}N^\mu) = \int C[f]\pi_m, \tag{4.58}$$

$$\frac{1}{\sqrt{-g}}\frac{\partial}{\partial x^\mu}(\sqrt{-g}T^{\nu\mu}) = -\Gamma^\nu{}_{\rho\mu}T^{\rho\mu} + \int C[f]\mathcal{L}^\nu{}_{\hat{\nu}}p^{\hat{\nu}}\pi_m. \tag{4.59}$$

for total neutrino number and 4-momentum, where N^μ and $T^{\mu\nu}$ are the neutrino number current and stress-energy tensor, respectively. In the above equations, $\pi_m = \sqrt{-g}\epsilon_{0ijk}dp^i \wedge dp^j \wedge dp^k//6|p_0|$ is the invariant volume element on a mass shell corresponding to the mass m in the tangent space (Cardall and Mezzacappa, 2002).

In this paper, we presented a detailed account of how one obtains a conservative differencing for the equation governing the evolution of the specific neutrino distribution function for spherically symmetric flows. Working with comoving frame neutrino four momenta, the conservation of total lab frame energy required a matching of the finite differencing of the radial, angular, and energy advection terms in this equation. One can, of course, take a different approach. While the use of comoving frame four momenta for the neutrinos is a prerequisite given the neutrino interactions with the stellar core, which are most naturally and most simply expressed in terms of these four momenta, the use of comoving frame neutrino densities (number, energy, . . .) may not be the best choice. Given that the global lab frame energy is conserved, working with densities in this frame that are expressed in terms of comoving frame neutrino four momenta might be the right combination considering both local neutrino interactions and global energy conservation.

Nonetheless, however natural this latter approach might seem, we are still not free of the need to match the finite difference representations for various terms in the kinetic equations, as we now show.

Consider the relationship between equation (4.55) for neutrino number kinetics and the t component of equation (4.56) for neutrino 4-momentum kinetics. Specifically, we need to relate the spatial and momentum divergence terms of these equations to each other and identify terms that cancel.

First we relate the spatial and momentum divergence terms of equation (4.55) and the t component of equation (4.56). Their spatial divergence terms

$$N_x[f] \equiv \frac{1}{\sqrt{-g}}\frac{\partial}{\partial x^\mu}(\sqrt{-g}\mathcal{L}^\mu{}_{\hat{\mu}}p^{\hat{\mu}}f), \tag{4.60}$$

$$\left(T^t\right)_x[f] \equiv \frac{1}{\sqrt{-g}}\frac{\partial}{\partial x^\mu}(\sqrt{-g}\mathcal{L}^t{}_{\hat{\nu}}\mathcal{L}^\mu{}_{\hat{\mu}}p^{\hat{\nu}}p^{\hat{\mu}}f), \tag{4.61}$$

are related by

$$\mathcal{L}^t{}_{\hat{\nu}}p^{\hat{\nu}}\; N_x[f] = \left(T^t\right)_x[f] - f\; p^{\hat{\mu}}p^{\hat{\nu}}\mathcal{L}^\mu{}_{\hat{\mu}}\frac{\partial}{\partial x^\mu}\left(\mathcal{L}^t{}_{\hat{\nu}}\right). \tag{4.62}$$

Their momentum divergence terms

$$\begin{aligned} &N_p[f] \\ \equiv\;& -E(\mathbf{p})\left|\det\left[\frac{\partial \mathbf{p}}{\partial \mathbf{u}}\right]\right|^{-1}\frac{\partial}{\partial u^{\hat{i}}}\left(\frac{1}{E(\mathbf{p})}\left|\det\left[\frac{\partial \mathbf{p}}{\partial \mathbf{u}}\right]\right|\Gamma^{\hat{j}}{}_{\hat{\mu}\hat{\rho}}p^{\hat{\mu}}p^{\hat{\rho}}\frac{\partial u^{\hat{i}}}{\partial p^{\hat{j}}}f\right) \end{aligned} \tag{4.63}$$

$$\begin{aligned} &\left(T^t\right)_p[f] \\ \equiv\;& -E(\mathbf{p})\left|\det\left[\frac{\partial \mathbf{p}}{\partial \mathbf{u}}\right]\right|^{-1} \\ \times\;& \frac{\partial}{\partial u^{\hat{i}}}\left(\frac{1}{E(\mathbf{p})}\left|\det\left[\frac{\partial \mathbf{p}}{\partial \mathbf{u}}\right]\right|\Gamma^{\hat{j}}{}_{\hat{\mu}\hat{\rho}}p^{\hat{\mu}}p^{\hat{\rho}}\frac{\partial u^{\hat{i}}}{\partial p^{\hat{j}}}\mathcal{L}^t{}_{\hat{\nu}}p^{\hat{\nu}}f\right), \end{aligned} \tag{4.64}$$

are related by

$$\mathcal{L}^t{}_{\hat{\nu}}p^{\hat{\nu}}\; N_p[f] = \left(T^t\right)_p[f] + f\;\Gamma^{\hat{j}}{}_{\hat{\mu}\hat{\rho}}p^{\hat{\mu}}p^{\hat{\rho}}\frac{\partial u^{\hat{i}}}{\partial p^{\hat{j}}}\;\mathcal{L}^t{}_{\hat{\nu}}\;\frac{\partial p^{\hat{\nu}}}{\partial u^{\hat{i}}}. \tag{4.65}$$

Comparison of equations (4.62) and (4.65) with equation (4.55) and the t component of Eq. (4.56) shows that the following equation must be valid:

$$E_S + E_M = -\Gamma^t{}_{\mu\rho}\mathcal{L}^\mu{}_{\hat{\mu}}\mathcal{L}^\rho{}_{\hat{\rho}}p^{\hat{\mu}}p^{\hat{\rho}}, \tag{4.66}$$

where

$$E_S \equiv f\, p^{\hat{\mu}} p^{\hat{\nu}} \mathcal{L}^{\mu}{}_{\hat{\mu}} \frac{\partial}{\partial x^{\mu}} \left(\mathcal{L}^{t}{}_{\hat{\nu}} \right) \tag{4.67}$$

is the "extra" term of Eq. (4.62) relating the spatial divergence terms of the number and energy balance equations, and

$$E_M \equiv -f\, \Gamma^{\hat{j}}{}_{\hat{\mu}\hat{\rho}} p^{\hat{\mu}} p^{\hat{\rho}} \frac{\partial u^{\hat{i}}}{\partial p^{\hat{j}}} \mathcal{L}^{t}{}_{\hat{\nu}} \frac{\partial p^{\hat{\nu}}}{\partial u^{\hat{i}}} \tag{4.68}$$

is the "extra" term in Eq. (4.65) relating the momentum divergence terms of the number and energy balance equations.

In a computational approach to transport in which a neutrino number distribution equation is solved and a neutrino energy equation is used as a consistency check, care must be taken that equation (4.66) is satisfied numerically. In particular, a given finite difference representation of the spacetime divergence $N_x[f]$ implies a corresponding finite difference representation of position-dependent terms in E_S, and the position-dependent terms in E_M that cancel with their counterparts in E_S must have a matching finite difference representation. Similarly, a given finite difference representation of the momentum divergence $N_p[f]$ implies a corresponding finite difference representation of momentum-dependent terms in E_M, and the momentum-dependent terms in E_S that cancel with their counterparts in E_M must have a matching finite difference representation.

It is important to note that equations (4.55) and (4.56) are general; i.e., they can be expressed in any spatial coordinate system using any neutrino four-momenta. If one specializes these equations to the comoving frame spatial coordinates and four-momenta used in the spherically symmetric case detailed herein and to the $O(v/c)$ limit, one finds that $E_S + E_M = 0$. The finite difference matching constructed in Section 6.5 guarantees that this sum is satisfied numerically in this case. In the general case, the sum is not zero and the number of terms is significantly greater, but the basic task is the same. The essential point is that we now have the complete framework within which to carry it out.

7. Neutrino Mass and Mixing

It is now an experimental fact that neutrinos have mass and mix. For the corresponding experimentally determined mixing parameters, there may be significant consequences for supernova dynamics, nucleosynthesis, and terrestrial neutrino signatures. Thus, no discussion of neutrino transport in core collapse supernovae would be complete without a discussion of the transport of massive neutrinos in stellar cores. However, while a number of developments have taken place in this and other contexts, the underlying kinetic theory is not yet at the same stage of maturity and, consequently, neither are the numerical methods and simulation efforts. The fundamental challenge, of course, is to simulate both the coherent quantum mechanical phenomenon of flavor transformation and the classical phenomena of transport and interaction (absorption, scattering, pair creation and annihilation, neutrino–neutrino interaction) in stellar cores. This requires quantum kinetic theory for the neutrino density matrix, whose diagonal components correspond to the neutrino distribution functions, for which we currently solve, and whose off-diagonal elements correspond to the amplitudes for quantum mechanical flavor transformation, currently not included in the state of the art in supernova simulations [except in simplified ways (see, e.g., Fuller et al., 1992, Mezzacappa and Bruenn, 1999)].

8. Conclusion

It was our intent in this review to illustrate by example the theoretical and computational challenge presented by neutrino kinetics in stellar cores and core collapse supernovae. While spherically symmetric models have achieved a significant level of sophistication, multidimensional models are in their infancy. Thus, this subject is very much an open one, waiting more for fundamental leaps than refinements. It is our hope that what we have outlined here has not only illustrated the requirements of this task, but a promising path to its completion. It is, after all, a task we must accept if we are to understand core collapse supernovae, hypernovae, gamma-ray bursts , and our place in the Universe.

Acknowledgments

This work was supported by the Oak Ridge National Laboratory, managed by UT-Battelle, LLC, for the DoE under contract DE-AC05-00OR22725; the National Science Foundation under contract AST-9877130; the U.S. Department of Energy (DoE) HENP Scientific Discovery through Advanced Computing Program; a DoE HENP PECASE award; and the Joint Institute for Heavy Ion Research.

References

Arnett, W. D. (1977). *Astrophysical Journal*, 218:815.

Balescu, R. (1975). *Equilibrium and Nonequilibrium Statistical Mechanics*. New York: Wiley-Interscience.

Baron, E. A., Cooperstein, J., and Kahana, S. (1985). *Nuclear Physics*, A440:744.

Bethe, H. A. and Brown, G. (1985). *Scientific American*, 252:60.

Bethe, H. A. (1990). *Reviews of Modern Physics*, 62:801.

Bethe, H. A. and Wilson, J. R. (1985). *Astrophysical Journal*, 295:14.

Bowers, R. L. and Wilson, J. R. (1982). *Astrophysical Journal Supplement*, 50:115.

Brown, G., Bethe, H. A., and Baym, G. (1982). *Nucl. Phys. A*, 375:481.

Bruenn, S. W. (1985). *Astrophysical Journal Supplement*, 58:771.

Bruenn, S. W. (1993). In Guidry, M. W. and Strayer, M. R., editors, *First Symposium on Nuclear Physics in the Universe*, page 31. Bristol. IOP Publishing.

Bruenn, S. W., DeNisco, K. R., and Mezzacappa, A. (2001). *Astrophysical Journal*, 560:326.

Bruenn, S. W. and Dineva, T. (1996). *Astrophysical Journal Letters*, 458:L71.

Buras, R., Rampp, M., Janka, H.-T., and Kifonidis, K. (2003). *Physical Review Letters*, submitted (astro-ph/0303171).

Burrows, A. (1987). *Physics Today*.

Burrows, A. (2003). This volume.

Burrows, A. and Goshy, J. (1993). *Astrophysical Journal Letters*, 416:L75.

Burrows, A., Hayes, J., and Fryxell, B. A. (1995). *Astrophysical Journal*, 450:830.

Burrows, A., Young, T., Pinto, P., Eastman, R., and Thompson, T. A. (2000). *Astrophysical Journal*, 539:865.

Cardall, C. Y. and Mezzacappa, A. (2002). *Physical Review D*, in press (astro-ph/0212460).

Castor, J. (1972). *Astrophysical Journal*, 178:779.

Colgate, S. A. and White, R. H. (1966). *Astrophysical Journal*, 143:626.

Cooperstein, J., Van Den Horn, L. J., and Baron, E. A. (1986). *Astrophysical Journal*, 309:653.

Fryer, C. L. and Heger, A. (2000). *Astrophysical Journal*, 541:1033.

Fryer, C. L. and Warren, M. S. (2002). *Astrophysical Journal*, 574:L65.

Fuller, G. M., Mayle, R., Meyer, B. S., and Wilson, J. R. (1992). *Astrophysical Journal*, 389:517.

Herant, M., Benz, W., Hix, W. R., Fryer, C. L., and Colgate, S. A. (1994). *Astrophysical Journal*, 435:339.

Hillebrandt, W., Nomoto, K., and Wolff, R. (1984). *Astronomy and Astrophysics*, 133:175.

Janka, H.-T. (2003). This volume.

Janka, H.-T. and Müller, E. (1996). *Astronomy and Astrophysics*, 306:167.

Levermore, C. D. and Pomraning, G. C. (1981). *Astrophysical Journal*, 248:321.

Lewis, E. and Miller, W. (1984). *Computational Methods of Neutron Transport*. New York: Wiley-Interscience.

Liebendörfer, M., Mezzacappa, A., Thielemann, F., Messer, O. E., Hix, W. R., and Bruenn, S. W. (2001). *Physical Review D*, 63:103004.

Liebendoerfer, M., Rampp, M., Janka, H.-T., and Mezzacappa, A. (2003). In preparation.

Liebendörfer, M. (2000). PhD thesis, University of Basel, Basel, Switzerland.

MacFadyen, A. I. and Woosley, S. E. (1999). *Astrophysical Journal*, 524:262.

Mezzacappa, A. and Bruenn, S. W. (1993a). *Astrophysical Journal*, 405:669.

Mezzacappa, A. and Bruenn, S. W. (1993b). *Astrophysical Journal*, 405:637.

Mezzacappa, A. and Bruenn, S. W. (1999). In Spooner, N. J. C. and Kudryavtsev, V., editors, *The Identification of Dark Matter*, page 655. World Scientific, Singapore.

Mezzacappa, A., Calder, A. C., Bruenn, S. W., Blondin, J. M., Guidry, M. W., Strayer, M. R., and Umar, A. S. (1998a). *Astrophysical Journal*, 493:848.

Mezzacappa, A., Calder, A. C., Bruenn, S. W., Blondin, J. M., Guidry, M. W., Strayer, M. R., and Umar, A. S. (1998b). *Astrophysical Journal*, 495:911.

Mezzacappa, A., Liebendörfer, M., Messer, O. E., Hix, W. R., Thielemann, F., and Bruenn, S. W. (2001). *Physical Review Letters*, 86:1935.

Mihalas, D. and Mihalas, B.. (1975). *Foundations of Radiation Hydrodynamics*. New York: Oxford University Press.

Mueller, E. (1991). In de Loore, C., editor, *Late Stages of Stellar Evolution: Computational Methods in Astrophysical Hydrodynamics*, page 97. Heidelberg: Springer-Verlag.

Myra, E. S., Bludman, S. A., Hoffman, Y., Lichtenstadt, I., Sack, N., and Van Riper, K. A. (1987). *Astrophysical Journal*, 318:744.

Rampp, M. and Janka, H.-T. (2000). *Astrophysical Journal*, 539:L33.

Rampp, M. and Janka, H.-T. (2002). *Astronomy and Astrophysics*, 396:361.

Shapiro, S. and Teukolsky, S. (1983). *Black Holes, White Dwarfs, and Neutron Stars*. New York: Wiley-Interscience.

Swesty, F. and Lattimer, J. (1994). *Astrophysical Journal*, 425:195.

Symbalisty, E. M. D. (1984). *Astrophysical Journal*, 285:729.

Thompson, T., Burrows, A., and Pinto, P. (2002). *Astrophysical Journal*, in press (astro-ph/0211194).

Van Riper, K. A. and Lattimer, J. M. (1981). *Astrophysical Journal*, 249:270.

Wilson, J. R. (1985). In Centrella, J. M., LeBlanc, J. M., and Bowers, R. L., editors, *Numerical Astrophysics*, page 422. Boston. Jones and Bartlett.

Wilson, J. R. and Mayle, R. W. (1993). *Physics Reports*, 227:97.

Chapter 5

NEUTRINO-MATTER INTERACTION RATES IN SUPERNOVAE

The Essential Microphysics of Core Collapse

Adam Burrows
Steward Observatory, The University of Arizona, Tucson, AZ 85721
burrows@zenith.as.arizona.edu

and Todd A. Thompson*
Astronomy Department and Theoretical Astrophysics Center,
601 Campbell Hall, The University of California, Berkeley, CA 94720
thomp@astro.berkeley.edu

Abstract Neutrino-matter interaction rates are central to the core collapse phenomenon and, perhaps, to the viability of the mechanism of core-collapse supernova explosions. In this paper we catalog and discuss the major neutrino scattering, absorption, and production processes that together influence the outcome of core collapse and the cooling of protoneutron stars. These are the essential inputs into the codes used to simulate the supernova phenomenon and an understanding of these processes is a prerequisite to continuing progress in supernova theory.

Keywords: Supernovae, Neutrino Interactions, Neutrino Spectra, Protoneutron Stars, Kinetic Theory

1. Introduction

One of the key insights of the 20th Century was that most of the elements of nature are created by nuclear processes in stars. Supernova explosions are one major means by which these elements are injected into the interstellar medium and, hence, into subsequent generations of stars. Therefore, supernovae are

*Hubble Fellow

C.L. Fryer (ed.), Stellar Collapse, 133–174.

central to the chemical evolution and progressive enrichment of the universe. Core-collapse supernova explosions signal the death of a massive star and are some of the most majestic and awe-inspiring events in the cosmos. However, to fully understand the role of supernovae in the grand synthesis of creation, one must have a firm handle on the nuclear data. In particular, understanding the origin of iron-peak, r-process, and rp-process elements hinges upon improvements in our knowledge of the properties of exotic nuclei, some of which are far from the valley of beta stability.

The mechanism of core-collapse supernovae is thought to depend upon the transfer of energy from the core to the mantle of the inner regions of a massive star after it becomes unstable to collapse. Neutrinos seem to be the mediators of this energy transfer. In order to understand this coupling and the role of neutrinos in supernova explosions, one needs to master the particulars of the neutrino-matter scattering, production, and absorption rates. Since recently there has been some progress in understanding the associated microphysics, it is fitting to summarize the neutrino-matter cross sections and the production rates of neutrinos in the core-collapse context. To this end, we have assembled here a short précis of many of the relevant processes and physics. This contribution does not attempt to explain the hydrodynamics of supernova explosions, but does try to present the relevant neutrino processes that play a role. For the former, the reader is referred to Burrows, Hayes, and Fryxell (1995) and Burrows (2000).

In §2, we present a physical derivation of stimulated absorption (the Fermionic correlate to stimulated emission) and then in §3 we present the basic cross sections. In §4, we discuss the inelastic neutrino-electron and neutrino-nucleon scattering processes and energy redistribution. This is followed with a discussion of the alternate, more powerful formalism, for determining differential interaction rates in the many-body context and for handling redistribution, namely that of dynamical structure factors. Both the non-interacting and the interacting (in the context of a simple nuclear model) cases are explored, as well as collective excitations of the medium. Source terms for electron-positron annihilation (§6), neutrino-anti-neutrino annihilation (§7; Janka 1991), and nucleon-nucleon bremsstrahlung (§8) cap off our review of the major processes of relevance in core-collapse simulations. Nucleon–nucleon bremsstrahlung can compete with pair annihilation as a source for ν_μ, $\bar{\nu}_\mu$, ν_τ, and $\bar{\nu}_\tau$ neutrinos. Clearly, a mastery of these neutrino-matter processes is a prerequisite for progress in supernova theory and it is in that spirit that we have assembled this review.

2. Stimulated Absorption

The concept of stimulated emission for photons is well understood and studied, but the corresponding concept of stimulated *absorption* for neutrinos is not so well appreciated. This may be because its simple origin in Fermi blocking and the Pauli exclusion principle in the context of *net* emission is not often explained. The *net* emission of a neutrino is simply the difference between the emissivity and the absorption of the medium:

$$\mathcal{J}_{net} = \eta_\nu - \kappa_a I_\nu \ . \tag{5.1}$$

All absorption processes involving fermions will be inhibited by Pauli blocking due to final–state occupancy. Hence, η_ν in eq. (5.1) includes a blocking term, $(1-\mathcal{F}_\nu)$(Bruenn 1985). $\mathcal{F}_\nu$ is the invariant distribution function for the neutrino, whether or not it is in chemical equilibrium.

We can derive stimulated absorption using Fermi's Golden rule. For example, the net collision term for the process, $\nu_e n \leftrightarrow e^- p$, is:

$$\begin{aligned} \mathcal{C}_{\nu_e n \leftrightarrow e^- p} &= \int \frac{d^3\vec{p}_{\nu_e}}{(2\pi)^3} \int \frac{d^3\vec{p}_n}{(2\pi)^3} \int \frac{d^3\vec{p}_p}{(2\pi)^3} \int \frac{d^3\vec{p}_e}{(2\pi)^3} \left(\sum_s |\mathcal{M}|^2 \right) \\ &\times \ \Xi(\nu_e n \leftrightarrow e^- p) \ (2\pi)^4 \, \delta^4(\mathbf{p}_{\nu_e} + \mathbf{p}_n - \mathbf{p}_p - \mathbf{p}_e) \ , \end{aligned} \tag{5.2}$$

where $\mathbf{p}$ is a four-vector and

$$\Xi(\nu_e n \leftrightarrow e^- p) = \mathcal{F}_{\nu_e}\mathcal{F}_n(1-\mathcal{F}_e)(1-\mathcal{F}_p) - \mathcal{F}_e\mathcal{F}_p(1-\mathcal{F}_n)(1-\mathcal{F}_{\nu_e}) \ . \tag{5.3}$$

The final–state blocking terms in eq. (5.3) are manifest, in particular that for the ν_e neutrino. Algebraic manipulations convert $\Xi(\nu_e n \leftrightarrow e^- p)$ in eq. (5.3) into:

$$\begin{aligned} \Xi(\nu_e n \leftrightarrow e^- p) &= \mathcal{F}_n(1-\mathcal{F}_e)(1-\mathcal{F}_p) \left[\frac{\mathcal{F}^{\rm eq}_{\nu_e}}{1-\mathcal{F}^{\rm eq}_{\nu_e}} (1-\mathcal{F}_{\nu_e}) - \mathcal{F}_{\nu_e} \right] \\ &= \frac{\mathcal{F}_n(1-\mathcal{F}_e)(1-\mathcal{F}_p)}{1-\mathcal{F}'_{\nu_e}} \left[\mathcal{F}^{\rm eq}_{\nu_e} - \mathcal{F}_{\nu_e} \right] , \end{aligned} \tag{5.4}$$

where

$$\mathcal{F}^{\rm eq}_{\nu_e} = \left[e^{(\varepsilon_{\nu_e} - (\mu_e - \hat{\mu}))\beta} + 1 \right]^{-1} \tag{5.5}$$

is an equilibrium distribution function for the ν_e neutrino and it has been assumed that only the electron, proton, and neutron are in thermal equilibrium. $\hat{\mu}$ is the difference between the neutron and the proton chemical potentials. Note that in $\mathcal{F}^{\rm eq}_{\nu_e}$ there is no explicit reference to a neutrino chemical potential, though of course in beta equilibrium it is equal to $\mu_e - \hat{\mu}$. There is no need to construct or refer to a neutrino chemical potential in neutrino transfer.

We see that eq. (5.4) naturally leads to:

$$\mathcal{J}_{net} = \frac{\kappa_a}{1 - \mathcal{F}_\nu^{\rm eq}} (B_\nu - I_\nu) = \kappa_a^*(B_\nu - I_\nu) \ . \tag{5.6}$$

Of course, B_ν is the black body function for neutrinos. This expression emphasizes the fact that $C_{\nu_e n \leftrightarrow e^- p}$ and $\mathcal{J}_{net}$ are the same entity. If neutrinos were bosons, we would have found a $(1 + \mathcal{F}_\nu^{\rm eq})$ in the denominator, but the form of eq. (5.6) in which I_ν is manifestly driven to B_ν, the equilibrium intensity, would have been retained. From eqs. (5.4) and (5.6), we see that the stimulated absorption correction to κ_a is $1/(1 - \mathcal{F}_\nu^{\rm eq})$. By writing the collision term in the form of eq. (5.6), with κ_a corrected for stimulated absorption, we have a net source term that clearly drives I_ν to equilibrium. The timescale is $1/c\kappa_a^*$. Though the derivation of the stimulated absorption correction we have provided here is for the $\nu_e n \leftrightarrow e^- p$ process, this correction is quite general and applies to all neutrino absorption opacities.

Kirchhoff's Law, expressing detailed balance, is:

$$\kappa_a = \eta_\nu / B_\nu \text{ or } \kappa_a^* = \eta_\nu' / B_\nu \, , \tag{5.7}$$

where η_ν' is not corrected for final–state neutrino blocking. Furthermore, the net emissivity can be written as the sum of its *spontaneous* and *induced* components:

$$\eta_\nu = \kappa_a \left[\frac{B_\nu}{1 \pm \mathcal{F}_\nu^{\rm eq}} + \left(1 - \frac{1}{1 \pm \mathcal{F}_\nu^{\rm eq}} \right) I_\nu \right] \, , \tag{5.8}$$

where $+$ or $-$ is used for bosons or fermions, respectively. Eq. (5.7) can be used to convert the absorption cross sections described in §3 into source terms.

3. Neutrino Cross Sections

Neutrino–matter cross sections, both for scattering and for absorption, play the central role in neutrino transport. The major processes are the super–allowed charged–current absorptions of ν_e and $\bar{\nu}_e$ neutrinos on free nucleons, neutral–current scattering off of free nucleons, alpha particles, and nuclei (Freedman 1974), neutrino–electron/positron scattering, neutrino–nucleus absorption, neutrino–neutrino scattering, neutrino–antineutrino absorption, and the inverses of various neutrino production processes such as nucleon–nucleon bremsstrahlung and the modified URCA process ($\nu_e + n + n \rightarrow e^- + p + n$). Compared with photon–matter interactions, neutrino– matter interactions are relatively simple functions of incident neutrino energy. Resonances play little or no role and continuum processes dominate. Nice summaries of the various neutrino cross sections of relevance in supernova theory are given in Tubbs & Schramm (1975) and in Bruenn (1985). In particular, Bruenn (1985) discusses in detail neutrino–electron scattering and neutrino–antineutrino processes using the full energy redistribution formalism. He also provides a serviceable approximation to the neutrino–nucleus absorption cross section (Fuller 1980; Fuller, Fowler, & Newman 1982; Aufderheide et al. 1994). For a neutrino energy of $\sim$10 MeV the ratio of the charged–current cross section to the ν_e–electron scattering cross section is $\sim$100. However, neutrino–electron scattering does play a role, along with neutrino–nucleon scattering and nucleon–nucleon bremsstrahlung, in the energy equilibration of emergent ν_μ neutrinos (Thompson, Burrows, & Horvath 2000).

Below, we list and discuss many of the absorption and elastic scattering cross sections one needs in detailed supernova calculations. In §4 and §5, we provide some straightforward formulae that can be used to properly handle inelastic scattering. The set of these processes comprises the essential microphysical package for the simulation of neutrino atmospheres and core–collapse supernovae.

$\nu_e + n \rightarrow e^- + p$:

The cross section per baryon for ν_e neutrino absorption on free neutrons is larger than that for any other process. Given the large abundance of free neutrons in protoneutron star atmospheres, this process is central to ν_e neutrino transport. A convenient reference neutrino cross section is σ_o, given by

$$\sigma_o = \frac{4G^2(m_ec^2)^2}{\pi(\hbar c)^4} \simeq 1.705 \times 10^{-44}\, cm^2 \,. \tag{5.9}$$

The total $\nu_e - n$ absorption cross section is then given by

$$\sigma^a_{\nu_e n} = \sigma_o \left(\frac{1+3g_A^2}{4}\right) \left(\frac{\varepsilon_{\nu_e} + \Delta_{np}}{m_ec^2}\right)^2 \left[1 - \left(\frac{m_ec^2}{\varepsilon_{\nu_e} + \Delta_{np}}\right)^2\right]^{1/2} W_M \,, \tag{5.10}$$

where g_A is the axial–vector coupling constant (~ -1.26), $\Delta_{np} = m_nc^2 - m_pc^2 = 1.29332$ MeV, and for a collision in which the electron gets all of the kinetic energy $\epsilon_{e^-} = \varepsilon_{\nu_e} + \Delta_{np}$. W_M is the correction for weak magnetism and recoil (Vogel 1984) and is approximately equal to $(1 + 1.1\varepsilon_{\nu_e}/m_nc^2)$. At $\varepsilon_{\nu_e} = 20$ MeV, this correction is only $\sim$ 2.5%. We include it here for symmetry's sake, since the corresponding correction ($W_{\bar{M}}$) for $\bar{\nu}_e$ neutrino absorption on protons is $(1-7.1\varepsilon_{\bar{\nu}_e}/m_nc^2)$, which at 20 MeV is a large -15%. To calculate κ^*_a, $\sigma^a_{\nu_e n}$ must be multiplied by the stimulated absorption correction, $1/(1-\mathcal{F}'_{\nu_e})$, and final–state blocking by the electrons and the protons à la eq. (5.4) must be included.

$\bar{\nu}_e + p \rightarrow e^+ + n$:

The total $\bar{\nu}_e - p$ absorption cross section is given by

$$\sigma^a_{\bar{\nu}_e p} = \sigma_o \left(\frac{1+3g_A^2}{4}\right)\left(\frac{\epsilon_{\bar{\nu}_e} - \Delta_{np}}{m_e c^2}\right)^2 \left[1 - \left(\frac{m_e c^2}{\epsilon_{\bar{\nu}_e} - \Delta_{np}}\right)^2\right]^{1/2} W_{\bar{M}}\,, \tag{5.11}$$

where $\epsilon_{e^+} = \epsilon_{\bar{\nu}_e} - \Delta_{np}$ and $W_{\bar{M}}$ is the weak magnetism/recoil correction given in §3. Note that $W_{\bar{M}}$ is as large as many other corrections and should not be ignored. To calculate κ^*_a, $\sigma^a_{\bar{\nu}_e p}$ must also be corrected for stimulated absorption and final–state blocking. However, the sign of $\mu_e - \hat{\mu}$ in the stimulated absorption correction for $\bar{\nu}_e$ neutrinos is flipped, as is the sign of μ_e in the positron blocking term. Hence, as a consequence of the severe electron lepton asymmetry in core–collapse supernovae, both coefficients are very close to one. Note that the $\bar{\nu}_e + p \rightarrow e^+ + n$ process dominates the supernova neutrino signal in proton–rich underground neutrino telescopes on Earth, such as Super Kamiokande, LVD, and MACRO, a fact that emphasizes the interesting complementarities between emission at the supernova and detection in Čerenkov and scintillation facilities.

$\nu_e A \leftrightarrow A'e^-$

From Bruenn (1985) the total $\nu_e - A$ absorption cross section, is approximated by

$$\sigma^a_A = \frac{\sigma_o}{14}\, g_A^2\, N_p(Z)\, N_n(N) \left(\frac{\varepsilon_\nu + Q'}{m_e c^2}\right)^2 \left[1 - \left(\frac{m_e c^2}{\varepsilon_\nu + Q'}\right)^2\right]^{1/2} W_{block}\,, \tag{5.12}$$

where $W_{block} = (1 - f_{e^-})\, e^{(\mu_n - \mu_p - Q')\beta}$, $Q' = M_{A'} - M_A + \Delta \sim \mu_n - \mu_p + \Delta$, Δ is the energy of the neutron $1f_{5/2}$ state above the ground state and is taken to be 3 MeV (Fuller 1982), and the quantities $N_p(Z)$ and $N_n(N)$ are approximated by: $N_p(Z) = 0$, $Z - 20$, and 8 for $Z < 20$, $20 < Z < 28$, and $Z > 28$, respectively. $N_n(N) = 6$, $40 - N$, and 0 for $N < 34$, $43 < N < 40$, and $N > 40$, respectively. The opacity, corrected for stimulated absorption, is then

$$\kappa^*_a = X_H\, \rho N_A \sigma^a_A (1 - \mathcal{F}^{\rm eq}_{\nu_e})^{-1}. \tag{5.13}$$

Since $N_n(N) = 0$ for $N > 40$, this absorption and emission process plays a role only during the very early phase of collapse. Typically at densities near $\rho \sim 10^{12}$ g cm^{-3} $\kappa^*_a \rightarrow 0$.

$\nu_i + p \rightarrow \nu_i + p$:

The total $\nu_i - p$ scattering cross section for all neutrino species is:

$$\sigma_p = \frac{\sigma_o}{4}\left(\frac{\varepsilon_\nu}{m_e c^2}\right)^2 \left(4\sin^4\theta_W - 2\sin^2\theta_W + \frac{(1+3g_A^2)}{4}\right) , \tag{5.14}$$

where θ_W is the Weinberg angle and $\sin^2\theta_W \simeq 0.23$. In terms of $C'_V = 1/2 + 2\sin^2\theta_W$ and $C'_A = 1/2$, eq. (5.14) becomes (Schinder 1990):

$$\sigma_p = \frac{\sigma_o}{4}\left(\frac{\varepsilon_\nu}{m_e c^2}\right)^2 \left[(C'_V - 1)^2 + 3g_A^2(C'_A - 1)^2\right] . \tag{5.15}$$

The differential cross section is:

$$\frac{d\sigma_p}{d\Omega} = \frac{\sigma_p}{4\pi}(1 + \delta_p \mu) , \tag{5.16}$$

where

$$\delta_p = \frac{(C'_V - 1)^2 - g_A^2(C'_A - 1)^2}{(C'_V - 1)^2 + 3g_A^2(C'_A - 1)^2} . \tag{5.17}$$

Note that δ_p, and δ_n below, are negative ($\delta_p \sim -0.2$ and $\delta_n \sim -0.1$) and, hence, that these processes are backward–peaked.

The transport (or momentum-transfer) cross section is simply

$$\sigma_p^{tr} = \frac{\sigma_o}{6}\left(\frac{\varepsilon_\nu}{m_e c^2}\right)^2 \left[(C'_V - 1)^2 + 5g_A^2(C'_A - 1)^2\right] . \tag{5.18}$$

where

$$\sigma_i^{tr} = \int \frac{d\sigma_i}{d\Omega}(1-\mu)\, d\Omega = \sigma_i \left(1 - \frac{1}{3}\delta_i\right) . \tag{5.19}$$

$\nu_i + n \rightarrow \nu_i + n$:

The total $\nu_i - n$ scattering cross section is:

$$\sigma_n = \frac{\sigma_o}{4}\left(\frac{\varepsilon_\nu}{m_e c^2}\right)^2 \left(\frac{1+3g_A^2}{4}\right) . \tag{5.20}$$

The corresponding differential cross section is:

$$\frac{d\sigma_n}{d\Omega} = \frac{\sigma_n}{4\pi}(1 + \delta_n \mu) , \tag{5.21}$$

where

$$\delta_n = \frac{1 - g_A^2}{1 + 3g_A^2} . \tag{5.22}$$

The transport cross section is

$$\sigma_n^{tr} = \frac{\sigma_o}{4}\left(\frac{\varepsilon_\nu}{m_e c^2}\right)^2\left(\frac{1+5g_A^2}{6}\right) . \tag{5.23}$$

The fact that δ_p and δ_n are negative and, as a consequence, that σ_i^{tr} is greater than σ_i increases the neutrino–matter energy coupling rate for a given neutrino flux in the semi–transparent region.

Horowitz (2002) has recently derived expressions that include a weak magnetism/recoil correction analogous to those previously discussed for the charged-current absorption rates $\nu_e n \leftrightarrow pe^-$ and $\bar{\nu}_e n \leftrightarrow ne^+$. We take the following form for the weak magnetism/recoil correction, a fit to the actual correction factor for the transport cross sections:

$$\sigma_{n,p}^{tr} \rightarrow \sigma_{n,p}^{tr}(1 + C_{W_M}\varepsilon_\nu/m_{n,p}), \tag{5.24}$$

where, for neutrino-neutron scattering $C_{W_M} \simeq -0.766$, for neutrino-proton scattering $C_{W_M} \simeq -1.524$, for anti-neutrino-neutron scattering $C_{W_M} \simeq -7.3656$, and for anti-neutrino-proton scattering $C_{W_M} \simeq -6.874$.

In fact, neutrino-nucleon scattering is slightly inelastic and when this is germane, as with mu and tau neutrinos, the more general formalisms of §5.3 and §5 are necessary.

$\nu_i + A \rightarrow \nu_i + A$:

In the post–bounce phase, nuclei exist in the unshocked region exterior to the shock. At the high entropies in shocked protoneutron star atmospheres there are very few nuclei. There are alpha particles, but their fractional abundances are generally low, growing to interesting levels due to reassociation of free nucleons just interior to the shock only at late times. However, nuclei predominate on infall and neutrino-nucleus scattering (Freedman 1974) is the most important process during the lepton trapping phase.

The differential $\nu_i - A$ neutral–current scattering cross section may be expressed as:

$$\frac{d\sigma_A}{d\Omega} = \frac{\sigma_o}{64\pi}\left(\frac{\varepsilon_\nu}{m_e c^2}\right)^2 A^2 \left\{\mathcal{W} C_{FF} + C_{LOS}\right\}^2 \langle S_{ion}\rangle (1+\mu) , \tag{5.25}$$

where

$$\mathcal{W} = 1 - \frac{2Z}{A}(1 - 2\sin^2\theta_W) , \tag{5.26}$$

Z is the atomic number, A is the atomic weight, and $\langle S_{ion}\rangle$ is the ion–ion correlation function, determined mostly by the Coulomb interaction between the nuclei during infall. $\langle S_{ion}\rangle$, in eq. (5.25) was investigated by Horowitz

(1997) who approximated it with the expansion

$$\langle S_{ion}(\epsilon)\rangle = \left[1 + exp\left(-\sum_{i=0}^{6} \beta_i(\Gamma)\epsilon^i\right)\right]^{-1} , \tag{5.27}$$

where

$$\Gamma = \frac{(Ze)^2}{a}\frac{1}{kT} \quad , \quad \epsilon_i = \frac{\varepsilon_{\nu_i}}{\hbar ca} \quad , \quad a = \left(\frac{3}{4\pi n_{ion}}\right)^{1/3} , \tag{5.28}$$

a is the interparticle spacing, n_{ion} is the number density of ions, Γ is the ratio of Coulomb potential between ions to the thermal energy in the medium, and β_i are specified functions of Γ for each neutrino species.

Leinson et al. (1988) have investigated the electron polarization correction, $\mathcal{C}_{LOS}$, and find that

$$\mathcal{C}_{LOS} = \frac{Z}{A}\left(\frac{1 + 4\sin^2\theta_W}{1 + (kr_D)^2}\right) , \tag{5.29}$$

where the Debye radius is

$$r_D = \sqrt{\frac{\pi\hbar^2 c}{4\alpha p_F E_F}} , \tag{5.30}$$

$k^2 = |\mathbf{p} - \mathbf{p}'|^2 = 2(\varepsilon_\nu/c)^2(1 - \mu)$, p_F and E_F are the electron Fermi momentum and energy, and α is the fine–structure constant ($\simeq 137^{-1}$). Note that $r_D \sim 10\hbar/p_F$ in the ultra–relativistic limit ($p_F >> m_e c$). The $\mathcal{C}_{LOS}$ term is important only for low neutrino energies, generally below $\sim$ 5 MeV.

Following Tubbs & Schramm (1975) and Burrows, Mazurek, & Lattimer (1981), the form factor term, $\mathcal{C}_{FF}$, in eq. (5.25) can be approximated by:

$$\mathcal{C}_{FF} = e^{-y(1-\mu)/2} , \tag{5.31}$$

where

$$y = \frac{2}{3}\varepsilon_\nu^2\langle r^2\rangle/(\hbar c)^2 \simeq \left(\frac{\varepsilon_\nu}{56\,\text{MeV}}\right)^2\left(\frac{A}{100}\right)^{2/3} ,$$

and $\langle r^2\rangle^{1/2}$ is the *rms* radius of the nucleus. $\mathcal{C}_{FF}$ differs from 1 for large A and ε_ν, when the de Broglie wavelength of the neutrino is smaller than the nuclear radius.

When $\langle S_{ion}\rangle = \mathcal{C}_{FF} = \mathcal{C}_{LOS} + 1 = 1$, we have simple coherent Freedman scattering reedman scattering. The physics of the polarization, ion–ion correlation, and form factor corrections to coherent scattering is interesting in its own right, but has little effect on supernovae (Bruenn & Mezzacappa 1997). The total and transport scattering cross sections for $\nu_i - \alpha$ scattering ($Z = 2; A = 4$) are simply

$$\sigma_\alpha = \frac{3}{2}\sigma_\alpha^{tr} = 4\,\sigma_o\left(\frac{\varepsilon_\nu}{m_e c^2}\right)^2 \sin^4\theta_W . \tag{5.32}$$

4. Inelastic Neutrino Scattering

Many authors have studied inelastic neutrino-electron scattering as an important energy redistribution process which helps to thermalize neutrinos and increase their energetic coupling to matter in supernova explosions (Bruenn 1985; Mezzacappa & Bruenn 1993abc). Comparatively little attention has been paid to inelastic neutrino-nucleon scattering. Thompson, Burrows, & Horvath (2000) and Raffelt (2001) showed that, at least for mu and tau neutrinos, this process cannot be ignored. Here, we review the Legendre expansion formalism for approximating the angular dependence of the scattering kernel, detail our own implementation of scattering terms in the Boltzmann equation, and include a discussion of neutrino-nucleon energy redistribution. In §5, we present an alternate approach involving dynamical structure factors that is more easily generalized to include many-body effects.

The general collision integral for inelastic scattering may be written as

$$\mathcal{L}_\nu^{\rm scatt}[f_\nu] = (1-f_\nu)\int \frac{d^3p'_\nu}{c(2\pi\hbar c)^3} f'_\nu\, R^{\rm in}(\varepsilon_\nu,\varepsilon'_\nu,\cos\theta) \tag{5.33}$$

$$-f_\nu\int \frac{d^3p'_\nu}{c(2\pi\hbar c)^3}(1-f'_\nu)\, R^{\rm out}(\varepsilon_\nu,\varepsilon'_\nu,\cos\theta) \tag{5.34}$$

$$= \tilde{\eta}_\nu^{\rm scatt} - \tilde{\chi}_\nu^{\rm scatt} f_\nu \tag{5.35}$$

where $\cos\theta$ is the cosine of the scattering angle, ε_ν is the incident neutrino energy, and ε'_ν is the scattered neutrino energy. Although we suppress it here, the incident and scattered neutrino phase space distribution functions (f_ν and f'_ν, respectively) have the following dependencies: $f_\nu = f_\nu(r,t,\mu,\varepsilon_\nu)$ and $f'_\nu = f'_\nu(r,t,\mu',\varepsilon'_\nu)$. μ and μ' are the cosines of the angular coordinate of the zenith angle in spherical symmetry and are related to $\cos\theta$ through

$$\cos\theta = \mu\mu' + [(1-\mu^2)(1-\mu'^2)]^{1/2}\cos(\phi-\phi'). \tag{5.36}$$

The only difference between f_ν and $\mathcal{F}_\nu$ in §2 is that here f_ν has explicit μ and ε_ν dependencies. $R^{\rm in}$ is the scattering kernel for scattering into the bin (ε_ν, μ) from any bin (ε'_ν, μ') and $R^{\rm out}$ is the scattering kernel for scattering out of the bin (ε_ν, μ) to any bin (ε'_ν, μ'). The kernels are Green functions that connect points in energy and momentum space. One may also write $R(\varepsilon_\nu,\varepsilon'_\nu,\cos\theta)$ as $R(q,\omega)$, where $\omega(=\varepsilon_\nu-\varepsilon'_\nu)$ is the energy transfer and $q(=[\varepsilon_\nu^2+\varepsilon_\nu'^2-2\varepsilon_\nu\varepsilon'_\nu\cos\theta]^{1/2})$ is the momentum transfer, so that the kernel explicitly reflects these dependencies (§5).

An important simplification comes from detailed balance, a consequence of the fact that these scattering rates must drive the distribution to equilibrium.

One obtains: $R^{\rm in} = e^{-\beta\omega} R^{\rm out}$, where $\beta = 1/T$. Therefore, we need deal only with $R^{\rm out}$. The scattering kernels for inelastic neutrino-nucleon and neutrino-electron scattering depend in a complicated fashion on scattering angle. For this reason, one generally approximates the angular dependence of the scattering kernel with a truncated Legendre series (Bruenn 1985). We take

$$R^{\rm out}(\varepsilon_\nu, \varepsilon'_\nu, \cos\theta) = \sum_{l=0}^{\infty} \frac{2l+1}{2} \Phi(\varepsilon_\nu, \varepsilon'_\nu, \cos\theta) P_l(\cos\theta), \tag{5.37}$$

where

$$\Phi_l(\varepsilon_\nu, \varepsilon'_\nu) = \int_{-1}^{+1} d(\cos\theta)\, R^{\rm out}(\varepsilon_\nu, \varepsilon'_\nu, \cos\theta) P_l(\cos\theta). \tag{5.38}$$

In practice, one expands only to first order so that

$$R^{\rm out}(\varepsilon_\nu, \varepsilon'_\nu, \cos\theta) \sim \frac{1}{2}\Phi_0(\varepsilon_\nu, \varepsilon'_\nu) + \frac{3}{2}\Phi_1(\varepsilon_\nu, \varepsilon'_\nu)\cos\theta. \tag{5.39}$$

Substituting into the first term on the right-hand-side of eq. (5.34) (the source) gives

$$\tilde{\eta}_\nu^{\rm scatt} = (1 - f_\nu) \int_0^\infty \frac{d\varepsilon'_\nu \varepsilon'^2_\nu}{c(2\pi\hbar c)^3} e^{-\beta\omega} \int_{-1}^{+1} d\mu' f'_\nu \int_0^{2\pi} d\phi' \left[\frac{1}{2}\Phi_0 + \frac{3}{2}\Phi_1 \cos\theta\right] \tag{5.40}$$

Substituting for $\cos\theta$ using eq. (5.36) and using the definitions

$$\tilde{J}_\nu = \frac{1}{2}\int_{-1}^{+1} d\mu f_\nu \tag{5.41}$$

and

$$\tilde{H}_\nu = \frac{1}{2}\int_{-1}^{+1} d\mu \mu f_\nu \tag{5.42}$$

we have that

$$\tilde{\eta}_\nu^{\rm scatt} = (1 - f_\nu) \frac{4\pi}{c(2\pi\hbar c)^3} \int_0^\infty d\varepsilon'_\nu \varepsilon'^2_\nu e^{-\beta\omega} \left[\frac{1}{2}\Phi_0 \tilde{J}'_\nu + \frac{3}{2}\Phi_1 \mu \tilde{H}'_\nu\right]. \tag{5.43}$$

Integrating over μ to get the source for the zeroth moment of the transport equation,

$$\frac{1}{2}\int_{-1}^{+1} d\mu\, \tilde{\eta}_\nu^{\rm scatt} = \frac{4\pi}{c(2\pi\hbar c)^3} \int_0^\infty d\varepsilon'_\nu \varepsilon'^2_\nu e^{-\beta\omega} \left[\frac{1}{2}\Phi_0 \tilde{J}'_\nu (1 - \tilde{J}_\nu) - \frac{3}{2}\Phi_1 \tilde{H}_\nu \tilde{H}'_\nu\right]. \tag{5.44}$$

Similarly, we may write the sink term of the Boltzmann equation collision term (second term in eq. 5.34), employing the Legendre expansion

$$\tilde{\chi}_\nu^{\rm scatt} = \frac{4\pi}{c(2\pi\hbar c)^3} \int_0^\infty d\varepsilon'_\nu \varepsilon'^2_\nu \left[\frac{1}{2}\Phi_0 (1 - \tilde{J}'_\nu) - \frac{3}{2}\Phi_1 \mu \tilde{H}'_\nu\right]. \tag{5.45}$$

In eqs. (5.57-5.59), the ξ_n's are differences between polylogarithmic integrals; $\xi_n = \mathrm{Li}_n(-\alpha_1) - \mathrm{Li}_n(-\alpha_2)$, where

$$\mathrm{Li}_n(y) = \int_0^y \frac{\mathrm{Li}_{n-1}(x)}{x}\, dx, \tag{5.61}$$

and $\mathrm{Li}_1(x) = \ln(1-x)$. The arguments necessary for computing the integrals are $\alpha_1 = \exp[\beta(e_- + \omega) - \eta_e]$ and $\alpha_2 = \exp(\beta e_- - \eta_e)$. Tables for computation of $\mathrm{Li}_n(y)$ and the I_ns can be provided by Sanjay Reddy.

Figure (5.1) shows the full scattering kernel for $\varepsilon_\nu = 20$ MeV and $\varepsilon'_\nu = 2$, 10, and 16 MeV as a function of the cosine of the scattering angle, $\cos\theta$. Note that although the absolute value of the energy transfer ($|\varepsilon_\nu - \varepsilon'_\nu|$) is the same for both $\varepsilon'_\nu = 16$ MeV and $\varepsilon'_\nu = 24$ MeV, the absolute value of $R^{\rm out}(20, 16, \cos\theta)$ is more than twice that of $R^{\rm out}(20, 24, \cos\theta)$, reflecting the fact that at this temperature the incoming neutrino is more likely to downscatter than upscatter. Figure (5.2) shows the scattering kernel for the same conditions as Fig. (5.1), but also includes both the first-order (short dashed lines) and second-order (long dashed lines) approximations to $R^{\rm out}$. We generally employ the former. The latter is included to illustrate the improvement in including higher-order terms. In fact, the actual degree of expansion necessary to capture accurately the physics can only be ascertained by running full transport calculations. We have run dynamical simulations with only the zeroth-order and first-order terms in the Legendre expansion and find little or no difference between the emergent spectra and detailed thermodynamical evolution in the models we have studied. Smit (1998) and Smit & Cernohorsky (1995) have explored the importance of including the second-order term ($\propto \cos^2\theta$, shown here) and find it negligible. The scattering-angle-averaged kernel, also the zeroth-order term in the Legendre series for $R^{\rm out}$, is shown in Fig. (5.3) for ν_e−electron scattering for a matter temperature (T) of 6 MeV and with an electron degeneracy factor $\eta_e = \mu_e/T = 20$ as a function of ε'_ν for various incoming neutrino energies, ε_νs.

In a full time-dependent simulation it is numerically costly to compute the Legendre moments of the scattering kernel ($\Phi_0(\varepsilon_\nu, \varepsilon'_\nu)$ and $\Phi_1(\varepsilon_\nu, \varepsilon'_\nu)$) via eq. (5.38) at each point in the computational domain. For this reason we tabulate $\Phi_0(\varepsilon_\nu, \varepsilon'_\nu)$ and $\Phi_1(\varepsilon_\nu, \varepsilon'_\nu)$ for each ε_ν and ε'_ν pair on a grid in temperature and η_e. Because the vector and axial-vector coupling constants for neutrino-electron scattering and the neutrino energy grouping differ between ν_e, $\bar{\nu}_e$, and ν_μ neutrinos, we construct separate tables for each species. Gauss-Legendre quadratures (16-point) are used to evaluate the angular integrals over $\cos\theta$ for $l = 0$ and $l = 1$ in eq. (5.38). At a given temperature/density/composition point the equation of state returns η_e and we then perform a six-point bivariant interpolation in T-η_e space, for the given ε_ν-ε'_ν combination, to obtain $\Phi_0(\varepsilon_\nu, \varepsilon'_\nu)$ and $\Phi_1(\varepsilon_\nu, \varepsilon'_\nu)$. The integrals over ε'_ν for each energy, which yield $\tilde{\eta}_\nu^{\rm scatt}$ and

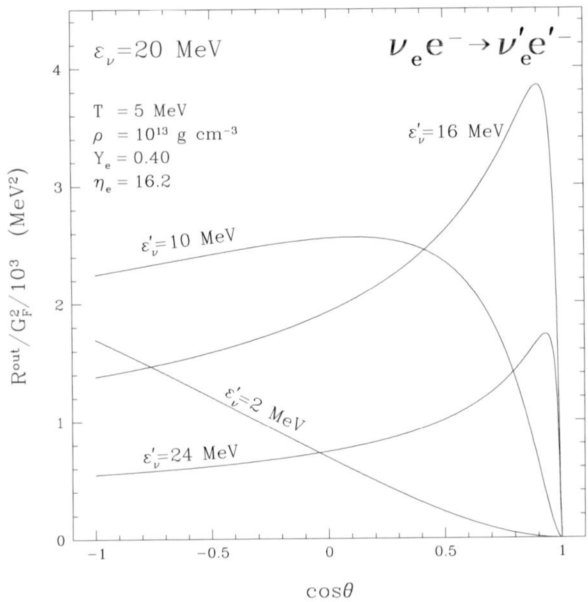

Figure 5.1. The scattering kernel $R^{\mathrm{out}}(\varepsilon_\nu, \varepsilon'_\nu, \cos\theta)$ for ν_e—electron scattering as a function of $\cos\theta$ for $\varepsilon_\nu = 20$ MeV and $\varepsilon'_\nu = 2$, 10, 16, and 24 MeV, at a representative thermodynamic point ($T = 5$ MeV, $\rho = 10^{13}$ g cm^{-3}, $Y_e = 0.4$).

$\tilde{\chi}_\nu^{\mathrm{scatt}}$, are then computed using simple trapezoidal rule integration. In practice, we use 40 energy groups (n_f), 30 temperature points (N_T), and 30 η_e points (N_η). The tables are then $l \times n_f \times n'_f \times N_T \times N_\eta$ in size, with $l = 2$ (Φ_0 and Φ_1), or approximately 50 Megabytes. The source and sink at each energy ($\tilde{\eta}_\nu^{\mathrm{scatt}}$ and $\tilde{\chi}_\nu^{\mathrm{scatt}}$) are then included *explicitly* in a manner analogous to any of the emission or absorption processes. Using this method, our calculations including neutrino-electron scattering, are just 10-15% slower than our calculations ignoring this important equilibration process (Thompson, Burrows, & Pinto 2002).

This calculation of $\tilde{\eta}_\nu^{\mathrm{scatt}}$ and $\tilde{\chi}_\nu^{\mathrm{scatt}}$ uses only values of the neutrino energy density and flux from the previous timestep and, hence, we introduce an explicit timescale into the energy and electron fraction updates returned by our transport algorithm. For this reason, when the scattering rate is large we may encounter a numerical instability. Because the largest scattering rates are encountered when the neutrino phase-space distribution function is in local thermodynamical equilibrium and the scattering off electrons is unimportant, we simply divide the source and sink by a large factor (typically 100 above $\rho = 10^{14}$ g cm^{-3}), thus

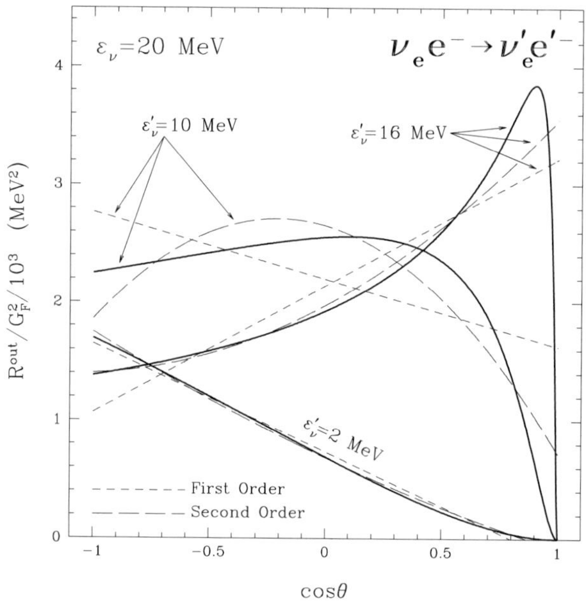

Figure 5.2. For the same thermodynamic point as used for Fig. (5.1), the scattering kernel ($R^{\rm out}$, thick solid lines) for ν_e−electron scattering as a function of $\cos\theta$, for $\varepsilon_\nu = 20$ MeV and $\varepsilon'_\nu = 2$, 10, and 16 MeV. Short dashed lines show the first-order Legendre series expansion approximation to $R^{\rm out}$, which is linear in $\cos\theta$; $R^{\rm out} \sim (1/2)\Phi_0 + (3/2)\Phi_1 \cos\theta$. The long dashed line shows the improvement in going to second order in $\cos\theta$ by taking $R^{\rm out} \sim (1/2)\Phi_0 + (3/2)\Phi_1 \cos\theta + (5/2)\Phi_2(1/2)(3\cos^2\theta - 1)$.

circumventing the problem of introducing a short and explicit timescale. Again, because $f_\nu = f_\nu^{\rm eq}$ at these high densities, this approximation is acceptable.

Neutrino-Nucleon Scattering

The kernel for inelastic neutrino-nucleon scattering can be related to the non-relativistic structure function employed in the thermalization studies of Thompson, Burrows, & Horvath (2000):

$$R^{\rm out}(\varepsilon_\nu, \varepsilon'_\nu, \cos\theta) = G^2 S(q,\omega)[(1+\cos\theta)V^2 + (3-\cos\theta)A^2], \quad (5.62)$$

where $S(q,\omega)$ is given in terms of the imaginary part of the polarization function, analogous to neutrino-electron scattering (§4).

The neutrino-nucleon scattering kernels, while larger in absolute magnitude than the corresponding neutrino-electron scattering kernels are, at most points in energy space and thermodynamic space, much less broad. In fact, in thermodynamic regimes most relevant for the formation of the various species' spectra, the kernel is quite sharply peaked in energy. That is, for a given ε_ν the distribu-

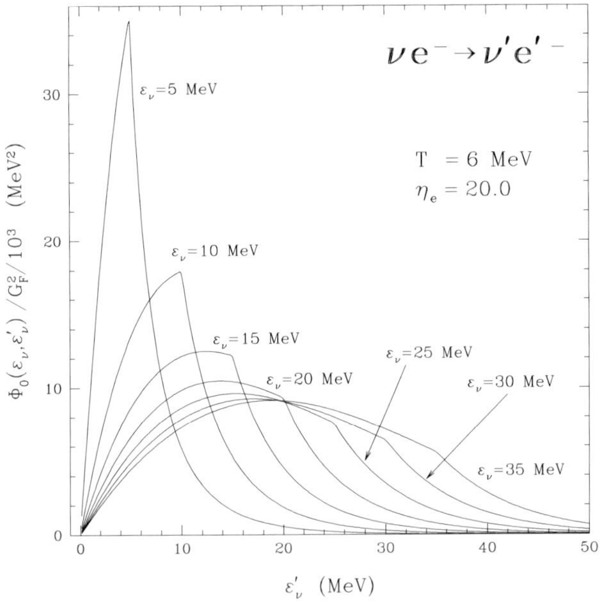

Figure 5.3. The $l = 0$ term in the Legendre expansion of the ν_e−electron scattering kernel, $\Phi_0(\varepsilon_\nu, \varepsilon'_\nu)$ (eq. 5.38), for $T = 6$ MeV and $\eta_e = 20$ as a function of ε'_ν for $\varepsilon_\nu = 5$, 10, 15, 20, 25, and 35 MeV. Note that for any ε_ν, the neutrino is predominantly downscattered. The magnitude of $\Phi_0(\varepsilon_\nu, \varepsilon'_\nu)$ and sign of $\langle\omega\rangle$ are to be compared with those in Fig. (5.5).

tion of ε'_ν's is tightly centered on ε_ν because the ratio of the neutrino energy to the nucleon mass is small. $R^{\rm out}$ as a function of $\cos\theta$, for neutrino-neutron scattering, is shown in Fig. (5.4) at a representative thermodynamic point. This plot is analogous to Fig. (5.1) for neutrino-electron scattering. By comparing the $\varepsilon'_\nu = 19$ MeV line with that for $\varepsilon'_\nu = 21$ MeV, one sees that the former is larger and, hence, downscattering is preferred. In addition, the overall magnitude is much larger than in the neutrino-electron scattering case. Figure (5.5) shows Φ_0 (eq. 5.38) for neutrino-neutron scattering as a function of ε'_ν for several ε_ν's at the same thermodynamic point used in Fig. (5.3), the corresponding figure for neutrino-electron scattering. Note that while downscatting is strong for the $\varepsilon_\nu = 35$ MeV kernel, there is almost equal upscattering at $\varepsilon_\nu = 5$ MeV. In order to explore the effect of this process on the emergent spectra in dynamical simulations we must first deal with a technical problem.

In a typical simulation, we employ 40 energy groups for all neutrino species with $1\,\mathrm{MeV} \leq \varepsilon_\nu \leq 320\,\mathrm{MeV}$ for electron-type neutrinos and $1\,\mathrm{MeV} \leq \varepsilon_\nu \leq 100\,\mathrm{MeV}$ for anti-electron and muon neutrinos. The grouping is generally logarithmic for the ν_es and linear for $\bar{\nu}_e$ and ν_μ. Neutrino-electron scattering and neutrino-nucleon scattering are most important as thermalization mechanisms

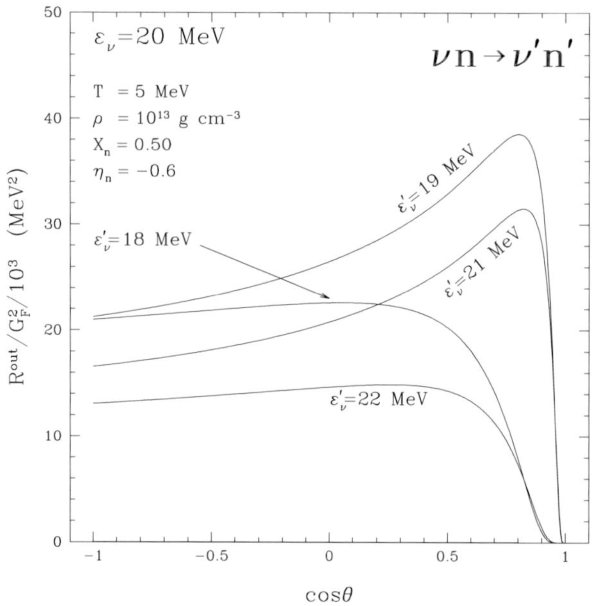

Figure 5.4. The scattering kernel $R^{\rm out}(\varepsilon_\nu, \varepsilon'_\nu, \cos\theta)$ for ν_e−neutron scattering as a function of $\cos\theta$ for $\varepsilon_\nu = 20$ MeV and $\varepsilon'_\nu = 18$, 19, 21, and 22 MeV, at a representative thermodynamic point ($T = 5$ MeV, $\rho = 10^{13}$ g cm^{-3}, $X_n = 0.5$). Note that although the absolute value of the energy transfer ($|\varepsilon_\nu - \varepsilon'_\nu|$) is the same for both $\varepsilon'_\nu = 19$ MeV and $\varepsilon'_\nu = 21$, the absolute value of $R^{\rm out}(20, 19, \cos\theta)$ is greater than that of $R^{\rm out}(20, 21, \cos\theta)$, reflecting the fact that at this temperature the incoming neutrino is more likely to downscatter than upscatter.

at energies below $\sim$ 60 MeV, where the phase space distribution of all neutrino species is largest. One can see clearly from Fig. (5.5) that a trapezoidal rule integration of Φ_0 over ε'_ν as it appears in eq. (5.44) and eq. (5.46) may grossly overestimate $\tilde{\eta}_\nu^{\rm scatt}$ and $\tilde{\chi}_\nu^{\rm scatt}$. In fact, with logarithmic energy grouping one may even calculate upscattering when there is none because the energy groups become larger with increasing energy.

For neutrino-electron scattering, we are able to employ a simple trapezoidal rule and adequately capture the qualities of the kernel. This implies that we use a linear interpolation for $\tilde{J}'_\nu$ and $\tilde{H}'_\nu$ in each energy bin. As Fig. (5.5) shows, however, in order to get an accurate integral over $\Phi_0(\varepsilon_\nu, \varepsilon'_\nu)$, we must do better than simple trapezoidal rule with linear interpolation. For neutrino-nucleon scattering, in order to increase the accuracy of our scheme without compromising computational efficiency, for a given energy grouping we pre-compute a grid of integrals over ε'_ν. We assume that during the dynamical calculation and the computation of $\tilde{\eta}_\nu^{\rm scatt}$ and $\tilde{\chi}_\nu^{\rm scatt}$, both $\tilde{J}'_\nu$ and $\tilde{H}'_\nu$ are proportional to $A\varepsilon'_\nu + B$ over an energy interval $\varepsilon_{\nu,i} \leq \varepsilon'_\nu \leq \varepsilon_{\nu,i+1}$. Given this assumption and

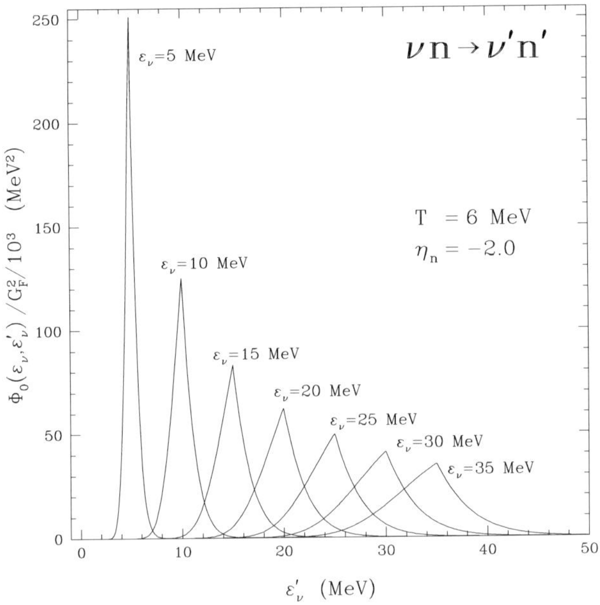

Figure 5.5. The $l = 0$ term in the Legendre expansion of the neutrino-nucleon scattering kernel, $\Phi_0(\varepsilon_\nu, \varepsilon'_\nu)$ (eq. 5.38), for $T = 6$ MeV and $\eta_n = -2$ as a function of ε'_ν for $\varepsilon_\nu = 5$, 10, 15, 20, 25, and 35 MeV. Note that for $\varepsilon_\nu = 5$ MeV the neutrino is predominantly upscattered, while for $\varepsilon_\nu = 35$ MeV the neutrino is predominantly downscattered. The magnitude of $\Phi_0(\varepsilon_\nu, \varepsilon'_\nu)$ and sign of $\langle\omega\rangle$ are to be compared with those in Fig. (5.3).

both $\Phi_0(\varepsilon_\nu, \varepsilon'_\nu)$ and $\Phi_1(\varepsilon_\nu, \varepsilon'_\nu)$ at a given T and $\eta_{n,p}$, we tabulate the following integrals:

$$\int_{\varepsilon_{\nu,\,i}}^{\varepsilon_{\nu,\,i+1}} d\varepsilon'_\nu \varepsilon'^2_\nu \Phi_l(\varepsilon_\nu, \varepsilon'_\nu),$$

$$\int_{\varepsilon_{\nu,\,i}}^{\varepsilon_{\nu,\,i+1}} d\varepsilon'_\nu \varepsilon'^2_\nu \varepsilon'_\nu \Phi_l(\varepsilon_\nu, \varepsilon'_\nu),$$

$$\int_{\varepsilon_{\nu,\,i}}^{\varepsilon_{\nu,\,i+1}} d\varepsilon'_\nu \varepsilon'^2_\nu e^{-\beta\omega} \Phi_l(\varepsilon_\nu, \varepsilon'_\nu),$$

and

$$\int_{\varepsilon_{\nu,\,i}}^{\varepsilon_{\nu,\,i+1}} d\varepsilon'_\nu \varepsilon'^2_\nu \varepsilon'_\nu e^{-\beta\omega} \Phi_l(\varepsilon_\nu, \varepsilon'_\nu),$$

where $l = 0, 1$ and each i is an individual energy group, set up at the beginning of the calculation. The integrals over ε'_ν are computed using 16-point Gauss-Legendre quadrature, each nested with another 16-point Gauss-Legendre quadrature for the integral over $\cos\theta$ necessary for each $\Phi_l(\varepsilon_\nu, \varepsilon'_\nu)$. These integrals are tabulated at 30 temperature and 30 $\eta_{n,p}$ points, analogous to the case for neutrino-electron scattering. For example, the integral

$$\int_0^\infty d\varepsilon'_\nu \varepsilon'^2_\nu \tilde{J}'_\nu \Phi_0(\varepsilon_\nu, \varepsilon'_\nu) = \sum_{i=1}^{n_f} A_i \int_{\varepsilon_{\nu,\,i}}^{\varepsilon_{\nu,\,i+1}} d\varepsilon'_\nu \varepsilon'^2_\nu \varepsilon'_\nu \Phi_0(\varepsilon_\nu, \varepsilon'_\nu) + \sum_{i=1}^{n_f} B_i \int_{\varepsilon_{\nu,\,i}}^{\varepsilon_{\nu,\,i+1}} d\varepsilon'_\nu \varepsilon'^2_\nu \Phi_0(\varepsilon_\nu, \varepsilon'_\nu), \quad (5.63)$$

where $A_i = (\tilde{J}^{i\prime}_\nu - \tilde{J}^{i+1\prime}_\nu)/(\varepsilon^{i\prime}_\nu - \varepsilon^{i+1\prime}_\nu)$ and $B_i = \tilde{J}^{i\prime}_\nu - A_i \varepsilon^{i\prime}_\nu$. The total source and sink are given by integrals analogous to eq. (5.63) with appropriate changes to A_i and B_i, depending on if the term in the source or sink is over $\tilde{J}'_\nu$ or $\tilde{H}'_\nu$. In practice, for a given temperature and density, we calculate the necessary integrals at the four nearest neighbor $T - \eta_{n,p}$ points saved in the table and then interpolate the solution using a four-point bivariant interpolation scheme. In this scheme, the energy integral over the kernel is reproduced extremely well and the primary uncertainty in calculating the total source and sink is due to the linear interpolation of $\tilde{J}'_\nu$ and $\tilde{H}'_\nu$ – the same as in the neutrino-electron scattering case. Also interesting is that because the neutrino-nucleon scattering kernel is so sharply peaked around ε_ν and drops off so quickly with ε'_ν, most terms in the sum in eq. (5.63) are zero. When we fill the table, we note the index i of the lowest and highest $\varepsilon'_{\nu,i}$ intervals that contribute significantly (to a part in 10^4) to the total ε'_ν integral over Φ_l. Typically, only four to five energy groups must be included in the final sum. This decreases both the size of the table and the amount of time needed to calculate $\tilde{\eta}^{\rm scatt}_\nu$ and $\tilde{\chi}^{\rm scatt}_\nu$. Although the vector and axial-vector couplings for neutrino-neutron scattering are independent of neutrino species, we calculate separate tables for ν_e, $\bar{\nu}_e$, and ν_μ so as to allow for arbitrary energy grouping for each species.

The Elastic Limit: With the inelastic formalism in hand, it is instructive to construct the elastic limit. Note that in the limit of zero energy transfer, $S(q,\omega) \rightarrow S(0) = 2\pi n_{n,p}\delta(\omega)$, where $n_{n,p}$ is the neutron or proton number density (see eq. 5.80). The scattering term can then be written as

$$\mathcal{L}^{\rm scatt}_{(0)}[f_\nu] = (1 - f_\nu)\frac{C n_{n,p}}{c(2\pi\hbar c)^3}\int_0^\infty d\varepsilon'_\nu \varepsilon'^2_\nu e^{-\beta\omega}\delta(\omega)\int_{-1}^{+1} d\mu' f'_\nu \int_0^{2\pi} d\phi' M - f_\nu \frac{C n_{n,p}}{c(2\pi\hbar c)^3}\int_0^\infty d\varepsilon'_\nu \varepsilon'^2_\nu \delta(\omega)\int_{-1}^{+1} d\mu'(1 - f'_\nu)\int_0^{2\pi} d\phi' M, \quad (5.64)$$

where $M = [(1 + \cos\theta)V^2 + (3 - \cos\theta)A^2]$. Using eqs. (5.36), (5.41), and (5.42) we find that

$$\mathcal{L}^{\rm scatt}_{(0)}[f_\nu] = \frac{4\pi C n_{n,p}}{c(2\pi\hbar c)^3}\int_0^\infty d\varepsilon'_\nu \varepsilon'^2_\nu \delta(\omega)[(1 - f_\nu)e^{-\beta\omega}\Xi_{\rm in} - f_\nu \Xi_{\rm out}], \quad (5.65)$$

where $\Xi_{\rm in} = V^2(\tilde{J}'_\nu + \mu\tilde{H}'_\nu) + 3A^2(\tilde{J}'_\nu - \mu\tilde{H}'_\nu/3)$ and $\Xi_{\rm out} = V^2(1 - \tilde{J}'_\nu - \mu\tilde{H}'_\nu) + 3A^2(1 - \tilde{J}'_\nu + \mu\tilde{H}'_\nu/3)$. Integrating over ε'_ν using the delta function, we have that

$$\begin{aligned}\mathcal{L}^{\rm scatt}_{(0)}[f_\nu] &= 4\pi C\varepsilon_\nu^2(V^2+3A^2)\left[(\tilde{J}_\nu - f_\nu) + \left(\frac{V^2-A^2}{V^2+3A^2}\right)\mu\tilde{H}_\nu\right] \\ &= \sigma_{n,p}n_{n,p}\left[(\tilde{J}_\nu - f_\nu) + \delta_{n,p}\mu\tilde{H}_\nu\right] \qquad (5.66)\end{aligned}$$

where $\sigma_{n,p} = (G^2/\pi c)\varepsilon_\nu^2(V^2 + 3A^2)$, $\delta_{n,p}$ is the scattering asymmetry for neutrino-neutron or neutrino-proton scattering, G is the weak coupling constant, and c is the speed of light. The result presented in eq. (5.66) is to be compared with the full scattering part of the collision term:

$$-\kappa_s f_\nu + \frac{\kappa_s}{4\pi}\int \Phi(\mathbf{\Omega},\mathbf{\Omega}')\, f_\nu(\mathbf{\Omega}')\, d\Omega'. \qquad (5.67)$$

Performing a first-order Legendre series expansion of the integral elastic scattering source term, and combining the scattering sink, $-\kappa_s f_\nu$, with the scattering source, $\kappa_s\tilde{J}_\nu + \kappa_s\delta\mu\tilde{H}_\nu$, we obtain eq. (5.66) and an expression for $\delta_{n,p}$.

Note that taking the zeroth moment of $\mathcal{L}^{\rm scatt}_{(0)}[f_\nu]$ yields zero, as it should in the elastic scattering limit. Further, note that the first moment is non-zero:

$$\begin{aligned}\frac{1}{2}\int_{-1}^{+1} d\mu\mu\,\mathcal{L}^{\rm scatt}_{(0)}[f_\nu] &= \frac{G^2}{\pi c}n_{n,p}\varepsilon_\nu^2\left[\frac{1}{3}\tilde{H}_\nu(V^2-A^2) - \tilde{H}_\nu(V^2+3A^2)\right] \\ &= -\frac{G^2}{\pi c}n_{n,p}\varepsilon_\nu^2\tilde{H}_\nu(V^2+3A^2)\left[1-\frac{1}{3}\delta_{n,p}\right] \\ &= -\sigma_{n,p}n_{n,p}\tilde{H}_\nu\left[1-\frac{1}{3}\delta_{n,p}\right]. \qquad (5.68)\end{aligned}$$

This expression is to be compared with the elastic scattering momentum source term on the right-hand side of the first moment of the transport equation. In fact, the quantity $\sigma_{n,p}n_{n,p}(1-\delta_{n,p}/3)$ defines the elastic transport cross section (see eq. 5.19).

5. Dynamic Structure Factors for Neutrino–Nucleon Interactions

An alternate formalism for handling inelastic neutrino-nucleon scattering that more straightforwardly than in §5.3 generalizes to include nucleon-nucleon correlations, whether due to fermi blocking or nuclear interactions, involves the dynamical structure factor $S(q,\omega)$. Our discussion here follows closely that found in Burrows and Sawyer (1998,1999). Recent explorations into the effects of many–body correlations on neutrino–matter opacities at high densities have revealed that for densities above 10^{14} gm cm^{-3} both the charged–current and the neutral–current interaction rates are decreased by a factor of perhaps 2 to 3, depending on the density and the equation of state (Burrows & Sawyer 1998,1999; Reddy, Prakash, & Lattimer 1998; Yamada 1999). Furthermore, it has been shown that the rate of energy transfer due to neutral–current scattering off of nucleons exceeds that due to ν_μ–electron scattering (Janka et al. 1996; Thompson, Burrows, & Horvath 2000). Previously, it had been assumed that neutrino–nucleon scattering was elastic (Lamb & Pethick 1977). However, these recent reappraisals reveal that the product of the underestimated energy transfer per neutrino–nucleon scattering with cross section exceeds the corresponding quantity for neutrino–electron scattering. Since ν_e and $\bar{\nu}_e$ neutrinos participate in super–allowed charged–current absorptions on nucleons, neutrino–nucleon scattering has little effect on their rate of equilibration. However, such scattering is important for ν_μ and ν_τ equilibration. Since the many–body correlation suppressions appear only above neutrinosphere densities ($\sim 10^{11} - 10^{13}$ gm cm^{-3}), it is only the kinematic effect, and not the interaction effect, that need be considered when studying the emergent spectra. In the following we adhere closely to the approach and formal development in Burrows and Sawyer (1998,1999). Without interactions, the relevant dynamical structure factor, $S(q,\omega)$, for neutrino–nucleon scattering is simply

$$S(q,\omega) = 2\int \frac{d^3p}{(2\pi)^3}\mathcal{F}(|\mathbf{p}|)(1-\mathcal{F}(|\mathbf{p}+\mathbf{q}|))2\pi\delta(\omega+\epsilon_{\mathbf{p}}-\epsilon_{\mathbf{p}+\mathbf{q}})\,, \quad (5.69)$$

where $\mathcal{F}(|\mathbf{p}|)$ is the nucleon Fermi–Dirac distribution function, $\epsilon_{\mathbf{p}}$ is the nucleon energy, ω is the energy transfer to the medium, and $\mathbf{q}$ is the momentum transfer. The magnitude of $\mathbf{q}$ is related to ω and E_1, the incident neutrino energy, through the neutrino scattering angle, θ, by the expression (see text in §4 after eq. 5.36),

$$q = [E_1^2 + (E_1-\omega)^2 - 2E_1(E_1-\omega)\cos\theta]^{1/2}\,. \quad (5.70)$$

In the elastic limit and ignoring final–state nucleon blocking, $S(q,\omega) = 2\pi\delta(\omega)$ n_n, the expected result, where n_n is the nucleon's number density.

The neutral current scattering rate off of either neutrons or protons is (Burrows & Sawyer 1998),

$$\frac{d^2\Gamma}{d\omega d\cos\theta} = (4\pi^2)^{-1}G_W^2(E_1-\omega)^2[1-\mathcal{F}_\nu(E_1-\omega)]\mathcal{I}_{\rm NC}\,, \quad (5.71)$$

where

$$\mathcal{I}_{\rm NC} = \Big[(1+\cos\theta)V + (3-\cos\theta)A\Big]S(q,\omega) \quad (5.72)$$

and

$$S(q,\omega) \;=\; 2{\rm Im}\Pi^{(0)}(1-{\rm e}^{-\beta\omega})^{-1}\,. \quad (5.73)$$

V and A are the applicable vector and axial–vector coupling terms (see §3 and §3) and $\beta = 1/kT$. The free polarization function, $\Pi^{(0)}$, contains the full kinematics of the scattering, as well as blocking due to the final–state nucleon, and the relevant imaginary part of $\Pi^{(0)}$ is given by:

$$\mathrm{Im}\Pi^{(0)}(q,\omega) = \frac{m^2}{2\pi q\beta}\log\left[\frac{1+e^{-Q_+^2+\beta\mu}}{1+e^{-Q_+^2+\beta\mu-\beta\omega}}\right], \tag{5.74}$$

where

$$Q_\pm = \left(\frac{m\beta}{2}\right)^{1/2}\left(\mp\frac{\omega}{q}+\frac{q}{2m}\right), \tag{5.75}$$

μ is the nucleon chemical potential, and m is the nucleon mass. The dynamical structure factor, $S(q,\omega)$, contains all of the information necessary to handle angular and energy redistribution due to scattering. The corresponding term on the right–hand–side of the transport equation is:

$$\mathcal{S}[\mathcal{F}_\nu] = (2\pi)^{-3}G_W^2\int d^3p'_\nu\,\mathcal{I}_{\rm NC}\,\Xi_{SF} \tag{5.76}$$

where

$$\Xi_{SF} = [1-\mathcal{F}_\nu(E_1)]\mathcal{F}'_\nu(E_1-\omega)e^{-\beta\omega} - \mathcal{F}_\nu(E_1)[1-\mathcal{F}'_\nu(E_1-\omega)] \tag{5.77}$$

and p'_ν is the final state neutrino momentum.

In the non-degenerate nucleon limit, eq. (5.74) can be expanded to lowest order in Q_+^2 to obtain, using eq. (5.73), an approximation to the dynamical structure factor:

$$S(q,\omega) = \frac{n(2\pi m\beta)^{1/2}}{q}e^{-Q_+^2}, \tag{5.78}$$

where n is the nucleon number density. This says that for a given momentum transfer the dynamical structure factor is approximately a Gaussian in ω.

For charged–current absorption process, $\nu_e + n \rightarrow e^- + p$, $\mathrm{Im}\Pi^{(0)}(q,\omega)$ is given by a similar expression:

$$\mathrm{Im}\Pi^{(0)}(q,\omega) = \frac{m^2}{2\pi\beta q}\log\left[\frac{1+e^{-Q_+^2+\beta\mu_n}}{1+e^{-Q_+^2+\beta\mu_p-\beta\omega}}\right]. \tag{5.79}$$

Eq. (5.79) inserted into eq. (5.73) with a $(1-e^{-\beta(\omega+\hat{\mu})})$, as is appropriate for the charged–current process, substituted for $(1-e^{-\beta\omega})$, results in an expression that is a bit more general than the one employed to date by most practitioners (Bruenn 1985), *i.e.*, $S = (X_n - X_p)/(1-e^{-\hat{\mu}/T})$. In the non-degenerate nucleon limit, the structure factor for the charged–current process can be approximated by eq. (5.78) with $n = n_n$. Note that for the structure factor of a charged-current interaction one must distinguish between the initial– and the final–state nucleons and, hence, between their chemical potentials. To obtain the structure factor for the $\bar{\nu}_e$ absorption process, one simply permutes μ_n and μ_p in eq. (5.79) and substitutes $-\hat{\mu}$ for $\hat{\mu}$ in the $(1 - e^{-\beta(\omega+\hat{\mu})})$ term.

In general, the widths of the structure factors are larger than might have been expected for scattering off of "heavy" particles. This is because in the past people thought that the neutrino could lose in ν–nucleon scattering an energy equal to only about $-E_1^2/m_n c^2$, *i.e.* that the fractional energy lost is of order $p_\nu/m_n c$ ($\sim$1%). However, this assumes that the nucleons are stationary. In fact, they are thermal and, the fractional energy they can in a collision transfer to the neutrino is of order $p_n/m_n c$. Since the nucleons have such a large mass, if they and the neutrino have the same energy, $p_n/m_n c$ is much larger than $p_\nu/m_n c$, at incident neutrino energies of 10–30 MeV by as much as an order of magnitude. The formalism above incorporates the kinematics of such a collision, a realistic Fermi–Dirac energy distribution for the nucleons, and final–state nucleon blocking. The upshot is broad distributions. Including many–body effects further flattens and broadens the distribution, while lowering the central values of $d\sigma/d\omega$, as well as the total integral over ω (Burrows & Sawyer 1998,1999).

Aside: Static Structure Factors

In the limit of heavy nucleons, when we perform the integration in eq. (5.69) over a range of ωs and evaluated the inner factors at $\omega = 0$, we can express this limit as,

$$(2\pi)^{-1}S(q,\omega) \rightarrow (2\pi)^{-1}\delta(\omega)\int d\omega' S(q,\omega') \equiv \delta(\omega)S(q), \qquad (5.80)$$

where $S(q)$ is the *static* structure factor. At the high densities and temperatures achieved in the supernova context, the $\omega = 0$ (elastic) limit is not particularly accurate (Burrows & Sawyer 1998; Reddy, Prakash, & Lattimer 1998).

$S(q)$ is merely the Fourier transform of the thermally–averaged density–density correlation function. This is the classic result that scattering off of a medium is in reality scattering off of the *fluctuations* in that medium. Also of interest is the long–wavelength limit, $q \rightarrow 0$, justified when the neutrino wavelength is much bigger than the interparticle separation. Statistical mechanics provides two useful and equivalent expressions for the long–wavelength limit, $S(0)$, the first (Landau & Lifschitz 1969) in terms of the isothermal compressibility of the medium K_T $(= -\frac{\partial \log V}{\partial P}|_T)$,

$$S(0) = \bar{n}^2 \beta^{-1} K_T = \bar{n} \frac{K_T}{K_0}, \tag{5.81}$$

where K_0 is the ideal gas compressibility and $\bar{n}$ is the average nucleon density, and the second in terms of the derivative of the density with respect to the chemical potential of the nucleons, μ,

$$S(0) = \beta^{-1} \frac{\partial \bar{n}}{\partial \mu}. \tag{5.82}$$

In the ideal gas limit of no correlation between particles, eqs. (5.81) and (5.82) show that $S(q)$ is simply equal to the number density, $\bar{n}$, as expected from eq. (5.69), without blocking. Eq. (5.81) reveals that if K_T is small because the matter is stiff, in the long–wavelength limit the neutrino–matter cross sections are *suppressed*. In general, when we acknowledge that the neutrino-matter interaction has axial–vector current and nucleon isospin terms, we require separate correlation functions for the neutron and the proton, as well as for spin correlations. These depend upon susceptibilities that are different from the compressibility, but we find suppression in these terms as well (Burrows & Sawyer 1998; Reddy, Prakash, & Lattimer 1998).

Eq. (5.82), equivalent to eq. (5.81) by a thermodynamic identity, is a powerful result of great generality. In standard approximation schemes for the many–body problem, the distribution function for a nucleon species is given by the Fermi–Dirac distribution in which the chemical potential μ is replaced by $\mu - v(\bar{n})$, where $v(\bar{n})$ is the average energy of interaction of the nucleon with the other nucleons and is a function of the density. Thus, the density is given implicitly by

$$\bar{n} = 2 \int \frac{d^3 p}{(2\pi)^3} [1 + e^{\beta[(p^2/(2m) - \mu + v(\bar{n})]}]^{-1}. \tag{5.83}$$

The expression (5.83) holds in the Hartree approximation; it holds in approaches that introduce mean meson fields instead of nuclear potentials; it holds in the Landau Fermi liquid theory (FLT), subject to the proviso that we use only results in which the derivative of the potential v (with respect to the $\bar{n}$) enters; and it holds in approaches using the Skyrme potential.

Differentiating (5.83), we can solve for $\frac{\partial \bar{n}}{\partial \mu}$ and $S(0)$,

$$S(0) = \beta^{-1} \frac{\partial \bar{n}}{\partial \mu} = h(\mu)[1 + h(\mu) \frac{\partial v}{\partial \bar{n}}]^{-1}, \tag{5.84}$$

where

$$h(\mu) = 2\int \frac{d^3p}{(2\pi)^3} \frac{e^{\beta[p^2/(2m)-\mu+v]}}{[1+e^{\beta[p^2/(2m)-\mu+v]}]^2} = 2\int \frac{d^3p}{(2\pi)^3}\mathcal{F}(p)(1-\mathcal{F}(p)) \tag{5.85}$$

and $\mathcal{F}(p)$ is the Fermi–Dirac function, but with the chemical potential displaced by v. If we regard particle densities as inputs to our calculations, then the displacement of the chemical potential by the nuclear potential is irrelevant, since the same difference, $\mu - v$, enters the calculation of the density in terms of the chemical potential. Thus, the numerator of (5.84) contains no more than the familiar Pauli blocking effects (for the case $q = 0$); the denominator contains all of the effect of the interactions.

As an example, consider a two–nucleon potential $V(r)$. In the Hartree approximation, the average potential seen by a single nucleon is given by $v = \bar{n}U$, where $U = \int d^3xV(x)$, and (5.84) becomes

$$S(0) = h(\mu)[1 + h(\mu)U]^{-1}, \tag{5.86}$$

the potential providing an enhancement, if negative, and a suppression, if positive. The latter is the case for high–density nuclear matter.

Procedure for Calculating ν–nucleon Structure Functions for Neutral-Current Scattering Including Interactions

Now, taking only the neutron part of the vector–current coupling, the differential neutrino-nucleon scattering rate is given by,

$$\frac{d^2\Gamma}{d\omega\, d\cos\theta} = (4\pi^2)^{-1}G_W^2E_2^2[1-f_\nu(E_2)]\Big[(1+\cos\theta)(C_V^n)^2S_{nn}(q,\omega) \\ +(3-\cos\theta)g_A^2[S_{pp}^A(q,\omega)+S_{nn}^A(q,\omega)-2S_{pn}^A(q,\omega)\Big], \tag{5.87}$$

where E_2=$E_1 - \omega$.

The structure functions, S (Fermi) and S^A (Gamow-Teller; axial), are elements of separate 2×2 symmetric matrices. For the vector dynamic structure function, S, we have

$$S(q,\omega) = \begin{pmatrix} S_{pp}(q,\omega) & S_{pn}(q,\omega) \\ S_{pn}(q,\omega) & S_{nn}(q,\omega) \end{pmatrix}.$$

The structure function matrix is given by,

$$S(q,\omega) = 2\mathrm{Im}\Big[\Pi^{(0)}(q,\omega)[1-v(q)\Pi^{(0)}(q,\omega)]^{-1}\Big](1-e^{-\beta\omega})^{-1} \tag{5.88}$$

where

$$\Pi^{(0)}(q,\omega) = \begin{pmatrix} \Pi_p^{(0)}(q,\omega) & 0 \\ 0 & \Pi_n^{(0)}(q,\omega) \end{pmatrix}$$

and $\Pi_p^{(0)}$ and $\Pi_n^{(0)}$ are given by the polarization function and evaluated with the proton and neutron chemical potentials, respectively (eq. 5.74). The potential matrix is,

$$v = \begin{pmatrix} v_1 + v_2 + 4\pi e^2(q^2 + q_{TF}^2)^{-1} & v_1 - v_2 \\ v_1 - v_2 & v_1 + v_2 \end{pmatrix},$$

where the v's were defined in terms of Fermi liquid parameters and the term containing q_{TF} is the Thomas-Fermi screened Coulomb potential ($q_{TF}^2 = 4e^2\pi^{1/3}(3\bar{n}_p)^{2/3}$). Following Burrows and Sawyer (1998,1999) and for simplicity, we use here FLT and Landau parameters, in lieu of a more developed nuclear interaction model.

In a real calculation, in all the kinematic expressions the nucleon mass (m) is to be replaced by m^*. Unfortunately, the relation of Landau parameters to experimental results depends upon the effective mass in model–dependent ways. Taking $m^* = 0.75m_n$ as our fiducial value for the effective mass, we use parameters from Backman, Brown, & Niskanen (1985) and Brown and Rho (1981): $F_0 = -0.28; F_0' = 0.95; G_0 = 0; G_0' = 1.7$ and $\lambda = 2.63 \times 10^{-5}\text{MeV}^{-2}$, obtaining,

$$\begin{aligned} v_1 &= -7.4 \times 10^{-6}\,\text{MeV}^{-2} \\ v_2 &= 2.5 \times 10^{-5}\,\text{MeV}^{-2} \\ v_3 &= 0 \\ v_4 &= 4.5 \times 10^{-5}\,\text{MeV}^{-2}. \end{aligned} \tag{5.89}$$

For other values of the effective mass, we keep these potentials at the same value, which is to say we assume that the Landau parameters are proportional to m^*/m.

The form for the Gamow–Teller matrix, $S^A(q,\omega)$, is the same as that for S, except that the potential matrix is replaced by v^A

$$v^A = \begin{pmatrix} v_3 + v_4 & v_3 - v_4 \\ v_3 - v_4 & v_3 + v_4 \end{pmatrix}.$$

Taking the matrix inverses leads to the following forms for the combinations of structure functions that appear in (5.87)

$$S_{nn}(q,\omega) = 2\text{Im}[\Pi_n^{(0)} D_V^{-1}](1 - e^{-\beta\omega})^{-1}, \tag{5.90}$$

where

$$D_V = 1 - (v_1 + v_2)\Pi_n^{(0)} - (v_1 - v_2)^2 \Pi_n^{(0)} \Pi_p^{(0)} Q_V{}^{-1} . \tag{5.91}$$

Q_V is given by the expression:

$$Q_V = 1 - 4\pi e^2 (q^2 + q_{TF}^2)^{-1} \Pi_p^{(0)} - (v_1 + v_2)\Pi_p^{(0)} . \tag{5.92}$$

If, as in (5.89), we take $v_3 = 0$, we obtain the simple result for the axial–current terms,

$$S_A(q,\omega) = 2\mathrm{Im}\Big[\frac{\Pi_p^{(0)}(q,\omega) + \Pi_n^{(0)}(q,\omega)}{1 - v_4[\Pi_p^{(0)}(q,\omega) + \Pi_n^{(0)}(q,\omega)]}\Big](1 - e^{-\beta\omega})^{-1} . \tag{5.93}$$

For the Fermi term, since $C_V^{(p)} = 1/2 - 2\sin^2\theta_W \sim 0$, we drop the proton structure function in (5.87). Furthermore, we use the potential parameters given in eq. (5.89), and in eq. (5.91) we drop the third term. This term would have been significant had it not been for the Coulomb term in the denominator, an illustration of the importance of the explicit inclusion of Coulomb forces, even for the neutron density correlations. Since the v_is are all real, we obtain for the structure factors used in (5.87),

$$S_F(q,\omega) = 2\mathrm{Im}\Pi_n^{(0)}(1 - e^{-\beta\omega})^{-1} \mathcal{C}_V{}^{-1} , \tag{5.94}$$

where

$$\mathcal{C}_V = (1 - v_F \mathrm{Re}\Pi_n^{(0)})^2 + v_F^2 (\mathrm{Im}\Pi_n^{(0)})^2 , \tag{5.95}$$

and

$$S_A(q,\omega) = 2\Big[\mathrm{Im}\Pi_p^{(0)}(q,\omega) + \mathrm{Im}\Pi_n^{(0)}(q,\omega)\Big](1 - e^{-\beta\omega})^{-1} \mathcal{C}_A{}^{-1} , \tag{5.96}$$

where

$$\mathcal{C}_A = \mathcal{C}_{A1} + \mathcal{C}_{A2} . \tag{5.97}$$

$\mathcal{C}_{A1}$ and $\mathcal{C}_{A2}$ are given by the expressions:

$$\mathcal{C}_{A1} = \Big[1 - v_{GT}(\mathrm{Re}\Pi_p^{(0)}(q,\omega) + \mathrm{Re}\Pi_n^{(0)}(q,\omega))\Big]^2 \tag{5.98}$$

and

$$\mathcal{C}_{A2} = v_{GT}^2 \Big[\mathrm{Im}\Pi_p^{(0)}(q,\omega) + \mathrm{Im}\Pi_n^{(0)}(q,\omega)\Big]^2 . \tag{5.99}$$

The F in $S_F(q,\omega)$ and the A in $S_A(q,\omega)$ stand for Fermi and Gamow–Teller (axial) and v_F and v_{GT} equal (v_1+v_2) and v_4, respectively, in Fermi Liquid Theory. $S_A(q,\omega)$ in eq. (5.96 is now the entire axial term in eq. (5.87). $\mathcal{C}_{V,A}$ is the correction factor due to many–body effects for a given momentum transfer (or scattering angle) and energy transfer. A similar procedure is employed for calculating the many-body corrections to the charged-current rates (Burrows & Sawyer 1999).

Collective Excitations of the Medium

Following Burrows and Sawyer (1998,1999), we note that for most regions of phase space, $\mathcal{C}_V$ and $\mathcal{C}_A$ in eqs. (5.95) and (5.97) are greater than one and represent suppression in the scattering rates. Their effects on the integrals over ω and θ are always suppressive. However, the terms containing the real parts have roots; these roots represent collective excitations. For the Fermi term, zero sound in the medium can be generated if the scattering has a (ω, q) pair that satisfies the mode's dispersion relation, *i.e.*, if it hits the resonance. Similarly, for the Gamow–Teller term, spin waves in the protons and the neutrons (related by a set phase) can be generated. These modes are the traveling–mode equivalents of the Gamow–Teller resonance in nuclei (a standing wave). The zero sound of the Fermi part is analogous to the Giant–Dipole resonance in nuclei. The resonances increase the structure function when the scattering transfer ratio, ω/q, equals the ratio of the collective excitation's angular frequency and wavenumber. For a given scattering angle, one can plot the differential cross section in ω and $\cos\theta$ as a function of ω/q to see the resonances. In Figure (5.6), we display this for five different angles between 15° and 180°, an incident neutrino energy of 20 MeV, a temperature of 5 MeV, a density of 3×10^{14} g cm^{-3}, and an electron fraction, Y_e, of 0.3. We see in Figure 5.6 that the resonances in both the forward and the backward directions line up at the same values of ω/q, as expected for a collective mode, and we can straightforwardly calculate the mode's dispersion relation. This is akin to the Čerenkov effect. Note that the Gamow–Teller term dominates the Fermi term, so that in Figure 5.6 we are really seeing the spin waves related to the Gamow–Teller resonance. However, the dispersion relations for zero sound and these spin waves are generally similar. In fact, recalling the classic result of Fetter and Walecka (1971) that in the weak–coupling limit, the speed of zero sound in a degenerate system is $\sim v_{fermi}$, where v_{fermi} is the Fermi velocity, and recalling that for nucleons in nuclei v_{fermi} is $\sim 0.3c$, the calculated resonance value of ω/q is not unexpected. In Figure 5.7, we plot the Gamow–Teller structure function versus ω/q for various values of ω, m^*, and two values of the density. At $m^*=m_n$, for each value of the density we obtain a sharp resonance, but at two different speeds, reflecting the crude $\rho^{1/3}$–dependence expected for v_{fermi}. For a given density, the mode speed is seen in Figure 5.7 to be inversely proportional to the effective mass. The width of the resonance is determined by the magnitude of the imaginary part of the polarization function.

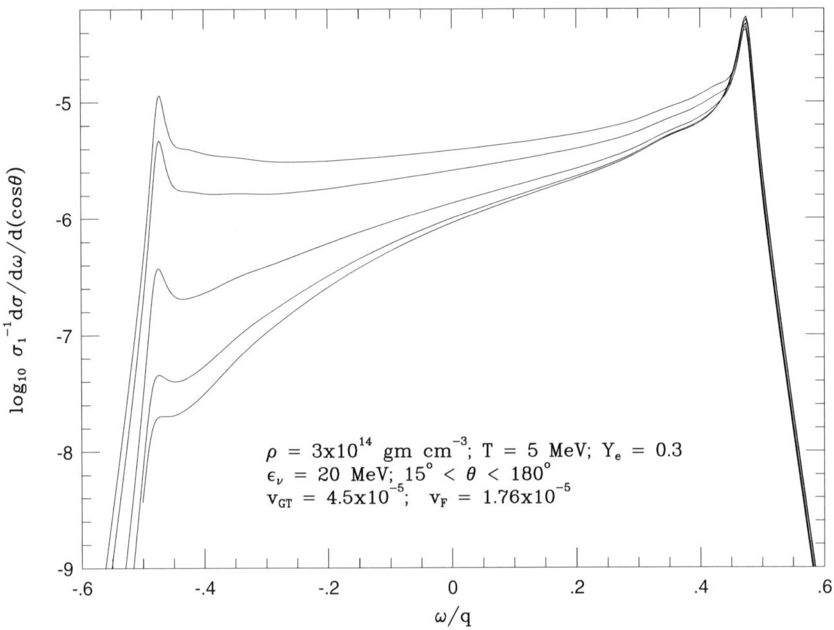

Figure 5.6. Log$_{10}$ of the doubly–differential cross section for neutral–current neutrino–nucleon scattering versus ω/q for scattering angles 15°, 45°, 90°, 135°, and 180° . The calculations were performed at a temperature of 5 MeV, a Y_e of 0.3, a ρ of 3×10^{14} g cm^{-3}, and an incident neutrino energy of 20 MeV. The default potentials ($v_{GT} = 4.5 \times 10^{-5}$ and $v_F = 1.76 \times 10^{-5}$) and effective mass ($m^* = 0.75\, m_n$) were employed. The differential cross section is divided by the total scattering cross section (σ_1) in the non–interacting, no–nucleon–blocking, $\omega = 0$ limit. (Figure taken from Burrows and Sawyer 1998.)

6. e^+e^- Annihilation

Ignoring phase space blocking of neutrinos in the final state and taking the relativistic limit ($m_e \rightarrow 0$), the total electron–positron annihilation rate into neutrino–antineutrino pairs can be written in terms of the electron and positron

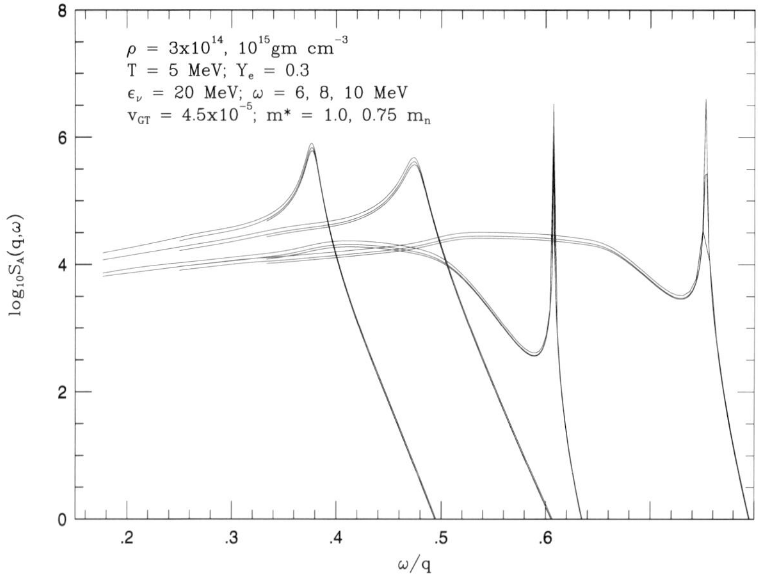

Figure 5.7. Log_{10} of the Gamow–Teller structure function versus ω/q for an incident neutrino energy of 20 MeV, energy transfers, ω, of 6, 8, and 10 MeV, two values of the effective mass ($m^* = [0.75m_n, 1.0m_n]$) and two values of the density ($\rho = 3 \times 10^{14}$ and 10^{15} g cm^{-3}). A temperature of 5 MeV and a Y_e of 0.3 were used, as was the default v_{GT} ($= 4.5 \times 10^{-5}$). (Figure taken from Burrows and Sawyer 1998.)

phase space densities (Dicus 1972):

$$Q_{\nu_e\bar{\nu}_e} = K_i \left(\frac{1}{m_e c^2}\right)^2 \left(\frac{1}{\hbar c}\right)^6 \int\int \mathcal{F}_{e^-}\mathcal{F}_{e^+}(\varepsilon_{e^-}^4\varepsilon_{e^+}^3 + \varepsilon_{e^-}^3\varepsilon_{e^+}^4)\, d\varepsilon_{e^-}\, d\varepsilon_{e^+} \ , \tag{5.100}$$

where $K_i = (1/18\pi^4)c\sigma_o(C_V^2 + C_A^2)$. Again, $C_V = 1/2 + 2\sin^2\theta_W$ for electron types, $C_V = -1/2 + 2\sin^2\theta_W$ for mu and tau types, and $C_A^2 = (1/2)^2$. Rewriting eq. (5.100) in terms of the Fermi integral $F_n(\eta)$, we obtain:

$$Q_{\nu_e\bar{\nu}_e} = K_i\,(kT) \left(\frac{kT}{m_e c^2}\right)^2 \left(\frac{kT}{\hbar c}\right)^6 [F_4(\eta_e)F_3(-\eta_e) + F_4(-\eta_e)F_3(\eta_e)] \ , \tag{5.101}$$

where $\eta_e \equiv \mu_e/kT$ and

$$F_n(\eta) \equiv \int_0^\infty \frac{x^n}{e^{x-\eta}+1}\,dx\ . \tag{5.102}$$

Integrating eq. (5.100), we obtain

$$Q_{\nu_e\bar{\nu}_e} \simeq 9.7615 \times 10^{24} \left[\frac{kT}{\mathrm{MeV}}\right]^9 f(\eta_e)\ \mathrm{ergs\,cm^{-3}s^{-1}}\,, \tag{5.103}$$

where

$$f(\eta_e) = \frac{F_4(\eta_e)F_3(-\eta_e)+F_4(-\eta_e)F_3(\eta_e)}{2F_4(0)F_3(0)}\ . \tag{5.104}$$

For $\nu_\mu\bar{\nu}_\mu$ and $\nu_\tau\bar{\nu}_\tau$ production combined,

$$Q_{\nu_{\mu,\tau}\bar{\nu}_{\mu,\tau}} \simeq 4.1724 \times 10^{24} \left[\frac{kT}{\mathrm{MeV}}\right]^9 f(\eta_e)\ \mathrm{ergs\,cm^{-3}s^{-1}}\,. \tag{5.105}$$

One can easily derive the spectrum of the total radiated neutrino energy (ε_T) by inserting a delta function ($\int\delta(\varepsilon_T-\varepsilon_{e^-}-\varepsilon_{e^+})d\varepsilon_T$) into eq. (5.100). Recall that the total energy of the neutrinos in the final state is equal to the sum of the electron and positron energies in the initial state. Integrating first over ε_{e^+} to annihilate the delta function and then over ε_{e^-} to leave a function of ε_T, one obtains:

$$\begin{aligned}\frac{dQ}{d\varepsilon_T} &= K_i\left(\frac{1}{m_ec^2}\right)^2\left(\frac{1}{\hbar c}\right)^6\int_0^{\varepsilon_T}\varepsilon_T(\varepsilon_T-\varepsilon_{e^-})^3 \\ &\quad \varepsilon_{e^-}^3\mathcal{F}_{e^-}[\varepsilon_{e^-}]\mathcal{F}_{e^+}[\varepsilon_T-\varepsilon_{e^-}]\,d\varepsilon_{e^-}\ .\end{aligned} \tag{5.106}$$

The numerical evalution of eq. (5.106) is straightforward. The average of ε_T is equal to:

$$\langle\varepsilon_T\rangle = \left(\frac{F_4(\eta_e)}{F_3(\eta_e)}+\frac{F_4(-\eta_e)}{F_3(-\eta_e)}\right)T\,, \tag{5.107}$$

which near $\eta_e \sim 0$ is $\sim 8T$ and for $\eta_e >> 1$ is $\sim 4T(1+\eta_e/5)$.

However, while the total energy loss rate (eq. 5.103) and the spectrum of ε_T pose no great mathematical problems, the production spectrum of an individual neutrino is not so easily reduced to a simple integral or to an analytic expression. This is due primarily to the awkward integration of the angular phase space terms, while subject to the momentum conservation delta function, and to the explicit dependence of the matrix elements on the electron/neutrino angles. From Dicus (1972), averaging over initial states and summing over final states, the matrix element for the $e^+e^- \rightarrow \nu\bar{\nu}$ process in the $m_e = 0$ limit is:

$$\frac{1}{4}\sum_s |\mathcal{M}|^2 = 16G^2[(C_V + C_A)^2 \mathbf{p}\cdot\mathbf{q}_{\bar{\nu}}\,\mathbf{p}'\cdot\mathbf{q}_\nu + (C_V - C_A)^2 \mathbf{p}\cdot\mathbf{q}_\nu\,\mathbf{p}'\cdot\mathbf{q}_{\bar{\nu}}]\ , \tag{5.108}$$

where p and p' are the four-momenta of the electron and positron, respectively, and q_ν and $q_{\bar{\nu}}$ are the four-momenta of the neutrino and antineutrino, respectively. Using the formalism of Bruenn (1985) and Fermi's Golden rule, expanding the production kernel in the traditional truncated Legendre series, performing the trivial angular integrals, taking the non–trivial angular integrals from Bruenn (1985), and ignoring final–state neutrino blocking, we obtain for the single–neutrino source spectrum due to e^+e^- annihilation:

$$\frac{dQ}{d\varepsilon_\nu} = \frac{8\pi^2}{(2\pi\hbar c)^6}\,\varepsilon_\nu^3 \int_0^\infty d\varepsilon_{\bar{\nu}}\,\varepsilon_{\bar{\nu}}^2\,\Phi_0^p(\varepsilon_\nu, \varepsilon_{\bar{\nu}})\ , \tag{5.109}$$

where

$$\Phi_0^p(\varepsilon_\nu, \varepsilon_{\bar{\nu}}) = \frac{G^2}{\pi}\int_0^{\varepsilon_\nu+\varepsilon_{\bar{\nu}}} d\varepsilon_{e^-}\,\mathcal{F}_{e^-}[\varepsilon_{e^-}]\mathcal{F}_{e^+}[\varepsilon_\nu + \varepsilon_{\bar{\nu}} - \varepsilon_{e^-}]\,H_0(\varepsilon_\nu, \varepsilon_{\bar{\nu}}, \varepsilon_{e^-})\ , \tag{5.110}$$

and

$$H_0(\varepsilon_\nu, \varepsilon_{\bar{\nu}}, \varepsilon_{e^-}) = (C_V + C_A)^2\,J_0^I(\varepsilon_\nu, \varepsilon_{\bar{\nu}}, \varepsilon_{e^-}) + (C_V - C_A)^2\,J_0^{II}(\varepsilon_\nu, \varepsilon_{\bar{\nu}}, \varepsilon_{e^-})\ . \tag{5.111}$$

The J_0s in eq. (5.111) come from the more obdurate angular integrals required by the dot products in eq. (5.108) and the momentum delta function and have the symmetry:

$$J_0^I(\varepsilon_\nu, \varepsilon_{\bar{\nu}}, \varepsilon_{e^-}) = J_0^{II}(\varepsilon_{\bar{\nu}}, \varepsilon_\nu, \varepsilon_{e^-})\ . \tag{5.112}$$

From eqs. (5.109) and (5.111), we see that the differences between the spectra of the ν_e and ν_μ neutrinos flow solely from their correspondingly different values of $(C_V + C_A)^2$ and $(C_V - C_A)^2$. One can use 4–point Gauss–Legendre integration to calculate eq. (5.110) and 16–point Gauss–Laguerre integration to calculate eq. (5.109).

At small η_e, the e^+e^- annihilation spectra and total energy loss rates for the ν_e and $\bar{\nu}_e$ neutrinos are similar, as are the average emitted ν_e and $\bar{\nu}_e$ neutrino energies. However, as η_e increases, both the total energy radiated in $\bar{\nu}_e$ neutrinos and the average $\bar{\nu}_e$ energy start to lag the corresponding quantities for the ν_e neutrinos. This is true despite the fact that the total number of ν_e and $\bar{\nu}_e$ neutrinos radiated is the same. If final–state blocking is ignored, $\langle\varepsilon_i\rangle/T$ is a function of η_e alone, becoming linear with η_e at high η_e and one half of eq. (5.107) ($\sim$4.0) at low η_e. Note also that $\langle\varepsilon_{\nu_\mu}\rangle/T$ and $\langle\varepsilon_{\bar{\nu}_\mu}\rangle/T$ are closer to one another than are $\langle\varepsilon_{\nu_e}\rangle/T$ and $\langle\varepsilon_{\bar{\nu}_e}\rangle/T$. The individual production spectra vary in peak strength, in peak energy, and in low–energy shape, but they are quite similar on the high–energy tail. Due to the parity–violating matrix element for the $e^+e^- \rightarrow \nu\bar{\nu}$ process and the fact that η_e is positive, the antineutrino spectra of all species are softer than the neutrino spectra. The pair sums of the integrals under these curves are given by eqs. (5.103) and (5.105). For $\eta_e = 0$, 50% of the pair energy emission of electron types is in $\bar{\nu}_e$ neutrinos, but at $\eta_e = 10$ only 42% of this total energy is in $\bar{\nu}_e$ neutrinos. However, at $\eta_e = 10$, the $\bar{\nu}_\mu$ neutrinos still constitute 48.5% of the $\nu_\mu/\bar{\nu}_\mu$ pair emission. These differences reflect differences in the corresponding coupling constants C_V and C_A.

7. $\nu_i\bar{\nu}_i$ Annihilation

In the limit of high temperatures and ignoring electron phase space blocking, the $\nu_i\bar{\nu}_i$ annihilation rate into e^+e^- pairs can be written (Janka 1991):

$$Q_{\nu_i\bar{\nu}_i} = 4K_i\pi^4\left(\frac{1}{m_ec^2}\right)\left(\frac{4\pi}{c}\right)^2\int\int \Phi'\, J_{\nu_i}J_{\bar{\nu}_i}(\epsilon_{\nu_i}+\epsilon_{\bar{\nu}_i})\, d\epsilon_{\nu_i}\, d\epsilon_{\bar{\nu}_i}\ , \tag{5.113}$$

where J_ν is the zeroth moment of the radiation field, ε_ν is the neutrino energy, K_i is defined as before (*i.e.*, $K_i = (1/18\pi^4)c\sigma_o(C_V^2 + C_A^2)$), and

$$\begin{aligned}\Phi'\left(\langle\mu_{\nu_i}\rangle, \langle\mu_{\bar{\nu}_i}\rangle, p_{\nu_i}, p_{\bar{\nu}_i}\right) &= \frac{3}{4}\Big[1 - 2\langle\mu_{\nu_i}\rangle\langle\mu_{\bar{\nu}_i}\rangle + p_{\nu_i}p_{\bar{\nu}_i} \\ &+ \frac{1}{2}(1-p_{\nu_i})(1-p_{\bar{\nu}_i})\Big]\ ,\end{aligned} \tag{5.114}$$

where the flux factor $\langle\mu_{\nu_i}\rangle = H_\nu/J_\nu$ and the Eddington factor $p_\nu = \langle\mu_{\nu_i}^2\rangle = P_\nu/J_\nu$. Eq. (5.113) can be rewritten in terms of the invariant distribution functions $\mathcal{F}_\nu$:

$$Q_{\nu_i\bar{\nu}_i} = K_i\left(\frac{1}{m_ec^2}\right)^2\left(\frac{1}{\hbar c}\right)^6\int\int \Phi'\,\mathcal{F}_{\nu_i}\mathcal{F}_{\bar{\nu}_i}(\varepsilon_{\nu_i}^4\varepsilon_{\bar{\nu}_i}^3 + \varepsilon_{\nu_i}^3\varepsilon_{\bar{\nu}_i}^4)\, d\varepsilon_{\nu_i}\, d\varepsilon_{\bar{\nu}_i}\ . \tag{5.115}$$

Note that when the radiation field is isotropic ($\Phi' = 1$) and when $\eta_e = 0$ the total rate for e^+e^- annihilation given in eq. (5.100) equals that for $\nu_i\bar{\nu}_i$ annihilation given in eq. (5.115), as expected. Buras et al. (2002) have addressed the related and interesting process of $\nu_i\bar{\nu}_i \rightarrow \nu_j\bar{\nu}_j$. We refer to that paper for a discussion of the relevance and rates of this process.

8. Nucleon–Nucleon Bremsstrahlung

A production process for neutrino/anti-neutrino pairs that has recently received attention in the supernova context is neutral-current nucleon–nucleon bremsstrahlung ($n_1 + n_2 \rightarrow n_3 + n_4 + \nu\bar{\nu}$). It importance in the cooling of old neutron stars, for which the nucleons are quite degenerate, has been recognized for years (Flowers 1975), but only in the last few years has it been studied for its potential importance in the quasi-degenerate to non-degenerate atmospheres of protoneutron stars and supernovae (Suzuki 1993; Hannestad & Raffelt 1998; Burrows et al. 2000; Thompson, Burrows, & Horvath 2000). Neutron–neutron, proton–proton, and neutron–proton bremsstrahlung are all important, with the latter the most important for symmetric matter. As a source of ν_e and $\bar{\nu}_e$ neutrinos, nucleon–nucleon bremsstrahlung can not compete with the charged–current capture processes. However, for a range of temperatures and densities realized in supernova cores, it may compete with e^+e^- annihilation as a source for ν_μ, $\bar{\nu}_\mu$, ν_τ, and $\bar{\nu}_\tau$ neutrinos ("ν_μ"s). The major obstacles to obtaining accurate estimates of the emissivity of this process are our poor knowledge of the nucleon–nucleon potential, of the degree of suitability of the Born Approximation, and of the magnitude of many–body effects (Hannestad & Rafflet 1998; Raffelt & Seckel 1998; Brinkmann & Turner 1988). Since the nucleons in protoneutron star atmospheres are not degenerate, we present here a calculation of the total and differential emissivities of this process in that limit and assume a one-pion exchange (OPE) potential model to calculate the nuclear matrix element. For the corresponding calculation for arbitrary nucleon degeneracy, the reader is referred to Thompson, Burrows, & Horvath (2000). The formalism we employ has been heavily influenced by those of Brinkman and Turner (1988) and Hannestad and Raffelt (1998), to which the reader is referred for details and further explanations.

Our focus is on obtaining a useful single–neutrino final–state emission spectrum, as well as a final–state pair energy spectrum and the total emission rate. For this, we start with Fermi's Golden Rule for the total rate per neutrino species:

$$Q_{nb} = (2\pi)^4 \int \Big[\prod_{i=1}^{4} \frac{d^3\vec{p}_i}{(2\pi)^3}\Big] \frac{d^3\vec{q}_\nu}{(2\pi)^3 2\omega_\nu} \frac{d^3\vec{q}_{\bar{\nu}}}{(2\pi)^3 2\omega_{\bar{\nu}}}\, \omega \sum_s |\mathcal{M}|^2 \delta^4(\mathbf{P})\, \Xi_{brems},$$

where

$$\Xi_{brems} = \mathcal{F}_1\mathcal{F}_2(1-\mathcal{F}_3)(1-\mathcal{F}_4), \tag{5.116}$$

$\delta^4(\mathbf{P})$ is four–momentum conservation delta function, ω is the energy of the final–state neutrino pair, (ω_ν,$\vec{q}_\nu$) and ($\omega_{\bar{\nu}}$,$\vec{q}_{\bar{\nu}}$) are the energy and momentum of the neutrino and anti–neutrino, respectively, and $\vec{p}_i$ is the momentum of nucleon i. Final–state neutrino and anti–neutrino blocking have been dropped.

The necessary ingredients for the integration of eq. (5.116) are the matrix element for the interaction and a workable procedure for handling the phase space terms, constrained by the conservation laws. We follow Brinkman and Turner (1988) for both of these elements. In particular, we assume for the $n + n \rightarrow n + n + \nu\bar{\nu}$ process that the matrix element is:

$$\sum_s |\mathcal{M}|^2 = \frac{64}{4} G^2 (f/m_\pi)^4 g_A^2 \Big[(\frac{k^2}{k^2+m_\pi^2})^2 + \ldots \Big] \frac{\omega_\nu \omega_{\bar\nu}}{\omega^2}$$
$$= A \frac{\omega_\nu \omega_{\bar\nu}}{\omega^2}, \qquad (5.117)$$

where the 4 in the denominator accounts for the spin average for identical nucleons, G is the weak coupling constant, f (~ 1.0) is the pion–nucleon coupling constant, g_A is the axial–vector coupling constant, the term in brackets is from the OPE propagator plus exchange and cross terms, k is the nucleon momentum transfer, and m_π is the pion mass. In eq. (5.117), we have dropped $\vec{q}_\nu \cdot \vec{k}$ terms from the weak part of the total matrix element. To further simplify the calculation, we set the "propagator" term equal to a constant ζ, a number of order unity, and absorb into ζ all interaction ambiguities.

Recently, Hanhart, Phillips, & Reddy (2001) have addressed these momentum terms in the context of axion emission and $\nu_\mu \bar\nu_\mu$ production in supernovae. In an effort to make contact with the approximation to the matrix element we present here, they plot ζ as a function of average relative thermal nucleon momentum ($\bar{p}$; Phillips, private communication). The function peaks for $\zeta(\bar{p})$ between $150 - 200$ MeV at $\zeta \simeq 0.47$. At $\bar{p} = 50$ MeV $\zeta \simeq 0.08$ and at $\bar{p} = 500$ MeV $\zeta \simeq 0.27$. We are most interested in the region around the ν_μ neutrinospheres, where the emergent spectrum might be most affected by nucleon-nucleon bremsstrahlung. Mass densities and temperatures in this region might be $10^{12} - 10^{13}$ g cm^{-3} and $5 - 10$ MeV, respectively. We estimate $\bar{p}$ in this regime to be ~ 175 MeV and take $\zeta = 0.5$ for all thermodynamical points. The constant A in eq. (5.117) remains.

Inserting a $\int \delta(\omega - \omega_\nu - \omega_{\bar\nu}) d\omega$ by the neutrino phase space terms times $\omega\omega_\nu\omega_{\bar\nu}/\omega^2$ and integrating over $\omega_{\bar\nu}$ yields:

$$\int \omega \frac{\omega_\nu \omega_{\bar\nu}}{\omega^2} \frac{d^3\vec{q}_\nu}{(2\pi)^3 2\omega_\nu} \frac{d^3\vec{q}_{\bar\nu}}{(2\pi)^3 2\omega_{\bar\nu}} \rightarrow \frac{1}{(2\pi)^4} \int_0^\infty \int_0^\omega \frac{\omega_\nu^2 (\omega - \omega_\nu)^2}{\omega} d\omega_\nu d\omega\,, \qquad (5.118)$$

where again ω equals $(\omega_\nu + \omega_{\bar\nu})$. If we integrate over ω_ν, we can derive the ω spectrum. A further integration over ω will result in the total volumetric energy emission rate. If we delay such an integration, after the nucleon phase space sector has been reduced to a function of ω and if we multiply eq. (5.116) and/or eq. (5.118) by ω_ν/ω, an integration over ω from ω_ν to infinity will leave the emission spectrum for the single final–state neutrino. This is of central use in multi–energy group transport calculations and with this differential emissivity and Kirchhoff's Law (§2) we can derive an absorptive opacity.

Whatever our final goal, we need to reduce the nucleon phase space integrals and to do this we use the coordinates and approach of Brinkman and Turner

(1988). We define new momenta: $p_+ = (p_1 + p_2)/2$, $p_- = (p_1 - p_2)/2$, $p_{3c} = p_3 - p_+$, and $p_{4c} = p_4 - p_+$, where nucleons 1 and 2 are in the initial state. Useful direction cosines are $\gamma_1 = p_+ \cdot p_-/|p_+||p_-|$ and $\gamma_c = p_+ \cdot p_{3c}/|p_+||p_{3c}|$. Defining $u_i = p_i^2/2mT$ and using energy and momentum conservation, we can show that:

$$\begin{aligned} d^3p_1 d^3p_2 &= 8d^3p_+ d^3p_- \\ \omega &= 2T(u_- - u_{3c}) \\ u_{1,2} &= u_+ + u_- \pm 2(u_+u_-)^{1/2}\gamma_1 \\ u_{3,4} &= u_+ + u_{3c} \pm 2(u_+u_{3c})^{1/2}\gamma_c\,. \end{aligned} \tag{5.119}$$

In the non–degenerate limit, the $\mathcal{F}_1\mathcal{F}_2(1 - \mathcal{F}_3)(1 - \mathcal{F}_4)$ term reduces to $e^{2y}e^{-2(u_+ + u_-)}$, where y is the nucleon degeneracy factor. Using eq. (5.119), we see that the quantity $(u_+ + u_-)$ is independent of both γ_1 and γ_c. This is a great simplification and makes the angle integrations trivial. Annihilating d^3p_4 with the momentum delta function in eq. (5.116), noting that $p_i^2 dp = \frac{(2mT)^{3/2}}{2} u_i^{1/2} du_i$, pairing the remaining energy delta function with u_-, and integrating u_+ from 0 to ∞, we obtain:

$$dQ_{nb} = \frac{Am^{4.5}}{2^8 \times 3 \times 5\pi^{8.5}} T^{7.5} e^{2y} e^{-\omega/T} (\omega/T)^4 \Big[\int_0^\infty e^{-x}(x^2 + x\omega/T)^{1/2} dx\Big] d\omega\,. \tag{5.120}$$

The variable x over which we are integrating in eq. (5.120) is equal to $2u_{3c}$. That integral is analytic and yields:

$$\int_0^\infty e^{-x}(x^2 + x\omega/T)^{1/2} dx = \eta e^{\eta} K_1(\eta)\,, \tag{5.121}$$

where K_1 is the standard modified Bessel function of imaginary argument, related to the Hankel functions, and $\eta = \omega/2T$. Hence, the ω spectrum is given by:

$$\frac{dQ_{nb}}{d\omega} \propto e^{-\omega/2T} \omega^5 K_1(\omega/2T)\,. \tag{5.122}$$

It can easily be shown that $\langle\omega\rangle = 4.364T$. Integrating eq. (5.120) over ω and using the thermodynamic identity in the non–degenerate limit:

$$e^y = \Big(\frac{2\pi}{mT}\Big)^{3/2} n_n/2\,, \tag{5.123}$$

where n_n is the density of neutrons (in this case), we derive for the total neutron–neutron bremsstrahlung emissivity of a single neutrino pair:

$$Q_{nb} = 1.04 \times 10^{30} \zeta (X_n \rho_{14})^2 (\frac{T}{\mathrm{MeV}})^{5.5} \,\mathrm{ergs\,cm^{-3}\,s^{-1}}\,, \tag{5.124}$$

where ρ_{14} is the mass density in units of 10^{14} gm cm^{-3} and X_n is the neutron mass fraction. Interestingly, this is within 30% of the result in Suzuki (1993), even though he has substituted, without much justification, $(1 + \omega/2T)$ for the integral in eq. (5.120). ($[1 + (\pi\eta/2)^{1/2}]$ is a better approximation.) The proton–proton and neutron–proton processes can be handled similarly and the total bremsstrahlung rate is then obtained by substituting $X_n^2 + X_p^2 + \frac{28}{3}X_nX_p$ for X_n^2 in eq. (5.124) (Brinkmann and Turner 1988). At $X_n = 0.7$, $X_p = 0.3$, $\rho = 10^{12}$ gm cm^{-3}, and T = 10 MeV, and taking the ratio of augmented eq. (5.124) to eq. (5.105), we obtain the promising ratio of $\sim 5\zeta$. Setting the correction factor ζ equal to ~ 0.5 (Hanhart, Phillips, and Reddy 2001), we find that near and just deeper than the ν_μ neutrinosphere, bremsstrahlung is larger than classical pair production.

If in eq. (5.118) we do not integrate over ω_ν, but at the end of the calculation we integrate over ω from ω_ν to ∞, after some manipulation we obtain the single neutrino emissivity spectrum:

$$\frac{dQ'_{nb}}{d\omega_\nu} = 2C\Big(\frac{Q_{nb}}{T^4}\Big)\omega_\nu^3 \int_{\eta_\nu}^{\infty} \frac{e^{-\eta}}{\eta} K_1(\eta)(\eta - \eta_\nu)^2 d\eta \tag{5.125}$$

$$= 2C\Big(\frac{Q_{nb}}{T^4}\Big)\omega_\nu^3 \int_1^{\infty} \frac{e^{-2\eta_\nu\xi}}{\xi^3}(\xi^2 - \xi)^{1/2} d\xi\,, \tag{5.126}$$

where $\eta_\nu = \omega_\nu/2T$, C is the normalization constant equal to $\frac{3\times5\times7\times11}{2^{11}}$ ($\cong 0.564$), and for the second expression we have used the integral representation of $K_1(\eta)$ and reversed the order of integration. In eq. (5.126), Q_{nb} is the emissivity for the pair.

Eq. (5.126) is the approximate neutrino emission spectrum due to nucleon–nucleon bremsstrahlung. A useful fit to eq. (5.126), good to better than 3% over the full range of important values of η_ν, is:

$$\frac{dQ'_{nb}}{d\omega_\nu} \cong \frac{0.234 Q_{nb}}{T}\Big(\frac{\omega_\nu}{T}\Big)^{2.4} e^{-1.1\omega_\nu/T}\,. \tag{5.127}$$

Thompson, Burrows, and Horvath (2000) should be consulted for a detailed discussion of nucleon-nucleon bremsstrahlung for arbitrary nucleon degeneracy.

9. Conclusion

The processes that have been described above are essential elements of the neutrino-driven supernova explosion mechanism. Coupling these with radiation-hydrodynamics codes, an equation of state, beta-decay and electron capture microphysics, and nuclear rates, one explores the viability of various scenarios for the explosion of the cores of massive stars (Liebendörfer et al. 2001ab; Rampp & Janka 2000). Recently, Thompson, Burrows, and Pinto (2002) have incorporated this neutrino microphysics into simulations of 1D (spherical) core collapse and have investigated the effects on the dynamics, luminosities, and emergent spectra of weak magnetism/recoil, nucleon-nucleon bremsstrahlung, inelastic neutrino-electron scattering, and a host of the cross section corrections described above. Figures 5.8 and 5.9 depict some of the resultant luminosity spectra and their temporal evolution for a representative simulation. The character of the spectra reflect the opacities and sources described in this paper. In particular, the energy hardness hierarchy from ν_e (softer) to ν_μ (harder) neutrinos is clearly demonstrated on these plots, as is the distinction between the ν_e pre-breakout and post-breakout spectra (Fig. 5.8). (See the figure captions for further details.)

To date, none of the detailed 1D simulations that have been performed explodes and it may be that multi-dimensional effects play a pivotal role in the explosion mechanism (Herant et al. 1994; Burrows, Hayes, & Fryxell 1995; Fryer et al. 1999; Janka & Müller 1996). Be that as it may, an understanding of neutrino-matter interactions remains central to unraveling one of the key mysteries of the nuclear universe in which we live.

Acknowledgments

We would like to thank Ray Sawyer, Sanjay Reddy, and Jorge Horvath for fruitful discussions and/or collaboration on some of the more thorny aspects of neutrino-matter interactions. Support for this work was provided by the Scientific Discovery through Advanced Computing (SciDAC) program of the DOE, grant number DE-FC02-01ER41184, and by NASA through Hubble Fellowship grant #HST-HF-01157.01-A awarded by the Space Telescope Science Institute, which is operated by the Association of Universities for Research in Astronomy, Inc., for NASA, under contract NAS 5-26555.

References

Aufderheide, M., Fushiki, I., Fuller, G., and Weaver, T. 1994, *Ap. J.*, 424, 257
Backman, S.-O., Brown, G.E., & Niskanen, J.A. 1985, Physics Reports, 124, 1
Bowers, R. L. & Wilson, J. R. 1982, ApJS, 50, 115
Brinkmann, R. P. & Turner, M. S. 1988, Phys. Rev. D, 38, 8, 2338
Brown, G.E. & Rho, M. 1981, Nucl. Phys., A 372, 397
Bruenn, S. W. 1985, ApJS, 58, 771
Bruenn, S.W. and A. Mezzacappa 1997, *Phys. Rev.* D, 56, 7529
Buras, R., Janka, H.-Th., Keil, M. Th., & Raffelt, G. G., Rampp, M. 2002, submitted to ApJ, (astro-ph/0205006)
Burrows, A., Mazurek T. J., & Lattimer, J. M. 1981, ApJ, 251, 325

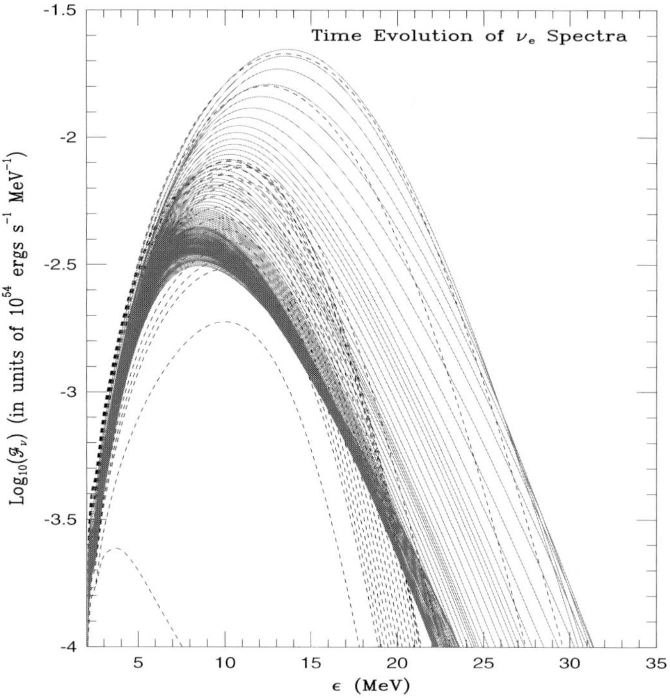

Figure 5.8. A collection of emergent ν_e spectra at various times during the core-collapse, bounce, and shock-stagnation phases of the core of an 11 $M_\odot$ progenitor. The luminosity spectrum (logarithm base ten) is in units of 10^{54} ergs s^{-1} MeV^{-1} and the neutrino energy (abscissa) is in units of MeV. The dashed curves cover the collapse phase (of duration $\sim$200 milliseconds) until just before the peak luminosity around shock breakout is achieved and the solid curves are for the subsequent cooling and deleptonization phases after the peak. The last curve is at $\sim$110 milliseconds after bounce. The lowest dashed curve is at a time early during collapse. Note the relative softness of the spectrum then. As the figure shows, the transition from the dashed to the solid curve happens close to the time when the ν_e spectrum is hardest.

Burrows, A., Hayes, J., & Fryxell, B. A. 1995, ApJ, 450, 830

Burrows, A. and Sawyer, R.F. 1998, *Phys. Rev.* C, 58, 554

Burrows, A. and Sawyer, R.F. 1999, *Phys. Rev.* C, 59, 510

Burrows, A. 2000, Nature, 403, 727

Burrows, A., Young, T., Pinto, P.A., Eastman, R., and Thompson, T. 2000, *Ap. J.*, 539, 865

Dicus, D.A. 1972, *Phys. Rev.* D, 6, 941

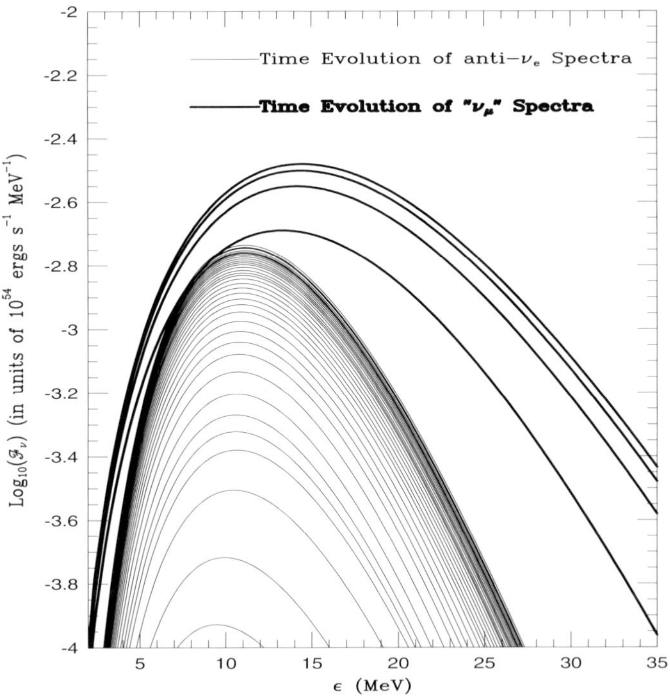

Figure 5.9. This figure shows the $\bar{\nu}_e$ (thin lines) and "ν_μ" (thick lines) emergent luminosity spectra for the 11 $M_\odot$ progenitor evolution depicted in Fig. 5.8. The luminosity spectra (logarithm base ten) are in units of 10^{54} ergs s^{-1} MeV^{-1} and the neutrino energy (abscissa) is in units of MeV. There is no appreciable flux prior to shock breakout for these species. To avoid clutter, we here depict only a few ν_μ spectra to $\sim$50 milliseconds after bounce. (These curves represent the sum of the ν_μ, $\bar{\nu}_\mu$, ν_τ, and $\bar{\nu}_\tau$ luminosity spectra.) However, the $\bar{\nu}_e$ spectra are shown until about 110 milliseconds after bounce. During the phases shown, both sets of luminosities are always increasing. Note that the ν_μ spectra are significantly harder than either the $\bar{\nu}_e$ or the ν_e spectra. This is a consequence of of the fact that the ν_μs do not have appreciable charged-current cross sections (eqs. 5.10 and 5.11), enabling one to probe more deeply into the hot core with these species.

Fetter, A.L. & Walecka, J.D. 1971, *Quantum Theory of Many Particle Systems* (New York: McGraw-Hill)
Flowers, E., Sutherland, P., and Bond, J.R. 1975, *Phys. Rev.* D, 12, 316
Freedman, D.Z. 1974, *Phys. Rev.* D, 9, 1389
Fryer, C. L., Benz, W., Herant, M., & Colgate, S. 1999, ApJ, 516, 892
Fuller, G. 1982, *Ap. J.*, 252, 741

Fuller, G. M., Fowler, W. A., & Newman, M. J. 1982, ApJ, 252, 715
Hanhart, C., Phillips, D. & Reddy, S. 2001, Phys. Lett. B, 499, 9
Hannestad, S. and Raffelt, G. 1998, *Ap. J.*, 507, 339
Herant, M., Benz, W., Hix, W. R., Fryer, C. L., Colgate, S. A. 1994, ApJ, 435, 339
Horowitz, C. J. 1997, Phys. Rev. D, 55, no. 8
Horowitz, C. J. 2002, Phys Rev. D, 65, 043001
Janka, H.-T. 1991, *Astr. Ap.*, 244, 378
Janka, H.-Th. & Müller, E. 1996, A&A, 306, 167
Janka, H.-T., Keil, W., Raffelt, G., and Seckel, D. 1996, *Phys. Rev. Lett.*, 76, 2621
Lamb, D. & Pethick, C. 1976, ApJL, 209, L77
Landau, L.D. and Lifschitz, E.M. 1969, *Statistical Physics*, 2'nd edition (Pergamon Press; New York) p.352
Leinson, L.B., Oraevsky, V.N., & Semikoz, V.B. 1988, Phys. Lett. B, 209, 1
Liebendörfer, M., Mezzacappa, A., Thielemann, F.-K., Messer, O. E. B., Hix, W. R., & Bruenn, S. W. 2001, PRD, 63, 103004
Liebendörfer, M., Mezzacappa, A., Thielemann, F.-K. 2001, PRD, 63, 104003
Mezzacappa, A. & Bruenn, S. W. 1993a, ApJ, 410, 637
Mezzacappa, A. & Bruenn, S. W. 1993b, ApJ, 410, 669
Mezzacappa, A. & Bruenn, S. W. 1993c, ApJ, 410, 740
Raffelt, G. & Seckel, D. 1998, Phys. Rev. Lett., 69, 2605
Raffelt, G. 2001, ApJ, 561,890
Rampp, M. & Janka, H.-Th. 2000, ApJL, 539, 33
Reddy, S., Prakash, M., & Lattimer, J. M. 1998, PRD, 58, 013009
Schinder, P.J. 1990, ApJS, 74, 249
Smit, J.M. 1998, Ph.D. Thesis, Universiteit van Amsterdam
Smit, J.M. & Cernohorsky, J. 1996, A&A, 311, 347
Suzuki, H. in *Frontiers of Neutrino Astrophysics*, ed. Suzuki, Y. & Nakamura, K. 1993, (Tokyo: Universal Academy Press), 219
Thompson, T. A., Burrows, A., & Horvath, J. E. 2000, PRC, 62, 035802
Thompson, T. A., Burrows, A., & Pinto, P.A. 2002, in preparation
Tubbs, D. L. & Schramm, D. N. 1975, ApJ, 201, 467
Vogel, P. 1984, Phys. Rev. D, 29, 1918
Yamada, S., Janka, H.-T., and Suzuki, H. 1999, *Astr. Ap.*, 344, 533

Chapter 6

PROTONEUTRON STAR WINDS

Todd A. Thompson*
Astronomy Department and Theoretical Astrophysics Center,
601 Campbell Hall, The University of California, Berkeley, CA 94720
thomp@astro.berkeley.edu

Abstract Neutrino-driven winds are thought to accompany the Kelvin-Helmholtz cooling phase of nascent protoneutron stars in the first seconds after a core-collapse supernova. These outflows are a likely candidate as the astrophysical site for rapid neutron-capture nucleosynthesis (the r-process). In this chapter we review the physics of protoneutron star winds and assess their potential as a site for the production of the heavy r-process nuclides.

We show that spherical transonic protoneutron star winds do not produce robust r-process nucleosynthesis for 'canonical' neutron stars with gravitational masses of $1.4\,M_\odot$ and coordinate radii of 10 km. We further speculate on and review some aspects of neutrino-driven winds from protoneutron stars with strong magnetic fields.

Keywords: nuclear reactions, nucleosynthesis, abundances — stars: magnetic fields — stars: winds, outflows — stars: neutron — supernovae: general

1. Introduction

r-Process Nucleosynthesis

A complete and self-consistent theory of the origin of the elements has been the grand program of nuclear astrophysics since the field's inception. Of the several distinct nuclear processes which combine to produce the myriad of stable nuclei and isotopes we observe, none has generated more speculation than r-process nucleosynthesis (Wallerstein 1997). The r-process, or rapid neutron-capture process, originally identified in Burbidge et al. (1957) and Cameron (1957), is a mechanism for nucleosynthesis by which seed nuclei

*Hubble Fellow

C.L. Fryer (ed.), Stellar Collapse, 175–202.

capture neutrons on timescales much shorter than those for β^- decay. The rapid interaction of neutrons with heavy, neutron-rich, seed nuclei allows a neutron capture-disintegration equilibrium to establish itself among the isotopes of each element. The nuclear flow proceeds well to the neutron-rich side of the valley of β-stability and for sufficient neutron-to-seed ratio ($\gtrsim$100) the r-process generates the heaviest nuclei (e.g., Eu, Dy, Th, and U), forming characteristic abundance peaks at $A \sim 80$, 130, and 195 (e.g. Burbidge et al. 1957; Meyer & Brown 1997; Wallerstein et al. 1997).

The neutron-to-seed ratio is the critical parameter in determining if r-process nucleosynthesis succeeds in producing nuclei up to and beyond the third abundance peak. If the neutron-to-seed ratio is too small, the r-process may only nuclei up to the first or second abundance peak. In a given hydrodynamical flow, the neutron-to-seed ratio is set primarily by the entropy entropy($s_{\rm a}$; 'a' here stands for 'asymptotic') of the flow, the neutron richness of the matter, and the dynamical timescale for expansion. The neutron richness is generally quantified by the the electron fraction ($Y_e^{\rm a}$, the number density of electrons per baryon). The dynamical timescale ($\tau_{\rm dyn}$) is a characteristic time for expansion, i.e. the e-folding time for temperature or density at a given temperature. Thus, $s_{\rm a}$, $Y_e^{\rm a}$, and $\tau_{\rm dyn}$ set the neutron-to-seed ratio. The higher the entropy, the lower the electron fraction, and the shorter the dynamical timescale, the larger the neutron-to-seed ratio and the higher in A the nucleosynthetic flow will proceed (e.g. Hoffman, Woosley, & Qian 1997; Meyer & Brown 1997).

Observational Motivation: A Remarkable Concordance

Recent observations of neutron-capture elements in ultra-metal-poor ([Fe/H] $\lesssim -2.5$] halo stars (Sneden et al. 1996; Burris et al. 2000; McWilliam et al. 1995a,b; Cowan et al. 1996; Westin et al. 2000; Hill et al. 2001) show remarkable agreement with the scaled solar r-process abundance pattern for $A \gtrsim 135$. Particularly for atomic numbers between $N = 55$ and $N = 75$, the distribution of abundances in these halo stars is identical with solar. Prototypical of this class are the stars CS 22892-052, BD +17°3248 and HD 115444 (Cowan & Sneden 2002). Figure 6.1 shows the scaled solar r-process abundances (solid line) together with those from the ultra-metal-poor halo star CS 22892-052 (points with error bars) as a function of atomic number. Note both the tight correspondence between the sun and CS 22892-052 above the second r-process peak and the large discrepancies at lower atomic number, in particular the elements Sn, Ag, and Y. The fact that a class of very old halo stars share the same relative abundance of r-process elements above the second peak suggests a universal mechanism for producing these nuclei, which must act early in the chemical enrichment history of the galaxy.

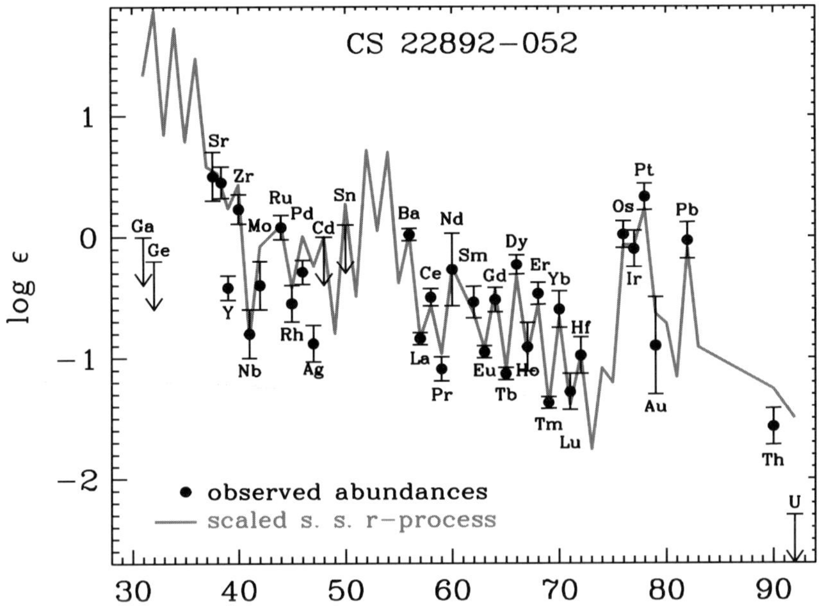

Figure 6.1. The scaled solar r-process abundances (solid line) and the abundances from CS 22892-052 (dots with error bars) as a function of atomic number (Chris Sneden, private communication).

The fact that observations of ultra-metal-poor halo stars show significant and scattered deviations from the scaled solar r-process abundance pattern below $A \sim 130$ has implied that there are perhaps two or more r-process sites (Qian, Vogel, & Wasserburg 1998; Wasserburg & Qian 2000; Qian & Wasserburg 2000). In these models of r-process enrichment, distinct astrophysical sites account for the observed r-process abundances below and above $A \sim 130$. Important in these considerations are the constraints set on the site or sites of the r-process by the integrated galactic r-process budget. For example, if all supernovae produce neutron stars that are accompanied by protoneutron star winds that produce a robust r-process, then, given a supernova rate of one every 50-100 years, 10^{-5}–$10^{-6}\,M_\odot$ of r-process material must be injected into the interstellar medium per event (Qian 2000). These numbers are to be contrasted with those from any other potential r-process site. For example, if the material possibly ejected to infinity in the formation of a black hole from neutron-star/neutron-star mergers accounts for the total galactic r-process budget (Freiburghaus, Rosswog, & Thielemann 1999, and references therein; Rosswog et al. 1999), then 10^{-1}–$10^{-2}\,M_\odot$ of r-process material must be ejected per event, given an event rate of one every 10^5 years (Kalogera et al. 2001; Qian 2000 and references therein).

Because the total r-process budget of the galaxy is fairly well known, any potential site must do more than just match the entropy, electron fraction, and dynamical timescales required for a high neutron-to-seed ratio. Simply attaining the necessary physical conditions for a neutron-to-seed ratio above $\sim$100 is not sufficient. The astrophysical site, given an event rate, must be able to consistently produce the robust r-process abundance pattern in ultra-metal-poor halo stars in accordance with the total galactic r-process budget. Thus, if $M_{\rm ej}^{\rm r}$ is the total mass of r-process material produced per event, the process of assessing an astrophysical site's potential for the r-process is merely a matter of mapping the region in $s_{\rm a} - Y_e^{\rm a} - \tau_{\rm dyn} - M_{\rm ej}^{\rm r}$ space physically accessible to the potential site. In what follows here we review the results for just-post-supernova, neutrino-driven protoneutron star winds .

Protoneutron Star Winds

The successful two-dimensional supernova explosion obtained by Burrows, Hayes, & Fryxell (1995) shows clearly a post-explosion neutrino-driven wind emerging into the evacuated region above the newly formed protoneutron star and behind the rapidly expanding supernova shock. Although no study was made of this outflow as a function of progenitor, such a wind phase might naturally accompany the post-explosion cooling epoch in many core-collapse supernovae.

A multiple of 10^{53} erg of binding energy will be lost as the nascent neutron star cools and contracts over its Kelvin-Helmholtz cooling time ($\sim 10 - 20$ seconds). This energy will be carried away predominantly by all species of neutrino. A small fraction of that energy will be deposited in the surface layers of the protoneutron star, ablating material from its upper atmosphere and driving a hydrodynamical wind. Whether this wind succeeds in escaping to infinity, or is prevented by fallback and reverse shocks as the supernova shock encounters the density stratifications of the overlying stellar mantle is an important future area of study. In addition, the actual emergence of the wind - how it overcomes the non-zero pressure of the matter exterior to the protoneutron star - has not been fully explored. In this work, however, we assume the existence of such outflows as evidenced by the calculation of Burrows, Hayes, & Fryxell (1995) as well as those of Janka & Müller (1995), and Takahashi, Witti, & Janka (1994). Our goal is to explore the basic physics of neutrino-driven outflows and assess this site as a candidate for r-process nucleosynthesis.

The basic scenario is as follows: as the supernova shock propagates outward in a successful explosion, the pressure in the region between the protoneutron star and the shock decreases and the wind, powered by neutrino heating, emerges into the post-shock ejecta. The surface of the protoneutron star is hot (temperatures of ~ 5 MeV) and has a low electron fraction (typically, $Y_e \sim 0.1$). The matter there is composed of relativistic charged leptons, free nucleons, and trapped photons. As the wind is driven outward, the matter descends a gradient in density and temperature. The wind material is heated only within the first $\sim$50 km, primarily via the charged-current absorption processes on free nucleons $\nu_e n \to pe^-$ and $\bar{\nu}_e p \to ne^+$, and its entropy increases concomitantly. As the temperature of the matter drops below ~ 1 MeV, nucleons combine into alpha particles, neutrino heating ceases and the material expands adiabatically with entropy $s_{\rm a}$. These charged-current processes also set the asymptotic electron fraction. As the matter expands away from the protoneutron star surface and the chemical equilibrium between ν_e and $\bar{\nu}_e$ neutrinos obtained near the neutrino decoupling radius (the neutrinosphere) is broken, the luminosity and energy density of the electron-type neutrino species determine the electron fraction. Typically, within 10 km of the protoneutron star surface Y_e asymptotes to $Y_e^{\rm a}$. At a temperature of $\sim$0.5 MeV (a radius in most models of $\sim$100 km) alpha particles combine with the remaining free neutrons in an α-process to form seed nuclei (Woosley & Hoffman 1992). At this location in the temperature profile, it is the steepness of the density or temperature gradient that sets the dynamical timescale of the wind. Simply stated, short dynamical timescales are favored for the r-process because for faster expansions there is less time to build seed nuclei, and, hence, the neutron-to-seed ratio is preferentially larger, all else being equal. When the wind material finally reaches a temperature of $\sim$0.1 MeV (at a radius of several hundred kilometers) the r-process may begin if the neutron-to-seed ratio, as set by $Y_e^{\rm a}$, $s_{\rm a}$, and $\tau_{\rm dyn}$, is sufficiently high.

Because the success of the r-process is so dependent on the neutron-to-seed ratio and, hence, the electron fraction, entropy, and dynamical timescale of the nucleosynthetic environment, we focus on these quantities and their sensitivity to various parameters of the protoneutron star. In addition to establishing correlations between these quantities in neutrino-driven wind environments, we must also fold in the constraint on the total mass ejected. We will show in §4 that this constraint on $M_{\rm ej}^{\rm r}$ significantly limits the $s_{\rm a} - Y_e^{\rm a} - \tau_{\rm dyn}$ space of relevance for winds.

Previous Work

Duncan, Shapiro, & Wasserman (1986) were the first to explore the physics of neutrino-driven winds. They identified a number of important scaling relations and the basic systematics and dependencies of the problem. In addition, they explored the relative importance of the neutrino and photon luminosities in determining the resulting hydrodynamics. Recent investigations have focused on the potential of these outflows for r-process nucleosynthesis as suggested in

Woosley & Hoffman (1992). Witti, Janka, & Takahashi (1994) and Takahashi, Witti, & Janka (1994) showed that although interesting α-process nuclei were created in their models of protoneutron star winds, conditions for a successful r-process fell short in entropy by a factor of ~ 5. Qian & Woosley (1996) made analytical estimates of the fundamental wind properties and systematics and compared their scalings to numbers from hydrodynamical simulations. They further explored interesting variations to their models such as inserting artificial heating sources and applying an external boundary pressure. These results were followed by nucleosynthetic calculations, taking wind trajectories as the time-history of Lagrangean mass elements in Hoffman, Woosley, & Qian (1997). These efforts also showed that with reasonable variations in the input physics protoneutron star winds do not achieve high enough entropy for the dynamical timescales and electron fractions derived.

Cardall & Fuller (1997) extended the analytical work of Qian & Woosley (1996) to general-relativistic flows. They found that significant enhancements in entropy might be obtained from compact neutron stars with large M/R. Several general-relativistic treatments of the problem followed, including the work of Otsuki et al. (2000), Sumiyoshi et al. (2000), Wanajo et al. (2001), and Thompson, Burrows, & Meyer (2001). In the last of these, purely transonic steady-state wind solutions were derived in general relativity, with a simultaneous solution for the evolution of the electron fraction in radius and a careful treatment of the boundary conditions.

Note that we distinguish here between models of protoneutron star *winds* and *bubbles*. The former is typified by those works already discussed, in which the wind velocity approaches the local speed of sound, or a large fraction of the speed of sound somewhere in the flow. Such winds were realized in the self-consistent supernova calculations in Burrows, Hayes, & Fryxell (1995). In contrast, in the work of Woosley et al. (1994) the protoneutron star outflow reached speeds of only a very small fraction of the sound speed in the region between the protoneutron star and the expanding supernova shock. Approximately 18 seconds after collapse and explosion, Woosley et al. (1994) obtained entropies of $\sim$400 (throughout, we quote entropy in units of k_B per baryon), long dynamical timescales, and electron fraction in the range $0.36-0.44$. However, in their model the supernova shock reached only 50,000 km at these late times. In turn, this external boundary caused the wind material to move slowly. It remained in the heating region for an extended period, thus raising the entropy above what any simulation or analytical calculation has since obtained. Although the r-process proceeded to the third abundance peak in their calculation, nuclei in the mass range near $A \sim 90$ (particularly, ^{88}Sr, ^{89}Y, and ^{90}Zr) were overproduced by more than a factor of 100.

This Review

In §2, we review the fundamental and general equations for time-independent energy-deposition-driven winds in Newtonian gravity and in general relativity. We also discuss the integrals of the flow and our numerical procedure for solving the relevant equations. In §3 we discuss some of the particulars of modeling protoneutron star winds, including the neutrino heating function, the equation of state, and the evolution of the electron fraction. We further critically examine several of the underlying assumptions that must accompany any such model. §4 summarizes our results for spherical winds. In §5 we consider the possible effects of magnetic fields and speculate on a number of issues in need of more thorough investigation. In §6 we summarize and conclude.

2. Hydrodynamics

The Newtonian Wind Equations

Assuming time-independent wind solutions, the equation for mass conservation is simply

$$\nabla \cdot (\rho \mathbf{v}) = 0, \tag{6.1}$$

implying that the mass-outflow rate ($\dot{M}$) of a wind is a constant in radius. In spherical symmetry, $\dot{M} = 4\pi r^2 \rho v = \mathrm{constant}$. This expression yields a differential equation for the evolution of the matter velocity in radius,

$$\frac{1}{v}\frac{dv}{dr} = -\frac{1}{\rho}\frac{d\rho}{dr} - \frac{2}{r}. \tag{6.2}$$

The equation for momentum conservation, neglecting the mass of the wind itself, is simply

$$v\frac{dv}{dr} = -\frac{1}{\rho}\frac{dP}{dr} - \frac{GM}{r^2} + F_\nu, \tag{6.3}$$

where M is the total mass of the protoneutron star. Although we include it here for completeness, the radiation force due to the neutrinos (F_ν) can be safely neglected. This approximation is justified because the neutrino Eddington luminosity ($L_\nu^{\mathrm{Edd}} = 4\pi GMc/\kappa_\nu$) is much larger than the neutrino luminosities that accompany the protoneutron star cooling/wind evolutionary phase. κ_ν is the total neutrino opacity and is dominated by $\nu_e n \to pe^-$ and $\bar{\nu}_e p \to ne^+$ for the electron and anti-electron neutrino, respectively. The μ- and τ-neutrino opacity is dominated by neutral-current scattering off free nucleons, the wind heating region being unpopulated by nuclei. Including these processes, one finds that $L_\nu^{\mathrm{Edd}} \gtrsim 10^{55}$ erg s^{-1}. We are thus safe in taking $F_\nu = 0$ because we consider winds with only $L_\nu^{\mathrm{tot}} < 5 \times 10^{52}$ erg s^{-1}.

Because neutrinos contribute heating and cooling to the flow, we must couple to these equations to the first law of thermodynamics. We define the net specific heating rate, $\dot{q} = \text{Heating} - \text{Cooling}$, so that

$$\frac{d\epsilon}{dt} = \dot{q} + \frac{P}{\rho^2}\frac{d\rho}{dt}, \tag{6.4}$$

where $d/dt = [\partial/\partial t + v \cdot \nabla]$. In the steady state, $\partial/\partial t = 0$ and we obtain

$$\dot{q} = Tv\frac{ds}{dr} = C_V v\frac{dT}{dr} - \frac{vT}{\rho^2}\left.\frac{\partial P}{\partial T}\right|_\rho \frac{d\rho}{dr}. \tag{6.5}$$

We choose to reduce eqs. (6.2), (6.3), and (6.5) to a set of coupled differential equations for dv/dr, $d\rho/dr$, and dT/dr that make the physics of the wind solution manifest and the solution to the problem more easily obtained. We start by eliminating the pressure. Expanding P differentially in ρ and T, we have that

$$dP = \left.\frac{\partial P}{\partial \rho}\right|_T \delta\rho + \left.\frac{\partial P}{\partial T}\right|_\rho \delta T. \tag{6.6}$$

Defining (D) as

$$D = \frac{T}{\rho}\left.\frac{\partial P}{\partial T}\right|_\rho, \tag{6.7}$$

we obtain

$$c_s^2 = c_T^2 + \frac{D^2}{C_V T}. \tag{6.8}$$

where C_V is the specific heat at constant volume, $c_s (= \partial P/\partial\rho|_s)$ is the adiabatic sound speed, and $c_T (= \partial P/\partial\rho|_T)$ is the isothermal sound speed. Taking eq. (6.6) and dropping F_ν, we can rewrite eq. (6.3) as

$$v\frac{dv}{dr} = -\frac{1}{\rho}\left[\left.\frac{\partial P}{\partial \rho}\right|_T \frac{d\rho}{dr} + \left.\frac{\partial P}{\partial T}\right|_\rho \frac{dT}{dr}\right] - \frac{GM}{r^2} \tag{6.9}$$

We can now eliminate dT/dr using eq. (6.5) so that

$$v\frac{dv}{dr} = -\frac{1}{\rho}\left\{\left.\frac{\partial P}{\partial \rho}\right|_T \frac{d\rho}{dr} + \left.\frac{\partial P}{\partial T}\right|_\rho \left[\frac{T}{C_V\rho^2}\left.\frac{\partial P}{\partial T}\right|_\rho \frac{d\rho}{dr} + \frac{\dot{q}}{C_V v}\right]\right\} - \frac{GM}{r^2} \tag{6.10}$$

The terms $d\rho/dr$ can be eliminated using eq. (6.2). Using eq. (6.8) and combining terms proportional to dv/dr, we obtain an expression for dv/dr in terms of thermodynamic quantities returned by the equation of state (D, P, C_V, c_s, etc.), the basic dependent hydrodynamical variables (ρ, v, and T), and the neutrino energy deposition function ($\dot{q}$):

$$\frac{dv}{dr} = \frac{v}{2r}\left(\frac{v_e^2 - 4c_s^2}{c_s^2 - v^2}\right) + \frac{D}{C_V T}\frac{\dot{q}}{c_s^2 - v^2}, \tag{6.11}$$

where $v_e = (2GM/r)^{1/2}$ is the escape velocity. Combining eq. (6.11) with eq. (6.2) and then with eq. (6.5), we obtain expressions for $d\rho/dr$ and dT/dr, respectively;

$$\frac{d\rho}{dr} = \frac{2\rho}{r}\left(\frac{v^2 - v_e^2/4}{c_s^2 - v^2}\right) - \frac{\rho}{v}\frac{D}{C_V T}\frac{\dot{q}}{c_s^2 - v^2} \tag{6.12}$$

and

$$\frac{dT}{dr} = \frac{2}{r}\frac{D}{C_V}\left(\frac{v^2 - v_e^2/4}{c_s^2 - v^2}\right) + \frac{\dot{q}}{C_V v}\left(\frac{c_T^2 - v^2}{c_s^2 - v^2}\right). \tag{6.13}$$

The General-Relativistic Wind Equations

The time-independent hydrodynamical equations for flow in a Schwarzschild spacetime can be written in the form (Thorne, Flammang, & Zytkow 1981; Flammang 1982; Nobili, Turolla, and Zampieri 1991)

$$\frac{1}{vy}\frac{d(vy)}{dr} + \frac{1}{\rho}\frac{d\rho}{dr} + \frac{2}{r} = 0, \tag{6.14}$$

$$\frac{1}{y}\frac{dy}{dr} + \frac{1}{\varepsilon + P}\frac{dP}{dr} = 0, \tag{6.15}$$

and

$$\frac{d\varepsilon}{dr} - \frac{\varepsilon + P}{\rho}\frac{d\rho}{dr} + \rho\frac{\dot{q}}{vy} = 0, \tag{6.16}$$

where $u_r(= vy)$ is the radial component of the fluid four-velocity, v is the velocity of the matter measured by a stationary observer,

$$y = \left(\frac{1 - 2GM/rc^2}{1 - v^2/c^2}\right)^{1/2}, \tag{6.17}$$

$\varepsilon\,(= \rho c^2 + \rho\epsilon)$ is the total mass-energy density, ρ is the rest-mass density, P is the pressure, ϵ is the specific internal energy, and $\dot{q}$ is the energy deposition rate per unit mass. These equations assume that the mass of the wind is negligible, as in the Newtonian derivation. Although not readily apparent in the form above, eqs. (6.14)–(6.16) exhibit a critical point when v equals the local speed of sound. In order to make the solution to this system tractable and the critical point manifest we recast the equations in the same form as eqs. (6.11-6.13):

$$\frac{dv}{dr} = \frac{v}{2r}\left[\frac{v_e^2}{y^2}\left(\frac{1 - c_s^2/c^2}{c_s^2 - v^2}\right) - 4c_s^2\left(\frac{1 - v^2/c^2}{c_s^2 - v^2}\right)\right]$$

$$+\frac{D}{C_V T}\frac{\dot{q}}{y}\left(\frac{1-v^2/c^2}{c_s^2-v^2}\right), \tag{6.18}$$

$$\frac{d\rho}{dr}=\frac{2\rho}{r}\left(\frac{v^2-v_e^2/4y^2}{c_s^2-v^2}\right)-\frac{\rho}{(vy)}\frac{D}{C_V T}\frac{\dot{q}}{c_s^2-v^2}, \tag{6.19}$$

and

$$\frac{dT}{dr}=\frac{2}{r\rho}\frac{D}{C_V}\frac{(P+\varepsilon)}{c^2}\left(\frac{v^2-v_e^2/4y^2}{c_s^2-v^2}\right)$$

$$+\frac{\dot{q}}{C_V(vy)}\left(\frac{(1-D/c^2)c_T^2-v^2}{c_s^2-v^2}\right). \tag{6.20}$$

In the above expressions M is the protoneutron star gravitational mass and

$$D=c^2\frac{T}{\varepsilon+P}\left.\frac{\partial P}{\partial T}\right|_\rho . \tag{6.21}$$

Note that by taking the limits $v/c \ll 1$ and $c_s/c \ll 1$, we recover the Newtonian wind equations in critical form. Note that by taking the limits $v/c \ll 1$ and $c_s/c \ll 1$, we recover the Newtonian wind equations in critical form.

Conservation and Numerics

There are two integrals of the flow which we use to gauge the accuracy of our solution to the Newtonian or general-relativistic wind equations. The first is the mass outflow rate, obtained from direct integration of the continuity equation (eq. 6.14), which yields the eigenvalue of the steady-state wind problem, $\dot{M} = 4\pi r^2 \rho v y$ ($y = 1$ in the Newtonian limit). The second is the Bernoulli integral, modified by energy deposition. In the Newtonian case,

$$\dot{M}\Delta\left(\epsilon+\frac{1}{2}v^2+\frac{P}{\rho}-\frac{GM}{r}\right)=\int_{R_\nu}^{r} d^3r'\,\rho\,\dot{q}=Q(r), \tag{6.22}$$

where R_ν is the coordinate radius of the protoneutron star surface and the Δ expresses the change in the quantity in parentheses between R_ν and r. We retain the subscript ν on R to emphasize that throughout this work the coordinate radius of the protoneutron star is assumed to coincide with the radius of decoupling for all neutrino species. In general relativity , with $\dot{q} = 0$, $\gamma h\sqrt{-g_{00}}$ is a constant. Here, γ is the Lorenz factor and h is the specific enthalpy. With a source term, the differential change in neutrino luminosity is given by

$$e^{-2\phi}\frac{\partial}{\partial\mu}(L_\nu e^{2\phi})=-\dot{q}, \tag{6.23}$$

where $d\mu/dr = 4\pi r^2 \rho\, e^{\Lambda}$. The expression $ds^2 = -e^{2\phi}dt^2 + e^{2\Lambda}dr^2 + r^2 d\Omega$ defines the metric. The total energy deposition rate is then,

$$Q = 4\pi \int_{R_\nu}^{\infty} dr\, r^2\, \rho\, \dot{q}\, e^{\Lambda}\, e^{2\phi}. \tag{6.24}$$

We solve the system of wind equations in critical form using a two-point relaxation algorithm between R_ν and the sonic point on an adaptive radial mesh (see Thompson, Burrows, & Meyer 2001; Press et al. 1992; London & Flannery 1982). We employ physical boundary conditions for the transonic wind problem, enforcing thermal ($\dot{q} = 0$) and chemical equilibrium ($dY_e/dr = 0$; see §3) at R_ν and $v = c_s$ at the outer edge of the computational domain. The last required boundary condition, which closes the system of equations, is an integral condition on the electron-neutrino optical depth (τ): $\tau_{\nu_e}(R_\nu) = 2/3$. This condition is combined with the other two at R_ν in a triple Newton-Raphson algorithm so that all boundary conditions are satisfied simultaneously. For modest radial zoning we typically maintain constant $\dot{M}$ and consistent Bernoulli integral to better than 1%, having imposed neither as a mathematical constraint on the system of wind equations.

3. Particulars of Protoneutron Star Winds

Electron Fraction Evolution

The charged-current electron-type neutrino interactions on free nucleons – $\nu_e n \leftrightarrow e^- p$ and $\bar{\nu}_e p \leftrightarrow e^+ n$ – affect the evolution of the electron fraction in radius;

$$(vy)\frac{dY_e}{dr} = X_n[\Gamma_{\nu_e n} + \Gamma_{e^+ n}] - X_p[\Gamma_{\bar{\nu}_e p} + \Gamma_{e^- p}], \tag{6.25}$$

where X_n and X_p are the neutron and proton fraction, respectively. The Γ's are the number rates for emission and absorption, taken from the approximations of Qian & Woosley (1996). The subscripts denote initial-state particles. Because the resulting nucleosynthesis in a given wind model is so sensitive to the Y_e at the start of the r-process, we solve the coupled system of the wind equations with dY_e/dr simultaneously. Ignoring the details of transport and neutrino decoupling near the neutrinospheres, the asymptotic electron fraction $Y_e^{\rm a}$ is determined by both the luminosity ratio $L_{\bar{\nu}_e}/L_{\nu_e}$ and the energy ratio $\langle\varepsilon_{\bar{\nu}_e}\rangle/\langle\varepsilon_{\nu_e}\rangle$, where $\langle\varepsilon_\nu\rangle = \langle E_\nu^2\rangle/\langle E_\nu\rangle$, and E_ν is the neutrino energy. To rough approximation (Qian et al. 1993; Qian & Woosley 1996),

$$Y_e^{\rm a} \simeq \frac{\Gamma_{\nu_e n}}{\Gamma_{\nu_e n} + \Gamma_{\bar{\nu}_e p}} \simeq \left(1 + \frac{L_{\bar{\nu}_e}}{L_{\nu_e}} \frac{\langle\varepsilon_{\bar{\nu}_e}\rangle - 2\Delta + 1.2\Delta^2/\langle\varepsilon_{\bar{\nu}_e}\rangle}{\langle\varepsilon_{\nu_e}\rangle + 2\Delta + 1.2\Delta^2/\langle\varepsilon_{\nu_e}\rangle}\right)^{-1}, \tag{6.26}$$

where $\Delta (= m_n - m_p \simeq 1.293$ MeV) is the energy threshold for the $\bar{\nu}_e$ neutrino absorption process, $\bar{\nu}_e p \rightarrow n e^+$. There are several important effects in protoneutron star winds which act to increase $Y_e^{\rm a}$: (1) The Threshold Effect : even if $\langle\varepsilon_{\bar{\nu}_e}\rangle/\langle\varepsilon_{\nu_e}\rangle$ is constant in time, if both $\langle\varepsilon_{\bar{\nu}_e}\rangle$ and $\langle\varepsilon_{\nu_e}\rangle$ decrease, $Y_e^{\rm a}$ must increase as a result of Δ and (2) The Alpha Effect: as the flow cools in moving away from the protoneutron star and α-particles are preferentially formed, residual excess neutrons capture ν_e neutrinos, thus increasing $Y_e^{\rm a}$ (Fuller & Meyer 1995; McLaughlin, Fuller, & Wilson 1996). . The α-effect is particularly important for flows with long dynamical timescales.

Neutrino Energy Deposition

The Charged Current Processes: In the protoneutron star wind context, the charged-current processes ($\nu_e n \leftrightarrow e^- p$ and $\bar{\nu}_e p \leftrightarrow e^+ n$) compete with neutrino-electron/positron scattering as the dominant energy deposition mechanisms. Ignoring final-state blocking and assuming relativistic electrons and positrons, the charged-current specific cooling rate can be written as

$$C_{\rm cc} \simeq 2.0 \times 10^{18}\, T^6 \left[X_p \frac{F_5(\eta_e)}{F_5(0)} + X_n \frac{F_5(-\eta_e)}{F_5(0)}\right], \tag{6.27}$$

where $F_n(y) = \int_0^\infty x^n (e^{x-y}+1)^{-1}\, dx$, T is in MeV, and $\eta_e = \mu_e/T$. The heating rate is

$$H_{\rm cc} \simeq 9.3 \times 10^{18} R_{\nu 6}^{-2} \left[X_n\, L_{\nu_e}^{51}\, \langle\varepsilon_{\nu_e}^2\rangle + X_p\, L_{\bar{\nu}_e}^{51}\, \langle\varepsilon_{\bar{\nu}_e}^2\rangle\right] \Phi^6\, \Xi(r), \tag{6.28}$$

where $R_{\nu 6}$ is the neutrinosphere radius in units of 10^6 cm, $L_\nu(= 10^{51} L_\nu^{51})$ and $\langle\varepsilon_\nu^2\rangle$ are defined at R_ν, $\Phi = [(1 - 2GM/R_\nu c^2)/(1 - 2GM/rc^2)]^{1/2}$ is the gravitational redshift, and $\Xi(r)$ is the spherical dilution function. In the vacuum approximation and assuming a sharp neutrinosphere,

$$\Xi(r) = 1 - \sqrt{1 - (R_\nu/r)^2/\Phi^2}. \tag{6.29}$$

The redshift term, Φ, appearing in eq. (6.29), accounts for the amplification of the heating processes due to the bending of null geodesics in general relativity (Salmonson & Wilson 1999). Cardall & Fuller (1997) showed that although this amplification is important, the dominant general-relativistic effect on the heating rates is due to the gravitational redshift of the neutrino energy and luminosity (note the Φ^6 in eq. 6.28).

With eqs. (6.27) and (6.28), the specific energy deposition rate (erg g^{-1} s^{-1}) due to the charged-current processes is simply $\dot{q}_{cc} = H_{cc} - C_{cc}$.

Inelastic Neutrino Scattering: At high entropies, electron-positron pairs are produced in abundance. The energy transfer associated with a single neutrino-electron or positron scattering event is well approximated by $\omega_i \simeq (\varepsilon_{\nu_i} - 4T)/2$, where ω is the energy transfer and i labels the neutrino species (Bahcall 1964). Note that ω allows for both net heating or net cooling, depending upon the local temperature and the neutrino spectral characteristics.

A number of researchers have dropped the '$4T$' cooling part from ω (Qian & Woosley 1996; Otsuki et al. 2000). This leads to larger net energy deposition and significant modifications to the wind solution. We have made a comparison between models with and without this term and find that omitting cooling leads to a $40 - 60\%$ increase in $\dot{M}$ and a $10 - 25\%$ decrease in $\tau_{\rm dyn}$. The asymptotic mechanical luminosity, $\dot{M}v_\infty^2/2$, is increased by as much as 80% and the asymptotic entropy is decreased at the 5% level. The largest deviations are in the models with the lowest luminosity and the highest $s_{\rm a}$. This is expected because as the luminosity decreases and the entropy of the flow increases, neutrino-electron scattering contributes more to the total energy deposition profile. Because the shape of the overall heating profile is not significantly affected by omitting the '$4T$' cooling term from ω, the asymptotic entropy is only modified slightly.

For inelastic neutrino-electron scattering, the net specific energy deposition rate can be approximated by $\dot{q} \simeq c n_e n_{\nu_i} \langle \sigma_{\nu_i e}\, \omega \rangle$, where n_e and n_{ν_i} are the number density of electrons and neutrinos, respectively, and $\sigma_{\nu_i e} \simeq \kappa_i T \varepsilon_{\nu_i}$ (Tubbs and Schramm 1975), $\kappa_i = \sigma_o \Lambda_i / 2m_e^2$ is a neutrino species dependent constant, where m_e is the mass of the electron in MeV, $\sigma_o \simeq 1.71 \times 10^{-44}$ cm^2, and Λ_i is the appropriate combination of vector and axial-vector coupling constants for neutrino species i. Averaging properly, we find that

$$\begin{aligned} \dot{q}_{\nu_i e} &= \frac{c}{\rho}\left(\frac{T^3}{(\hbar c)^3}\frac{F_2(\eta_e)}{\pi^2}\right)\frac{L_\nu}{4\pi r^2 c \langle\varepsilon_\nu\rangle\langle\mu\rangle}\Phi^2 \ \text{erg g}^{-1}\ \text{s}^{-1} \\ &\times \left[\frac{\kappa}{2}\langle\varepsilon_\nu\rangle\frac{F_4(\eta_\nu)}{F_3(\eta_\nu)}T\left(\langle\varepsilon_\nu\rangle\Phi\frac{F_2(\eta_\nu)}{F_3(\eta_\nu)} - 4T\frac{F_3(\eta_\nu)}{F_4(\eta_\nu)}\right)\right], \end{aligned} \tag{6.30}$$

where η_ν is an effective neutrino degeneracy parameter (Janka and Hillebrandt 1989), and $\langle\mu\rangle$ is the flux factor, which is related to eq. (6.29) by $\langle\mu\rangle = R_\nu^2/2\Phi^2\Xi(r)r^2$. In order to obtain the contribution to the net heating from neutrino-positron scattering, $\eta_e \to -\eta_e$ and one must also make appropriate changes to Λ_i.

Electron/Positron Annihilation and its Inverse: Also at high entropies, cooling and heating due to $e^+e^- \leftrightarrow \nu_i\bar{\nu}_i$ must be included. Assuming relativistic electrons and positrons, and ignoring Pauli blocking in the final state, the cooling rate is

$$C \simeq 1.4 \times 10^{17} T^9 \rho_8^{-1}\, f(\eta_e)\ \text{ergs g}^{-1}\,\text{s}^{-1}, \tag{6.31}$$

where $f(\eta_e) = [F_4(\eta_e)F_3(-\eta_e) + F_4(-\eta_e)F_3(\eta_e)]/2F_4(0)F_3(0)$, ρ_8 is the mass density in units of 10^8 g cm^{-3} and T is in MeV. The specific heating rate due to the inverse process, $\nu\bar{\nu} \rightarrow e^+e^-$, is simply (Qian & Woosley 1996)

$$H \simeq 1.6 \times 10^{19} \frac{\Psi(x)}{\rho_8\, R_{\nu 6}^4} \Phi^9 \left[L_{\bar{\nu}_e}^{51} L_{\nu_e}^{51} (\langle \varepsilon_{\bar{\nu}_e} \rangle + \langle \varepsilon_{\nu_e} \rangle) + \frac{6}{7} (L_{\nu_\mu}^{51})^2 \langle \varepsilon_{\nu_\mu} \rangle \right], \tag{6.32}$$

where $\Psi(x) = (1-x)^4(x^2+4x+5)$, and $x = (1-(R_\nu/r)^2/\Phi^2)^{1/2}$.

Equation of State

At the temperatures and densities encountered in protoneutron star winds, exterior to the radius of neutrino decoupling, to good approximation, free neutrons, protons, and alpha particles may be treated as non-relativistic ideal gases. We also include photons and a fully general electron/positron equation of state is employed (Evonne Marietta, private communication). Sumiyoshi et al. (2000) have found that using a general electron/positron EOS can decrease the dynamical timescale in the nucleosynthetic region of the wind by as much as a factor of two. Such a modification is important when considering the viability of the neutrino-driven wind as a candidate site for the r-process. For this reason, a general electron/positron EOS is essential. Also of importance is the inclusion of alpha particles. The formation of alpha particles effectively terminates energy deposition via the processes $\nu_e n \leftrightarrow e^- p$ and $\bar{\nu}_e p \leftrightarrow e^+ n$. Failure to include alpha particles results in more heating and a broader energy deposition profile. The entropy of the flow is thereby higher. In low luminosity, late-time, high entropy ($s_{\rm a} \sim 200$) the difference in $s_{\rm a}$ is ~ 20 units.

4. Results: Spherical Models

For a given M, R_ν, and L_ν, the solution to eqs. (6.18)-(6.20) yields radial profiles of temperature, density, electron fraction, and velocity. From these quantities, one obtains $s_{\rm a}$, $Y_e^{\rm a}$, and $\tau_{\rm dyn}$ – the critical parameters in determining the neutron-to-seed ratio and, hence, the resulting r-process. By comparing these numbers with r-process nucleosynthesis survey calculations like those of Hoffman, Woosley, & Qian (1997) or Meyer & Brown (1997), one can quickly see if a given wind solution inhabits a point in $s_{\rm a}$–$Y_e^{\rm a}$–$\tau_{\rm dyn}$ space where a robust 3$^{\rm rd}$-peak r-process is likely.

In order to map the wind solution space, we have constructed evolutionary tracks from our steady-state models. As the supernova commences, we expect the protoneutron star to have large radius and high neutrino luminosity. As the cooling epoch proceeds, the protoneutron star will contract to its final radius (perhaps 10 km) and the luminosity may decrease as a power-law or quasi-exponentially in time (Burrows & Lattimer 1986; Pons et al. 1999). The actual time dependence of L_ν and R_ν depends sensitively on the equation of state of dense nuclear matter and the details of transport and deleptonization of the protoneutron star by neutrinos.

Figure 6.2 summarizes the results of Thompson, Burrows, & Meyer (2001) for protoneutron stars with gravitational masses of 1.4, 1.6, 1.8, and 2.0 $M_\odot$. Each thin solid line is a sequence of steady-state general-relativistic wind models in the plane of $\tau_{\rm dyn}$ versus $s_{\rm a}$. Each track, for a given M, starts with $R \simeq 20$ km and $L^{51}_{\bar{\nu}_e} = 8.0$, corresponding to a total neutrino luminosity of 3.7×10^{52}, where $L^{51}_\nu = L_\nu/10^{51}$. The models with largest L_ν and R_ν have the lowest $s_{\rm a}$ ($\sim 50-70$) and moderate $\tau_{\rm dyn}$ (~ 9 ms). We take $L_\nu \propto t^{-0.9}$ and $R_\nu(t)$ such that the protoneutron star radius decreases linearly in time from 20 km to 10 km in one second (thin solid lines labeled, 'Fast Contraction'). For comparison, we also include a model that has $R_\nu(t) \propto t^{-1/3}$ for $M = 1.4\,M_\odot$ (labeled, 'Slow Contraction'). In the 'Fast' cases, all models move to much higher $s_{\rm a}$ as R_ν goes from 20 km to 10 km. The $s_{\rm a}$ reached at each L_ν is set in part by $M/R_\nu(t)$, with the 2.0 $M_\odot$ model reaching $s_{\rm a} \simeq 150$ when R_ν reaches 10 km. Once the protoneutron star reaches 10 km, R_ν is fixed and each track makes a sharp turn toward much longer $\tau_{\rm dyn}$ and only moderately higher $s_{\rm a}$. Due to the relatively slow contraction, the model with $R_\nu(t) \propto t^{-1/3}$ never exhibits such a sharp turn in the $s_{\rm a}$–$\tau_{\rm dyn}$ plane and eventually joins the 'Fast' 1.4 $M_\odot$ evolutionary track at $\tau_{\rm dyn} \sim 0.015$ seconds. Note that after R_ν has reached 10 km, the evolutionary tracks evolve along characteristic curves in the $s_{\rm a}$–$\tau_{\rm dyn}$ plane. We find that these two r-process parameters approximately follow the power law

$$s_{\rm a} \propto \tau_{\rm dyn}^{0.2}. \tag{6.33}$$

In their analytic and Newtonian exploration of protoneutron star winds, Qian & Woosley (1996) found that $s_{\rm a} \propto \tau_{\rm dyn}^{1/6}$ at constant R_ν and M. This $1/6^{\rm th}$ scaling and that found in eq. (6.33), from our general-relativistic wind solutions, are to be compared with the analytic work of Hoffman, Woosley, & Qian (1997). They find (see their eqs. 20a & 20b) that the $s_{\rm a}$ required to achieve 3rd-peak r-process nucleosynthesis as a function of $\tau_{\rm dyn}$, at constant $Y_e^{\rm a}$, is given by

$$s_{\rm a} \sim 730 \left(\frac{Y_e^{\rm a}}{0.50}\right)\left(\frac{\tau_{\rm dyn}}{0.10\ {\rm s}}\right)^{1/3} \ {\rm k}_B\ {\rm baryon}^{-1} \tag{6.34}$$

for $Y_e^{\rm a} > 0.38$. In Fig. 6.2 we show (thick solid lines) results from eq. (6.34) for $Y_e^{\rm a} = 0.48$ and 0.38. Because all of the several hundred wind models in Fig. 6.2 have $0.45 \lesssim Y_e^{\rm a} \lesssim 0.495$, we conclude that models with $M = 1.4\,M_\odot$ fall short of the required entropy by at least a factor of ~ 3. These lines are meant only to delineate the relevant range of $s_{\rm a}$ and $\tau_{\rm dyn}$ required for 3rd-peak nucleosynthesis. Actual nucleosynthetic calculations in the wind profiles themselves are preferred to the simple comparison on this plot. Thompson, Burrows, & Meyer (2001) did just this, carrying out the full r-process calculation in the 1.4 $M_\odot$ evolutionary track. They found that nucleosynthesis did not proceed beyond $A \sim 100$, in accordance with the predictions of the survey calculations of Meyer & Brown (1997) and Hoffman, Woosley, & Qian (1997).

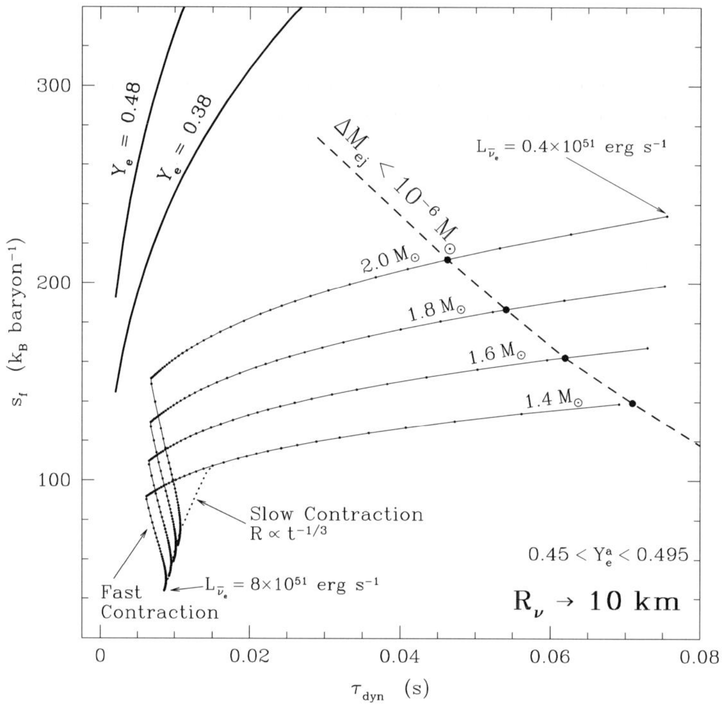

Figure 6.2. Evolutionary tracks of steady-state wind models showing the various correlations between s_a and $\tau_{\rm dyn}$. The solid lines show models for 1.4, 1.6, 1.8, and 2.0 $M_\odot$ protoneutron stars (from Thompson, Burrows, & Meyer 2001). The thick solid lines show an analytic approximation from Hoffman, Woosley, & Qian (1997) for the s_a required for 3rd-peak r-process nucleosynthesis for $Y_e^a = 0.48$ and 0.38. If s_a goes above this line then production of the 3rd-peak nuclides is likely. The thick dashed line denotes the $\tau_{\rm dyn}$ beyond which, for given M, $\dot M$ is too small for the wind to contribute significantly to the total galactic r-process budget.

That the exponent $1/3$ appears in eq. (6.34) and 0.2 appears in eq. (6.33) implies that once R_ν is set, for a given M, the wind cannot evolve into a region of the $s_{\rm a}$–$\tau_{\rm dyn}$ plane where the r-process can take place. Put another way, the nearly horizontal lines of constant R_ν, for a given M, in Fig. 6.2 cannot cross the lines of eq. (6.34) at arbitrarily long $\tau_{\rm dyn}$. Thus, we conclude, along with Takahashi, Witti, & Janka (1994), Qian & Woosley (1996), Sumiyoshi et al. (2000), Wanajo et al. (2000), and Otsuki et al. (2000), that winds from protoneutron stars with 'canonical' parameters $M = 1.4\,{\rm M}_\odot$ and $R_\nu = 10\,{\rm km}$ fail to produce robust r-process nucleosynthesis up to and beyond the 3rd r-process peak. Although the effects of general relativity are important in determining the dynamical timescales and entropy of the wind as predicted in Cardall & Fuller (1997), for reasonable $Y_e^{\rm a}$'s , the entropy falls short of that required for 3rd-peak nucleosynthesis by a large factor ($\sim$ 3, slightly better than the factor of $\sim$ 5 found by Takahashi, Witti, & Janka 1994).

Mass Loss

Having solved for $\dot{M}$ at every point along these evolutionary tracks, and assuming that $L_\nu(t) \propto t^{-0.9}$, we can calculate the total mass ejected as a function of time:

$$M_{\rm ej}(t) = \int_0^t \dot{M}(t')\,dt'. \tag{6.35}$$

Observations of ultra-metal-poor halo stars (see §1) suggest that the astrophysical site for production of the heavy r-process nuclides is universal and acts early in the chemical enrichment history of the galaxy. The fact that if all supernovae produce this r-process signature, then $10^{-5} - 10^{-6}\,{\rm M}_\odot$ of r-process material must be ejected per event (Qian 2000) allows us to constrain the space of relevant wind solutions with eq. (6.35). The heavy dashed line in Fig. (6.2), assuming $L_\nu(t) \propto t^{-0.9}$, shows the point along each evolutionary track in the $s_{\rm a}$–$\tau_{\rm dyn}$ plane beyond which (to longer $\tau_{\rm dyn}$) $\dot{M}$ is simply too small to generate the required total mass loss so as to contribute significantly to the total galactic r-process budget. If r-processing begins to the right of this line, less than $10^{-6}\,{\rm M}_\odot$ will be ejected. Although the position of this $\Delta M_{\rm ej}$ line must change for different $L_\nu(t)$ and $R_\nu(t)$, such a bound must exist for any cooling model. Thus, although the wind may eventually evolve to arbitrarily long dynamical timescales, we conclude that the range of $\tau_{\rm dyn}$ relevant for r-process nucleosynthesis is significantly constrained by consideration of $\dot{M}$ and $M_{\rm ej}$. For example, the track for $M = 1.4\,{\rm M}_\odot$ reaches $s_{\rm a} \sim 400$ only when $\tau_{\rm dyn}$ is several seconds and $\dot{M}$ is of order $10^{-11}\,{\rm M}_\odot\,{\rm s}^{-1}$. Hence, even if the r-process could commence in this epoch, it would need to persist for $\sim 10^5$ seconds in order to contribute significantly to the total galactic r-process budget. Conservatively, then, if transonic protoneutron star winds are the primary site for the r-process, this constraint on the amount of mass ejected per supernova implies that the epoch of r-process nucleosynthesis must occur for $\tau_{\rm dyn}$ less than $\sim 0.07 - 0.1$ seconds.

Conclusions from Spherical Models

Models of transonic winds from neutron stars with $M \simeq 1.4\,{\rm M}_\odot$ and $R_\nu \simeq 10\,{\rm km}$ fail to produce the heavy r-process nuclides. Models with much higher gravitational mass and even smaller coordinate radii, with large neutrino luminosities can achieve 3rd-peak nucleosynthesis. We find that with $M \simeq 2.0\,{\rm M}_\odot$

and $R_\nu \simeq 9\,\mathrm{km}$ that some 3rd-peak nuclides are produced. It is difficult to understand how such massive and compact objects might be created in standard supernova scenarios. This has led Thompson, Burrows, & Meyer (2001) to speculate that the near environments of collapsars, black holes caused by stellar collapse, surrounded by a thick accretion disk, might generate outflows much like the wind solutions described here, but benefiting from the general relativistic effects as with a $M \simeq 2.0\,\mathrm{M}_\odot$ and $R_\nu \simeq 9\,\mathrm{km}$ protoneutron star. Barring these possibilities, however, we are left with possible modifications to the physics described here which might lower Y_e^{a} , increase s_{a}, or decrease τ_{dyn}.

Some possibilities for decreasing Y_e^{a} include (1) different ν_e and $\bar{\nu}_e$ spectral characteristics (see eq. 6.26), (2) neutrino oscillations (Qian & Fuller 1995a,b), and (3) neutrino transport effects. Of relevance for (1), changes to the high-density nuclear equation of state may interestingly effect the electron and anti-electron neutrino spectra. Modifications to the neutrino energy-deposition profile may effect both s_{a} and τ_{dyn}. An increase in $\dot{q}(r)$ at fairly large radius $(50 - 100\,\mathrm{km})$ can increase s_{a} and decrease τ_{dyn} (Qian & Woosley 1996). Such a modification to $\dot{q}(r)$ could be caused by non-standard neutrino physics or even by magnetic field reconnection (see §5).

Finally, one may suspect that the assumption of sphericity is a fundamental problem – that protoneutron star winds and their ejecta simply cannot be understood fully in one spatial dimension. In particular, one may add additional degrees of freedom and break spherical symmetry, by considering the effects of rotation and magnetic fields. As a start to the very complex problem of full magnetohydrodynamic and general-relativistic outflows, in the following we attempt to quantify some of the basic numbers and scalings.

5. Magnetic Protoneutron Star Winds

The solar wind cannot be explained in detail without consideration of magnetic effects on the outflow. This, coupled with the fact that a class of neutron stars are observed to have very high surface magnetic field strengths (*magnetars*, Kouveliotou et al. 1999; Duncan & Thompson 1992), motivates an examination of MHD effects on neutrino-driven protoneutron star winds and their nucleosynthetic ejecta.

To date, such effects have received little attention.

Qian & Woosley (1996) speculated qualitatively on the role of magnetic fields in their wind solutions, noting that tangled field topologies might impede the flow in escaping to infinity, but the effects discussed were not quantified. Nagataki & Kohri (2001) considered a monopole-like magnetic field with rotation in one dimension by restricting their attention to the equatorial plane. The formulation was directly analogous to that of Weber & Davis (1967). However, because of the complex critical point topology encountered in this MHD wind problem, they were unable to assess the importance of field strengths above $\sim 10^{11}$ G, although they were able to consider a variety neutron star rotation periods. Recently, Cameron (2001) argued qualitatively that core collapse, rotation, and magnetic fields conspire to form jets and a post-collapse accretion disk that feeds these jet outflows.

Below we briefly discuss a small subset of the possible effects expected from neutrino-driven MHD winds.

Non-Spherical Expansion

In a strong magnetic field, the character of the neutrino-driven outflow may be significantly modified by the non-spherical divergence of open field lines, along which the wind is channeled. Kopp & Holzer (1976) first to considered these effects in their models of the solar wind in coronal hole regions. In this case, eq. (6.1) becomes

$$\frac{d}{ds}(A\rho v) = 0, \tag{6.36}$$

where ds is the differential line element along the magnetic field and $A(s)$ is an arbitrary area function. The derivation of equations analogous to eqs. (6.11)-(6.13), starting with eq. (6.36), is straightforward. The solution to those equations, however, is complicated by the possibility that more than one critical point may exist in the flow if $A(s)$ changes rapidly (Kopp & Holzer 1976; Bailyn, Rosner, & Tsinganos 1985). Indeed, standing shocks, connecting physical solutions, may exist in the flow for rapid areal divergence (Habbal & Tsinganos 1983; Bailyn, Rosner, & Tsinganos 1985; Leer & Holzer 1990). For smoothly and modestly changing $A(s)$, however, the solution proceeds as in the spherical case – the one dimensional problem now along ds instead of dr. Charboneau & Hundhausen (1996) have constructed quasi-two-dimensional models of flow in the field lines of the open region in a helmet streamer configuration (see also Pneuman & Kopp 1971; Low 1986). Helmet streamer/coronal hole magnetic field configurations in the context of the sun have been the focus of considerable theoretical effort (e.g. Mestel 1968; Pneuman & Kopp 1970; Pneuman & Kopp 1971; Steinolfson, Suess, & Wu 1982; Usmanov et al. 2000; Lionello et al. 2002). In these models, pressure forces, inertia, gravity, and a strong ordered dipole magnetic field conspire to produce a region of closed magnetic field lines close to the central star, at latitudes near the magnetic equator. At the magnetic poles, the flow is radial. At intermediate latitudes, between the pole and the closed zone, open magnetic field lines bend toward the equator close to the star and then extend radially. In these models $A(s)$ exhibits smooth variations and only for streamlines emerging from latitudes very near the closed zone do large deviations from purely radial flow exist (Charboneau & Hundhausen 1996).

Taking $A(r) = r^2 f(r)$ and $f(r) = (f_{\rm max}\exp[(r - R_1)/\sigma] + f_1)/(\exp[(r - R_1)/\sigma]+1)$, where $f_1 = 1-(f_{\rm max}-1)\exp[(R_\nu - R_1)/\sigma]$, as in Kopp & Holzer (1976), we have computed several Newtonian wind models for comparison with purely spherical expansion. This function varies most rapidly near R_1, with the change in $f(r)$ over a radial distance $R_1 \pm \sigma$. With $f_{\rm max} = 4$ so that $A(r)$ is four times as large at a given radius as spherical expansion and with $R_1 = 2R_\nu$ and $\sigma = R_\nu$ so that the divergence is smooth, we find that $\tau_{\rm dyn}$ increases by $\sim 60\%$, $\dot{M}$ decreases by $\sim 65\%$, $s_{\rm a}$ increases from 68 to 74 $k_{\rm B}$ baryon^{-1}, and that the asymptotic mechanical luminosity ($P_{\rm mech} = \dot{M}v_\infty^2/2$) drops by more than a factor of three. In contrast, constricting the flow with $f_{\rm max} = 1/4$ yields a much faster wind. In this case, $\tau_{\rm dyn}$ decreases from 3.2 milliseconds to 1.7 milliseconds. $s_{\rm a}$ decreases from 68 to 66.5 $k_{\rm B}$ baryon^{-1}. $\dot{M}$ and $P_{\rm mech}$ increase by a factor of ~ 3 and ~ 5.5, respectively. Clearly the quality of the areal divergence can significantly influence the properties of the flow. Although $s_{\rm a}$ was not affected by more than a few percent, the changes in $\tau_{\rm dyn}$ evidenced by this simple comparison imply that a more thorough investigation is warranted. We save a detailed exploration of these effects for a future work.

Closed Zones & Trapping

The ideas of this section have recently been set down in Thompson (2003). Here, we follow the discussion of Thompson (2003) closely.

Figure 6.3 shows profiles of temperature, entropy, energy deposition rate, and pressure for a high neutrino luminosity protoneutron star wind model with $M = 1.4\,{\rm M}_\odot$ and $R_\nu = 10$ km. Also shown (thick dotted line) is the magnetic energy density $B^2/8\pi$. For simplicity we take $B = B_0(R_B/r)^3$, where R_B is a reference radius for the magnetic field footpoints. Because of the exponential near-hydrostatic atmosphere in these wind models (note sharp drop in P in Fig. 6.3), we take $R_B = 11$ km. The surface magnetic field strength B_0 is here set to 1.5×10^{15} G. We define the quantity $\beta = P/(B^2/8\pi)$ and R_β as the radius where $\beta = 1$. Here, $R_\beta \sim 46.5$ km. From this figure it is clear that a $\sim 10^{15}$ G field can dominate the matter pressure during the wind epoch. Note that this figure is only for a single wind model, with a single neutrino luminosity. As L_ν drops, the pressure profile drops everywhere so that R_β moves out in radius for constant B_0. Because $P(r)$ drops everywhere as L_ν decays, at any instant in time, a lower B_0 is required such that $\beta \sim 1$ at some radius. This implies that even though a 10^{15} G field may not dominate the wind dynamics at early times, if the protoneutron star somehow maintains this field strength as L_ν drops, the field will eventually dominate as $P(r)$ decreases (Thompson 2003).

From this admittedly limited comparison, we conclude that neutrino-driven winds from protoneutron stars with magnetar-like surface field strengths may be significantly affected by the presence of such a field.

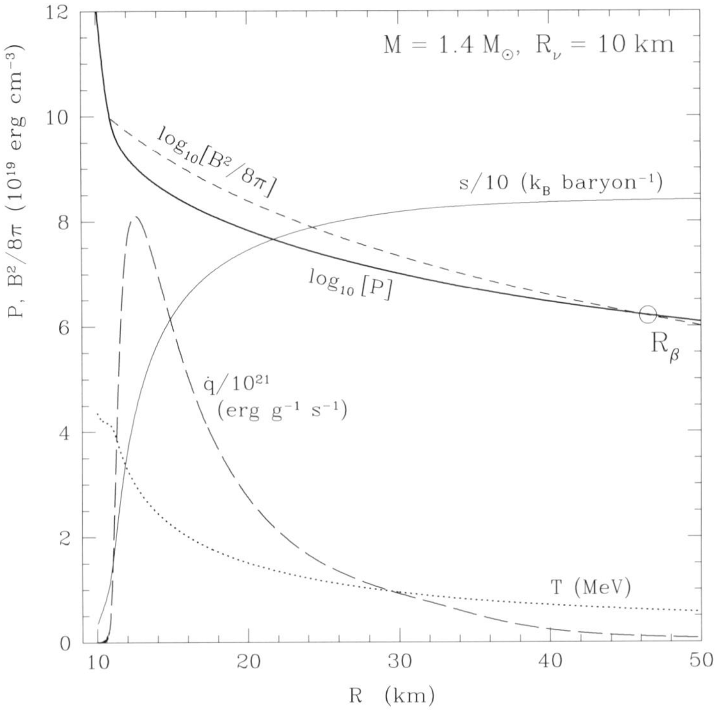

Figure 6.3. Temperature T (MeV, dotted line), entropy s (10 k$_{\rm B}$ baryon^{-1}, thin solid line), energy deposition rate $\dot{q}$ (10^{21} erg g s^{-1}, long dashed line), log of the pressure P (10^{19} erg cm^{-3}, thick solid line), and log of the magnetic energy density $B^2/8\pi$ (10^{19} erg cm^{-3}, short dashed line) for a protoneutron star wind model with $M = 1.4$ M$_\odot$ and $R_\nu = 10$ km. Here, we take $B = B_0(R_B/r)^3$, where $R_B = 11$ km and $B_0 = 1.5 \times 10^{15}$ G.

From MHD models and observations of the solar wind, we expect that a strong magnetic field, that dominates the wind pressure inside R_β, may form a closed zone at these radii, at latitudes near the magnetic equator. The configuration of the flow would then be analogous to the helmet streamer described in §5 (Steinolfson, Suess, & Wu 1982; Usmanov et al. 2000; Lionello et al. 2002). If heating and cooling balance so that $\dot{q} = 0$ throughout the closed zone, this structure may be stable and the matter in this regime permanently trapped. However, if heating dominates then the pressure of the trapped matter will increase. Thus, if the material is trapped inside R_β with $P \ll B^2/8\pi$, net neutrino energy deposition in the closed zone (see Fig. 6.3) must increase P to $\sim B^2/8\pi$. If this happens, we expect the matter to escape dynamically. In this way, the closed magnetic field structures that form where $\beta < 1$ are unstable. Importantly for the r-process, because the pressure increase of the matter is caused by neutrino heating, it is necessarily accompanied by an increase in the matter entropy (Thompson 2003).

Very roughly, the matter will be trapped for a time set by P, $B^2/8\pi$, and $\dot{q}$;

$$\tau_{\rm trap} \sim [B^2/8\pi - P]/[\dot{q}\rho]. \tag{6.37}$$

Assuming that T, ρ, and $\dot{q}$ do not change significantly in $\tau_{\rm trap}$, the entropy amplification associated with such an increase in pressure is then (Thompson 2003)

$$\Delta s \sim \dot{q}\tau_{\rm trap}/T. \tag{6.38}$$

When eqs. (6.37) and (6.38) are evaluated at a characteristic radius for energy deposition (say the half-asymptotic-entropy radius), they yield an order of magnitude estimate for Δs. For very high B_0 and low $P(r)$ (slow winds with low L_ν), Thompson 2003 found that there is a radius (R_q) inside of which cooling balances heating before P approaches $B^2/8\pi$. Therefore, in this simple picture, the matter interior to R_q is permanently trapped (barring MHD instabilities that might very well arise). Importantly, for any B_0, R_q is always less than R_β so that the trapped matter between these two radii can escape with Δs set by eq. (6.38).

Figure 6.4 shows $s_{\rm a}$ versus $\tau_{\rm dyn}$ (analogous to Fig. 6.2) for a large set of protoneutron star wind models (Thompson 2003). The thick solid line shows spherical, steady-state models as described in §4 for constant $R_\nu = 10$ km and $M = 1.4\,{\rm M}_\odot$. The dashed line shows again the analytical results of Hoffman, Woosley, & Qian (1997) for the $s_{\rm a}$ required, at a given $\tau_{\rm dyn}$, for 3$^{\rm rd}$-peak r-process nucleosynthesis (see eq. 6.34; compare with Fig. 6.2). Above this dashed line, for $Y_e^{\rm a} = 0.48$, the neutron-to-seed ratio is high enough for a robust r-process. The thin solid lines are constructed from the non-magnetic wind models (thick solid line), by applying eqs. (6.37) and (6.38) for $B_0 = 2\times10^{14}$, 2×10^{14}, 4×10^{14}, 6×10^{14}, 8×10^{14}, 1×10^{15}, and 2×10^{15} G. From Fig. 6.4 it is clear both that the spherical non-magnetic wind models fall short of the entropy required for 3$^{\rm rd}$-peak r-process and that for $B_0 \sim 10^{15}$ G, the entropy enhancements caused by trapping may be sufficient to account for this deficit.

dynamics?

There are a number of effects that might decrease the entropy enhancements discussed here as a result of trapping in closed magnetic field structures. Any physical effect that globally disrupts the closed zone on a timescale much less than $\tau_{\rm trap}$ would significantly undermine the entropy enhancements estimated in this scenario. Such effects might include MHD instabilities, differential rotation, and rapid motion of the magnetic field footpoints due to convection (Thompson 2003; see also Duncan & Thompson 1992; Thompson & Duncan 1993; Thompson & Murray 2001). It is worth noting that the very early configuration of the protoneutron star magnetic field is highly uncertain and may

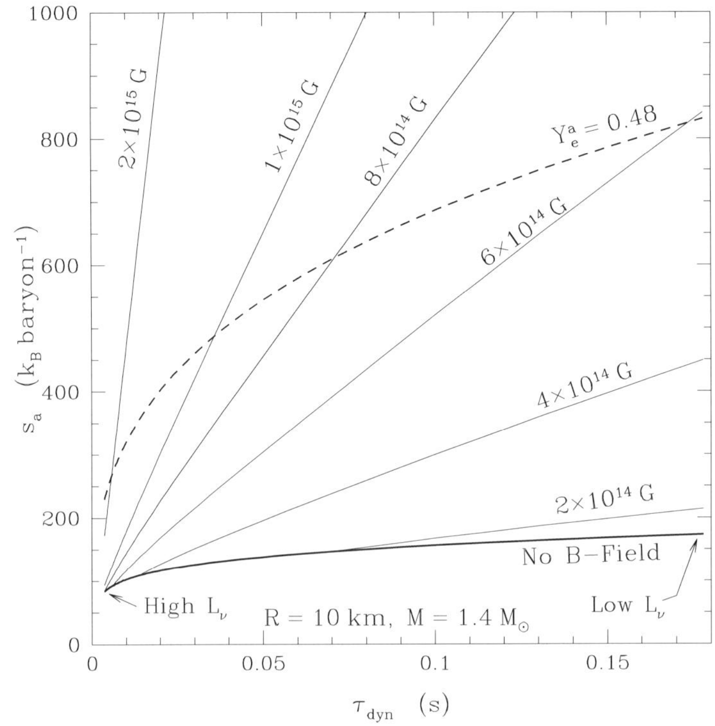

Figure 6.4. Non-magnetic, spherically symmetric wind models (thick solid line) in the plane of s_a versus τ_{dyn} for protoneutron star wind models with $M = 1.4\,M_\odot$ and $R_\nu = 10$ km, for a large range in neutrino luminosities. Thin solid lines show s_a as a function of τ_{dyn}, employing the entropy enhancement as in eq. (6.38) for $2 \times 10^{14} \leq B_0 \leq 2 \times 10^{15}$ G. The dashed line shows the results of eq. (6.34) for $Y_e^a = 0.48$ (from Hoffman, Woosley, & Qian 1997). Figure from Thompson (2003).

be a complex of high-order multipoles. The large-scale closed zone desceribed in Thompson (2003) might not then obtain. However, in this case many closed zones may exist and eq. (6.38) may be used to estimate the entropy enhancement locally at many sites on the surface of the protoneutron star as closed regions with a variety of β are generated and then opened by neutrino heating. If the field is very complex, twisted, or sheared reconnection may deposit energy in the flow, as first suggested by Qian & Woosley (1996) in this context. Studies of extra energy deposition show that this may increase or decrease the entropy of the flow, depending crucially on where the energy is deposited (Qian & Woosley 1996; Thompson, Burrows, & Meyer 2001).

6. Summary, Conclusions, & Implications

The subject of protoneutron star winds is relatively young, born only in the early-1990s. Much of the physics attending the emergence and evolution of these outflows in the just-post-supernova environment is uncertain and intimately tied with other outstanding issues in neutron star birth: rotation and magnetic fields. Many of the phenomena intensively investigated in the context of the sun – reconnection, flares, coronal mass ejections, closed loops, prominences, flux emergence, coronal holes, streamers – may play important roles in determining the nucleosynthetic consequences of the wind/cooling epoch.

The results presented in §4 from Thompson, Burrows, & Meyer (2001) as well as the results of Takahashi, Witti, & Janka (1994), Qian & Woosley (1996), Sumiyoshi et al. (2000), Otsuki et al. (2000), and Wanajo et al. (2000) indicate that spherical steady-state winds from canonical neutron stars cannot attain the requisite entropy for robust r-process nucleosynthesis. Figure 6.2 shows, however, that very compact, massive, and luminous neutron stars may realize a short dynamical timescale, modest entropy r-process. In addition, it may also be that the actual electron fraction of protoneutron star winds is much lower than that derived in Thompson, Burrows, & Meyer (2001) and implied by the neutron star cooling calculations of Pons et al. (1999). If Y_e^{a} could be made to be ~ 0.3 in the models of Fig. 6.2, the $1.4\,\mathrm{M}_\odot$ evolutionary track might naturally generate 3rd-peak nuclides. Then again, perhaps $1.4\,\mathrm{M}_\odot$ neutron stars are responsible for production of only the r-process elements below $A \sim 130$ as in Fig. (6.1) (see §1).

Section 5 shows that magnetic effects can change the entropy and dynamical timescale of a given flow solution considerably for surface field strengths of order 10^{15} G. It may be that only neutron stars born with magnetar-like field strengths produce robust r-process signatures. In any case, the need for multi-dimensional magnetohydrodynamic simulations of wind emergence and evolution are required to address many of these still open questions more fully.

Acknowledgments

I am indebted to Adam Burrows, Brad Meyer, Eliot Quataert, Jon Arons, and Anatoly Spitkovsky for helpful conversations. Special thanks go to Evonne Marietta for making her tabular electron/positron EOS available and to Chris Sneden for providing Figure 6.1. Support for this work is provided in part by NASA through a GSRP fellowship (while the author was at the University of Arizona, Tucson) and by NASA through Hubble Fellowship grant #HST-HF-01157.01-A awarded by the Space Telescope Science Institute, which is operated by the Association of Universities for Research in Astronomy, Inc., for NASA, under contract NAS 5-26555.

References

Arzoumanian, Z. 1995, PhD Thesis, Princeton University

Bahcall, J. N. 1964, Physical Review, 136, B1164

Bailyn, C., Rosner, R., & Tsinganos, K. 1985, ApJ, 296, 696

Bruenn, S. W., De Nisco, K. R., & Mezzacappa, A. 2001, preprint (astro-ph/0101400)

Burbidge, E. M., Burbidge, G. R., Fowler, W. A., & Hoyle, F. 1957, Rev. Mod. Phys., 29, 547

Burris, D. L., Pilachowski, C. A., Armandroff, T. E., Sneden, C., Cowan, J. J., & Roe, H. 2000, ApJ, 544, 302

Burrows, A. & Lattimer, J. M. 1986, ApJ, 307, 178

Burrows, A. 1988, ApJ, 334, 891

Burrows, A., Hayes, J., & Fryxell, B. A. 1995, ApJ, 450, 830

Cameron, A. G. W. 1957, PASP, 69, 201

Cameron, A. G. W. 2001, ApJ, 562, 456

Cardall, C. Y. & Fuller, G. M. 1997, ApJL, 486, 111

Charbonneau, P. & Hundhausen, A. J. 1996, Sol. Phys., 165, 237

Chevalier, R. 1989, ApJ, 346, 847

Cayrel, R., Hill, V., Beers, T. C., Barbuy, B., Spite, M., Spite, F., Plez, B., Andersen, J., Bonifacio, P., Francois, P., Molaro, P., Nordström, B., & Primas, F. 2001, Nature, 409, 691

Cowan, J. J., Sneden, C., Truran, J. W., & Burris, D. L. 1996, ApJL, 460, 115

Cowan, J. J. & Sneden, C. 2002, To appear in the Proceedings of the 3rd International Conference on *Fission and Properties of Neturon-Rich Nuclei*, astro-ph/0212149

Duncan, R. C., Shapiro, S. L., & Wasserman, I. 1986, ApJ, 309, 141

Duncan, R. C. & Thompson, C. 1992, ApJL, 392, 9

Flammang, R. A. 1982, MNRAS, 199, 833

Freiburghaus, C., Rosswog, S., & Thielemann, F.-K. 1999, ApJL, 525, 121

Freiburghaus, C., Rembges, J.-F., Rauscher, T., Kolbe, E., Thielemann, F.-K., Kratz, K.-L., Pfeiffer, B., & Cowan, J. J. 1999, ApJ, 516, 381

Fryer, C. L., Benz, W., Herant, M., & Colgate, S. 1999, ApJ, 516, 892

Fuller, G. M. & Meyer, B. S. 1995, ApJ, 453, 792

Fuller, G. M. & Qian, Y.-Z. 1996, Nuc. Phys. A, 606, 167

Habbal, S. R. & Tsinganos, K. 1983, J. Geophys. Res., 88(A3), 1965

Hill, V., Plez, B., Cayrel, R., & Beers, T. C. 2001, proceedings of Astrophysicaql Ages and Timescales, ASP Conference Series, preprint (astro-ph/0104172)

Hoffman, R. D., Woosely, S. E., Fuller, G. M., & Meyer, B. S. 1996, ApJ, 460, 478

Hoffman, R. D., Woosely, S. E., & Qian,Y.-Z. 1997, ApJ, 482, 951

Horowitz, C. J. & Li, G. 2000, Proceedings of Intersections Conference, Qubec, preprint (astro-ph/0010042)

Itoh, N., Hayashi, H., Nishikawa, A., & Kohyama, Y. 1996, ApJS, 102, 411

Janka, H.-Th. & Müller, E. 1995, ApJL, 448, 109

Janka, H.-T. 1991, A&A, 244, 378

Janka, H.-Th. & Hillebrandt, W. 1989, AA, 224, 49

Kajino, T., Otsuki, K., Wanajo, S., Orito, M., & Mathews, G. 2001, to appear in Few-Body Systems Suppl., preprint (astro-ph/0006079)

Kalogera, V., Narayan, R., Spergel, D. N., Taylor, J. H. 2001, ApJ, 556, 340

Kopp, R. A. & Holzer, T. E. 1976, Solar Phys., 49, 43

Kouveliotou, C., Strohmayer, T., Hurley, K., van Paradijs, J., Finger, M. H., Dieters, S., Woods, P., Thompson, C., Duncan, R. C. 1999, ApJL, 510, 115

Lamers, H. J. G. L. M. & Cassinelli, J. P., *Introduction to Stellar Winds* (Cambridge University Press, Cambridge, 1999)

Lattimer, J. M. & Prakash, M. 2001, ApJ, 550, 426

Leer, E. & Holzer, T. E. 1990, ApJ, 358, 680

Liebendörfer, M., Mezzacappa, A., Thielemann, F.-K., Messer, O. E. B., Hix, W. R., & Bruenn, S. W. 2000, PRD, submitted, preprint (astro-ph/0006418)

Lionello, R., Linker, J. A., & Mikíc, Z. 2001, ApJ, 546, 542

London, R. A. & Flannery, B. P. 1982, ApJ, 258, 260

Low, B. C. 1986, ApJ, 310, 953

MacFadyen, A. I. & Woosley, S. E. 1999, ApJ, 524, 262

Mayle, R., Wilson, J. R., & Schramm, D. N. 1987, ApJ, 318, 288

Mclaughlin, G., Fuller, G. M., & Wilson, J. R. 1996, ApJ, 472, 440

McWilliam, A., Preston, G. W., Sneden, C., & Searle, L. 1995a, AJ, 109, 2757

McWilliam, A., Preston, G. W., Sneden, C., & Shectman, S. 1995b, AJ, 109, 2757

Mestel, L. 1968, MNRAS, 138, 359

Meyer, B. S. & Brown, J. S. 1997, ApJS, 112, 199

Meyer, B. S., Howard, W. M., Mathews, G. J., Woosley, S. E., & Hoffman, R. D. 1992, ApJ, 399, 656

Mezzacappa, A., Liebendörfer, M., Messer, O. E. B., Hix, W. R., Thielemann, F.-K., & Bruenn, S. W. 2001, PRL, 86, 1935

Myra, E. S. & Burrows, A. 1990, ApJ, 364, 222

Nagataki, S. & Kohri, K. 2001, PASJ, 53, 547

Nobili, L., Turolla, R., & Zampieri, L. 1991, ApJ, 383 250

Otsuki, K., Tagoshi, H., Kajino, T., & Wanajo, S.-Y. 2000, ApJ, 533, 424

Pneuman, G. W. & Kopp, R. A. 1971, 18, 258

Pons, J. A., Reddy, S., Prakash, M., Lattimer, J. M., & Miralles, J. A. 1999, ApJ, 513, 780

Press, W. H., Teukolsky, S. A., Vetterling, W. T., & Flannery, B. P. 1992, Numerical Recipes in Fortran 77 (2d Ed.; New York, NY: Cambridge University Press)

Qian, Y.-Z., Fuller, G. M., Mathews, G. J., Mayle, R. W., Wilson, J. R., Woosley, S. E. 1993, PRL, 71, 1965

Qian, Y.-Z. 2000, ApJL, 534, 67

Qian, Y.-Z. & Fuller, G. M. 1995a, PRD, 51, 14, 1479

Qian, Y.-Z. & Fuller, G. M. 1995b, PRD, 52, 2, 656

Qian, Y.-Z. & Wasserburg, G. J. 2000, Phys. Reps., 333, 77

Qian, Y.-Z. & Woosley, S. E. 1996, ApJ, 471, 331

Qian, Y.-Z., Vogel, P., & Wasserburg, G. J. 1998, ApJ, 494, 285

Rampp, M. & Janka, H.-Th. 2000, ApJL, 539, 33

Reddy, S., Prakash, M., & Lattimer, J. M. 1998, PRD, 58, 013009

Riley, P., Linker, J. A., Mikíc, Z., Odstrcil, D., Pizzo, V. J., & Webb, D. F. 2002, ApJ, 578, 972

Rosswog, S., Liebendörfer, M., Thielemann, F.-K., Davies, M., Benz, W., &Piran, T. 1999, A&A, 341, 499

Salmonson, J. D. & Wilson, J. R. 1999, ApJ, 517, 859

Sneden, C., McWilliam, A., Preston, G. W., Cowan, J. J., Burris, D. L., & Armosky, B. J. 1996, ApJ, 467, 819

Steinolfson, R. S., Suess, S. T., & Wu, S. T. 1982, ApJ, 255, 730

Sumiyoshi, K., Suzuki, H., Otsuki, K., Teresawa, M., & Yamada, S. 2000, PASJ, 52, 601

Surman, R., Engel, J., Bennett, J. R., & Meyer, B. S. 1992, PRL, 79, 1809

Takahashi, K., Witti, J., & Janka, H.-T. 1994, A&A, 286, 857

Thompson, C. & Duncan, R. C. 1993, ApJ, 408, 194

Thompson, C. & Murray, N. 2001, ApJ, 560, 339

Thompson, T. A., Burrows, A., & Meyer, B. S. 2001, ApJ, 562, 887

Thompson, T. A. 2003, ApJL, 585, L33

Usmanov, A. V., Goldstein, M. L., Besser, B. P., & Fritzer, J. M. 2000, Journal of Geophysical Research, 105, 12675

Wallerstein, G., Iben, I., Parker, P., Boesgaard, A. M., Hale, G. M., Champagne, A. E., Barnes, C. A., Käppeler, F., Smith, V. V., Hoffman, R. D., Timmes, F. X., Sneden, C., Boyd, R. N., Meyer, B. S., & Lambert, D. L. 1997, Rev. Mod. Phys., 69, 995

Wanajo, S., Kajino, T., Mathews, G. J., & Otsuki, K. 2001, accepted to ApJ

Wang, A.-H., Wu, S. T., Suess, S. T., & Poletto, G. 1993, Solar Phys., 147, 55

Wasserburg, G. J. & Qian, Y.-Z. 2000, ApJL, 529, 21

Weber, E. J. & Davis, L. 1967, ApJ, 148, 217

Westin, J., Sneden, C., Gustafsson, B., & Cowan, J. J. 2000, ApJ, 530, 783

Witti, J., Janka, H.-T., & Takahashi, K. 1994, A&A, 286, 841

Woosley, S. E. & Hoffman, R. D. 1992, ApJ, 395, 202

Woosley, S. E., Wilson, J. R., Mathews G. J., Hoffman, R. D., & Meyer, B. S. 1994, ApJ, 433, 209

Woosley, S. E. & Weaver, T. A. 1995, ApJS, 101, 181

II

ASYMMETRIES IN COLLAPSE, BEYOND THE BASIC SUPERNOVA MECHANISM

Introduction

Chris Fryer

In Part I of this book, we discussed the basic neutrino-driven mechanism for supernovae. However, with the observation of SN 1998bw (a.k.a GRB 980425), we discovered a new type of outburst that may be produced by stellar collapse: Gamma-Ray Bursts. Studies of energetic outbursts may make up 1% of all "supernova" explosions. These outbursts have been termed hypernovae or jet-driven supernovae.

Part II of this book deals with these rare, but highly interesting explosions which arise from stellar collapse. This field is still in its infancy, and some of what is stated here will not hold true in the future. Most of the models are essentially using the same simplified physics routines used by theorists studying the neutrin-driven mechanism back in the 1960s. But the chapters here will give a flavor of the range of ideas that have been proposed to explain these objects.

C.L. Fryer (ed.), Stellar Collapse, 205.

Chapter 7

RADIOACTIVE DECAY IN CORE-COLLAPSE SUPERNOVAE

Motivation for Asymmetry

Aimee L. Hungerford

Transport Methods, Los Alamos National Laboratory, Los Alamos, NM, 87545

University of Arizona, Tucson, AZ, 85721

Abstract High energy emission from supernovae provide a direct window into the quantity and distribution of radioactive elements produced in these explosions. This information has proven to be of tremendous importance in inferring details of the shock evolution in core-collapse supernovae. This chapter highlights the theoretical advances in core-collapse explosion simulations which have been motivated by the many surprises in γ-ray observations. In particular, the presence of asymmetries in the explosion, as inferred from the high energy spectra, are emphasized. Recent simulations combining supernova explosion calculations with 3D Monte Carlo γ-ray transport are presented. The diagnostic capability of the high energy emission to distinguish mild explosion asymmetries is demonstrated.

Keywords: Core-Collapse, Supernovae, gamma-rays, radioactive decay

Introduction

Core-collapse supernovae are among the most energetic explosions in the universe since the Big Bang. They mark the catastrophic end of stars more massive than 8 $M_\odot$ leaving behind compact remnants such as neutron stars and stellar mass blackholes. The nucleosynthesis in these massive stars, and their subsequent explosion, is responsible for most of the heavy element enrichment in our galaxy. So naturally, any attempt to address human origins must begin with an understanding of core-collapse supernovae. Among the heavy elements synthesized during the explosive burning phase are a set of unstable isotopes such as ^{56}Ni and ^{44}Ti. The daughter products from these radioactive decays are often left in excited nuclear states and must de-excite through the emission

C.L. Fryer (ed.), Stellar Collapse, 207–236.

of γ-radiation. This radiation is of primary importance to the observations of supernovae evolution, as it is the photons from these decays which deposit sufficient energy to power the lightcurve, making supernovae visible across the universe.

Although the deposited γ-radiation is of utmost importance in powering the magnificent light show of the supernova, it is the γ-rays that escape with little or no interaction that provide the most direct diagnostic of the underlying physics in the explosion itself. Compared with observations at other wavelengths, the MeV γ-radiation is much less complicated in terms of its emission and interaction processes. The emission of radioactive decay lines depends only on the decay rate of the parent nucleus and is not sensitive to excitation conditions in the ejecta material. In addition, the dominant interaction affecting these energetic photons is Compton scatter off electrons. Unlike the complicated line opacities present in the optical, infrared and X-ray bands, this scattering opacity varies smoothly with wavelength and is only weakly dependent on composition. Furthermore, since the decay lines arise from the de-excitation of an excited nuclear state, their emission provides abundance information about specific isotopes of an element. For these reasons, the hard X- and γ-ray spectra from core-collapse events present a relatively unprocessed message regarding the underlying structure of a supernova's nucleosynthetic yields.

The mean lifetimes of the dominant radioactive isotopes span a range of values and thus probe supernova evolution over a variety of timescales. In fact, these isotopes fall primarily into three lifetime categories: 1) Short-lived isotopes with mean decay lifetimes of order 1 year (^{56}Ni, ^{56}Co), 2) Intermediate-lived isotopes with lifetimes of order 100 years (^{44}Ti) and 3) Long-lived isotopes with lifetimes of order 10^6 years (^{26}Al, ^{60}Fe). The average time between supernova events in our galaxy is $\sim$ 50 years (for a supernova rate of $\sim$ 2 century^{-1}). This implies that decay emission from isotopes in the first two lifetime categories will be observed on timescales consistent with a single supernova event. In contrast, emission from the long-lived isotopes is contributed by thousands of supernova events spanning the million year lifetime of the isotope. The information conveyed by these observations is a statistical average over an entire population of core-collapse events. In this review, we will concentrate on short-lived and intermediate-lived radioactive isotopes, and the insight they provide for the early explosion and extended remnant phases of the supernova. In particular, we will concentrate on the γ-ray work pertaining to SN 1987A , which is the hallmark example of what can be learned from short-lived radioisotopic observations, and SNR Cassiopeia A, which highlights the power of intermediate-lived isotopes to probe the details of the explosion mechanism even centuries after the explosion itself.

1. Supernova 1987A

Supernova 1987A marked the spectacular death of the massive star Sanduleak -69 202 (see review by Arnett et al. 1989). The proximity of this supernova (located at 50 kpc in the Large Magellanic Cloud) allowed for the first direct detection of nuclear decay lines from a supernova explosion. As early as 1957 (Burbidge, Burbidge, Fowler & Hoyle), it was believed that radioactive decay from unstable elements was primarily responsible for powering the energetic lightcurves of supernovae. By the late sixties, researchers had predicted that it was ^{56}Ni (synthesized during an explosive nucleosynthesis phase in the supernova) and its decay products ^{56}Co and ^{56}Fe which were responsible for providing the luminosity in the lightcurve. The detection of the ^{56}Co 847 and 1238 keV decay lines from SN 1987A was a phenomenal verification of that prediction. (Matz et al. 1988, Cook et al. 1988, Mahoney et al. 1988, Tueller et al. 1990)

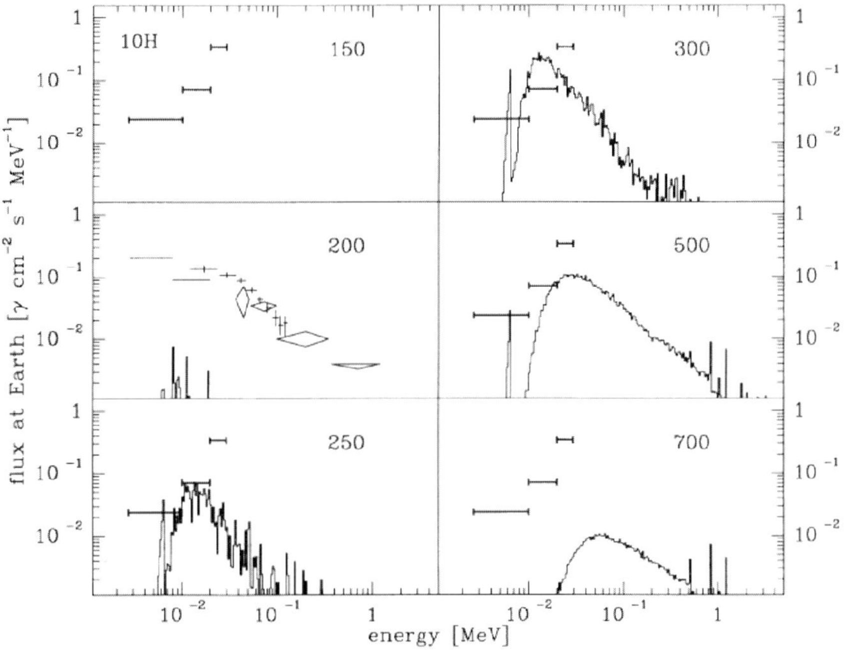

Figure 7.1. High energy spectrum from FASTGAM simulation using progenitor model 10H (from Pinto & Woosley 1988b).

However, the γ-line detections came in nearly 6 months earlier than theorists had predicted (based on simulations using spherically symmetric explosion models.) Figure 7.1 (Pinto & Woosley 1988) shows an example of the simulated high energy spectra at different times for progenitor model 10H (Woosley 1988). There are 3 key features of the simulated spectrum (solid lines) which should be noted: 1) the decay lines from the unstable radioisotopes can be seen as the forest of lines around 1 MeV, 2) the linear power law continuum below the decay lines arises from down scattered decay photons which have suffered several Compton scatters and 3) the abrupt fall off at energies around 20 keV results from photoelectric absorption of the Comptonized photons. Also pictured in Figure 7.1 are the hard X-ray observations from the HEXE (crosses) and PULSAR X-1 (diamonds) instruments on board the MIR satellite (Sunyaev et al. 1987). It is clear from the figure that theoretical predictions of the hard X-ray flux for model 10H (a favored progenitor model from bolometric light curve constraints alone), fell significantly short of the actual hard X-ray observations (Dotani et al. 1987, Sunyaev et al 1987). The most straightforward explanation for this discrepancy between observation and theory was that the paradigm of onion-skin stellar structure was not preserved during the explosion. By assuming that nickel, synthesized in the inner layers of the explosion, was mixed to larger radii in the SN ejecta, better fits to the high energy observations were obtained (Sunyaev et al. 1990, Pinto & Woosley 1988).

The early emergence of hard X- and γ-rays was not the only indication that a break in spherical symmetry was needed for the explosion phase. Line profiles from the ^{56}Co decay emission, as well as the FeII IR line emission, were broadened to velocities of order 3500 km/s (Spyromilio, Meikle & Allen 1990, Haas 1990). This implied that the Cobalt and Iron, both daughter products of ^{56}Ni decay, had managed to mix all the way into the hydrogen envelope of the supernova ejecta. In addition, as can be seen in Figure 7.2, both the γ-ray lines and FeII IR lines show a redshift of order 500 km/s. This has been difficult to reproduce in the simulated spectra since optical depth effects tend to cause a blueward shift of the line centroid. We will return to this puzzle in Section 3 when we discuss recent models of γ-ray emission from axisymmetric supernova explosion simulations. As we will see, these models also seem incapable of achieving redshifts along any viewing angle, which suggests a need for asymmetries of a non-axisymmetric nature.

2. Cassiopeia A Supernova Remnant

It is widely accepted that Cassiopeia A is the expanding remnant of a core-collapse supernova, possibly observed by J. Flamsteed in 1680 A.D (Ashworth 1980). With an age of roughly 325 years and a distance of $\sim$3.4 kpc (Reed et al. 1995), it is the nearest and youngest supernova remnant in the galaxy. This has made it a primary target for remnant studies at a variety of wavelengths,

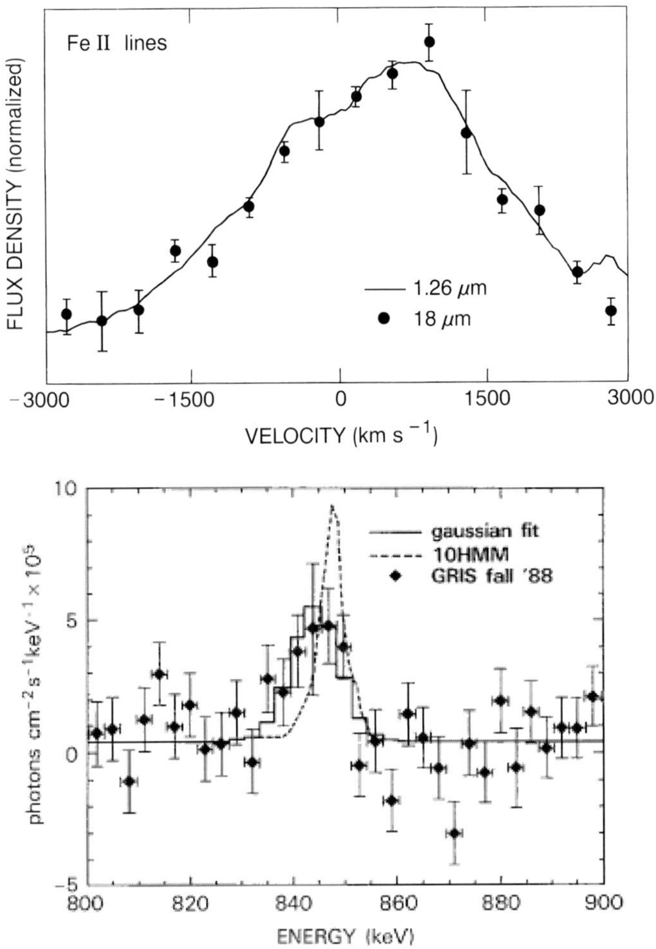

Figure 7.2. Line profiles for FeII 18 μm and 2 μm lines (top panel) and the ^{56}Co 847 keV line (bottom panel). Both lines exhibit a redshift of roughly 500 km/s relative to the systemic velocity of the LMC (280 km/s). (from Spyromilio, Meikle, & Allen 1990 and Tueller et al. 1990).

including X-rays and γ-rays. Indeed, the first light target for the Chandra X-ray satellite was a 5000 s observation towards Cas A. The resulting image revealed further evidence for a break with the onion-skin paradigm of stellar explosions, just as the γ-ray observations of SN 1987A had required. Figure 7.3 shows the false color composite image of Cas A from the Chandra satellite first light observations (Hughes et al. 2000). Three wavelength band images have been combined to produce this image, with each band dominated by line emission from a different set of elements. Of particular importance in this image is

capture occurs primarily through the capture of inner shell electrons, sufficient ionization of the isotope blocks its decay, effectively altering its half life. In this way, the mass of ^{44}Ti inferred from line flux observations and a laboratory determined decay lifetime will not correctly reflect the mass of initially synthesized ^{44}Ti. Indeed, for the remnant age of Cas A, Mochizuki et al. (1999) show that taking this ionization effect into account brings the observationally inferred masses into closer agreement with theoretical predictions. In order to achieve the best agreement between theory and observations, the assumption of inhomogeneous clumping of titanium associated with iron-group enriched material was required, though as we have discussed, there is significant evidence for the existence of such inhomogeneity.

An alternative for boosting the ^{44}Ti emission has been proposed by Nagataki et al. (1998b) and addresses the conditions under which the ^{44}Ti is synthesized. The primary formation scenario for this isotope is through the "α-rich freeze-out". This occurs when material in NSE (nuclear statistical equilibrium) at low densities cools so rapidly that the free α particles do not have time to burn via the inefficient triple-α reaction. This leaves the heavy nuclei to cool in the presence of a large abundance of α particles, the capture of which synthesizes heavier nuclei like ^{44}Ti. The abundance of α particles is particularly sensitive to the entropy of the gas, with high entropy leading to a larger abundance of αs. In this way, higher entropy supernova explosions can enhance the ^{44}Ti/^{56}Ni abundance ratio. Nagataki et al. (1998b) have demonstrated this effect using 2D hydrodynamical explosion models with asymmetric initial explosion velocities. The asymmetry (ranging from axis ratios of 2:1 to 8:1 pole versus equator) create a supernova shock with higher entropy along the polar direction. For the particular case of a Cas A model, they find enhancements by a factor of 3-8 for their range of explosion asymmetries.

It seems likely that both the ionization effect proposed by Mochizuki et al. (1999), and the asymmetry effects of Nagataki et al. (1998b) are playing some role in the ^{44}Ti emission from Cas A. However, it is worth mentioning that for the case of SN 1987A, constraints on the ^{44}Ti abundance have been inferred from late-time bolometric light curve observations. In that case as well, observations suggest the presence of a greater abundance of ^{44}Ti than can be accounted for by progenitor models (Kumagai et al. 1993). For the current age of SN 1987A, the ionization effect put forth by Mochizuki et al. (1999) should cause an overall reduction in the observed emission. Perhaps SN 1987A is a stronger case for the argument that asymmetry effects must be included in models of core-collapse supernova remnants.

3. 3D γ-Ray Simulations

Both objects discussed above argue for a closer inspection of the role a global asymmetry may play in the explosion mechanism. In fact, the evidence for asymmetries in core-collapse supernovae goes well beyond that offerred by the γ-rays alone. Large optical polarization has been observed toward several core-collapse supernovae over the past decade (Leonard & Filippenko 2001; Wang et al. 2001). These observations show increasing polarization with time, suggesting that the explosion driving these supernovae is inherently asymmetric (Höflich 1991). An equally compelling argument for global asymmetry arises from attempts to understand the high space velocities of neutron stars. The high observed velocities of pulsars and the formation scenarios of neutron star binaries both suggest that neutron stars are given strong kicks at birth. These kicks are most easily explained by some asymmetry in the supernova explosion (see Fryer, Burrows, & Benz 1996 for a review).

In fact, the most likely explanation for the polarization and neutron star kicks is that the explosion asymmetries are imprinted by the explosion mechanism itself. For instance, even if the collapsing star is initially spherically symmetric, some asymmetry can be produced due to convection taking place in, and above, the proto-neutron star (Herant et al. 1994; Burrows, Hayes, & Fryxell 1995; Janka & Müller 1996). To date, these asymmetries are not extreme enough in the theoretical models to explain the mixing. Large asymmetries may occur if the collapsing star is asymmetric due to nuclear burning (Burrows & Hayes 1996; Lai 2000), however, sufficiently large departures from spherical symmetry have only been produced by assuming extremely asymmetric collapsing cores (Burrows & Hayes 1996). Alternatively, rotation can produce significant asymmetries in the supernova explosion (Mönchmeyer & Müller 1989; Janka & Mönchmeyer 1989; Fryer & Heger 2000, Khokhlov et al. 1999). The nature of these asymmetries depends upon the angular momentum profile of the collapsing star and, although most calculations predict jet-like explosions along the rotation axis, some calculations imply that an equatorial explosion could occur (Mönchmeyer & Müller 1989).

By comparison to the asymmetric explosions discussed in the chapter by Höflich, Wang & Khokhlov (these proceedings), the explosion asymmetries which result from rotation are relatively mild (velocity axis ratio of order 2:1 at roughly 2 seconds after bounce; Fryer & Heger 2000). In this section, we'll discuss results from 3-dimensional supernova simulations (Hungerford et al. 2003) investigating explosions with the mild asymmetries implied by rotating core-collapse models. Since γ-ray observations have played an important role in motivating theoretical investigations in multi-dimensions, we will place particular emphasis on the high energy spectral signatures from these rotating models.

Explosion Simulations

In the hydrodynamic simulations described below, we employed the 15 $M_\odot$ progenitor (s15s7b) by Weaver & Woosley (1993). This star was evolved with a piston-driven explosion to 100 s after bounce, producing 0.24 $M_\odot$ of ^{56}Ni. The total energy of this model was roughly 1.5×10^{51} erg with roughly 1.0×10^{51} erg in kinetic energy. As the explosion moves through a progenitor star, the shock passes through composition boundaries where strong entropy gradients exist. When the shock hits these boundaries, Rayleigh-Taylor instabilities develop, causing the star to mix (Chevalier & Klein 1978; Weaver & Woosley 1980). Our simulations have modeled this mixing and the effects that asymmetries have on it.

Numerical Schemes. The simulations presented here (Table 1) have been calculated using a 3-dimensional smooth particle hydrodynamics code (see Warren et al. 2003, in preparation, for details) based on the parallel oct-tree algorithm developed by Warren & Salmon (1993). This lagrangian code tracks the composition exactly. For 2 of our simulations, we included the energy injection from ^{56}Ni and ^{56}Co decay (and hence also trace the abundances of ^{56}Co and ^{56}Fe). For these 2 models, we assume that all of the decay energy is injected into the gas. At late times, this will overestimate the total energy deposited, as a fraction of this energy will escape, but since we would like an upper limit on the effects of ^{56}Ni decay, and because the injection of energy is less important to the explosion dynamics at late times, this assumption is adequate. For decay energy, we assume that the total energy from decay is 9.3×10^{16} erg g^{-1} with 33% being released from ^{56}Ni decay with a 6.1 d half-life and the other 67% being released from ^{56}Co decay with a 77 d half-life (Colgate, Petschek, & Kriese 1980). This is similar to what Herant & Benz (1992) and Herant & Woosley (1994) used for their models. For an equation of state, we use the same "low-density" equation of state developed by Blinnikov, Dunina-Barkovskaya, & Nadyozhin (1996) that is used in our progenitor model.

We mapped the Weaver & Woosley (1993) model with its 100 s long, spherically symmetric explosion into our 3D SPH code with 2.2 million variably-massed particles. We model the entire sphere, so there are no axis boundaries in this simulation and the neutron star mass at the center is added through an external gravitational force. Asymmetries are added artificially to the velocities only and, for most of the explosions, are added in such a way to insure that the total kinetic energy of the explosion is conserved. The two asymmetries we model are jet explosions:

$$v_{\rm radial} = (\alpha + \beta \times |z|/r) v_{\rm radial}^{\rm sym} \tag{7.1}$$

and equatorial explosions:

$$v_{\rm radial} = (\alpha - \beta \times |x|/r) v_{\rm radial}^{\rm sym} \tag{7.2}$$

where $v_{\rm radial}^{\rm sym}$ is the velocity from the one dimensional calculation, $v_{\rm radial}$ is the radial velocity for the asymmetric setup, and x, z, r are the x-position, z-position, and radius of the particle. The values of α and β for each model are given in Table 1, along with the initial kinetic energy of each model. The magnitude of the asymmetries are guided by the results of Fryer & Heger (2000), who found that their rotating core-collapse simulations in 2-dimensions had velocities that were a factor of 2 higher in the pole than in the equator 1.5 s after bounce. We do not use the extreme asymmetric explosions of Khokhlov et al. (1999) which concentrate the explosion energy into a narrow jet. As we shall show (and as Nagataki 2000 argued), such extreme asymmetries are not required to explain outward mixing of nickel in supernovae. Although at the end of the Fryer & Heger simulation, the asymmetry in polar vs. equatorial velocities is growing, the trend as the shock moves outward will be to spherize the shock. So at this point, it is not clear whether the shock at 100 s will be more or less asymmetric than what Fryer & Heger (2000) found at the end of their simulation.

After mapping these models into our 3-dimensional SPH code, we then run the explosion out to 1 year. When the shock reaches the edge of the star, we assume it is moving through a vacuum. In reality, such a star will have a stellar wind atmosphere surrounding it, but the density of a 15 $M_\odot$ wind is so low, that for the purposes of our simulation, zero density material is appropriate. In addition, photon transport is not modeled in our calculations. However, prior to shock breakout, the photons are essentially trapped in the shock. By the time the shock breaks out, most of the internal energy has already been converted to kinetic energy, so although including photon transport will change the ionization state of the exploding stellar material, it does not affect the kinematics significantly.

The smooth particle hydrodynamic technique automatically adds a level of perturbation in the code. With our initial conditions, these perturbations are random with a maximum $1 - \sigma$ deviation in a shell of 5-7%. Such large deviations only occur at composition boundaries, specifically the silicon and oxygen shells, and match well the deviations arising from explosive oxygen and silicon flashes prior to collapse (e.g. Bazan & Arnett 1998).

Explosion and Nickel Distribution. Although we use the same 15 $M_\odot$ progenitor that was used by Herant & Woosley (1994), it has a higher explosion energy and we map this model onto our 3-dimensional grid 200 s earlier than

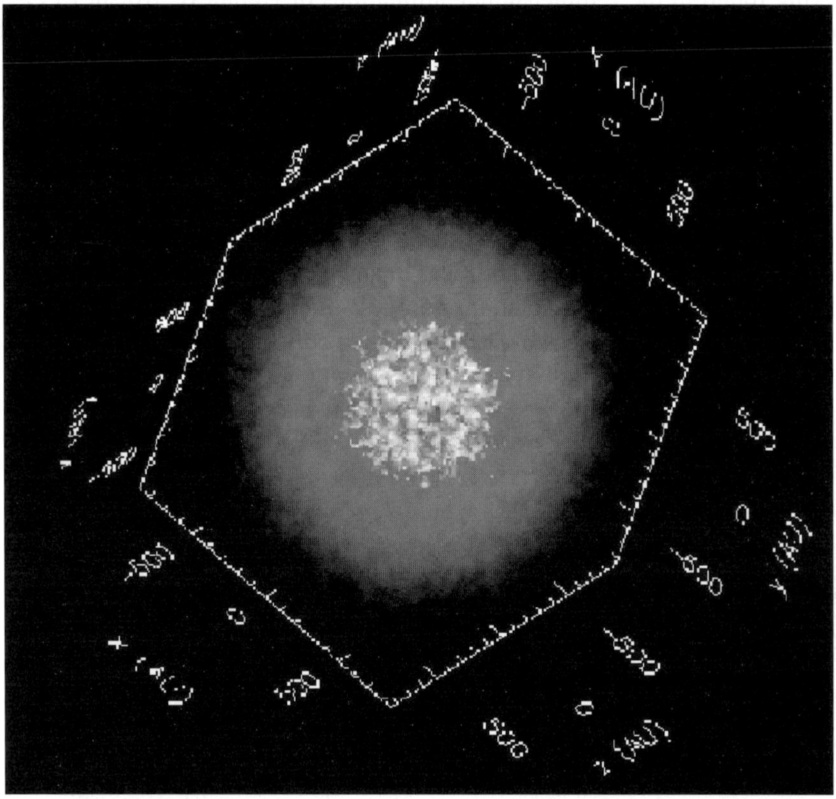

Figure 7.7. 3-dimensional simulation of the symmetric explosion 1 year after the shock launch. The contours represent the cobalt distribution with a number density of 10^{-5}. The colors denote the density distribution. Note that although the explosion is symmetric, Rayleigh-Taylor instabilities mix out the nickel.

the red supergiant envelope (Chevalier & Soker 1989), the distribution of ^{56}Co (the decay product of ^{56}Ni) retains a large asymmetry. We discuss the effects of these asymmetries on the γ ray emission in §3.2.

Like Nagataki et al. (1998a) and Nagataki (2000), we find that the asymmetries broaden the velocity profile of ^{56}Ni (Figure 7.10). However, in our simulations, mild asymmetries ($v_{\rm Pole}/v_{\rm Equator} = 2$) led to only a small increase in the maximum nickel velocity from 2500 km s^{-1} to 2900 km s^{-1}. For Nagataki et al. (1998a), such mild asymmetries increase the maximum nickel velocity from 2200 km s^{-1} to 3200 km s^{-1}! This difference could be progenitor dependent, an effect of 3-dimensional vs. 2-dimensional convection, or the

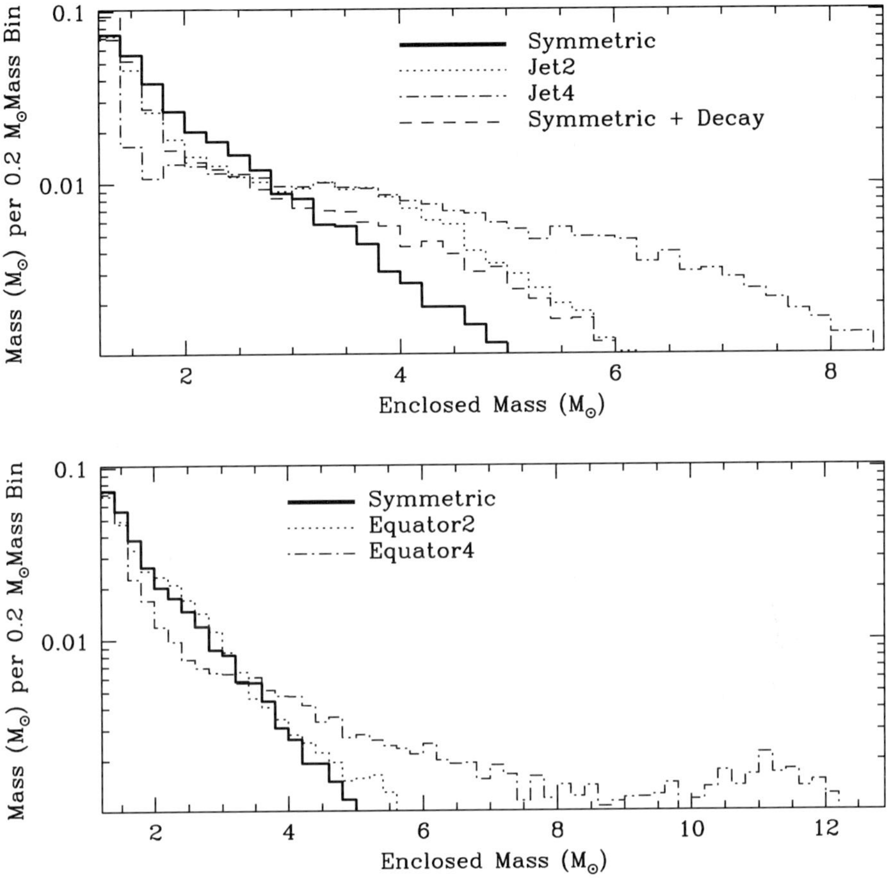

Figure 7.8. Distribution of the nickel ejecta in mass, comparing jet explosions with a symmetric explosion (top panel) and equatorial explosions with a symmetric explosion (bottom panel). Note that as we increase the degree of asymmetry (Jet4 versus Jet2, Equator4 versus Equator2), the mixing increases dramatically, placing nickel well into the hydrogen envelope of the star. The dashed line (top panel) shows the extent of mixing if all of the nickel/cobalt decay energy is deposited into the nickel ejecta and produces almost as much mixing in a symmetric explosion as the Jet2 model.

lack of resolution in our 3-dimensional models. Extracting the true cause of this difference awaits future calculations with similar initial conditions.

However, increasing the amount of asymmetry by another factor of 2 (Models Jet4, Eq4) causes some nickel to be ejected at very high velocities. The amount of mixing in these cases reaches extremes with the nickel well into the hydrogen envelope. Such mixing has decided signatures in both the emergence of the γ-

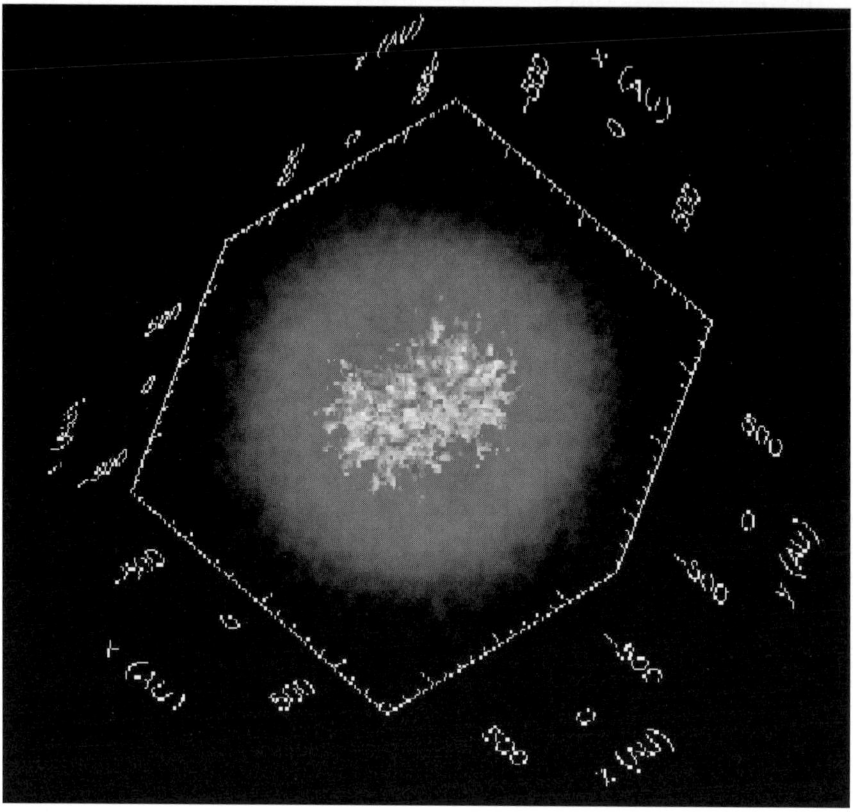

Figure 7.9. 3-dimensional simulation of the Jet2 asymmetric explosion 1 year after shock launch. As in Fig. 3, the contours represent the cobalt distribution with a number density of 10^{-5}. The colors denote the density distribution. The nickel is mixed out extensively in the polar direction where the explosion was strongest. However, the density distribution did not gain large asymmetries and remained fairly symmetric.

ray line flux and the shape of the γ-ray lines (§3.2). Note that the energy released from the decay of nickel also helps to mix out the nickel (on par with the effects of mild asymmetries). Clearly, the energy released from the decay of nickel cannot be neglected in any accurate mixing calculation.

This mixing also has important repercussions for nucleosynthetic yields and the mass-cut for the remnant mass. Most black holes are formed in stars which produce supernova explosions that are too weak to throw off all of the stellar envelope and the subsequent fallback produces a black hole (Fryer & Kalogera 2001). In our models (Jet2, Sym+Decay, etc.), more than 10% of the nickel produced is ejected well beyond the helium core (Figure 7.11). If this trend

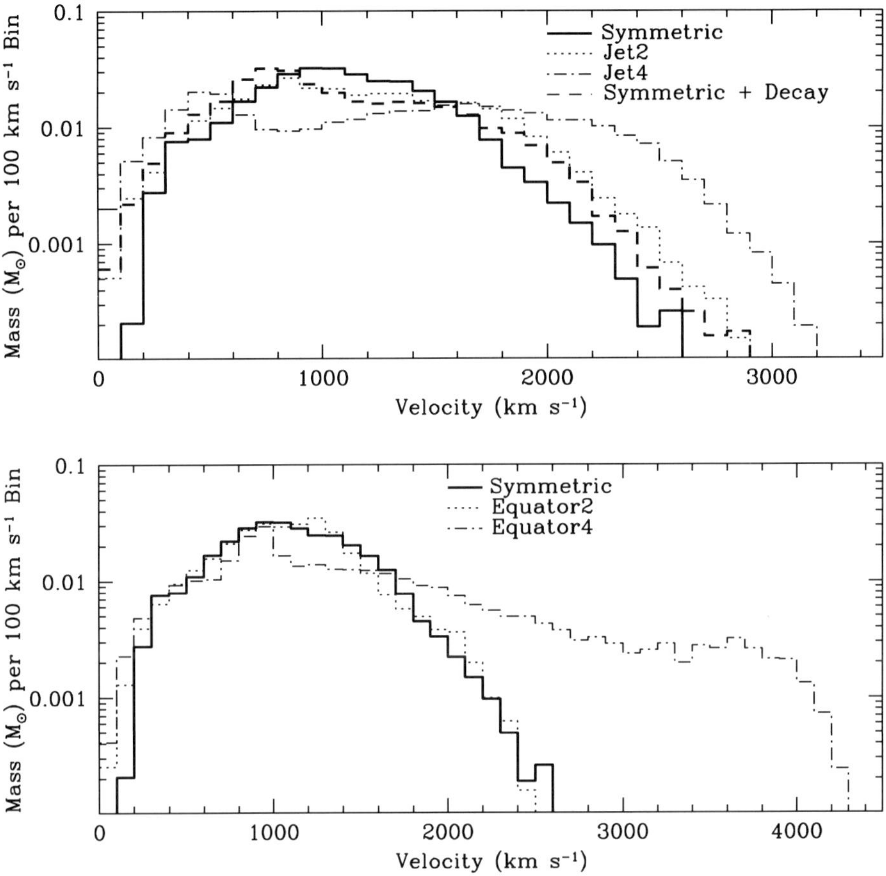

Figure 7.10. Distribution of the nickel ejecta versus velocity comparing jet explosions with a symmetric explosion (top panel) and equatorial explosions with a symmetric explosion (bottom panel). Note that as we increase the degree of asymmetry (Jet4 versus Jet2, Equator4 versus Equator2), the mixing increases dramatically, producing nickel velocities in excess of 3000 km s^{-1}. The dashed line (top panel) shows the velocities achieved if all of the nickel/cobalt decay energy is deposited into the nickel ejecta and produces almost as much mixing in a symmetric explosion as the Jet2 model.

holds for more massive stars such as the progenitor of SN 1997D (Turatto et al. 1998), then the entire helium core of such a star ($>8\,M_\odot$) could fall back and still enough nickel would escape to power the observed light curve! Bear in mind, however, that weaker explosions may well produce less mixing, so adapting the results of our simulations to supernovae like 1997D must be taken with some caution.

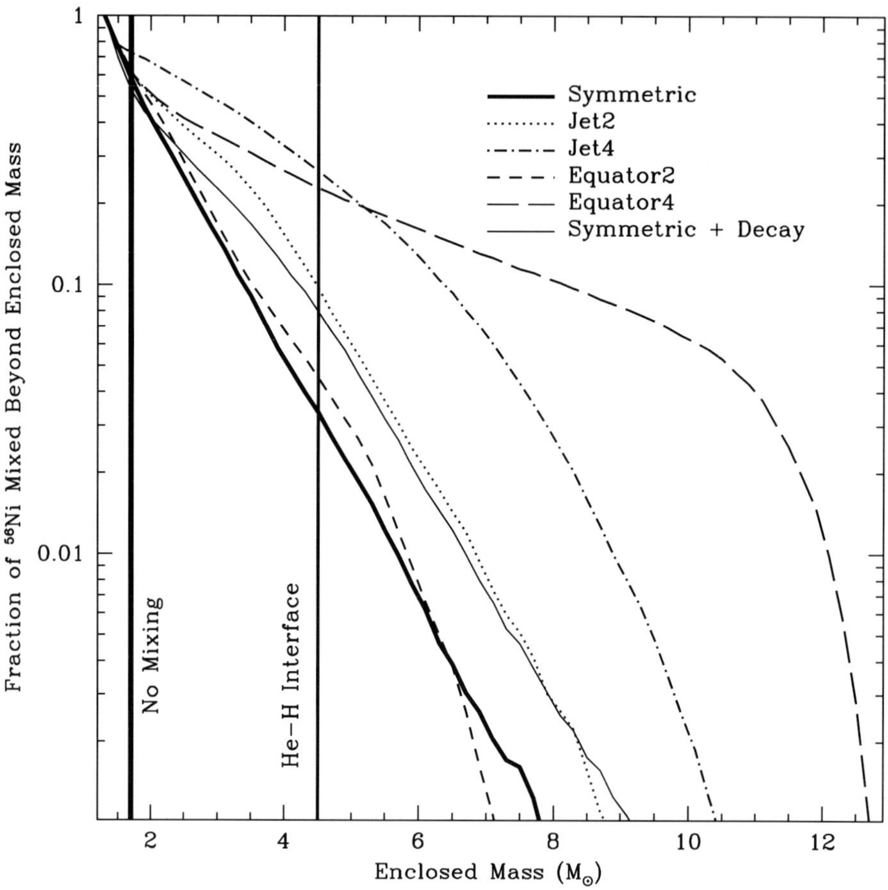

Figure 7.11. Fraction of nickel ejecta mixed out into the star for our set of simulations. Note that for mild asymmetries (Jet2) or if decay energy is included in a symmetric explosion (Symmetric+Decay), nearly 10% of the nickel mass is injected into the hydrogen envelope. If this amount of mixing occurs in weak explosions, an explosion that resulted in a 4.5 $M_\odot$ black hole remnant would still eject a moderate amount of nickel.

High Energy Spectral Calculations

For the spectral calculations, we have used data from the 3-dimensional explosion simulation discussed in §3.1. We input ejecta material properties from five different snapshots in time at 150, 200, 250, 300 and 365 days after explosion. Spectral calculations were carried out for both the Jet2 and Symmetric explosion models. Our analysis of these model spectra concentrates on the differences in total luminosity and line profile shape with the introduction of realistic explosion asymmetries. Since the progenitor star used as input to our simulations was a 15 $M_\odot$ red supergiant, we are unable to directly compare our calculated spectra with the observations of SN 1987A discussed in §1. However, we discuss how our models compare to various spectral trends observed from SN 1987A.

Numerical Schemes. We used a Monte Carlo technique, similar to that described in Ambwani & Sutherland 1988, for modelling γ-ray transport in 3-dimensions. Input models of the supernova ejecta (element abundances, density and velocities) were taken from the "Jet2" and "Symmetric" SPH explosion simulations and mapped onto a $140 \times 140 \times 140$ cartesian grid. Escaping photons were tallied into 250 coarse energy bins, with finer binning at the decay line energies to provide line profile information. The emergent photons were also tallied into 11 angular bins ($\Delta\theta = 10°$) along the polar axis (the models investigated in this work are axisymmetric, alleviating the need to tally in azimuthal angle as well.)

The decay of the radioactive species (predominantly ^{56}Ni and its decay product: ^{56}Co) in the supernova ejecta gives rise to the γ-ray line emission. As in Ambwani & Sutherland 1988, we assign the energy of the emitted photon packets according to decay probabilities (Lederer & Shirley 1978) for the various radioactive species (^{56}Ni, ^{56}Co, ^{57}Co, ^{44}Ti, ^{44}Sc, and ^{22}Na). We include a total of 56 decay lines from these species, but for the explosion times considered, the packets fall predominantly into $\sim$ 15 decay lines. Roughly 10^9 Monte Carlo photon bundles were generated for each input model in proportion to the mass of radioactive material distributed throughout the ejecta. The material properties of the ejecta were not evolved with photon flight time. However, we found that 99 % of the photons contributing to the observed model spectra have escape times of less than 2 days. This is sufficiently shorter than the timescales for change in the hydrodynamic models that our assumption of a fixed material background should be valid for the time slices considered here. The luminosity weight of each photon packet and the opacities seen by the packet were calculated in the comoving frame of the fluid, but all photon properties were boosted to the observer's frame before being tallied into spectral observables.

In all models, photoelectric and pair production opacities were calculated for the elements H, He, C, N, O, Ne, Mg, Si, S, Ar, Ca, Ti, Cr, Fe, Co, and Ni which correspond to the elements used in the nucleosynthesis calculations for the progenitor star from Weaver & Woosley (1993). The cross section data for these elements were taken from the LLNL Evaluated Nuclear Data Library (Plechaty, Cullen, and Howerton 1981, revised 1987). The angle- and energy-dependent Compton scattering opacities were calculated assuming that all electrons, bound and free, contribute to the total cross section. Daughter products from the absorption processes were not followed (e.g. positron annihilation photons from pair production and K-shell fluorescence photons from photoelectric absorption.) In order to test the validity of this last approximation, we compared the results of our 3-dimensional Monte Carlo transport code (Maverick) with the 1-dimensional code FASTGAM (Pinto & Woosley 1988a). Good agreement was found between the model spectra whether the daughter products of absorption were included in FASTGAM or not. This justified the decision to disregard the fluorescence and annihilation photons in Maverick.

Figure 7.12, shows comparison spectra with the 1-dimensional FASTGAM code for model 10HMM (Pinto & Woosley 1988b). This is a logarithmic plot of photon flux (photons/second/MeV/cm^2, assuming a distance of 60 kpc) across the energy range 0.3 keV - 4 MeV. The input model (10HMM) is an artificially mixed version of the Woosley 1988 10H model progenitor for SN 1987A. There is good agreement across the spectrum to within the uncertainties of the Monte Carlo calculation.

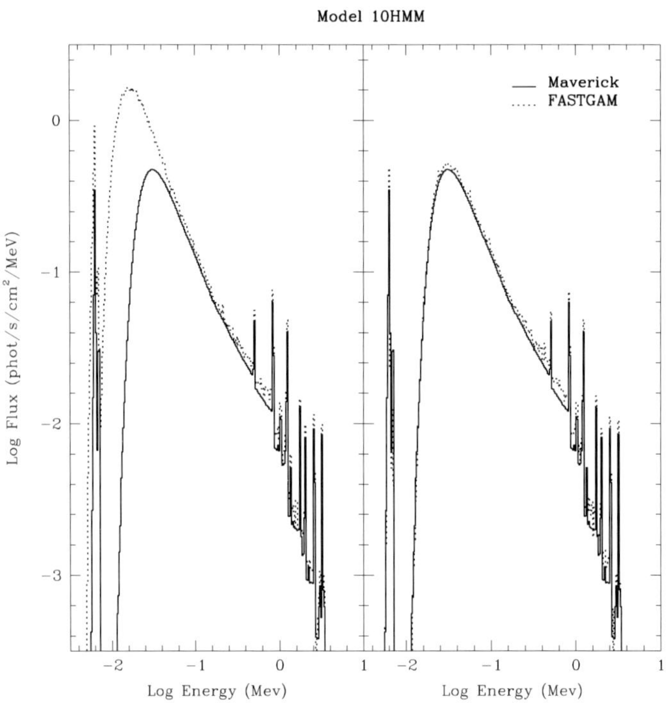

Figure 7.12. Total hard X- and γ-ray spectrum comparison between the 1D Monte Carlo transport code FASTGAM and the 3D code used in this work (Maverick). The left panel shows the two calculated spectra before a correction to the absorptive opacities in FASTGAM. The right panel shows the comparison once this correction had been made. The agreement between the two code results is quite good.

Hard X-ray and γ-ray Spectrum. Figure 7.13 is a logarithmic plot of photon flux in units of photons/second/MeV/cm^2 across the energy range investigated with these simulations (0.3 keV - 4 MeV). We have placed this object at the distance of the Large Magellanic Cloud (60 kpc) for easy comparison with flux data from SN 1987A observations. The 5 panels are spectra from the different time slices; in each panel, we plot the spectrum for the Symmetric model, along with polar and equatorial views of the Jet2 model. The effects of mixing are present in both these simulations, though at differing levels due to the differences in explosion asymmetry. It can be seen immediately that the hard X-rays emerge earlier from the ejecta with a global explosion asymmetry (Jet2 model). This holds regardless of viewing angle (pole versus equator) towards the explosion.

The fact that the hard X-ray flux in the aspherical explosion model is larger than the symmetric explosion, regardless of line of sight, can be understood in principle from optical depth arguments. In Figure 7.14, we show a contour plot of density (outer contour) and ^{56}Co number density (inner contour) for the Jet2 and Symmetric models at t = 150 days. Decay of ^{56}Co is the major source of γ-ray photons, so the inner contour essentially traces the surface of the emission region. The horizontal and vertical lines in Figure 7.14 represent lines of sight from the ejecta surface to the emission source and are labeled with the optical depth along that line of sight. The dominant opacity for the hard X- and γ-rays is Compton scattering off electrons and, since the density contours remain roughly spherical in both models, the optical depth from a given point to the ejecta surface is roughly constant. In the Jet2 model, the ^{56}Ni was mixed out to larger radii in the polar direction, so it makes perfect sense that we see enhanced emission over the Symmetric model spectrum for that viewing angle (the optical depth that the high energy photons must pass through is 10 in the symmetric model versus $\sim 6-7$ along the polar line-of-sight in the Jet2 model).

At a first glance, one might expect that, in the Jet2 explosion along the equatorial line-of-sight, the total flux should also be low (the optical depth from the nickel in the equator is also roughly 10). However, this material does not dominate the high energy emission seen along the equatorial line-of-sight. The material ejected along the poles has been mixed far enough out in the ejecta that the optical depth these high energy photons must travel through, even along the equatorial line-of-sight, is quite low ($\sim$ 7). It is this nickel which dominates the hard X-ray emission at all viewing angles. In fact, the optical depth from the "ends" of the ^{56}Co distribution does not differ very much between the polar view and the equator view (τ = 6 & 7 respectively), which explains why the overall hard X-ray flux depends only mildly on viewing angle.

For the later time slices, this mismatch in escaping emission from the "ends" versus the equatorial plane ejecta is less pronounced, and the equator view spectrum has comparable contributions from both regions.

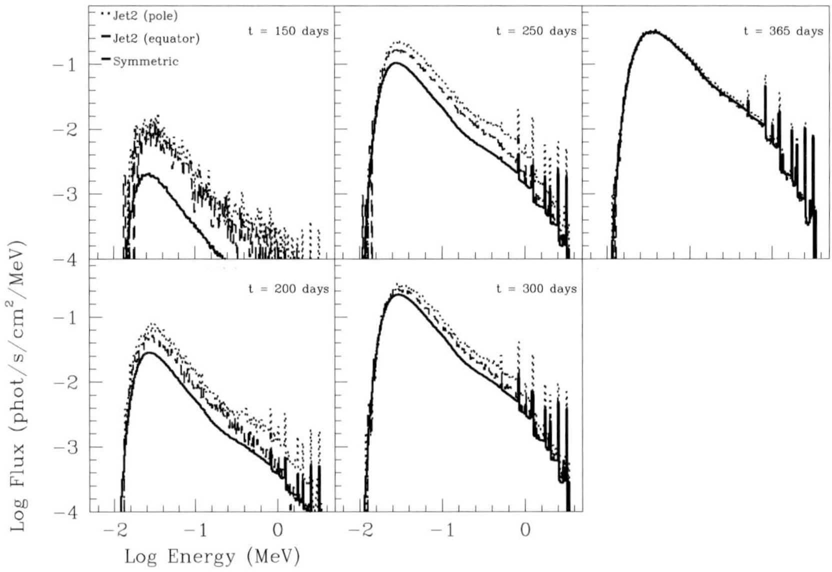

Figure 7.13. Total hard X- and γ-ray spectrum at 5 different times during the explosion (150,200,250,300,365 days) for symmetric (solid lines) and aspherical (Jet2) explosions (dotted and dashed lines). The flux is determined by assuming the object is 60 kpc from the observer. The dotted lines refer to an aspherical explosion where the jet is directed along the line-of-sight of the observer. The dashed lines refer to an explosion where the observer line-of-sight is directed 90° off of the jet axis, in the equatorial direction. Regardless of observer viewing angle, the aspherical explosion is brighter than the symmetric explosion.

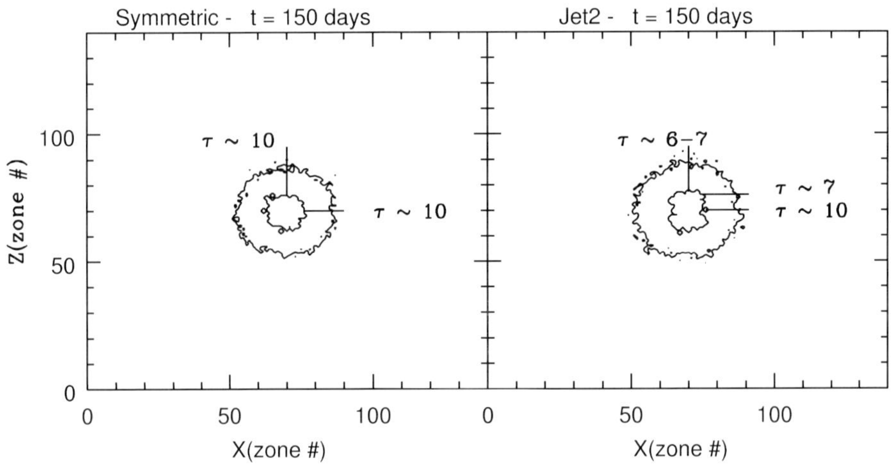

Figure 7.14. Contour plots in the xz-plane of the Symmetric and Jet2 explosion models at t = 150 days. Inner contour is for ^{56}Co number density which traces the surface of the γ-ray emitting region. Outer contour is for the mass density which follows electron density and thus traces the dominant opacity source (Compton scattering). The lines represent lines-of-sight through the ejecta for which the optical depth from emission region to ejecta surface has been calculated. Regardless of viewing angle, the optical depth of the ^{56}Co ejected along the poles in the Jet2 explosion remains quite low. Hence, it is this material that dominates the observed emission for all viewing angles in the aspherical explosion.

γ-ray Line Profiles. Although the overall hard X-ray emission shows little variation between pole and equator views, a detailed look at the γ-ray line profile shapes and strengths, for the 1.238 and 0.847 MeV ^{56}Co lines, does reveal trends with viewing angle. Figures 7.15 and 7.16 show line profiles of these two ^{56}Co lines for both the Symmetric and Jet2 explosion models. At these γ-ray energies, a single Compton scatter reduces the photon energy by roughly a factor of 2. This effectively removes the photon from the line profile region, leaving the γ-ray line profiles to be formed primarily by photons that escape with no interaction. In this way, the broadening of the line is caused entirely by Doppler velocity shifts resulting from the *spatial* distribution of radioactive nickel in the homologously expanding ejecta. The 4 panels shown are for days 200, 250, 300, and 365 after explosion. The three lines in the Jet2 spectra represent different viewing angles through the ejecta (along the pole, the equator and an intermediate angle $\sim 45°$.) For the Symmetric spectra, we have plotted these same viewing angles.

Both explosion scenarios (Symmetric and Jet2) show blue-shifted line profiles, though to varying degree. These differences can be best understood by examining the physical effects which dictate the formation of the line profile edges. The blue edge to the lines is set by the maximum observed line-of-sight velocity of the ^{56}Co in the ejecta. Since the expansion is basically homologous after 100 days, the line of sight velocity of a fluid element in the ejecta is proportional to its distance above the mid-plane of the explosion. Each spectral energy bin in the line profile can be mapped to a unique line of sight velocity in the ejecta, which can in turn be mapped to a specific height above the mid-plane. For example, defining the line of sight to be along the z-axis, the line profile shape should be proportional to the total mass of Cobalt summed in x and y as a function of z height in the ejecta. Therefore, the bluest edge of the line will arise from material that was mixed furthest out along the line of sight direction.

It is clear from Figure 7.14 that the nickel was mixed furthest out in the polar direction (and thus achieves higher velocities) for the asymmetric explosion. Following the discussion above, it is not suprising that the γ-ray line profiles viewed along the polar direction are much more blueshifted for the Jet2 model than the Symmetric model. Figure 7.14 does not show a very large difference in the extent of mixing along the equatorial direction between the two models. Correspondingly, the blue edge of the Symmetric lines and the equator view of the Jet2 lines are comparable.

The red edge of the lines is determined by the escaping emission from ^{56}Co with the smallest line-of-sight velocity in the ejecta. In a Symmetric model, this should be an indication of how deep into the ejecta we can see along a given viewing angle. However, there is a more pronounced effect at play in the asymmetric explosion models. As we mentioned in the previous sub-section,

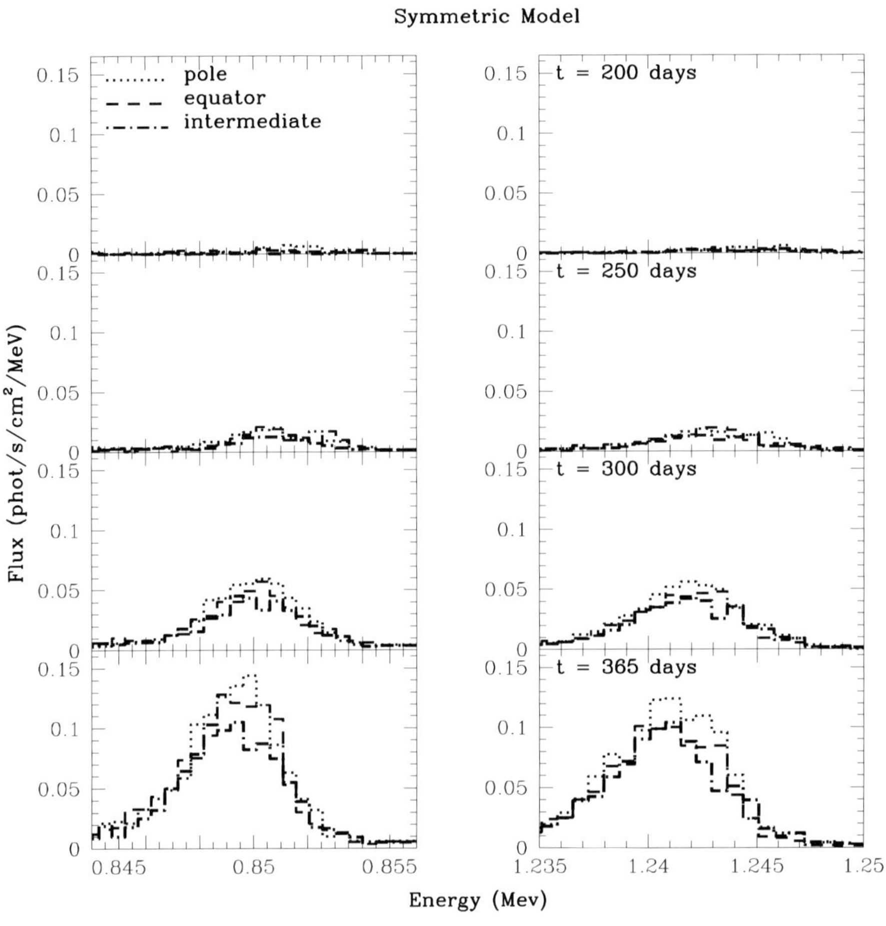

Figure 7.15. Line profiles of the ^{56}Co 1.238 and 0.847 MeV lines for the Symmetric model at 4 different times during the explosion (200, 250, 300, 365 days). 3 different viewing angles are shown: polar view (dotted lines), equatorial view (dashed lines) and an intermediate view angle of $\sim$ 45° (dash-dot lines). The line profiles do not show significant variation with viewing angle (as would be expected for a symmetric explosion.)

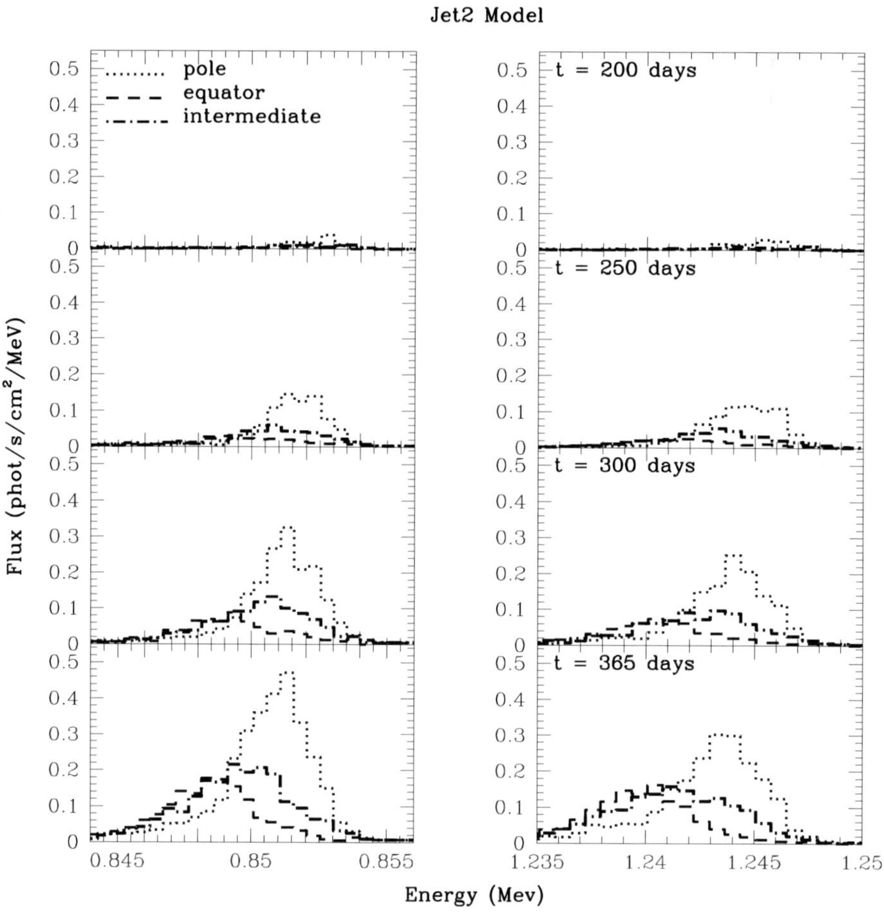

Figure 7.16. Line profiles of the ^{56}Co 1.238 and 0.847 MeV lines for the Jet2 model at 4 different times during the explosion (200, 250, 300, 365 days). 3 different viewing angles are shown: polar view (dotted lines), equatorial view (dashed lines) and an intermediate view angle of $\sim 45^{\circ}$ (dash-dot lines). The flux axis is scaled by a factor of 4 over the Symmetric model profiles shown in Figure 11. Significant variations in the line profiles with viewing angle are apparent, and can be explained by considering the velocity distribution (and thus radial distribution in a homologous expansion) of the ejecta responsible for the observed emission.

much of the γ-ray emission for the equatorial view arises from the "tips" of the ^{56}Co distribution. This material has a very low line-of-sight velocity for an equatorial observer, since it is being ejected predominantly in the polar direction. This allows for a significantly lower velocity red edge of the equator view lines, even though the optical depth profiles do not vary much between polar and equator viewing angles.

Also note that the line centroids become less blue-shifted with time. As the supernova expands, the opacity in the ejecta drops and emission from material located deeper into the ejecta (and thus at smaller radial velocities) becomes visible. This results in an overall broadening of the line, as well as a redward shift of the line centroid.

As mentioned in §1, the γ-ray line profiles from SN 1987A were in fact redshifted, a trend that is not obtained with these simulations. Although the γ-ray data uncertainties were quite high, this redshift was also observed in the far infrared forbidden lines of FeII, providing verification for the γ-ray line centroid measurements (see Figure 7.2). As we've discussed above, the spectral line shape is directly correlated with the total Cobalt mass at a given z-coordinate along the line of sight. With this in mind, the observed red-shifted line profiles towards SN 1987A imply, not only a break in spherical symmetry, but also a break in axisymmetry of the ejecta. Namely, there should be more Nickel/Cobalt mass on the far side of SN 1987A's ejecta as seen from our viewing angle. Pulsar velocity distributions also support the need for some non-axisymmetry in core-collapse supernova explosions. An interesting study that we have currently underway, is to link the magnitude of kick velocity imparted to a neutron star with the compositional asymmetry implied by the redshifted line profiles of SN 1987A.

4. Conclusions

γ-ray observations of core-collapse supernovae have played an integral role in motivating theoretical studies to go beyond the onion-skin paradigm of stellar structure. The early emergence of ^{56}Co decay lines observed from SN 1987A, and the broadened line profiles from those detections, were among the first observational signs that a break in spherical symmetry had occurred during the supernova explosion. Furthermore, attempts to explain the unexpectedly bright line flux from decay lines of ^{44}Ti in Cas A has also required the presence of inhomogeneities in the heavy element distribution, and possibly even global asymmetries in the explosion itself.

The 3D simulations presented here show that globally asymmetric explosions lead to extensive mixing of the supernova ejecta, placing the products of explosive burning well into the helium layer of the star. Even mild 2:1 asymmetries can mix 10% of the nickel out to the inner edge of the hydrogen layer.

If such mixing occurred in weak explosions, as well as the strong explosions presented in this paper, these mildly asymmetric supernova explosions could explain the extensive mixing required in population III stars (Umeda & Nomoto 2002) and black hole binary systems such as Nova Scorpii (Podsiadlowski et al. 2002).

As discussed in §1, the observed high energy spectrum of SN 1987A differed from the predictions of theoretical onion-skin models in two fundamental ways. Both the broad lines of nickel and the early emergence of the hard X-rays could be explained qualitatively by invoking a mixing argument. From a theoretical standpoint, including a 1D prescription for that mixing makes the assumption that both data points can be fit with one free parameter. However, the simulations in this work suggest that the addition of a global asymmetry will change the direct correlation between the emergence time and the degree of line broadening. That is to say, for a given hard X-ray flux, the Symmetric model will correspond to a single line profile, regardless of viewing angle. The Jet2 model, however, produces similar hard X-ray continua for different viewing angles, but the line profile varies significantly with viewing angle. In fact, the data for SN 1987A (the γ-line profiles and hard X-ray continuum) were not fit well by 1D models. In particular, the model 10HMM (Pinto & Woosley 1988), which was mixed sufficiently to account for the flux in the hard X-ray continuum observations, resulted in γ-line centroids that were shifted too far to the blue (Tueller et al. 1990). Although the uncertainties in this data were relatively large, this trend may be in the right direction to suggest a global asymmetry. That is to say, an asymmetric explosion scenario for SN 1987A could produce the same hard X-ray flux level, but with a redder line profile than the symmetric explosion scenario.

Notes

1. Note that the 2-dimensional simulations of Herant & Woosley (1994) have poor mass resolution and low resolution could be a problem in the 2-dimensional simulations as well as the 3-dimensional simulations

References

Ambwani, K. & Sutherland, P. 1988, ApJ, 325, 820

Arnett, W. D., Bahcall, J. N., Kirshner, R. P., & Woosley, S. E. 1989, ARA&A, 27, 629

Ashworth, W., 1980, Journal Hist. Astron., 11, 1

Bazan, G.,& Arnett, D 1998, ApJ, 496, 316

Blinnikov, S. I., Dunina-Barkovskaya, N. V., & Nadyozhin, D. K. 1996, ApJS, 106, 171

Burrows, A., Hayes, J., & Fryxell, B. A. 1995, ApJ, 450, 830

Burrows, A., & Hayes, J. 1996, PRL, 76, 352

Chevalier, R. A., & Klein, R. I. 1978, ApJ, 269, 281

Chevalier, R. A., & Soker, N. 1989, ApJ, 341, 867
Colgate, S. A., Petschek, A. G., Kriese, J. T. 1980, ApJ, 237, L81
Cook, W. R., Palmer, D. M., Prince, T. A., Schindler, S. M., Starr, C. H., & Stone, E. C. 1988, ApJ, 334, L87
Dotani, T., Hayashida, K., Inoue, H., Itoh, M., & Koyama, K. 1987, Nature, 330, 230
Fesen, R., Becker, R. & Blair, W., 1987, ApJ, 313, 378
Fryer, C.L., Burrows, A., & Benz, W. 1996, ApJ, 496, 333
Fryer, C. L., & Heger, A. 2000, ApJ, 541, 1033
Fryer, C. L., & Kalogera, V. 2001, ApJ, 554, 548
Haas, M. R., Erickson, E. F., Lord, S. D., Hollenbach, D. J., Colgan, S. W. J., & Burton, M. G. 1990, ApJ, 360, 257
Herant, M., & Benz, W. 1992, ApJ, 387, 294
Herant, M., & Woosley, S.E. 1994, ApJ, 425, 814
Herant, M., Benz, W., Hix, W.R., Fryer, C.L. & Colgate, S.A. 1994, ApJ, 435, 339
Höflich, P., 1991, A&A, 246, 481
Hughes, J., Rakowski, C., Burrows, D. & Slane, P., 2000, ApJ, 528, L109
Hungerford, A., Fryer, C. & Warren, M., ApJ, 594, 390
Iyudin, A., Diehl, R., Bloemen, H., Hermsen, W., Lichti, G., Morris, D., Ryan, J., Schoenfelder, V., Steinle, H., Varendorff, M., de Vries, C. & Winkler, C., 1994, A&A, 284, L1
Iyudin, A., 1997, Nuclear Phys. A, 654, 900c
Janka, H.-T. & Mönchmeyer, R., 1989, A&A, 209, L5
Janka, H.-Th., & Müller, E. 1996, A&A, 306, 167
Khokhlov, A.M., Höflich, P.A., Oran, E.S., Wheeler, J.C., Wang, L., Chtchelkanova, A. Yu. 1999, ApJ, 524, L107
Kumagai, S., Shigeyama, T., Nomoto, K., Hashimoto, M. & Itoh, S., 1993, A&A, 273, 153
Lai, D. 2000, ApJ, 540, 946
Lederer, C. M., & Shirley, V. S. 1978, *Table of the Isotopes* (7th ed., New York; Wiley)
Leonard, D. C., & Filippenko, A. V. 2001, PASP, 113, 920
Mahoney, W. A., Varnell, L. S., Jacobson, A. S., Ling, J. C., Radocinski, R. G., & Wheaton, Wm. A. 1988, ApJ, 334, L81
Matz, S. M., Share, G. H., Leising, M. D., Chupp, E. L., & Vestrand, W. T. 1988, Nature, 331, 416
Mochizuki, Y., Takahashi, K., Janka, H-T., Hillebrandt, W. & Diehl, R., 1999, A&A, 346, 831
Mönchmeyer, R. & Müller, E., 1989, in NATO ASI series, Timing Neutron Stars, ed. H. Ögelman & E.P.J. van den Heuvel (New York: ASI)
Nagataki, S., Shimizu, T.M., & Sato, K. 1998, ApJ, 495, 413

Nagataki, S., Hashimoto, M., Sato, K., Yamada, S. & Mochizuki, Y., 1998, ApJ, 492, L45

Nagataki, S. 2000, ApJS, 127, 141

Pinto, P. A., & Woosley, S. E. 1988a, ApJ, 329, 820

Pinto, P. A., & Woosley, S. E. 1988b, Nature, 333, 534

Plechaty, E., Cullen, D. & Howerton, R., 1981, LLNL Rep. UCRL-50400, Vol. 6(rev. 3[1987])

Podsiadlowski, P., Nomoto, K., Maeda, K., Nakamura, T., Mazzali, P., & Schmidt, B. 2002, ApJ, 567, 491

Reed, J., Hester, J, Fabian, A. & Winkler, P., 1995, ApJ, 440, 706

Spyromilio, J., Meikle, W. P. S., & Allen, D. A. 1990, MNRAS, 242, 669

Sunyaev, R., Kaniovskii, A., Efremov, V., Gilfanov, M., & Churazov, E. 1987, Nature, 330, 227

Sunyaev, R., Kaniovskii, A., Efremov, V., Grebenev, S., Kuznetsov, A., Engelhauser, J., Döbereiner, S., Pietsch, W., Reppin, C., Tuümper, J., Kendziorra, E., Maisack, M., Mony, B., & Staubert, R., 1990, Soviet Astron. Lett., 16(3), 171

Timmes, F., Woosley, S., Hartmann, D. & Hoffman, R., ApJ, 464, 332

Tueller, J., Barthelmy, S., Gehrels, N., Teegarden, B. J., Leventhal, M., & MacCallum, C. J. 1990, ApJ, 351, L41

Turatto, M., Mazzali, P. A., Young, T. R., Nomoto, K., Iwamoto, K., Benetti, S., Cappallaro, E., Danziger, I. J., de Mello, D. F., Phillips, M. M., Suntzeff, N. B., Clocchiatti, A., Piemonte, A., Leibundgut, B., Covarrubias, R., Maza, J., Sollerman, J. 1998, ApJ, 498, L129

Umeda, H. & Nomoto, K. 2002, ApJ, 565, 385

Vink, J., Laming, J. M., Kaastra, J., Bleeker, J., Bloemen, H. & Oberlack, U., 2001, ApJ, 560, L79

Wang, L., Howell, D. A., Höflich, P., Wheeler, J. C. 2001, ApJ, 556, 302

Warren, M. S. & Salmon, J. K. 1993, Supercomputing '93, IEEE Comp. Soc, pg. 12

Weaver, T. A., & Woosley, S. E. 1980, in AIP Conf. Proc. 63, Supernovae Spectra, ed. R. Meyerott & G. H. Gillespie (New York: AIP), 15

Weaver, T. A., & Woosley, S. E. 1993, Phys. Rep., 227, 65

Woosley, S. E. 1988, ApJ, 330, 218

Chapter 8

ASYMMETRIC SUPERNOVA EXPLOSIONS

P. Höflich

Department of Astronomy, University of Texas, Austin, TX 78681, USA

pah@astro.as.utexas.edu

L. Wang

Lawrence Berkeley Lab, 1 Cyclotron Rd, Berkeley, CA 94720, USA

A. Khokhlov

Naval Research Lab, Washington DC, USA

Abstract Core collapse supernovae (SN) are the final stages of stellar evolution in massive stars during which the central region collapses, forms a neutron star (NS) or black hole, and the outer layers are ejected. Recent explosion scenarios assume that the ejection is due to energy deposition by neutrinos into the envelope, but current models with detailed neutrino transport do not produce powerful explosions. There is new and mounting evidence for an asphericity and, in particular, for axial symmetry in several supernovae which may be hard to reconcile within the spherical picture. This evidence includes the observed high polarization and its variation with time, pulsar kicks, high velocity iron-group and intermediate-mass elements observed in remnants, and direct observations of the debris of SN1987A. Some of the new evidence is discussed in more detail. To be in agreement with the observations, any successful mechanism must invoke some sort of axial symmetry for the explosion. Based on models in literature, we expect no such asymmetries from neutrino driven explosions.

As a limiting case for aspherical explosions, we consider jet-induced/dominated explosions of "classical" core collapse supernovae. Bipolar outflows may be formed as a consequence of an accretion disk around the central object which is formed just after the core collapse, MHD mechanisms, or, maybe, some new instabilities within the neutrino picture. Our study is based on detailed 3-D hydrodynamical and radiation transport models. We demonstrate the influence of the jet properties and of the underlying progenitor structure on the final density and chemical structure. Our calculations show that low velocity, massive jets can explain the observations. Both asymmetric ionization and density/chemical

C.L. Fryer (ed.), Stellar Collapse, 237–258.

distributions have been identified as crucial for the formation of asymmetric photospheres. Even within the picture of jet-induced explosion, the latter effect alone fails to explain early polarization in core collapse supernovae with a massive, hydrogen-rich envelope such as SN1999em. The need for an asymmetric distribution of freshly formed ^{56}Ni may lend additional support for the idea that the explosion mechanism itself is asymmetric. Solving neutrino transport is an important 'component' to solve the SN problem but, apparently, not the complete solution. A successful model has to include all the effects, i.e. the core bounce, neutrino transport, convective flows and, in addition, significant effects due to rotation and, maybe, magnetic fields. Finally, we discuss observational consequences and tests.

1. Introduction

Supernovae (SNe) are among the most spectacular events in the universe, reaching the same brightness as an entire galaxy. This makes them good candidates to determine extragalactic distances and to measure the basic cosmological parameters. Moreover, they are thought to be the major contributors to the chemical enrichment of the interstellar matter with heavy elements. Energy injection by SNe into the interstellar medium, triggered star formation and feedback in galaxy formation are regarded as a key for our understanding of the formation and evolution of galaxies.

Core collapse supernovae are thought to be the final stages of the evolution of massive stars which live only 10^6 to 2×10^8 years. Such supernovae could be the brightest objects in the distant past when stars first began to form. A detailed understanding of core collapse is essential to probe the very early phases of the Universe right after the initial star forming period which occurs at redshifts $z \geq 3...5$. Understanding the mechanism of core collapse supernovae explosions is a problem that has challenged researchers for decades (Hoyle & Fowler 1960). In the general scenario for the explosion, the central region of a massive star collapses and forms a neutron star or, for massive progenitors, a black hole. Eventually, parts of the potential energy will cause the ejection of the envelope. Light emitted from the rapidly expanding envelope is observed as the supernova phenomenon. This general scenario has been confirmed by a wealth of observations including the direct detection of neutrinos in SN1987A and neutron stars in young supernovae remnants. However, the mechanisms which cause the ejection of the envelope are not well understood.

2. Models for Collapse Supernova

The mechanism of the energy deposition into the envelope is still debated. The process likely involves the bounce and the formation of the prompt shock (e.g. Van Riper 1978, Hillebrandt 1982), radiation of the energy in the form of neutrino (e.g. Bowers & Wilson 1982) and the interaction of the neutrino with the material of the envelope and various types of convective motions (e.g. Herant et al. 1994, Burrows et al. 1995, Müller & Janka 1997, Janka & Müller 1996), rotation (e.g. LeBlanc & Wilson 1970, Saenz & Shapiro S.L. 1981, Mönchmeyer et al. 1991) and magnetic fields (e.g. LeBlanc & Wilson 1970, Bisnovati-Kogan 1971).

Currently, the most favored mechanism invokes the neutrino deposition in an, overall, spherical configuration. The results depend critically on the progenitor structure, equation of state, neutrino physics, and implementation of the neutrino transport. Recent results indicate that spherical models fail to produce successful explosions even when using sophisticated, relativistic Boltzmann solvers for the neutrino transport and taking different flavors and neutrinos into account (Yamada et al. 1999, Ramp & Janka 2000, Mezzacappa et al. 2001). Multi-dimensional effects such as convection during the core collapse itself must be included but, still, it is an open question whether convection combined with the neutrino transport provides the solution to the supernova problem (Ramp et al. 1998, Fryer et al. 2002, and references therein). In the current calculations, the size and scale of the convective motions seem to be too small to explain the observed asymmetries in the envelope. The angular variability of the neutrino flux caused by the convection has been invoked to explain the neutron star kicks (Burrows et al. 1995, Janka & Müller 1994). Calculations give kick velocity up to $\simeq 100\,\mathrm{km\,s^{-1}}$ whereas NS with velocities of several $100\,\mathrm{km\,s^{-1}}$ are common. Even if successful, these models cannot explain the observed, large scale asymmetries (see below). Within the spherical core-collapse picture, additional mechanisms must be invoked which operate within the envelope itself.

Two such mechanisms have been studied in some detail. One is the Rayleigh-Taylor instability which causes mixing of the layers of different composition when the outgoing shock front passes through these layers (Müller et al. 1989, Benz & Thielemann 1990, Fryxell et al. 1991). This effect can explain mixing of the carbon, oxygen and helium-rich layers required for the SN1987A, but none of the simulations were able to account for the high velocity of Ni observed in SN1987A (Kifonidis et al. 2000). Rayleigh-Taylor mixing provides rather small-scale structures and can hardly account for the observed polarization which requires a global asymmetry of the expanding envelope (Höflich 1991). Another mechanism involves an explosion inside a rapidly and differentially rotating supernova progenitor (Steinmetz & Höflich 1992). With this mechanism it was possible to account for the polarization in SN1987A which originated from a blue supergiant. This mechanism may have difficulty accounting for the early polarization in some Type II supernovae (Wang et al. 2000) whose light curves indicate a red-giant progenitors. A strong differential rotation in red supergiants can hardly be expected due to their convective envelopes.

Attempts have also been made to include multi-dimensional effects into a model of collapse itself. The collapsing core becomes unstable due to the gradients of both electron mole fraction Y_e and entropy. The developing convection then affects both the neutrino flux and the energy deposition behind the stalled shock. Numerous studies have demonstrated the presence of this effect. It is still debated whether convection combined with the neutrino transport provides the solution to the supernova problem (Ramp et al. 1998 and references therein). In the current calculations, the size and scale of the convective motions seem to be too small to explain the observed asymmetries. The angular variability of the neutrino flux caused by the convection has been invoked to explain the neutron star kicks (Burrows et al. 1995, Janka & Müller 1994). Calculations give kick velocity up to $\simeq 100$ km s^{-1} whereas NS with velocities of several 100 km s^{-1} are common.

Rotation of the collapsing core may also be important. It tends to facilitate the explosion because the centrifugal barrier reduces the effective potential for the material moving in the equatorial plane and introduces an axial symmetry in the fluid motions. Simulations made so far indicate that the rotation alone has no or has only a weak effect on the explosion (e.g. Mönchmeyer et al. 1991, Zwerger & Müller 1997). In the latter case it induces a rather weak asymmetry of the explosion with more energy going along the rotational axis. As a consequence of stellar rotation, an accretion disk will form around the neutron star or black hole, and this may cause axial symmetric outflow or jets.

It has long been suggested that the magnetic field can play an important role in the explosion (LeBlanc & Wilson 1970; Ostriker & Gunn 1971, Bisnovati-Kogan 1971, Symbalisty 1984). Simulations by LeBlanc & Wilson showed the amplification of the magnetic field due to rotation and the formation of two oppositely directed, high-density, supersonic jets of material emanating from the collapsed core. Their simulations assumed a rather high initial magnetic field $\sim 10^{11}$ Gauss and produced very strong final fields of the order of $\sim 10^{15}$ Gauss which seemed to be unreasonable at the time. The recent discovery of pulsars with very high magnetic fields (Kouveliotou et al. 1998, Duncan & Thomson 1992) revived the interest in the role of rotating magnetized neutron stars in the explosion mechanism. It is not clear whether a high initial magnetic field required for the LeBlanc & Wilson mechanism is realistic. On the other hand, it may not be needed.

In conclusion, the current picture of the core collapse process is unsettled. A quantitative model of the core collapse must eventually include all the elements mentioned above.

Due to the difficulty of modeling core collapse from first principles, a very different line of attack on the explosion problem has been used extensively and proved to be successful in understanding of the supernova problem, SN1987A in particular (Arnett et al. 1990, Hillebrand & Höflich 1991). The difference of characteristic time scales of the core (a second or less) and the envelope (hours to days) allows one to divide the explosion problem into two largely independent parts - the core collapse and the ejection of the envelope. By assuming the characteristics of the energy deposition into the envelope during the core collapse, the response of the envelope can be calculated. Thus, one can study the observational consequences of the explosion and deduce characteristics of the core collapse and the progenitor structure. This approach has been extensively applied in the framework of the 1D spherically symmetric formulation. The major factors influencing the outcome have been found to be the explosion energy and the progenitor structure. The same approach can be applied in multi-dimensions to investigate the effects of asymmetric explosions. In this paper we study the effects and observational consequences of an asymmetric, jet-like deposition of energy inside the envelope of SN.

3. Observational Evidence for Asymmetry

In recent years, there has been a mounting evidence that the explosions of massive stars (core collapse supernovae) are highly aspherical. (1) The spectra of core-collapse supernovae (e.g., SN87A, SN1993J, SN1994I, SN1999em) are significantly polarized indicating asymmetric envelope with axis ratios up to 2 (Méndez et al. 1988; Höflich 1991; Jeffrey 1991; Wang et al. 1996; Wang et al. 2001b). The degree of polarization tends to vary inversely with the mass of the hydrogen envelope, being maximum for Type Ib/c events with no hydrogen (Wang et al. 2001b). For SN1999em (Fig. 8.1), Leonard et al. (2000) showed that the polarization (and thus, the asphericity) increases with time. Both trends suggest a connection of the asymmetries with the central engine. For supernovae with good time and wavelength coverage, the orientation of the polarization vector tends to stay constant both in time and with wavelength. This implies that there is a global symmetry axis in the ejecta (Leonard et al. 2001, Wang et al. 2001b). (2) Observations of SN 1987A showed that radioactive material was brought to the hydrogen rich layers of the ejecta very quickly during the explosion (Lucy 1988; Tueller et al. 1991). (3) The remnant of the Cas A supernova shows rapidly moving oxygen-rich matter outside the nominal boundary of the remnant and evidence for two oppositely directed jets of high-velocity material (Fesen & Gunderson 1997). (4). Recent X-ray observations with the Chandra satellite have shown an unusual distribution of iron and silicon group elements with large scale asymmetries in Cas A (Hughes et al. 2001). (5) After the explosion, neutron stars are observed with high velocities, up to 1000 $\mathrm{km\,s^{-1}}$ (Strom et al. 1995). (6) Direct HST-images from June 11,2000, are able to resolve the inner debris of SN1987A showing its prolate geometry with an axis ratio of ≈ 2 (Fig. 8.1). Both the ejecta and the inner ring around SN1987A show a common axis of symmetry. By combining the HST-images with spectral and early-time polarization data, Lifan Wang worked out the details of the chemical and density structure. By connecting the HST-images with the polarization data from earlier times, he demonstrated that the overall geometry of the entire envelope of SN1987A (including the H-rich envelope) was elongated all along the same symmetry axis, and that the distribution of the products of stellar burning (O, Ca, etc.) are concentrated in the equatorial plane (Wang et al. 2001a).

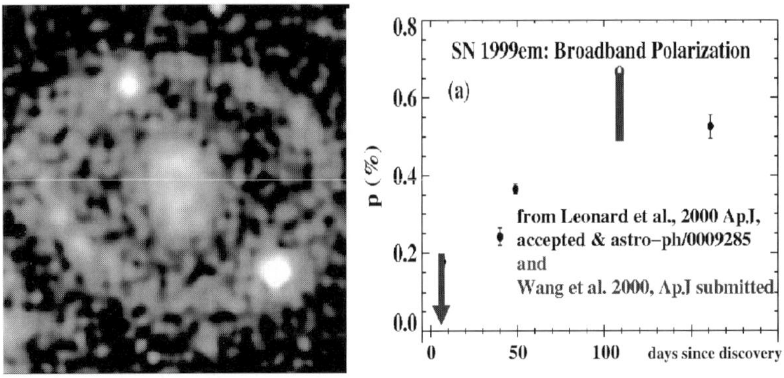

Figure 8.1. Observational evidence for asphericity in core collapse supernovae. The HST image of SN 1987A (left panel) shows the inner debris of the SN-ejecta with an axis ratio of ≈ 2 and the ring on June 11th, 2000 (from Wang et al. 2002b). Note that the inner ring has been formed during the stellar evolution about 10,000 years before the explosion. The right panel shows the evolution with time of the linear polarization P in the plateau supernovae SN1999em. Although P increased with time, the polarization angle remained constant with time and wavelength indicating a common axis of symmetry in the expanding envelope (Leonard et al. 2001, Wang et al. 2001c).

4. Numerical Methods

3-D Hydrodynamics: The explosion and jet propagation are calculated by a full 3-D code within a cubic domain of size D. The stellar material is described by the time-dependent, compressible, Euler equations for inviscid flow with an ideal gas equation with $\gamma = 5/3$ plus a component due to radiation pressure with $\gamma = 4/3$. The Euler equations are integrated using an explicit, second-order accurate, Godunov type, adaptive-mesh-refinement, massively parallel, Fully-Threaded Tree (FTT) program, ALLA (Khokhlov 1998). Euler fluxes are evaluated by solving a Riemann problem at cell interfaces. FTT discretization of the computational domain allowed the mesh to be dynamically refined or coarsened at the level of individual cells (Fig. 8.2). For more details, see Khokhlov (1998) and Khokhlov et al. (1999).

Hydrodynamical Radiation Transport: The computational tools for the radiation hydrodynamics were used to carry out many of the analyses of SNIa

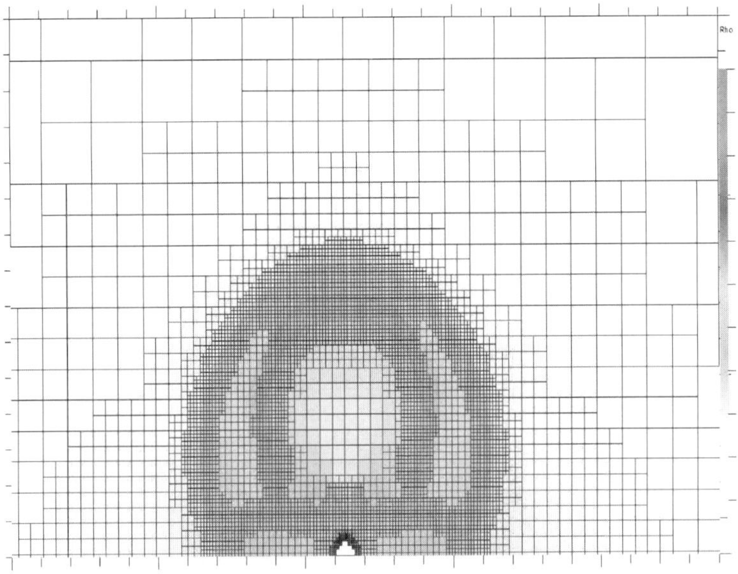

Figure 8.2. Jet-induced model for the explosion of a C/O core (SNIc) using a 3-D code with adaptive mesh refinement using symmetric multi-processing, i.e. the same equations are solved by all processors but for various sub-domains. The total computational domain is $1.5 \times 10^{11} cm$ with a resolution of $5 \times 10^{7} cm$. The snapshot shows the density distribution ($1.2 < log(\rho) < 7$, in CGS) after about 10 seconds for the explosion of a star with 2.5 $M_{\odot}$ triggered by low velocity jet ($\approx 11,000\, s^{-1}$, $E_{jet,total} = 1.6E50erg$).

and Core Collapse Supernovae (e.g. Höflich 1988, Höflich, Müller & Khokhlov 1993, Höflich, 1995, Howell et al. 2001). For the nomenclature we follow Mihalas & Mihalas (1984) if quantities are not defined explicitly. All components of the codes have been written or adopted in a modular form with well defined interfaces which allows an easy coupling (see Fig. 8.5) and code verification by exchanging physics modules (e.g. various radiation transport schemes) but keeping the remaining setup identical. The modules consist of physical units to provide a solution for the nuclear network, the statistical equations to determine the atomic level population, equation of states, the opacities, the hydro or the radiation transport problem. The individual modules are coupled explicitly. Consistency between the solutions is achieved iteratively by perturbation methods. Currently, not all modules can be combined simultaneously because not all iteration schemes have been implemented and because of requirements on CPU time: a) For full NLTE-spectra with large model atoms and a high frequency resolution we are restricted to the time-independent case based on a given hydrodynamical structure and, for 3-D models, reduced atomic atomic models (level-merging/super-levels) have to be used. b) Radiation hydrodynamics is restricted to a reduced frequency resolutions with reduced atomic levels

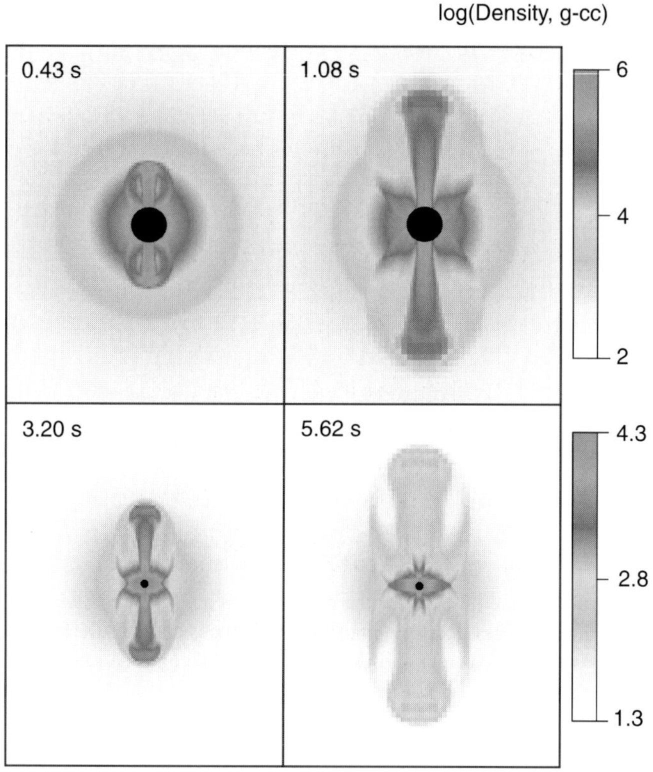

Figure 8.3. Logarithm of the density structure as a function of time for a helium core. The total mass of the ejecta is 2.6 $M_\odot$. The initial radius, velocity and density of the jet were taken to 1200 km 32,000 km s^{-1} and $6.5E5g/cm^3$, respectively. The shown domains 7.9, 9.0, 36 and 45 $\times 10^9 cm$. The total energy is about 9E50 erg. After about 4.5 seconds, the jet penetrates the star. The energy deposited in the stellar envelope by the jet is about 4E50 erg, and the final asymmetry is of the order of two.

(level-merging) and spherical geometry. In case of multi-dimensional radiation hydrodynamics, CPU-time requirements restrict applications even further. Currently, we can use a few frequencies to represent the fluxes e.g. in the Lyman and 'Balmer and higher continua', and 3-level atoms plus spherically symmetric velocity fields for the RT(Höflich, Khokhlov & Wang 2002), or the grey case for arbitrary field. Technically, we use the so called 'trivial' parallelization: Parallelization (via MPI and PVM) is achieved on the module basis, e.g. parallelization of the nuclear network over the grid points, using groups of photons for the Monte-Carlo radiation transport, and sub-domains with 'ghost-cells' with respect to the spacial and frequency coordinates for the hydro and the comoving frame radiation transport, respectively. In the remainder of this section, we want to describe the various modules.

Hydrodynamics: The structure of the expanding envelopes are obtained by three different modules which by a) assuming free expansion, by solving the hydro equations in b) the Lagrangian frame for spherical geometry including a front tracking scheme to resolve shock fronts (HWT98), or the Eulerian scheme for full 3-D using Cartesian coordinates based on PROMETHEUS (Fryxell, Arnett & Müller 1991). The hydro modules use an explicit Piecewise Parabolic Method (PPM) by Collella & Woodward (1984) to solve the compressible reactive flow equations. PPM is implemented as a step followed by separate remaps of the thermal and kinetic energy to avoid numerical generation of spurious pressure disturbances during propagation of reaction fronts (flames and detonations).

Nuclear network & high density EOS: Nuclear burning is taken into account using Thielemann's nuclear reaction-network library (Thielemann, Nomoto & Hashimoto 1994, and references therein). The main sources for experimental rates are Caughlan et al.(1985),Caughlan & Fowler(1988) and Wiescher, Görres & Thielemann (1990). Typically, between 20 and 618 isotopes are taken into account. For the equation of state, we use a relativistic Fermi-gas with Coulomb corrections and crystallization, and radiation.

Opacities and low density equation of state: For low densities, we solve the full rate equations (see below) to determine the level population, or assume LTE. Typically, about 500 to 600 discrete NLTE levels are included in full NLTE with about 20,000 to 40,000 line transitions. For the 3-D hydro, the effective number of levels is reduced by about a factor of 10 by level-merging (Höflich 1990), often referred to as the use of super-levels. Complete redistribution over each individual line both in frequency and in angle are assumed. This means that the relative populations within the sublevels or the merged levels are described by

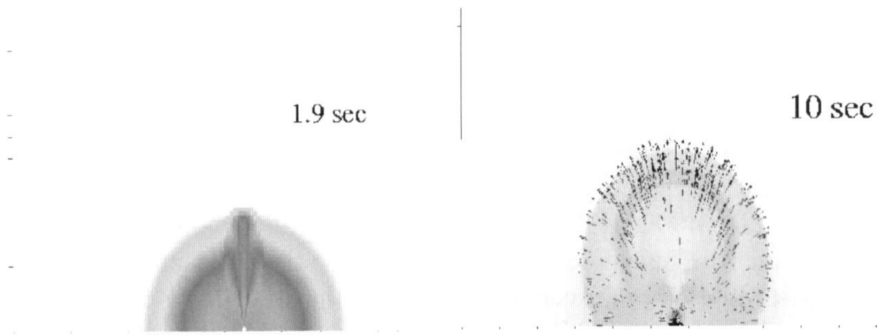

Figure 8.4. Same as Fig. 8.3 ($0.5 \leq log(\rho) \leq 5.7$) but for a jet velocity of 61,000 km s^{-1} and a total energy of 10 foe at $\approx 1.9sec$ (left), and 11,000 km s^{-1} and a total energy of 0.6 foe (right). The size of the presented domains are 5 (left) and 2 $10^{10}cm$ (right), respectively. For the high velocity jet, most of the energy is carried away by the jet. Only 0.9 foe are deposited in the expanding envelope. In case of a low velocity jet, the bow-shock still propagates through the star after the jet is switched off, and the entire jet energy is deposited in the expanding envelope.

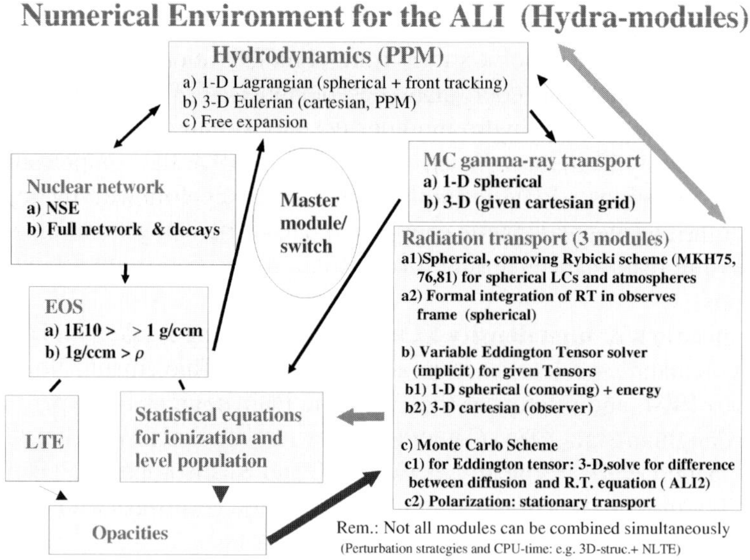

Figure 8.5. Block diagram of our numerical scheme to solve radiation hydrodynamical problems including detailed equation of state, nuclear and atomic networks. For specific problems, a subset of the modules is employed (see text). The code is still under active development. For its current status, see Höflich, 2002b).

a Maxwell-Boltzmann distribution. The data for the atomic line transitions are taken from the new compilation of Kurucz (Kurucz 1991; Kurucz 1995) and the opacity project (TOPBASE, e.g. Cunto & Mendoza 1992) from which about 500,000 -2,000,000 lines are extracted, depending on the temperature and density range. Photon scattering by free electrons is included in the Klein-Nishina limit. Free-free cross sections are treated in the hydrogen approximation with free-free Gaunt factors according to Seaton(1960) and Gayet(1970). Radiative bound-free cross sections are taken from the opacity project ($Z \leq 28$).

Gamma-ray transport and energy deposition: The γ-ray transport is computed in spherical and three-dimensions using a Monte Carlo method (Höflich, Khokhlov & Müller 1992, Höflich, 2002a) including relativistic effects and a consistent treatment of both the continua and line opacities. The interaction processes allowed are: Compton scattering according to the full angle-dependent Klein-Nishina formula, pair-production, and bound-free transitions (Ambwani & Sutherland 1988, Brown & Leventhal 1987).

Radiation transport: Several modules are available. For the spherical geometry and in the stationary case, the radiation transport equation is solved in the comoving frame (Mihalas, Kunacz & Hummer 1975, Mihalas, Kunacz & Hummer 1976, Mihalas, Kunacz & Hummer 1976b, hereafter MKH-methods). For the time dependent cases, we use variable Eddington Tensor solvers (implicit in time) (Mihalas & Mihalas 1984, Stone, Mihalas & Norman 1992, Höflich et al. 1993). We assume that the Eddington factors are constant during each time step. In the spherical case, we use the MKH methods. To obtain the correct solution for the Eddington tensor elements in 3-D, we use a Monte Carlo method to compute the difference between the solution of the non-equilibrium diffusion and full radiation transport equation. We calculate the difference between the solutions for computational accuracy and efficiency. In particular, the random element in MC avoids spatial fluctuations (and, thus, wiggles) in propagating, plane light fronts though, still, its speed in the free stream limit may differ from the light speed by 5 to 20 %. The same Monte Carlo solver is used which has been applied to compute γ-ray and polarization spectra for scattering dominated atmospheres (e.g. Höflich et al. 1992, Howell et al. 2001). The Monte Carlo method is appropriate for this problem because of its flexibility with respect to the geometrical and velocity structures.

AutomaticMeshRefinement for radiation transport by a MC torch: Automatic Mesh Refinement is a well established procedure in hydrodynamics and allows us to adjust the resolution to the requirements. The same holds true also for radiation transport. In stellar atmospheres, a logarithmic spacing of the optical depth τ is adopted to guarantee an appropriate resolution. In dynamical problems, extended atmospheres/envelopes with arbitrary morphologies the problem boils down to determine the region of the photosphere, or the 'skin' of an optically thick object. We start with a grid, equally spaced in mass (in comoving frame) or space (in Eulerian frame), and employ a 'Monte Carlo Torch' for the grid refinement. As a basic concept, the photosphere is defined as the region from which the photons can escape or, because of the symmetry of the problem, a photon can penetrate to when coming the outside. With the 'torch', we illuminate the object from the outside and calculate the path of the photons (with various energies) in random directions. If a photon is interacting in a computational cell, we increase its photon counter. Typically, the number of test photons is of the order of the number of cells n_R in a grid times the number of representative frequencies ν_{rep}. Typically, about 10^6 to 10^7 are used both in the spherical ($n_r = 912$ & $\nu_{rep} = 1000$)) and in the 3-D ($n_r \approx 5 \times 10^5$ & $10 \times \nu_{rep}$). Computational cells are divided in half if their photon count exceeds the average by a factor of 5 to 10. By illuminating the object from the edges of the domain, 3-D surfaces will be traced only if they can be seen from the domain boundaries. Concave structures such as the inner edges of Rayleigh Taylor fingers will be missed. Therefore, $\approx$ 10 additional photons are emitted from each grid cell. A photon triggers a counter only if it has crossed 2 or more cell boundaries to avoid rezoning of optically thick layers. The requirements of this approach depends on the physical situation. In our examples, we allow

for rezoning every 20 to 50 time steps. To test the accuracy, we increase the number of test photons by a factor of 10 about every 10 rezoning steps.
Accelerated Lambda Iteration: The radiation field couples the local, statistical equations and the hydro equations, resulting in complex systems of integro-differential equations of rank $DIM = n(x)n(y)n(z)n(\nu) \sum_{el} \sum_{ion(el)} \sum_{level(el,ion)} n(el, ion, level)$ where n(x,y,z) and n(ν) are the spatial and frequency coordinates, and n(el,ion,level) are the atomic level population of "level" for element "el" with the ionization "ion". The rank can easily exceed $10^{12...13}$ which prohibits a direct solution. However, the iterative solution provides an effective way to separate global and local quantities (Höflich, 2002b).

To overcome these problems, we use perturbation techniques. a) The method of accelerated lambda iteration is used to remove the global dependencies produced by the the radiation field.
b) The statistical equations and the energy terms in the hydro equations are solved by a partial linearization method (Mihalas & Mihalas 1984). When evaluating the mean intensity, J_ν, in the rate equations two extreme cases are considered in which the source function of the specific transition does or does not dominate the total source function. This approach further assumes that the variation of the total source function can be represented by the transition under consideration or by the change of the source function during the previous iteration step, respectively. The first case is quite similar to the assumptions of the equivalent two level approximation, but with the difference that the perturbation terms are used instead of the total rates.
c) The concept of "leading" elements is introduced. An appropriate ordering of elements or groups of elements allows a separation of the equations, which are solved by partial linearization. With this technique, the system of equations remains small, independently of the problem. This property is needed to achieve numerical stability and computing efficiency.
d) We use the concept of level-locking (Höflich 1990), often called super-levels for numerical efficiency. For stability, we use level locking when the system is far from convergence.
e) The scattering and thermal contributions to the source functions are separated by an equivalent two-level approximation for transitions from the ground levels. In effect, this allows to include the non-local nature of optically thick, scattering dominated transitions already during solution of the full radiation transport problem. In effect, this approach introduces an acceleration term for the convergence of ALI (Athay 1992, Avrett & Loeser 1992, Höflich, 1995).
f) Based on the explicit forms for the total opacity $\chi(\nu)$ and the source function $S(\nu)$, we reconstruct redistribution functions in ν for the photon, and limit the relative change between model iterations to $\approx 10\%$. For more details, see Höflich 1990, Höflich, 2002a, Höflich, 2002b, and references therein.

5. Results for Jet-Induced Supernovae

Jet propagation:

The Setup: The computational domain is a cube of size L with a spherical star of radius $R_{\rm star}$ and mass $M_{\rm star}$ placed in the center. The innermost part with mass $M_{\rm core} \simeq 1.6 M_\odot$ and radius $R_{\rm core} = 4.5 \times 10^8$ cm, consisting of Fe and Si, is assumed to have collapsed on a timescale much faster than the outer, lower-density material. It is removed and replaced by a point gravitational source with mass $M_{\rm core}$ representing the newly formed neutron star. The remaining mass of the envelope M_{env} is mapped onto the computational domain. At two polar locations where the jets are initiated at R_{core}, we impose an inflow with velocity v_j and a density ρ_j. At $R_{\rm core}$, the jet density and pressure are the same as those of the background material. For the first 0.5s, the jet velocity at $R_{\rm core}$ is kept constant at v_j. After 0.5s, the velocity of the jets at $R_{\rm core}$ was gradually decreased to zero at approximately 2 s. The total energy of the jets is E_j. These parameters are consistent with, but somewhat less than, those of the LeBlanc-Wilson model.

The reference model: As a baseline case, we consider a jet-induced explosion in a helium star. Jet propagation inside the star is shown in Fig. 8.3. As the jets move outwards, they remain collimated and do not develop much internal structure. A bow shock forms at the head of the jet and spreads in all directions, roughly cylindrically around each jet. The jet-engine has been switched off after about 2.5 seconds. The material of the bow shock continues to propagate through the star. The stellar material is shocked by the bow shock. Mach shocks travel towards the equator resulting in a redistribution of the energy. The opening angle of the jet depends on the ratio between the velocity of the bow shock to the speed of sound. For a given star, this angle determines the efficiency of the deposition of the jet energy into the stellar envelope. Here, the efficiency of the energy deposition is about 40 %, and the final asymmetry of the envelope has an axis ratio of about two.

Influence of the jet properties: Fig. 8.4 shows two examples of an explosion with with a low and a very high jet velocity compared to the baseline case (Fig. 8.3). Fig. 8.4 demonstrates the influence of the jet velocity on the opening angle of the jet and, consequently, on the efficiency of the energy deposition. For the low velocity jet, the jet engine is switched off long before the jet penetrates the stellar envelope. Almost all of the energy of the jet goes into the stellar explosion. On a contrary, the fast jet (61,000 km s^{-1}) triggers only a weak explosion of 0.9 foe although its total energy was $\approx$ 10 foe.

Influence of the progenitor: For a very extended star, as in case of 'normal' Type II Supernovae, the bow shock of a low velocity jet stalls within the envelope, and the entire jet energy is used to trigger the ejection of the stellar envelope. In our example (Fig. 8.10), the jet material penetrates the helium

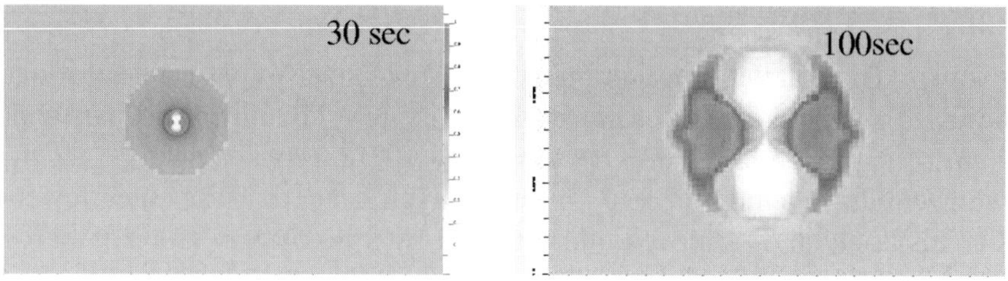

Figure 8.6. Same as Fig. 8.3 but helium abundance (between 0 to 1) for the explosion of a red supergiant with 207 $R_\odot$ and 7.6 $M_\odot$. The jet velocity of 11,000 km s^{-1} and a total energy of 2 foe has been taken. A domain of about 1.4×10^{12}cm is shown. After about 30 seconds, the material of the bow shock penetrates the Helium core, and, at about 250 seconds, the jet material 'stalls' in the hydrogen rich layers. Subsequently a almost spherical shock front propagates through the star. In the final configuration, asymmetries are restricted to the layers within the 2 to 3 solar masses of the h-rich envelope. After homologous expansion, the region corresponding to the stalled shock expands with about 4000 km s^{-1}. All of the jet energy is deposited in the expanding envelope.

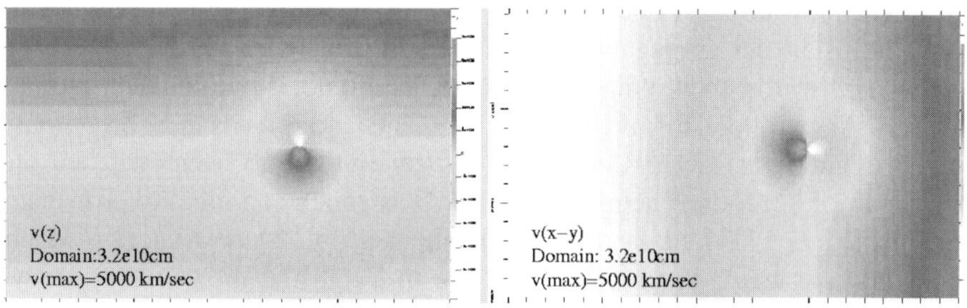

Figure 8.7. Same as Fig. 8.6 the velocity distribution in the xy- and yz plane for the very inner regions at about 250 sec. Note the qualitative difference between 1-D and multidimensional results. In 1-D, a large, almost empty cavity is found with expansion velocities of a about 4000 km s^{-1} for corresponding explosions. In multidimensional simulations shown here, this cavity is all but absent. Still, after in multidimensional simulations material can be found up with low velocities. Even infall can persist over a rather extended period of time.

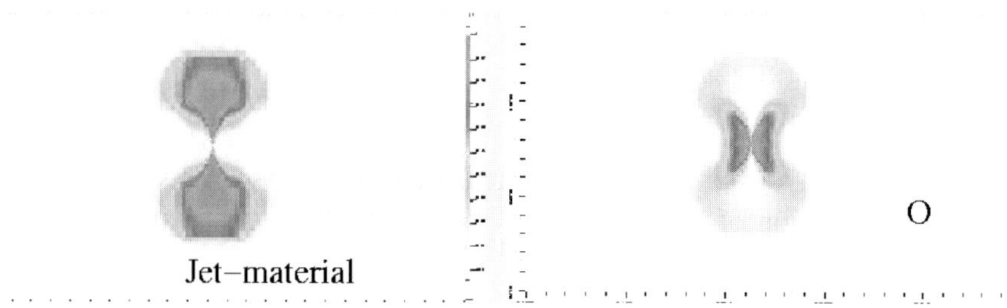

Figure 8.8. Same as Fig. 9.3 but the distribution of O and the jet material

core at about 100 seconds. After about 250 seconds the material of the jet stalls within the hydrogen rich envelope and after passing about 5 solar masses in the radial mass scale of the spherical progenitor. At this time, the isobars are almost spherical, and an almost spherical shock front travels outwards. Consequently, strong asphericities are limited to the inner regions. After about 385 seconds, we stopped the 3-D run and remapped the outer layers into 1-D structure, and followed the further evolution in 1-D. After about 1.8E4 seconds, the shock front reaches the surface. After about 3 days, the envelope expands homologously. The region where the jet material stalled, expands at velocities of about 4500 km s^{-1}.

Fallback: Jet-induced supernovae have very different characteristics with respect to fallback of material and the innermost structure. In 1-D calculations and for stars with main sequence masses of less than 20 $M_\odot$ and explosion energies in excess of 1 foe, the fallback of material remains less than 1.E-2 to 1.E-3 $M_\odot$ and an inner, low density cavity is formed with an outer edge of ^{56}Ni . For explosion energies between 1 and 2 foe, the outer edge of the cavity expands typically with velocities of about 700 to 1500 km s^{-1} (e.g. Woosley 1997, Höflich et al. 2000). In contrast, we find strong, continuous fallback of $\approx 0.2 M_\odot$ of O-rich matter in the the 3-D hydro models, and no lower limit for the velocity of the expanding material (Fig. 8.4). This significant amount of fallback must have important consequences for the secondary formation of a black hole. The exact amount and time scales for the final accretion on the neutron star will depend sensitively on the rotation and momentum transport.

Chemical Structure: The final chemical profiles of elements formed during the stellar evolution such as He, C, O and Si are 'butterfly- shaped' whereas the jet material fills an inner, conic structure (Fig. 8.10, upper, middle panel).

The composition of the jets must reflect the composition of the innermost parts of the star, and should contain heavy and intermediate-mass elements, freshly synthesized material such as ^{56}Ni and, maybe, r-process elements because, in our examples, the entropy at the bow shock region of the jet was as high as a few hundred. In any case, during the explosion, the jets bring heavy and intermediate mass elements into the outer H-rich layers.

Radiation Transport Effects

For the compact progenitors of SNe Ib/c, the final departures of the isodensity contours from sphericity are typically a factor of two. This will produce a linear polarization of about 2 to 3 % (Fig. 8.9) consistent with the values observed for Type Ib/c supernovae. In case of a red supergiant, i.e. SNe II, the asphericity is restricted to the inner layers of the H-rich envelope. There the iso-densities show an axis ratios of up to ≈ 1.3. The intermediate and outer H-rich layers remain spherical. This has strong consequences for the

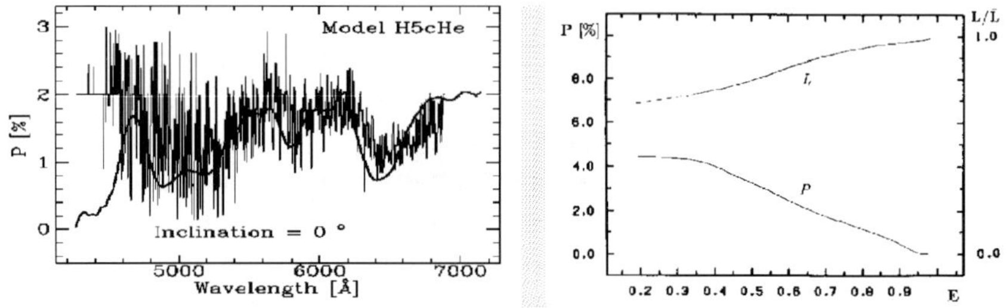

Figure 8.9. Polarization spectrum for SN1993J for an axis ratio of 1/2 for an oblate ellipsoid in comparison with observations by Trammell et al. (1993) are given in the left plot. On the right, the dependence of the continuum polarization (right) and directional dependence of the luminosity is shown as a function axis ratios for oblate ellipsoids seen from the equator (from Höflich, 1991 & Höflich et al. 1995).

observations, in particular, for polarization measurements. The polarization should be larger in SNe Ib/c compared to classical SNe II which is consistent with the observations by Wang et al. (2001b). Early on, we expect no or little polarization in supernovae with a massive, hydrogen rich envelope but that the polarization will increase with time to about 1% (Höflich 1991), depending on the inclination the SN is observed. This is also consistent both with the long-term time evolution of SN1987A (Jefferies 1991).

Recently, the plateau supernova SN1999em has been observed with VLT and Keck providing the best time coverage up to now of any supernovae (Wang et al. 2001b; Leonard et al. 2001). The basic trend has been confirmed which we expected from the hydro. Indeed, P is very low early on, and it rises when more central parts are seen. However, there are profound differences which point towards an additional mechanism to produce aspherical photospheres.

SN1999em is an extreme plateau supernovae with a plateau lasting for more than 100 days (IAUC 7294 to 7355). However, no detailed light curves have been published. Therefore, in Fig. 8.10, we show a theoretical LC which resembles SN 1999em with respect to the duration of the plateau and its brightness, and the typical expansion velocities. The light curves of SNe IIp show three distinct phases (Fig. 8.10). 1) Most of the envelope is ionized. This phase depends sensitively on the explosion energy, mixing of radioactive Ni, and the mass of the progenitor, e.g. either strong mixing or $E_{kin} \leq 1$ foe will cause a steep and steady increase in the luminosity (and in B and V); 2) The emitted energy is determined by the receding (in mass) of the H recombination front which is responsible for both the release of stored, thermal and the recombination energy. At the recombination front, the opacity drops by about 3 orders of magnitude when it changes from electron scattering dominated to bound-free/ free-free. This provides a self-regulating mechanism for the energy release. If too little energy is released, the opacity drops fast, causing an increase in the speed at which the photosphere is receding. In turn, this causes a larger energy release.

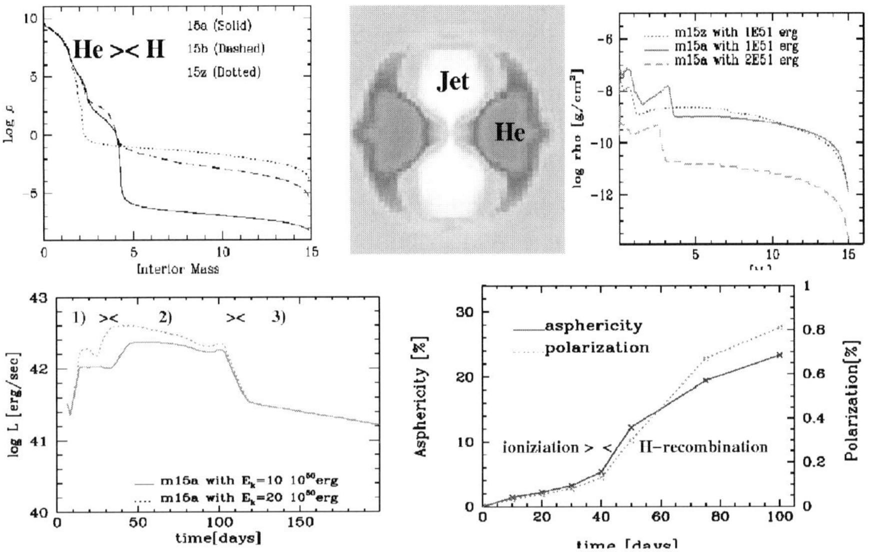

Figure 8.10. Polarization produced by an aspherical chemical distribution for a SN IIp model with $15M_{\odot}$ and an explosion energy $E_{exp} = 2 \times 10^{51} erg$. This model resembles SN 1999em (see above). The initial density profile is given for a star at the final stage of stellar evolution for metalicities Z of 0.02, 0.001 and 0 (models 15a, 15b, 15z, upper left panel, from Höflich et al. 2000 & Chieffi et al. 2001). The model for the Red Supergiant, 15a, has been used to calculate the jet-induced explosion 2). In the upper, middle panel, the chemical distribution of He is given at 250 sec for the He-rich layers after the jet material has stalled. The color-codes white, yellow, green, blue and red correspond to He mass fractions of 0., 0.18, 0.36, 0.72, and 1., respectively. The subsequent explosion has been followed in 1-D up to the phase of homologous expansion. In the upper, right panel, the density distribution is given at about 5 days after the explosion. The steep gradients in the density in the upper right and left panels are located at the interface between the He-core and the H-mantle. In the lower, left panel, the resulting bolometric late-curves are given for $E_{exp} = 2 \times 10^{51} erg$ (dotted line) and, for comparison, for $1 \times 10^{51} erg$, respectively. Based on full 3-D calculations for the radiation & γ-ray transport, we have calculated the location of the recombination front as a function of time. The resulting shape of the photosphere is always prolate. The corresponding axis ratio and the polarization seen from the equator are shown (lower, right panel). Note the strong increase of the asphericity after the onset of the recombination phase between day 30 to 40 (see also SN 1999em in Fig. 8.1).

Hydrogen recombines at a specific temperature at or just below the photosphere. Due to the flat density profiles of the expanding envelopes in the RSG case, the photospheric radius and, thus, the luminosity L stays almost constant. After the recombination front has passed through the H-rich envelope, the brightness drops fast. During phase 3), L is given by the instant energy release by radioactive decay of ^{56}Co . Obviously, the steep rise in P of SN1999em coincides with the transition from phase 1 to 2, pointing towards anisotropic excitation as a new mechanism for producing aspherical photospheres and, consequently, polarization. To quantify the effect, we have calculated the temperature and ionization structure of a SN IIp (Fig. 8.10). Starting from a spherical model, the initial chemical distribution has been taken from our 3-D jet simulation. As mentioned above, the chemical profile is frozen out after about 250 sec, and the expansion becomes spherical. The further evolution can be followed with our 1-D radiation code. For several moments of time, we have calculated the ionization structure and continuum polarization based of the 3-D chemical structure and the spherical density distribution under the assumption that the distribution of the radioactive Ni coincides with the jet-material. Note that the ^{56}Ni layers extend throughout the He-core ($\approx 5M_{\odot}$) in polar directions but they are confined to the very center along the equatorial plane (Fig. 8.10, upper, middle panel) leading to an increased transport of energy and, consequently, heating toward polar directions. Before the recombination phase, the opacity is dominated by Thomson scattering which does not depend on the temperature, and the shape of the photosphere remains almost spherical. However, during the recombination phase, the location of the photosphere depends sensitively on the heating, and the photosphere becomes prolate (Fig. 8.10, lower, right panel). There is a gradual increase in P and no jump in P because the optically thick H-rich layers below the photosphere redistribute the photons. The increase of P depends on the geometrical expansion and speed (in mass) of the receding recombination front.

6. Conclusions

We have numerically studied the explosion of Core Collapse supernovae caused by supersonic jets generated in the center of the supernova as a result of the core collapse into a neutron star. We simulated the process of the jet propagation through the star, the redistribution of elements, and radiation transport effects. A strong explosion and a high efficiency for the conversion of the jet energy requires low jet velocities or a low initial collimation of the jet. With increasing extension of the envelope, the conversion factor increases. Typically, we would expect higher kinetic energies in SNe II compared to SNe Ib/c if a significant amount of explosion energy is carried away by jets. Within the framework of jet-induced SN, the lack of this evidence suggests that the jets have low velocities.

The He, C, O and Si rich layers of the progenitor show characteristic, butterfly-shape structures. This morphology and pattern should be observable in supernova remnants, X-ray satellites such as Chandra observatory, despite some modifications and instabilities when the expanding medium interacts with the interstellar material.

During the explosion, the jets bring heavy and intermediate mass elements including ^{56}Ni into the outer layers. Due to the high entropies of the jet material close to the center, this may be a possible site for r-process elements. Spatial distribution of the jet material will influence the properties of a supernova. In our model for a SN II, the jet material stalled within the expanding envelope corresponding to a velocity of $\approx 4500\, s^{-1}$ during the phase of homologous expansion. In SN1987A, a bump in spectral lines of various elements has been interpreted as due to material excited by a clump of radioactive ^{56}Ni (Lucy 1988). Within our framework, this bump may be a measure of the region where the jet stalled. This could also explain the early appearance of X-rays in SN1987A which requires strong mixing of radioactive material into the hydrogen-rich layers (see above), and the overall distribution of elements and distribution of elements in the resolved HST-images of the inner debris of SN 1987A. We note that, if this interpretation is correct, the 'mystery spot' (Nisenson et al. 1988) would be unrelated. In contrast to 1-D simulations, we find in our models strong, continuous fallback over an extended period of time, and a lack of an inner, almost empty cavity. This significant amount of fallback during late times and the consequences for the secondary formation of a black hole shall be noted. Moreover, fallback and the low velocity material may alter the escape probability for γ-rays produced by radioactive decay of ^{56}Ni. In general, the lower escape probability is unimportant for the determination of the total ^{56}Ni production by the late light-curves because full thermalization can be assumed in core collapse SN during the first few years. However, in extreme cases such as SN98BW (e.g. Schaefer et al. 2000), only a small fraction of gamma's are trapped. Effects of multi-dimensionality will strongly alter the energy input by radioactive material and disallow a reliable estimate for the total ^{56}Ni mass.

Qualitatively, the jet-induced picture allows reproduction of the polarization observed in core collapse supernovae. Both asymmetric ionization and density/chemical distributions have been identified as crucial. Even within the picture of jet-induced explosion, the latter effect alone cannot (!) account for the high polarization produced in the intermediate H-rich layers of core-collapse SN with a massive envelope such as SN1999em. The former effect operates only during a recombination phase, and can be expected to dominate the polarization in core-collapse supernovae with massive H-rich envelopes during the first 1 to 2 months. Complete time series of polarization measurements are needed to test this suggestion.

Finally, we want to emphasize the limits of this study and some of the open questions which will be addressed in the future. We have assumed that jets are formed in the course of the formation of a neutron star, and have addressed observational consequences and constrains. However, we have not calculated the jet formation, we do not know if they really form, and, if they form, whether they form in all core-collapse supernovae. Qualitatively, the observational properties of core collapse supernovae are consistent with jet-induced supernovae and support strongly that the explosion mechanism is highly aspherical, but more detailed comparisons with individual objects must be performed as soon as the data become available. We cannot claim that the jets are the only mechanism that can explain asphericity in supernovae, but any competing mechanism must involve some sort of axial symmetry on large scales with a profound impact on the explosion such as rapid rotation. It remains to be seen whether asymmetry and axial symmetry are the 'smoking gun' for our understanding of the SN-mechanism.

Acknowledgments: We want to thank our colleagues for helpful discussions, in particular, D. Baade, E.S. Oran, J.C. Wheeler, Inzu Yi A., C. Mayers, J.C. Wilson, A. Chieffi, M. Limongi, and O. Straniero. This work is supported in part by NASA Grant LSTA-98-022.

References

hamb88 Ambwani, K., Sutherland, P.G. 1988, ApJ 325, 820

Arnett W.D., Bahcall J.N., Kirshner, R.P., Woosley, S.E. 1990, ARAA 27, 62

Athay R. 1972 *Radiation Transport in Spectral Lines*, Reidel Publ.C., Dordrecht

Avrett G., Loeser R. 1992, 7th Cambridge Workshop on Cool Stars, ASP Conference Series 26, 489

Bisnovatyi-Kogan G. 1971, Soviet Astronomy AJ, 14, 652

Bowers R.L., Wilson J.R. 1982 ApJS 50, 115

Brown, B.L., Leventhal, M. 1987, ApJ 319, 637

Burrows A., Hayes J., Fryxell B. 1995, ApJ 450

Caughlan G.R , Fowler A.M., Harris M.J., Zimmerman B.A. 1985, At. Data Nucl. Dat a Tables 32, 197

Caughlan G.R , Fowler A.M. 1988, Atomic Data Nucl. Table, 40, 283

Cunto, W., Mendoza, C. 1992 Topbase, IBM Venezuela Scientific Center

Colella, P., Woodward, P.R. 1984, J.Comp.Phys. 54, 174

Duncan, R. C. & Thompson, C. 1992, ApJ, 392, L9

Fesen, R. A. & Gunderson, K. S. 1996, ApJ, 470, 967

Fryer C.L., Warren M.S. 2002, ApJ, accepted & astro-ph/astro-ph/0206017

Fryxell B., Arnett D., Müller E. 1991, ApJ 367, 619

Gayet R. 1970, A&A 9, 312

Herant M., Benz W., Hix W.R., Fryer C.L., Colgate S.A. 1994, ApJ 435, 339

Hillebrandt W., Höflich 1991, Nuclear Physics B 19, 113

Hillebrandt W. 1982, ApJ 103, 147

Hoyle F., Fowler W. A. 1960, ApJ 132, 565

Höflich, P. 2002a, New Astronomy 46, 475 & astro-ph/0110098

Höflich, P. 2002b, in: Stellar Atmospheric Modelling, ed. Hubeny, Mihalas & Werner, ASP Conference Series, in press & astro-ph/0207103

Höflich, P., Khokhlov A.M., , Wang L. 2002, in: 20th Texas Converence

Höflich P., Straniero O., Limongi M. Dominguez I. Chieffi A. 2000, 7th TexMex-Conference, eds. W. Lee & S. Torres-Peimbert, UNAM-Publ., in press & astro-ph/005037

Höflich P., Wheeler, J.C., Thielemann F.-K. 1998. ApJ 495, 617

Höflich P., Wheeler, J.C., Hines, D., Trammell S. 1995, ApJ 459, 307

Höflich, P. 1995, ApJ 443, 89

Höflich, P., Müller, E., & Khokhlov, A. 1993, A&A, 268, 570

Höflich, P., Khokhlov, A., & Mueller, E. 1992, A&A, 259, 549

Höflich, P. 1990, *A quantitative analysis of type II supernovae atmospheres*, thesis submitted in partial fulfillment of the conditions for admission to the degree of Dr.rer.nat.habil., U München, MPA 563

Höflich, P., Müller E., Khokhlov A. 1993, A&A 268, 570

Höflich, P., Müller E., Khokhlov A. 1992, A&A 259, 243

Höflich, P. 1991 A&A 246, 481

Höflich, P. 1988, PASP 7, 434

Howell, D. A., Höflich, P., Wang, L., & Wheeler, J. C. 2001, ApJ, 556, 302

Hughes J.P., Rakowski C.E., Burrows D.N., Slane P.O. 2001, AJ, in press & astro-ph/9910474

Janka H.T. & Müller E., 1994 A&A 290, 496

Jeffrey D.J., 1991, ApJ, 375, 264

Khokhlov A.M., Höflich P. 2001, in: 1st KIAS Astrophysics, IAP-Publishing, ed. I. Yi, in press & astro-ph/0011023

Khokhlov A., Höflich P., Oran E.S., Wheeler J.C., P. Wang L., 1999, ApJ 524, L107

Khokhlov, A.M. 1998, J.Comput.Phys., 143, 519

Kifonidis K., Plewa T., Janka H.T., Müller E. 2000, ApJ 531, 123

Kouveliotou, C., Strohmayer, T., Hurley, K., Van Paradijs, J., Finger, M. H., Dieters, S., Woods, P., Thompson, C. & Duncan, R. C. 1998, ApJ, 510, 115

Kurucz R.L. 1995, CD-23, Center for Astrophysics

Kurucz R.L. 1991, CD-1, Center for Astrophysics

Lucy L.B. 1988, Proc. of the 4th George Mason conference, ed. by M. Kafatos, Cambridge University Press, p. 323

LeBlanc, J. M. & Wilson, J. R. 1970, ApJ, 161, 541

Leonard D.C., Filippenko, A.V., Barth A.J., Matheson T. 2000, ApJ 536, 239

Mendez R.H., Clocchiatti A., Benvenuto O.G., Feinstein C. Marraco H. 1977, ApJ 334, 295

Mezzacappa A., Liebendoerfer M., Bronson Messer O.E. Hix R., Thielemann F-K, Bruenn S.W. 2001, Phy.Rev.Let., accepted

Mihalas, D. & Mihalas, B. W. 1984, 'Foundation of Radiation Hydrodynamics', New York, Oxford University Press

Mihalas D., Kunasz R.B., Hummer D.G. 1975

Mihalas D., Kunasz R.B., Hummer D.G. 1976

Mihalas D., Kunasz R.B., Hummer D.G. 1976b
Mönchmeyer R., Schaefer G., Mueller E., Kates R.E. 1991 A&A 246, 417
Müller E., Janka H.T. 1997, A&A 317, 140
Müller E., Hillebrandt W., Orio M., Höflich P., Mönchmeyer R., Fryzell B.A. 1989, A&A 220, 167
Nisenson P., Papaliolios C., Karovska M., Noyes R. 1988, ApJ 324, 35
Ostriker, J. P. & Gunn, J. E. 1971, ApJ, 164, L95
Ramp M., Müller E., Ruffert M. 1998, A&A 332, 969
Ramp M. and Janka, H.-T. 2000, ApJ 593, L33
Saenz R.A., Shapiro S.L. 1981, ApJ 244, 1033
Schaefer B. 2000, ApJ 533, 21
Seaton M.J. 1960, Rep. Prog. Phys. 23, 313
Steinmetz M., Höflich P. 1992, A&A 257, 641
Stone, J. M., Mihalas, D., & Norman, M. L. 1992, ApJs, 80, 819
Strom R., Johnston H.M., Verbunt F., Aschenbach B. 1995, Nature, 373, 587
Symbalisty E.M.D. 1984, ApJ 285, 729
Thielemann F.-K., Nomoto K., Hashimoto M. 1994, in: Supernovae, Les Houcheseds. S. Bludman R., Mochkovitch J., Mochkovitch J., Zinn-Justin, Elsevier, Amsterdam, p 629
Trammell S., Hines D., Wheeler J.C. 1993, ApJ 414, 21
Tueller J., Barthelmy S., Gehrels N., Leventhal M., MacCallum C.J., Teegarden B.J. 1991, in: Supernovae, ed. S.E. Woosley, Springer Press, p. 278
Van Riper K.A. 1978, ApJ 221, 304
Wang L. et al. 2002a, ApJ, in press & astro-ph/0205337
Wang L., Howell A., Höflich P., Wheeler J.C. 2002b, ApJ, 550, 1030
Wang L. et al. 2001a, *The Bipolar Ejecta of SN1987A*, ApJ,
Wang, L., Wheeler, J. C., Li, Z. W., & Clocchiatti, A. 1996, ApJ, 467, 435
Wang, L., Wheeler, J.C., Höflich, P. 1997, ApJ, 476, 27
Wiescher M., Görres J., Thielemann, F.K. 1990, ApJ 363, 340
Yamada S., Janka H.T., Suzuki H. 1999, A&A 344, 533
Zwerger T., Müller E. 1997, A&A 320, 209

Chapter 9

MAGNETIC FIELDS IN SUPERNOVAE

Shizuka Akiyama
Astronomy department
University of Texas at Austin
1 University Station C1400
Austin, TX 78712
shizuka@astro.as.utexas.edu

J. Craig Wheeler
Astronomy department
University of Texas at Austin
1 University Station C1400
Austin, TX 78712
wheel@astro.as.utexas.edu

Abstract

A relatively modest value of the initial rotation of the iron core, a period of $\sim 6 - 31$ s, will give a very rapidly rotating protoneutron star and hence strong differential rotation with respect to the infalling matter. Under these conditions, a seed field is expected to be amplified by the magnetorotational instability (MRI) and to grow exponentially. Exponential growth of the field on the time scale Ω^{-1} by the MRI will dominate the linear growth process of field line "wrapping" with the same characteristic time. The shear is strongest at the boundary of the newly formed protoneutron star. Modest initial rotation velocities of the iron core result in sub–Keplerian rotation and a sub–equipartition magnetic field that nevertheless produce substantial magnetohydrodynamic (MHD) luminosity and hoop stresses: saturation fields of order $10^{15} - 10^{16}$ G develop ~ 300 msec after bounce with an associated MHD luminosity of $\sim 10^{49} - 10^{53}$ erg s^{-1}. Bi-polar flows driven by this MHD power can affect or even cause the explosions associated with core-collapse supernovae. If the initial rotation is too slow, then there will not be enough rotational energy to power the supernova despite the high luminosities. The MRI should be active and may qualitatively alter the flow if a black hole forms directly or after a fall-back delay.

C.L. Fryer (ed.), Stellar Collapse, 259–275.

1. The Magneto-Rotational Instability

Akiyama et al. (2002) simulated the collapse of a model iron core of a 15 $M_\odot$ progenitor with a one-dimensional flux-limited diffusion code (Myra et al. 1987). The evolution of the angular velocity profile $\Omega(r)$ was computed using the radial density profiles produced by the core collapse code assuming that the specific angular momentum of a given shell is constant. The magnetic field was obtained using the resulting $\rho(r)$, $\Omega(r)$, and $d\Omega/dr$ profiles according to the theory of the MRI. The MHD luminosity and hoop stress were estimated from the resulting magnetic field.

Angular Velocity Profile

The 15 $M_\odot$ model of Heger et al.(2000) attains an angular velocity of 10 rad s^{-1} (see their Fig. 8) in the center of the iron core at the precollapse stage. Their simulations did not include the effects of a magnetic field. It is possible that the iron core rotates slower if the effect of magnetic braking is included (Spruit & Phinney 1998; Heger & Woosley 2002). Fryer & Heger(2000) studied the rotational effects on pure hydrodynamic core collapse explosions with initial velocity profiles obtained by Heger et al.(2000) with a central rotational velocity of 4 rad s^{-1}. Akiyama et al. (2002) adopted the initial angular velocity profile, $\Omega_0(r)$, of Fryer & Heger(2000) (hereafter called the FH profile) as one case to study in addition to an analytic (MM profile) form (Möchmeyer & Müller 1989; Yamada & Sato 1994; Fryer & Heger 2000) and solid body profiles. The adopted profiles, characterized by the initial central value of the rotational frequency, $\Omega_{0,c}$ had small enough angular momentum that little departure from spherical geometry will occur.

It is inevitable that the collapsing core spins up and generates strong differential rotation for very general choices of the $\Omega_0(r)$ profile, since the inner regions collapse larger relative distances than the outer regions. A strong shear must form at the boundary of the protoneutron star (PNS). At bounce, the original homologous core has a positive gradient in $\Omega(r)$, and about 50 ms after bounce, the density profile is nearly identical to that of the initial iron core, giving a nearly flat rotation profile . After that, the density profile becomes somewhat more centrally condensed than the original iron core and the rotation profile decreases monotonically outward even deep within the PNS (Fig. 1, 2).

Ruderman et al.(2000) noted that the collapse of a white dwarf to a PNS gives a positive $\Omega(r)$ gradient since the relativistic degenerate core of the white dwarf has a steeper density profile than the PNS. The PNS will thus be relatively more compact for a given central density. There are two important differences in the calculations of Akiyama et al. (2002). The most critical is that the core collapsing is not in isolation as for the accretion-induced collapse scenario. Rather, the PNS forms within the massive star collapse ambience, and the PNS

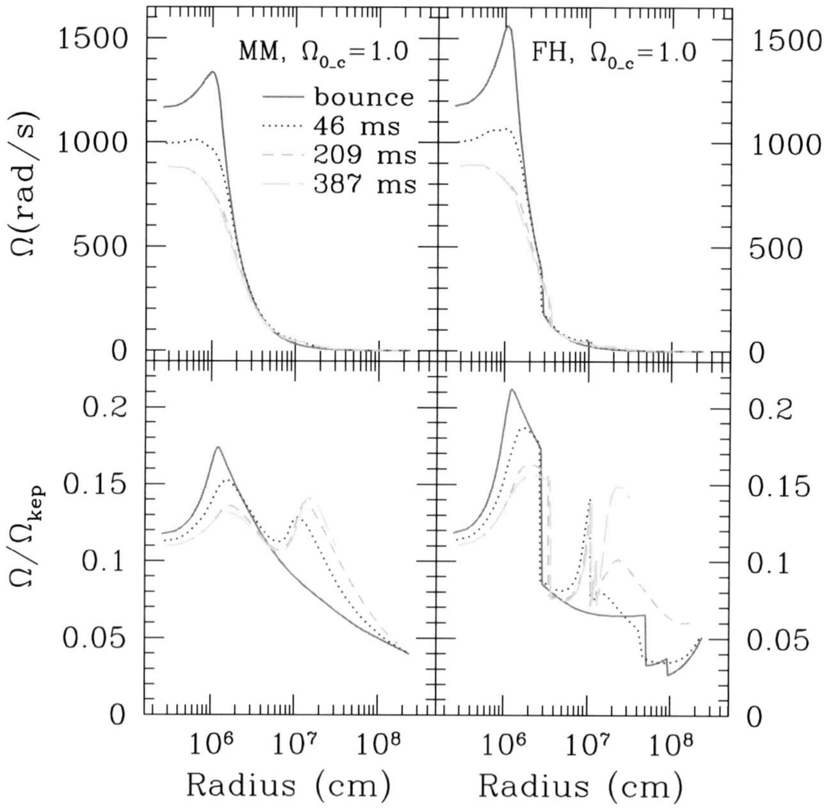

Figure 9.1. Rotational profiles and $\Omega/\Omega_{\rm kep}$ for the initial differential rotation cases (MM and FH) with $\Omega_{0,c} = 1.0$ rad s^{-1}. The collapse generates strong differential rotation at the boundary of the initial homologous core. The rotation is always sub–Keplerian.

must be strongly differentially rotating with respect to the still-infalling matter. This will generate a strong shear at the boundary of the PNS that would not pertain to a collapsing isolated white dwarf. Another, more subtle, difference is the equation of state. The equation of state of a partially degenerate iron core is not as different as for the highly relativistic white dwarf collapsing to a non-relativistic neutron star.

Magnetic Field

The MRI generates turbulence in a magnetized rotating fluid body that amplifies the magnetic field and transfers angular momentum. The MRI should pertain in this environment and amplify the magnetic field exponentially and

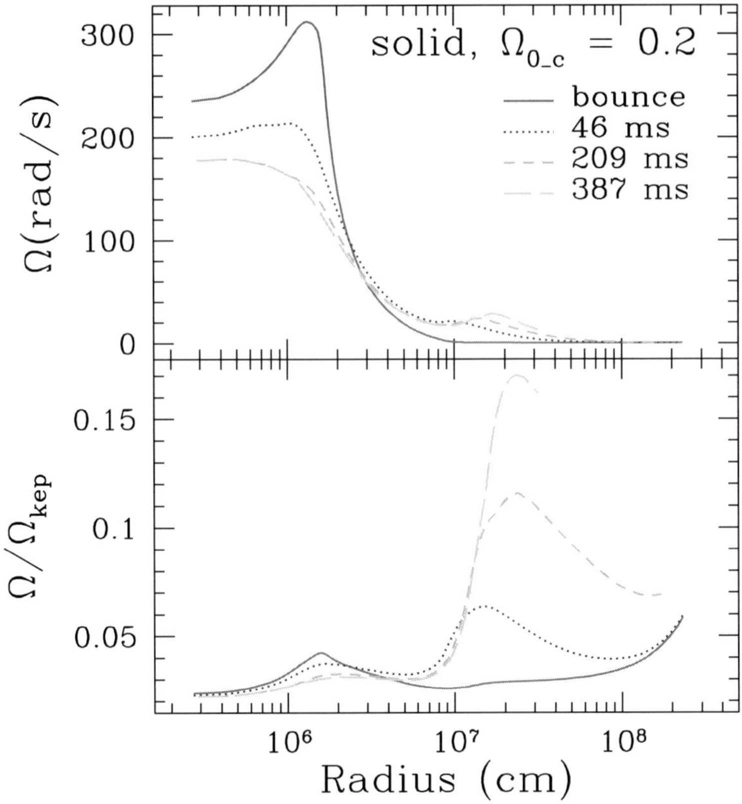

Figure 9.2. Rotational profiles and $\Omega/\Omega_{\rm kep}$ for the initial solid body rotation case with $\Omega_{0,c} = 0.2$ rad s^{-1}. The rotational profiles are very similar to those of initial differential rotation cases.

perhaps, in turn, power MHD bi-polar flow or jets. Key questions are the amplitude of the magnetic field and the effect on the dynamics.

Ignoring entropy gradients, the condition for the instability of the slow magnetosonic waves in a magnetized, differentially rotating plasma is (Balbus & Hawley 1991, 1998):

$$\frac{d\Omega^2}{d\ln r} + (\mathbf{k}\cdot\mathbf{v}_{\rm A})^2 < 0, \tag{9.1}$$

where

$$v_{\rm A} = \frac{B}{\sqrt{4\pi\rho}} \tag{9.2}$$

is the Alfvén velocity. When the magnetic field is very small, and/or the wavelength is very long, $(\mathbf{k}\cdot\mathbf{v}_{\rm A})^2$ is negligible, and the instability criterion for the MRI is simply that the angular velocity gradient be negative (Balbus & Hawley

1991, 1998), i.e.:

$$\frac{d\Omega^2}{d\ln r} < 0. \tag{9.3}$$

The growth of the magnetic field associated with the MRI is exponential with characteristic time scale of order Ω^{-1}. The time scale for the maximum growing mode is given by (Balbus & Hawley 1998):

$$\tau_{\rm max} = 4\pi \left|\frac{d\Omega}{d\ln r}\right|^{-1}. \tag{9.4}$$

We thus expect the MRI to dominate any process such as "wrapping of field lines" (cf. Wheeler et al. 2000 and references therein) that only grows linearly in time, even if on about the same time scale. The MRI will also operate under conditions of moderate rotation that are not sufficient to compete with the PNS convective time scales to drive the sort of $\alpha - \Omega$ dynamo invoked by, e.g., Duncan & Thompson (1992). The resulting unstable flow is expected to become non-linear, develop turbulence, and drive a dynamo that amplifies and sustains the field.

An order of magnitude estimate for the saturation field can be obtained by equating the shearing length scale $\ell_{\rm shear} \sim dr/d\ln\Omega$ to the characteristic mode scale $\ell_{\rm mode} \sim v_{\rm A} \cdot (d\Omega/d\ln r)^{-1}$. The resulting saturation magnetic field is given by:

$$B^2_{\rm sat} \sim 4\pi\rho r^2\Omega^2. \tag{9.5}$$

This is the same result as obtained by setting the Alfvén velocity equal to the local rotational velocity, $v_{\rm A} = r\Omega$.

The empirical value of the saturation field obtained by the numerical simulation of Hawley et al.(1996) is:

$$\begin{aligned} B_{\rm sim} &= \sqrt{\frac{\rho}{\pi}} r\Omega \\ &= \frac{1}{2\pi} \cdot B_{\rm sat}. \end{aligned} \tag{9.6}$$

This saturation field is achieved after turbulence is fully established, which takes about 20 rotations following the initial exponential growth (Hawley et al. 1996). For conditions of rotation at much less than Keplerian, these saturation fields are much less than the equipartition field for which $B^2/8\pi$ is comparable to the ambient pressure, i.e. for the current calculations, $c_{\rm s} \gg r\Omega \sim v_{\rm A}$.

When a vertical seed field exists, the maximum unstable growing mode (Balbus & Hawley 1998) implies a saturation field of:

$$B^2_{\rm max} = -4\pi\rho\lambda^2_{\rm max}\Omega^2 \cdot \left[\frac{1}{8\pi^2}\left(1 + \frac{1}{8}\frac{d\ln\Omega^2}{d\ln r}\right)\frac{d\ln\Omega^2}{d\ln r}\right], \tag{9.7}$$

where $\lambda_{\rm max}$ is the wavelength of the maximum growing mode which is not allowed to exceed the local radius r. With $\lambda_{\rm max} = r$, eq. (9.7) becomes:

$$B^2_{\rm max} = -B^2_{\rm sat} \cdot \left[\frac{1}{8\pi^2} \left(1 + \frac{1}{8} \frac{d\ln\Omega^2}{d\ln r} \right) \frac{d\ln\Omega^2}{d\ln r} \right]. \tag{9.8}$$

This expression for the saturation field depends on the shear explicitly as well as indirectly through the stability criterion. Note that for the maximum growing mode the expression for $B^2_{\rm max}$ acquires a negative value when

$$\begin{aligned} \frac{d\Omega^2}{d\ln r} &< -8\Omega^2 \text{ or },\\ \kappa^2 &< -4\Omega^2 < 0, \end{aligned} \tag{9.9}$$

where κ is the epicyclic frequency:

$$\kappa^2 = \frac{1}{r^3}\frac{d(r^4\Omega^2)}{dr} = 4\Omega^2 + \frac{d\Omega^2}{d\ln r}. \tag{9.10}$$

When eq. (9.9) is true, the epicyclic motion dominates over the MRI and prevents growth of the perturbation. Akiyama et al. (2002) turned off field amplification when this condition arose. In practice, the gradient of Ω may be reduced by mixing due to the epicyclic motion, and the MRI may eventually be active in a region in which it was at first suppressed by a strong negative gradient of $\Omega(r)$. Akiyama et al. (2002) also discuss the situation when the protoneutron star is convectively unstable.

Results

The shear is the strongest at the boundary of the initial homologous core. At bounce the shear is positive inside of the initial homologous core, and the region is stable against the MRI though convective instability can destabilize the structure. The solid body profile possesses a similar shear profile to the FH and MM profiles.

Even a relatively modest value of Ω_0 gives a very rapidly rotating PNS and hence strong differential rotation with respect to the infalling matter. At bounce, the peak of $\Omega/\Omega_{\rm kep}$ is at the boundary of the initial homologous core. At later times, however, the peak moves to the second hump which is located inside the stalled shock. This hump is at the same location as a maximum in entropy which is caused by shocked material with higher density.

For the given initial rotational profiles, the magnetic fields of eq. (9.5) and eq. (9.8) are amplified exponentially with the time scale of eq. (9.4). The resulting magnetic field for $B_{\rm sat}$ (eq. 9.5) and the ratio $\beta^{-1} \equiv P_{\rm mag}/P_{\rm gas}$ (where β is the conventional β in plasma physics) are presented in Fig. 3. Given the

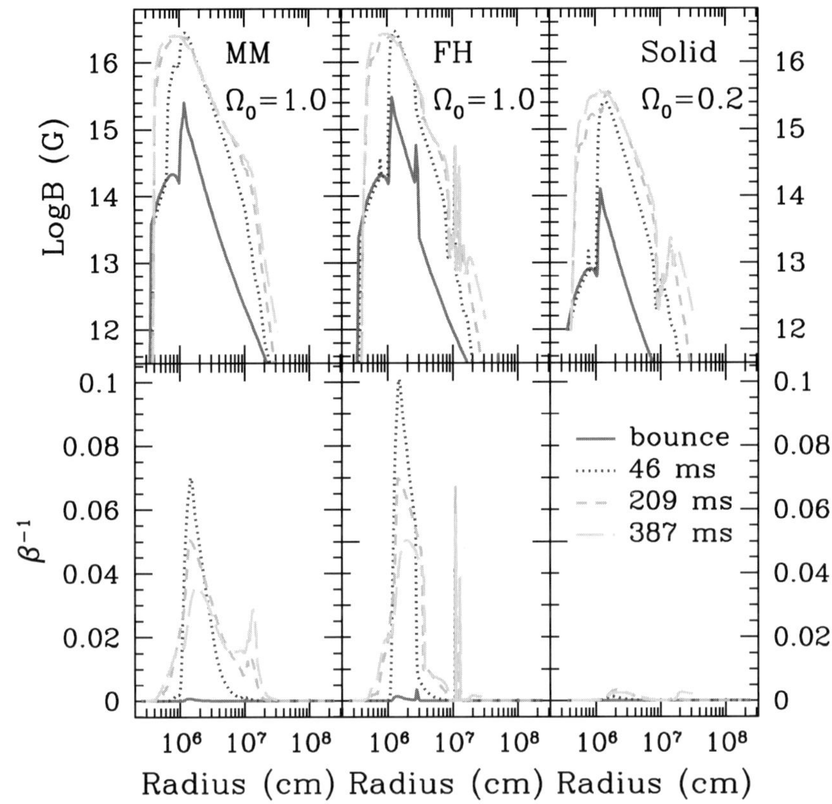

Figure 9.3. Magnetic field that of $B_{\rm sat}$ (eq. 9.5) and the ratio $\beta^{-1} = P_{\rm mag}/P_{\rm gas}$ for MM, FH, and solid body profiles.

limitation of the current calculations, we can only argue that these fields are roughly representative of what one would expect during core collapse.

For the cases with initial differential rotation, the peak values of the magnetic field at the end of our calculation at 387 ms after bounce are $B_{\rm sat} = 2.7 \times 10^{16}$ G and $B_{\rm max} = 2.5 \times 10^{15}$ G for the FH profile, $B_{\rm sat} = 2.5 \times 10^{16}$ G and $B_{\rm max} = 2.5 \times 10^{15}$ G for the MM profile, and $B_{\rm sat} = 3.7 \times 10^{15}$ G and $B_{\rm max} = 4.0 \times 10^{14}$ G for the solid body profile. The amplitude of the magnetic field is remarkably high and above the QED limit ($B_{\rm Q} = 4.4 \times 10^{13}$ G), but remains less than equipartition. For the case of $B_{\rm sat}$, $P_{\rm mag}$ is above 10% of $P_{\rm gas}$, and magnetic buoyancy may limit growth of the magnetic field (Wheeler, Meier, & Wilson 2002).

We expect that the magnetic field generated by the MRI will power MHD bi-polar outflow. The characteristic power of non-relativistic MHD outflow is given by Blandford & Payne (1982; see also Meier 1999, Wheeler, Meier & Wilson 2002):

$$L_{\rm MHD} = \frac{B^2 r^3 \Omega}{2}. \tag{9.11}$$

The outflow carries energy, angular momentum and mass. Employing this characteristic power of a Blandford & Payne type MHD outflow, the outflow luminosity $L_{\rm MHD}$ calculated for the three initial rotational profiles, MM, FH, and solid body. The profiles of this MHD luminosity mimic those of the magnetic field. For the cases with initial differential profiles, the peaks of the MHD luminosity are at the boundary of the PNS.

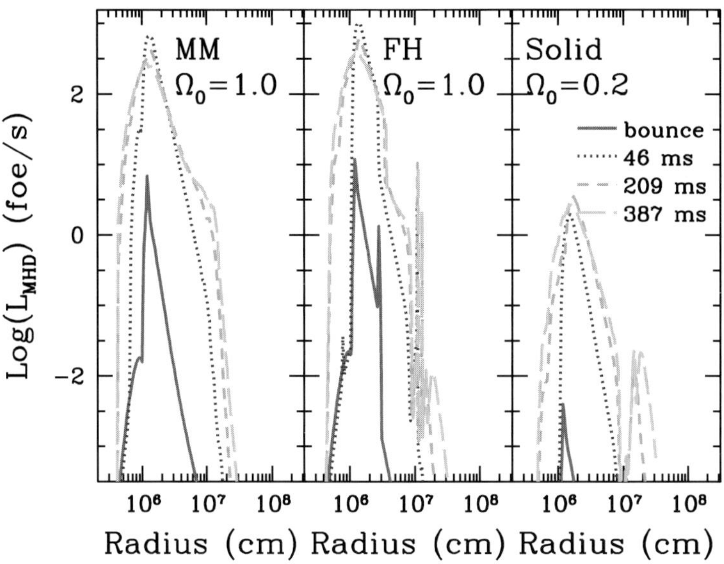

Figure 9.4. MHD jet luminosity in units of 1 foe ($= 10^{51}$ erg) corresponding to $B_{\rm sat}$. Sub-Keplerian rotation with MM, FH, and solid body initial rotation profiles results in high luminosity.

Our calculations are limited to sub–Keplerian rotation and sub-equipartition fields, and yet they potentially produce significant MHD luminosity (Fig. 4): for the saturation field $B_{\rm sat}$, the maximum values 387 ms after bounce are 3.8 $\times 10^{53}$ erg s^{-1} for FH, 2.5 $\times 10^{53}$ erg s^{-1} for MM, and 1.9 $\times 10^{51}$ erg s^{-1} for the initial solid rotation. For the saturation field $B_{\rm max}$, the peak values of MHD luminosity 387 ms after bounce are 5.5 $\times 10^{51}$ erg s^{-1} for FH, 4.2 $\times 10^{51}$ erg s^{-1} for MM, and 8.2 $\times 10^{49}$ erg s^{-1} for the initial solid rotation. The investigation of how the MHD luminosity can be turned into a bi-polar flow is left for future work, although we outline some possibilities in the discussion below.

2. Discussion and Conclusions

No one would doubt that the progenitors of core collapse supernovae rotate and possess some magnetic field. The question has always been whether rotation and magnetic fields would be incidental perturbations or a critical factor in understanding the explosion. Akiyama et al. (2002) have shown that with plausible rotation from contemporary stellar evolution calculations and any finite seed field with a component parallel to the rotation axis, the magnetorotational instability can lead to the rapid exponential growth of the magnetic field to substantial values on times of a fraction of a second, comparable to the core collapse time. Even a relatively modest value of initial rotation gives a very rapidly rotating PNS and hence strong differential rotation with respect to the infalling matter.

This result promises to be robust because the instability condition for the MRI is basically only that the gradient in angular velocity be negative. This condition is broadly satisfied in core collapse environments. Rotation can weaken supernova explosions without magnetic field (Fryer & Heger 2000); on the other hand, rotational energy can be converted to magnetic energy that can power MHD bi-polar flow that may promote supernova explosions. *The implication is that rotation and magnetic fields cannot be ignored in the core collapse context.*

As expected, the shear and hence the saturation fields are often highest at the boundary of the PNS where strong MHD activity is anticipated. Even artificially limiting the post-collapse rotation to sub–Keplerian values as done by Akiyama et al. (2002), we find fields in excess of 10^{15} G near the boundary of the neutron star are produced. While this field strength is sub–equipartition, the implied MHD luminosities are of order 10^{52} erg s^{-1}. This is a substantial luminosity and could, alone, power a supernova explosion if sustained for a sufficiently long time, a fraction of a second. As pointed out by Wheeler, Meier & Wilson (2002), the fields do not have to be comparable to equipartition to be important because they can catalyze the conversion of the large reservoir of rotational energy into buoyant, bi-polar MHD flow. Higher rates of initial rotation that are within the bounds of the evolutionary calculations could lead to even larger post-collapse rotation and even larger magnetic fields. If the initial rotation of the iron core proves to be substantially lower than we have explored here, then the MRI would be of little consequence to the explosion. The MHD luminosities derived here are comparable to the typical neutrino luminosities derived from core collapse, $\sim 10^{52}$ erg s^{-1}. One important difference is that the matter beyond the PNS is increasingly transparent to this neutrino luminosity, whereas the MHD power is deposited locally in the plasma. Another difference is that the neutrino luminosity is basically radial so it resists the inward fall of the collapse, the very source of the neutrino luminosity itself. In contrast, hoop stresses associated with the magnetic field (see below) will tend to pull inward and force matter selectively up the rotation axis.

We note that for complete self-consistency, one should apply the MRI to the evolution of rotating stars where even a weak field renders the Høiland dynamical stability criterion “all but useless" in the words of Balbus & Hawley (1998). Recent calculations by Heger & Woosley(2002) based on a prescription for magnetic viscosity by Spruit(2002) yield rapidly rotating iron cores. Heger & Woosley(2002) find PNS rotation rates of 4 to 8 ms, consistent with the values we have explored here. Clearly, much more must be done to understand the magnetorotational evolution of supernova progenitors.

The configuration of the magnetic field in a precollapse iron core is not well understood. In this calculation we have assumed there exists a seed vertical field to calculate the growth of the field due to the MRI; however, the MRI can amplify other components of the magnetic field. The final configuration of the magnetic field after collapse may be less uncertain since the system has a strongly preferred direction due to rotation. Most of the shear is in the radial direction, so the radial component is greatly amplified by the MRI and turned into toroidal field due to differential rotation (Balbus & Hawley 1998). The dominant component is most likely to be the toroidal field.

Another uncertainty is the rotational profile. It is not clear what profile to use in the PNS, since, we note, even the rotational profile of the Sun is not well understood. A full understanding of the rotational state of a PNS remains a large challenge.

We have assumed various prescriptions for the saturation field. All are variations on the theme that, within factors of order 2π, the saturation field will be given by the condition $v_{\rm A} \sim r\Omega$. In the numerical disk simulations, about 20 rotations are required to reach saturation. The region of maximum shear in these calculations, around 15 km, typically has an angular velocity of 500 rad s^{-1} or a period of about 0.013 s. That means that by the end of the current calculations at 0.387 s, there have been about 30 rotations. Although the prescriptions for the growth and saturation fields we use here are heuristic, this aspect of our results is certainly commensurate with the numerical simulations of the MRI.

The issues of the saturation field and the nature of astrophysical dynamos are still vigorously explored. Vishniac & Cho (2001) conclude that the MRI has the required properties for a dynamo, anisotropic turbulence in a shearing flow, to generate both disordered and ordered fields of large strength. The saturation limits we have adopted here are consistent with those found in numerical calculations of the MRI saturation, but this topic clearly deserves more study.

Both the magnetic pressure and the magnetic viscosity are small for the sub–Keplerian conditions explored here. For most cases β^{-1} is less than 0.1 for the conditions we have assumed, (the $B_{\rm sat}$ case for FH with $\Omega_{0,\rm c} = 1.0$ pushes this limit), so the direct dynamical effect of the magnetic field is expected to be small. The viscous time scale is $\tau_{\rm vis} \sim (\alpha\Omega)^{-1}(\rm r/h)^2$, where α is the viscosity parameter and h is the vertical scale height, with h $\sim$ r for our case. For a magnetically-dominated viscosity,

$$\alpha \sim \frac{B_{\rm r}B_\phi}{4\pi P} = \left(\frac{B_{\rm r}}{B_\phi}\right)\frac{B_\phi^2}{4\pi P} \sim 2\left(\frac{B_{\rm r}}{B_\phi}\right)\beta^{-1}. \tag{9.12}$$

With this expression for α, the viscous time becomes:

$$\tau_{\rm vis} \sim \frac{1}{2}\left(\frac{B_\phi}{B_{\rm r}}\right)\left(\frac{1}{\beta^{-1}\Omega}\right) \gg \Omega^{-1}, \tag{9.13}$$

where the final inequality follows from $B_\phi > B_{\rm r}$ and $\beta^{-1} < 1$. This prescription for viscosity is reasonable in the absence of convection. In the portions of the structure that are convective, the viscosity could be enhanced significantly.

We have not discussed the role of neutrinos here, although the processes of neutrino loss and de-leptonization are included in our calculation of the cooling PNS. It is possible that the neutrino flux affects the magnetic buoyancy (Thompson & Murray 2002) and that the magnetic fields affect the neutrino emissivity (Thompson & Duncan 1996) and interactions with the plasma (Laming 1999). The time scale for shear viscosity due to neutrino diffusion is much longer than the times of interest here, although magnetic fields and turbulence can make it shorter (Goussard et al. 1998). The MRI provides magnetic field and turbulence, so this issue deserves further study. In addition to affecting the shear, the neutrino viscosity might also affect the turbulence needed to make the MRI work.

An obvious imperative is to now understand the behavior of the strong magnetic fields we believe are likely to be attendant to any core collapse situation. The fields will generate strong pressure anisotropies that can lead to dynamic response even when the magnetic pressure is small compared to the isotropic ambient gas pressure. As argued in Wheeler, Meier & Wilson (2002), a dominant toroidal component is a natural condition to form a collimated magneto-centrifugal wind, and hence polar flow. A first example of driving a polar flow with the MRI is given by Hawley & Balbus (2002).

The MRI is expected to yield a combination of large scale and small scale magnetic fields. A key ingredient to force flow up the axis and to collimate it is the hoop stress from the resulting field. Hoop stresses and other aspects of the strongly anisotropic Maxwell stress tensor will tend to lead to enhanced flow inward on the equator and up the axis, thus promoting a jet. These hoop stresses will occur for a field with a large scale toroidal component, but also in cases with only a small scale, turbulent field, i.e. when $< B_\phi >= 0$ but $< B_\phi^2 >\neq 0$ (Ogilvie 2001, Williams 2002).

Akiyama et al. (2002) found that the acceleration implied by the hoop stresses of the saturation fields, $a_{\rm hoop} = B_\phi^2/4\pi\rho r$, was competitive with, and could even exceed, the net acceleration of the pressure gradient and gravity. The large scale toroidal field is thus likely to affect the dynamics by accelerating matter inward along cylindrical radii. The flow, thus compressed, is likely to be channeled up the rotation axes to begin the bi-polar flow that will be further accelerated by hoop and torsional stresses from the field, the "spring and fling" outlined in Wheeler, Meier & Wilson (2002).

While there has been some excellent work on the generation and propagation of jets through stars in the context of the "collapsar" model (Aloy et al. 1999, 2000; MacFadyen & Woosley 1999; MacFadyen, Woosley & Heger 2000; Zhang, Woosley & MacFadyen 2002) and for supernovae (Khokhlov et al. 1999; Khokhlov & Höflich 2001; Höflich, Wang, and Khokhlov 2001), none of this numerical work has taken explicit account of rotation and magnetic fields in the origin and propagation of the jet. The same is true for the associated analytic work on jet propagation (Tan, Matzner, & McKee 2001; Matzner 2002; Mészáros & Rees 2001; Ramirez-Ruiz, Celotti & Rees 2002).

The dynamics of MHD jets may depart substantially from pure hydrodynamical jets, since they will tend to preserve the flux in the Poynting flow and be subject to hoop stresses and other magnetic phenomena. In addition, reconnection can accelerate the matter (Spruit, Daigne & Drenkhahn 2001). The magnetic forces in the jet may affect the collimation of the jet and the efficiency with which the surrounding cocoon is heated and expands in a transverse manner. A sufficiently strong magnetic field could thus alter the efficiency with which the propagating jet deposits energy into the stellar envelope. The magnetic field can also affect the stability of the jet. Large scale helical fields tend to be unstable to the kink instability. Li (2002) has recently argued that the effective hoop stresses of small scale turbulent fields could collimate magnetic jets and stabilize the flow against kinking.

Understanding of the role of magnetic fields in supernovae may also shed light on the production of collimated jets and magnetic field in the more extreme case of γ-ray bursts . One of the outstanding questions associated with γ-ray bursts is the origin of the magnetic field that is implicit in all the modeling of synchrotron emission. The fields deduced from the modeling are comparable to, but substantially less than, equipartition. Such fields cannot arise simply from shock compression of the ambient field of the ISM. While some schemes for generating this field in the γ-ray burst shock have been proposed (Medvedev & Loeb 1999), there is no generally-accepted understanding of the origin of this strong field.

Jets arising from the rotation and magnetic fields of neutron stars are likely to be important in asymmetric core-collapse supernova explosions (Wheeler et al. 2000; Wheeler, Meier, & Wilson 2002; Akiyama et al. 2002). Rotation and magnetic fields are critical in current models for the origin of jets in everything from protostars to AGN (Meier et al. 2001) and are very likely to be involved in the rapidly rotating environment that must occur if core collapse to a black hole is to produce anything like a γ-ray burst in the collapsar scenario. If the magnetic field plays a significant role in launching a relativistic γ-ray burst jet from within a collapsing star, then the magnetic field may also play a role in the propagation, collimation, and stability of that jet within and beyond the star. These factors have not been considered quantitatively. If magnetic flux is carried out of the star in the jet, then the magnetic field required to explain the observed synchrotron radiation may already be present and will not have to be generated *in situ*.

The MRI can operate under conditions of moderate rotation. This means that the MRI will be at work even as the disk of material described by MacFadyen & Woosley (1999) begins to form and makes a transition from a non-Keplerian to quasi-Keplerian flow. The resulting unstable flow is expected to become non-linear, develop turbulence, and drive a dynamo that amplifies and sustains the field.

For a complete understanding of the physics in a core collapse supernova explosion, a combination of neutrino–induced and jet–induced explosion may be required. Understanding the myriad implications of this statement will be a rich exploration.

Acknowledgments

This work was supported in part by NASA Grant NAG59302 and NSF Grant AST-0098644.

References

Akiyama, S., Wheeler, J. C., Meier, D. L. & Lichtenstadt, I. 2002, ApJ, in press (astro-ph/0208128)

Aloy, M. A., Ibáñez, J. M., Martí, J. M., Gómez, J.-L., & Müller, E. 1999, *Ap. J. (Letters)*, 523, L125

Balbus, S. A. & Hawley, J. F. 1991, *Ap. J.*, 376, 214

Balbus, S. A. & Hawley, J. F. 1998, Review of Modern Physics, 70, 1

Bisnovatyi-Kogan, G. S. 1971, Soviet Astronomy AJ, 14, 652

Bisnovatyi-Kogan, G. S., & Ruzmaikin, A. A. 1976, Ap & Sp. Sci., 42, 401

Blandford, R. D., & Payne, D. G. 1982, MNRAS, 199, 883

Dubner, G., Giacani, E., Gaensler, B. M., Goss, W. M. & Green. 2002, in Neutron Stars in Supernova Remnants, eds. P. O. Slane & B. M. Gaensler (Astron. Soc. Pac.: Provo), in press (astro-ph/0112155)

Fesen, R. A., & Gunderson, K. S. 1996, *Ap. J.*, 470, 967

Fesen, R. A. 2001 *Ap. J. Suppl.*, 133, 161

Fryer, C. L. & Heger, A. 2000, *Ap. J.*, 541, 1033
Goussard, J-O., Haensel, P., & Zdunik, J. L. 1998, *Astr. Ap.*, 330, 1005
Hawley, J. F., Gammie, C. F., & Balbus, S. A. 1996, *Ap. J.*, 464, 690
Hawley, J. F. & Balbus, S. A. 2002, *Ap. J.*, in press (astro-ph/0203309)
Heger, A., Langer, N., & Woosley, S. E. 2000, *Ap. J.*, 528, 368
Heger, A. & Woosley, S. E. 2002, astro-ph/0206005
Helfand, D. J., Gotthelf, E. V., & Halpern, J. P. 2001, *Ap. J.*, 556, 380
Höflich, P., Khokhlov, A., & Wang, L. 2001, in Proc. of the 20th Texas Symposium on Relativistic Astrophysics, eds. J. C. Wheeler & H. Martel (New York: AIP), 459
Hughes, J. P., Rakowski, C. E., Burrows, D. N., & Slane, P. O. 2000, *Ap. J. (Letters)*, 528, L109
Hwang, U., Holt, S. S., & Petre, R. 2000, *Ap. J. (Letters)*, 537, L119 & Meyer-Hofmeister, E., & Thomas, H. C. 1970, *Astr. Ap.*, 5, 155
Khokhlov, A. M., Höflich, P., Oran, E. S., Wheeler, J. C., Wang, L., & Chtchelkanova, A. Yu. 1999, *Ap. J. (Letters)*, 524, L107
Khokhlov, A., & Höflich, P. 2001, in AIP Conf. Proc. No. 556, Explosive Phenomena in Astrophysical Compact Objects, eds. H.-Y, Chang, C.-H., Lee, & M. Rho (New York: AIP), 301
Koide, S., Meier, D. L., Shibata, L., & Kudoh, T. 2000, *Ap. J.*, 536, 668
Kundt, W. 1976, Nature, 261, 673
Laming, J. M. 1999, New Astronomy, 4, 389
Leonard, D. C., Filippenko, A. V., Barth, A. J., & Matheson, T. 2000, *Ap. J.*, 536, 239
Leonard, D. C., Filippenko, A. V., Ardila, D. R., & Brotherton, M. S. 2001, *Ap. J.*, 553, 861
Li, L. 2002, *Ap. J.*, 564, 108
MacFadyen, A. & Woosley, S. E. 1999, ApJ, 524, 262
MacFadyen, A., Woosley, S. E., & Heger, A. 2001, ApJ, 550, 410
Matzner, C. D. 2002, MNRAS, in press (astro-ph/0203085)
Medvedev, M. V. & Loeb, A. 1999, ApJ, 526, 697
Meier, D. L. 1999, *Ap. J.*, 522, 753
Meier, D. L., Koide, S., &, Uchida, Y. 2001, Science, 291, 84
Mészáros, P. & Rees, M. J. 2001, ApJ, 556, L37
Mönchmeyer, R. & Müller, E. 1989, in Timing Neutron Stars, ed. H. Ögelman & E. P. J. van den Heuvel (NATO ASI Ser. C, 262; New Tork: ASI), 549
Myra, E. S., Bludman, S. A., Hoffman, Y., Lichtenstadt, I., Sack, N., & van Riper, K. A. 1987, *Ap. J.*, 318, 744
Ogilvie, G. I. 2001, MNRAS, 325, 231
Ostriker, J. P., & Gunn, J. E. 1971, *Ap. J. (Letters)*, 164, L95
Pun, C. S. J., & Supernova INtensive Studies (SINS) Collaboration. 2001, AAS, 199, 9402

Ramirez Ruiz, E., Celotti, A. & Rees, M. J. 2002, MNRAS, in press (astro-ph/0205108)
Ruderman, M. A., Tao, L., & Kluźniak, W. 2000, *Ap. J.*, 542, 243
Spruit, H. C. 2002, *Astr. Ap.*, 381, 923
Spruit, H.C., & Phinney, E. S. 1998, Nature, 393, 139
Spruit, H. C., Daigne, F., & Drenkhahn, G. 2001, *Astr. Ap.*, 369, 694
Tan, J. C., Matzner, C. D., & McKee, C. F. 2001, ApJ, 551, 946
Thompson, C. & Duncan, R. C. 1996, *Ap. J.*, 473, 322
Thompson, C., & Murray, N. 2002, preprint (astro-ph/0105425)
Vishniac, E. T. & Cho, J. 2001, *Ap. J.*, 550, 752
Wang, L., Wheeler, J. C., Li, Z. W., & Clocchiatti, A. 1996, *Ap. J.*, 467, 435
Wang, L., Howell, D. A., Höflich, P., & Wheeler, J. C. 2001, *Ap. J.*, 550, 1030
Wang, L., Weeler, J.C., H"oflich, P., Khokhlov, A., Baade, D., Branch, D., Challis, P., Filippenko, A.V., Fransson, C., Garnavich, P., Kirshner, R.P., Lundqvist, P., McCray, R., Panaja, N., Pun, C.S.J., Phillips, M.M., Sonneborn, G., Suntzeff, N.B. 2002, *Ap. J.*, in press (astro-ph/0205337)
Weisskopf, M. C., Hester, J. J., Tennant, A. F., Elsner, R. F., Schulz, N. S., Marshall, H. L., Karovska, M., Nicholas, J. S., Swartz, D. A., Kolodziejczak, J. J., & O'Dell, S. L. 2000, *Ap. J. (Letters)*, 536, L81
Wheeler, J. C., Yi, I., Höflich, P., Wang, L. 2000, *Ap. J.*, 537, 810
Wheeler, J. C., Meier, D. L., & Wilson, J. R. 2002, *Ap. J.*, 568, 807
Williams, P. T., IAOC Workshop "Galactic Star Formation Across the Stellar Mass Spectrum," ASP Conference Series, ed. J. M. De Buizer, in press (astro-ph/0206230)
Woosley, S. E., & Heger, A. 2001, AAS, 198, 3801
Willingale, R., Bleeker, J. A. M., van der Heyden, K. J., Kaastra, J. S. & Vink, J. 2002, *Astr. Ap.*, 381, 1039
Yamada, S., & Sato, K. 1994, *Ap. J.*, 434, 268
Zhang, W., Woosley, S. E. & MacFadyen, A. I. 2002, ApJ, in press (astro-ph/0207436)

Chapter 10

HYPERNOVAE AND OTHER BLACK-HOLE-FORMING SUPERNOVAE

Ken'ichi Nomoto,[1,2] Keiichi Maeda,[1] Paolo A. Mazzali,[2,3] Hideyuki Umeda,[1] Jinsong Deng,[1,2] Koichi Iwamoto,[4]

[1] *Department of Astronomy, School of Science, University of Tokyo*

[2] *Research Cenetr for the Early Universe, School of Science, University of Tokyo*

[3] *Osservatorio Astronomico di Trieste, Italy*

[4] *Department of Physics, College of Science and Technology, Nihon University*

Abstract During the last few years, a number of exceptional core-collapse supernovae (SNe) have been discovered. Although their properties are rather diverse, they have the common feature that at least some of their basic parameters (kinetic energy of the explosion, mass of the ejecta, mass of the synthesized ^{56}Ni), and sometimes all of them, are larger, sometimes by more than an order of magnitude, than the values typically found for this type of SNe. Therefore, these SNe have been given the collective classification of 'Hypernovae'. The best known object in this class is SN 1998bw, which owes its fame to its likely association with the gamma-ray burst GRB 980425. In this paper, we first describe how the basic parameters of SN 1998bw can be derived from observations and modeling, and discuss the properties of other hypernovae individually. These hypernovae seem to come from rather massive stars, being more massive than $\sim$ 20 - 25 $M_\odot$ on the main-sequence, thus forming black holes. On the other hand, there are some examples of massive SNe with only a small kinetic energy. We suggest that stars with non-rotating black holes are likely to collapse "quietly" ejecting a small amount of heavy elements (Faint supernovae). In contrast, stars with rotating black holes are likely to give rise to very energetic supernovae (Hypernovae). We present distinct nucleosynthesis features of these two types of "black-hole-forming" supernovae. Nucleosynthesis in Hypernovae are characterized by larger abundance ratios (Zn,Co,V,Ti)/Fe and smaller (Mn,Cr)/Fe. Nucleosynthesis in Faint supernovae is characterized by a large amount of fall-back. We show that the abundance pattern of the recently discovered most Fe deficient star, HE0107-

C.L. Fryer (ed.), Stellar Collapse, 277–325.

5240, and other extremely metal-poor carbon-rich stars are in good accord with those of black-hole-forming supernovae, but not pair-instability supernovae. This suggests that black-hole-forming supernovae made important contributions to the early Galactic (and cosmic) chemical evolution. Finally we discuss the nature of First (Pop III) Stars.

Keywords: Supernovae, Hypernovae, Nucleosynthesis, Chemical Evolution, Gamma-Ray Bursts

1. Introduction

One of the most interesting recent developments in the study of supernovae (SNe) is the discovery of some very energetic supernovae, whose kinetic energy (KE) exceeds 10^{52} erg, about 10 times the KE of normal core-collapse SNe (hereafter $E_{51} = E/10^{51}$ erg). The most luminous and powerful of these objects, the Type Ic supernova (SN Ic) 1998bw, was probably linked to the gamma-ray burst GRB 980425 (Galama et al. 1998), thus establishing for the first time a connection between gamma-ray bursts (GRBs) and the well-studied phenomenon of core-collapse SNe. However, SN 1998bw was exceptional for a SN Ic: it was as luminous at peak as a SN Ia, indicating that it synthesized $\sim 0.5\ M_\odot$ of ^{56}Ni, and its KE was estimated at $E \sim 3 \times 10^{52}$ erg (Iwamoto et al. 1998; Woosley, Eastman, & Schmidt 1999). Because of its large KE, SN 1998bw was called a "Hypernova (HN)".

Subsequently, other "hypernovae" of Type Ic have been discovered or recognised, such as SN 1997ef (Iwamoto et al. 2000; Mazzali, Iwamoto & Nomoto 2000), SN 1997dq (Matheson et al. 2001), SN 1999as (Knop et al. 1999), and SN 2002ap (Mazzali et al. 2002). Although these SNe Ic did not appear to be associated with GRBs, most recent "hypernova" SN 2003dh is clearly associated with GRB 030329 (Stanek et al. 2003). Figures 10.1 and 10.2 show the near-maximum spectra and the absolute V-light curves of these hypernovae. These objects span a wide range of properties, although they all appear to be highly energetic compared to normal core-collapse SNe.

SN 1999as is the most luminous supernova ever discovered, reaching a peak magnitude $M_V < -21.5$, while the brightness of SN 2002ap appears to be similar to that of normal core collapse SNe. The analysis of these various objects suggests that the KE may be related to the progenitor's main-sequence mass, which was probably $\gtrsim 50\ M_\odot$ for SN 1999as, $\sim 40\ M_\odot$ for SN 1998bw, $\sim 30\ M_\odot$ for SN 1997ef, and $\sim 20 - 25\ M_\odot$ for SN 2002ap. Another possible hypernovae, although of Type IIn, were SN 1997cy and 1999E, which was also estimated to have a large mass ($\sim 25\ M_\odot$; Germany et al. 2000; Turatto et al. 2000; Rigon et al. 2003). These mass estimates place hypernovae at the

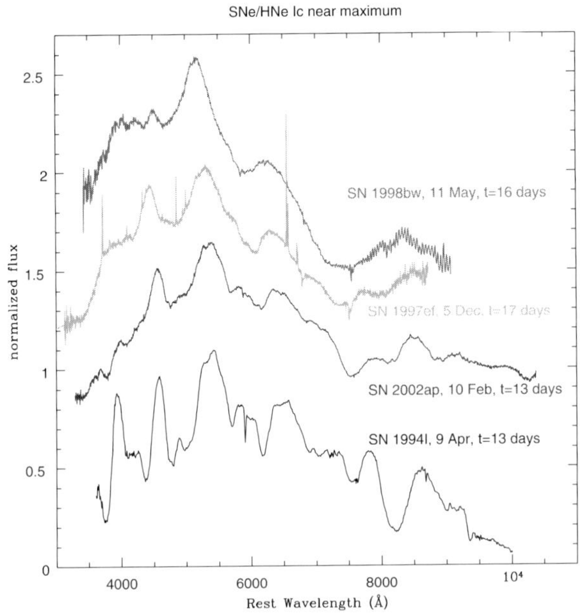

Figure 10.1. The near-maximum spectra of Type Ic SNe and hypernovae: SNe 1998bw, 1997ef, 2002ap, and 1994I (Mazzali et al. 2002).

high-mass end of SN progenitors, as they are always larger than the mass of the progenitors of normal core-collapse SNe, which is $\sim 15 - 20\ M_{\odot}$.

In the following sections, we examine the properties of these hypernovae as derived from optical light curves and spectra and discuss what may be the discriminating factor for the birth of a hypernova or the connection with a GRB. We then focus on nucleosynthesis in hypernovae, which is quite distinct from the case of ordinary supernovae, thus making a unique contribution to galactic chemical evolution.

2. SN 1998bw and GRB980425

In the early spectra of SN 1998bw, only few features were present. Absorption features were very broad, if present at all, while broad peaks were present at 4000, 5,000, 6,000 and 8,000 Å(Galama et al. 1998). The absence of any hydrogen lines, of any clear He I features, and of a strong Si II 6355 line indicated that SN1998bw should be classified as a Type Ic SN, despite some deviation from previously known objects of this class, mostly due to the extreme line width. SNe Ic are thought to be the result of core-collapse-induced explosions of C+O stars, which are massive stars that have lost their H-rich envelope and

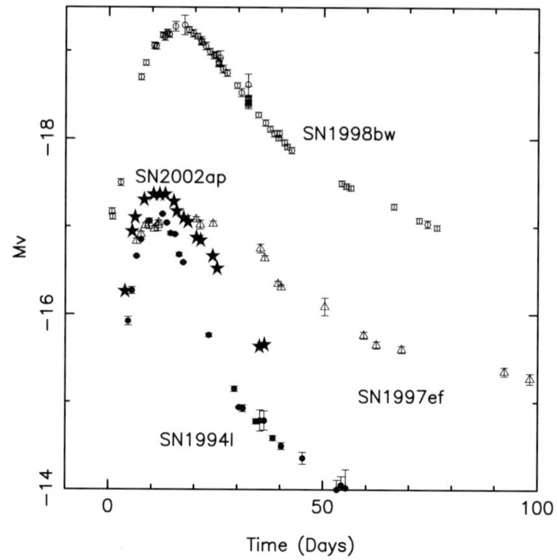

Figure 10.2. The observed V-band light curves of SNe 1998bw (*open circles*), 1997ef (*open triangles*), 2002ap (*stars*), and 1994I (*filled circles*) (Mazzali et al. 2002).

almost all of their He shell, through either a wind or binary interaction (Nomoto et al. 1994).

SN 1998bw is located in a spiral arm of the barred spiral galaxy ESO 184-G82, for which a distance is measured as $\sim$ 38 Mpc from its redshift velocity 2550 km s^{-1} (Galama et al. 1998) and a Hubble constant $H_0 = 65$ km s^{-1} Mpc^{-1}. Then, the peak absolute luminosity is estimated to be $\sim 10^{43}$ ergs sec^{-1}, which is about ten times brighter than typical core-collapse SNe (SNe Ib/Ic, or II). Assuming that the GRB was at the same distance as SN 1998bw, the gamma-ray fluence of GRB980425 reported by the BATSE (CGRO) group, $(4.4 \pm 0.4) \times 10^{-6}$ erg cm^{-2} (Kippen et al. 1998), corresponds to a burst energy $\sim (7.2 \pm 0.65) \times 10^{47}$ ergs (in gamma-rays). This is four orders of magnitude smaller than for average GRBs, implying that either GRB980425 might not be a typical burst or that it may not be related to SN 1998bw. However, the recent case of GRB030329/SN2003dh has provided a solid evidence of connection between the ordinary GRBs and Hypernovae (Stanek et al. 2003). Wang & Wheeler (1998) studied the correlation between SNe and GRBs in the literature systematically.

Extensive follow-up observations of SN 1998bw have brought us invaluable information on this peculiar SN and put rather strong constraints on corresponding theoretical models.

Early Light Curve

The early light curve of SN 1998bw has been modeled based on spherically symmetric explosions. Various investigations reached the similar conclusion that the early light curve can be successfully reproduced by the explosion of a massive C+O star with a kinetic energy more than ten times larger than a canonical SN explosion (Iwamoto et al. 1998; Woolsey et al. 1999). Figure 10.5 compares the light curve of SN 1998bw with those of other SNe Ic and their model light curves (Table 1).

SN 1998bw showed a very early rise. It had a luminosity of $\sim 10^{42}$ erg s^{-1} already at day 1. This rapid rise requires the presence of ^{56}Ni near the surface. Since spherically symmetric explosion models produce ^{56}Ni in deep inner layers of the ejecta, they can only reproduce this behaviour if extensive mixing is postulated to have occurred, dredging ^{56}Ni up to outer layers (Iwamoto et al. 1998; Nakamura et al. 2001a).

The light curve of SN 1998bw reaches the peak on about day 16, with the peak absolute magnitude being comparable to that of normal SNe Ia. We obtain a qualitative guess of model parameters such as ejecta mass M_{ej}, explosion kinetic energy $E_{\mathrm{k}} = E_{51} \times 10^{51}$ erg, and the ^{56}Ni mass M_{Ni} by employing an analytic solution by Arnett (1982). At early times, the bolometric luminosity L_{bol} of a "compact" SN that is powered by radioactive decay of ^{56}Ni is written as

$$L_{\mathrm{bol}} = \frac{\epsilon_{\mathrm{Ni}} M_{\mathrm{Ni}}}{\tau_{\mathrm{Ni}}} \Lambda(x, y), \tag{10.1}$$

where $\tau_{\mathrm{Ni}} = 8.8$ days and $\epsilon_{\mathrm{Ni}} = 2.96 \times 10^{16}$ erg g^{-1} are the decay time of ^{56}Ni and the energy deposited per gram of ^{56}Ni, respectively, and the function Λ is given by

$$\Lambda(x, y) = \exp(-x^2) \int_0^x \exp(-2zy + z^2) 2z dz. \tag{10.2}$$

It is assumed that ^{56}Ni is distributed homogeneously and that γ-rays are all trapped in the ejecta. This latter assumption is correct at the early phases we are considering. The dimensionless variable $x = t/\tau_c$, where t is the elapsed time and τ_c is the characteristic time of the light curve $\tau_c = (2\tau_h \tau_d)^{1/2}$, where τ_h and τ_d are the hydrodynamical timescale and the diffusion time scale of optical photons through the ejecta, respectively. The variable y is defined as $y = \tau_c/(2\tau_{\mathrm{Ni}})$. For the principal mode of diffusion (Arnett 1982; Pinto & Eastman 2000), τ_c turns out to be

$$\tau_c \sim 8 \text{ days} \left(\frac{\kappa}{0.05}\right)^{1/2} M_{\mathrm{ej},\odot}^{3/4} E_{\mathrm{k},51}^{-1/4} \tag{10.3}$$

where κ is the effective opacity. Hereafter, the symbol $M_{\mathrm{x},\odot}$ denotes $M_{\mathrm{x}}/M_{\odot}$. The ejecta are assumed to be a sphere of constant density expanding homologously, and thus M_{ej} and $E_{\mathrm{k}} = E_{\mathrm{k},51} \times 10^{51}$ erg are given by $M_{\mathrm{ej}} = (4\pi/3)R^3\rho$, $E_{\mathrm{k}} \sim (3/5)(1/2)M_{\mathrm{ej}}v_s^2$, where R, ρ, and v_s are the radius, density, and surface velocity of the ejecta, respectively.

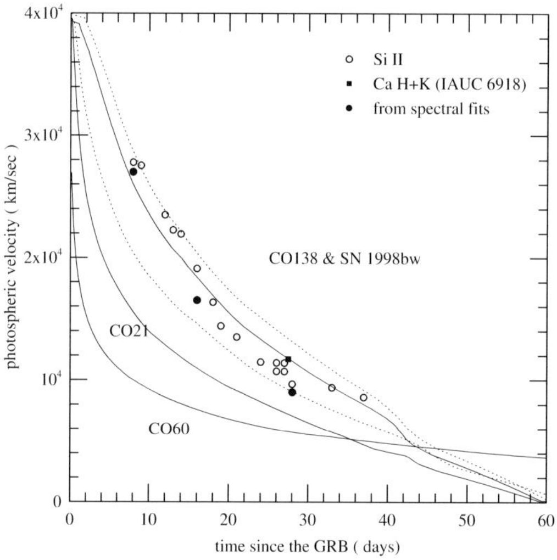

Figure 10.3. Photospheric velocities of SN 1998bw. The photospheric velocities obtained by spectral modeling (filled circles) and the observed velocities of Si II 6347 Å, 6371 Å lines measured at the absorption cores (open circles: Patat et al. 2001), and that of the Ca II H+K doublet derived from the spectrum of 23 May 1998 (filled square) are compared with a series of C+O star explosion models (Iwamoto et al. 1998).

Differentiating Λ with respect to x, it is found that Λ has a maximum $\Lambda_{\max} = \exp(-2xy)$. As given in Table 1 of Arnett (1982), the maximum occurs at $x_{\max}y = t_{\max}/(2\tau_{\mathrm{Ni}}) \sim 0.42 + 0.48y$ and $\Lambda_{\max} \sim 0.165/y$ (Arnett 2001). For SN 1998bw, $t_{\max} \sim 16$ days corresponds to $y \sim 1$ and $\Lambda_{\max} \sim 0.165$, thus $L_{\max} \sim 1.3 \times 10^{43} M_{\mathrm{Ni},\odot}\mathrm{erg\ s^{-1}}$. Given the fact that the ^{56}Co contribution doubles this luminosity at day ~ 16, the mass of ^{56}Ni is approximately given by

$$M_{\mathrm{Ni},\odot} \sim 0.38 L_{\max,43} \tag{10.4}$$

Model	$M_{\rm ms}$ $(M_\odot)$	$M_{\rm C+O}$ $(M_\odot)$	$M_{\rm ej}$ $(M_\odot)$	^{56}Ni mass $(M_\odot)$	$M_{\rm cut}$ $(M_\odot)$	$E_{\rm K}$ $(10^{51}$ erg)
CO138E50	~ 40	13.8	10	0.4	4	50
CO138E30	~ 40	13.8	10.5	0.4	3.5	30
CO138E7	~ 40	13.8	11.5	0.4	2.5	7
CO138E1	~ 40	13.8	12	0.4	2	1
CO100	~ 30	10.0	7.6	0.15	2.4	10
CO60	~ 25	6.0	4.4	0.15	1.4	1
CO21	~ 15	2.1	0.9	0.07	1.2	1

Table 10.1. Models and their parameters for SNe Ic. CO138E1 is an ordinary SN Ic model, in which a C+O star of $M_{\rm CO} = 13.8 M_\odot$ (which is the core of a $\sim 40 M_\odot$ main-sequence star) explodes with $E_{\rm K} = 1 \times 10^{51}$ ergs and $M_{\rm ej} = M_{\rm CO} - M_{\rm cut} \simeq 12 M_\odot$. $M_{\rm cut}$ (= 2 $M_\odot$ in this case) denotes the mass cut, which corresponds to the mass of the compact star remnant, either a neutron star or a black hole. CO138E50, CO138E30, and CO138E7 are hypernova models, in which the progenitor C+O star is the same as CO138E1 but explodes with different energies. The mass cut is chosen so that the ejected mass of ^{56}Ni is the value required to explain the observed peak brightness of SN 1998bw.

where $L_{\rm max,43}$ is the peak luminosity in units of 10^{43} erg s^{-1}. The observed peak luminosity of SN 1998bw, $L_{\rm max,43} \sim 1 \times (d/37.8{\rm Mpc})^2$ with d being the distance, translates into an estimated mass of ^{56}Ni $M_{\rm Ni,\odot} \sim 0.4(d/37.8{\rm Mpc})^2$, which is much larger than that of typical core-collapse SNe $M_{\rm Ni,\odot} \lesssim 0.1$.

A constraint for $M_{\rm ej,\odot}$ and $E_{\rm k,51}$ can be obtained from $y \sim 1$ with the use of equation (3) such that

$$M_{\rm ej,\odot}^3 / E_{\rm k,51} \sim 23 \left(\frac{\kappa}{0.05} \right)^{-2} \qquad (10.5)$$

Another equation is necessary to resolve the degeneracy of masses and energies in this equation. One useful quantity to use is the evolution of the photospheric velocity, as this has a different dependence on the parameters $M_{\rm ej,\odot}$ and $E_{\rm k,51}$. The photospheric velocity can either be obtained from observations, determining the approximate velocity of absorption lines, or it can be computed in the explosion models. An equivalent but more quantitative approach is to compare the observed and synthetic spectra.

In Figure 10.3, the observed line velocities and photospheric velocities of SN 1998bw are compared with calculated photospheric velocities for different models. The radius of the photosphere $r_{\rm ph}$ is defined by $\int_{r_{\rm ph}}^{\infty} \kappa \rho dr = 2/3$. For the constant density sphere model used in analytic light curves, the photosphere is located at a fraction $\sim 80\pi E_{\rm k}/(27\kappa M_{\rm ej}^2)t^2$ of the ejecta radius from the surface. Then, the photospheric velocity $v_{\rm ph}$ at early phases is given by

$$v_{\rm ph} \sim v_s \left(1 - \frac{80\pi E_{\rm k}}{27\kappa M_{\rm ej}^2}t^2\right) = \left(\frac{10E_{\rm k}}{3M_{\rm ej}}\right)^{1/2} \left(1 - \frac{80\pi E_{\rm k}}{27\kappa M_{\rm ej}^2}t^2\right). \quad (10.6)$$

Using the fact that on day 8 $v_{\rm ph} \sim 27,000$ km s^{-1}, we obtain from equation (6) another constraint:

$$E_{\rm k,51}/M_{\rm ej,\odot} \sim 4.3$$

which, combined with equation (5), results in

$$M_{\rm ej,\odot} \sim 10, \qquad E_{\rm k,51} \sim 43. \quad (10.7)$$

This very large energy, exceeding 10^{52} erg, led us to refer to SN 1998bw as a "Hypernova (HN)", an exceptional class of energetic SN explosion with kinetic energy $E_{\rm k,51} \gtrsim 10$ (Iwamoto et al. 1998; Woosley et al. 1999). The above estimate provides model parameters in good agreement with those given by detailed light curve models (Iwamoto et al. 1998; Woosley et al. 1999; Nakamura et al. 2001a). Nakamura et al. (2001a) selected a model (CO138E50) with $M_{\rm ej,\odot} = 10$, $E_{\rm k,51} = 50$, $M_{\rm Ni,\odot} = 0.4$ as the best model to reproduce the light curve and photospheric velocities using a distance 37.8Mpc and an extinction $A_{\rm V} = 0.05$ for SN1998bw (see Table 1 and Fig. 10.5).

Early Spectra

A more quantitative constraint on these basic model parameters and crucial diagnosis of chemical compositions of the ejecta can be obtained by detailed spectral modeling (Iwamoto et al. 1998; Branch 2001; Nomoto et al. 2001ab; Nakamura et al. 2001a; Mazzali et al. 2001).

Figure 10.4 shows early spectra of SN 1998bw for three epochs (Nakamura et al. 2001a; Patat et al. 2001). The spectra are dominated by broad absorption features. These line features are also seen in other SNe Ic. For SN 1998bw, however, they are exceptionally broad and blueshifted. Stathakis et al. (2000) found that absorption line minima are shifted 10-50 per cent blueward at day 15 in comparison with ordinary SNe Ic. These features shift significantly to the red over the three weeks covered by the spectra shown, indicating how the photosphere is receding to inner, lower velocity parts of the ejecta. The bluest of these features is likely due to Fe II lines, while the feature near 6000Å is

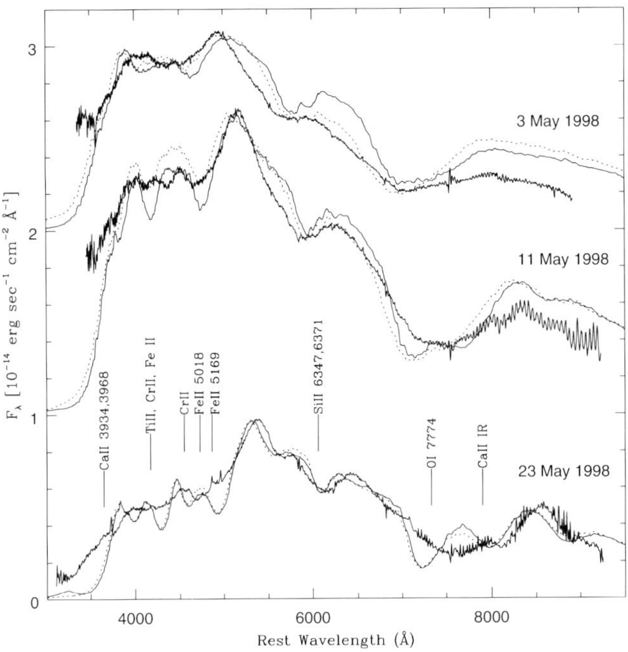

Figure 10.4. The observed spectra of SN 1998bw at three epochs: 3, 11, and 23 of May 1998 (day 8, 16, and 28, respectively: Patat et al. 2001) are compared with synthetic spectra (dashed lines) using model CO138E50 (Nakamura et al. 2001a). A distance modulus $\mu = 32.89$ mag and $A_V = 0.05$ are assumed, which corresponds to a distance of 37.8 Mpc with $H_0 = 65$ km s^{-1} Mpc^{-1}. The observed featureless spectra are the result of blending of many metal lines having large velocities and large velocity spreads. The apparent emission peaks are actually low opacity regions of the spectra in which photons can escape.

dominated by Si II and the redmost one is a blend of O I and Ca II. That the O I and Ca I lines merge into a single broad absorption is very unusual for any SN, and it indicates that the ejecta velocities are very large (the line separation is ~ 30000km/s).

In Fig. 10.4, the observed spectra are compared with synthetic spectra computed with a hypernova model CO138E50 (Nakamura et al. 2001a). A distance modulus of $\mu = 32.89$ mag and an extinction $A_V = 0.05$ are adopted. The assumption of low reddening is supported by the upper limit of 0.1 Å in the equivalent width of Na I D line obtained from high-resolution spectra (Patat et al. 2001).

The synthetic spectra of CO138E50 are in good agreement with the observed spectra. The Si II feature near 6,000Å and, in particular, the OI+CaII feature between 7,000 and 8,000Å, are as broad as the observations. Nevertheless, the blue sides of those absorptions are still too narrow, indicating that the new model

of CO138E30 shows a slower decline, in better agreement with SN1998bw, but it still declines too rapidly (Nakamura et al. 2001a; McKenzie & Schaefer 1999; Patat et al. 2001). All of these models (see Table 1) originate from a $13.8 M_\odot$ CO core, which is formed in a $\sim 40 M_\odot$ star. The explosion may have given rise to a Black Hole.

The early light curve of SN 1998bw is well reproduced by higher-energy models, but they deviate from the observations at late times. On the other hand, the late-time light curve is more easily reproduced by a lower-energy model with a smaller ^{56}Ni mass (model CO138E7), which is however too dim at early phases (Nakamura et al. 2001a). This difficulty may be overcome if the ejecta have multiple components with different characteristic velocities, suggesting that either the density distribution or the ^{56}Ni distribution in the ejecta or both are not spherically symmetric. A similar situation occurrs for SN 1997ef (Iwamoto et al. 2000; Mazzali et al. 2000) and SN 2002ap (Maeda et al. 2003).

As seen in Fig.10.5, the light curve of SN 1998bw seems to flatten after about day 400. The model light curves do not really follow this behavior. At $t \gtrsim 400$ days, the γ-ray deposition fraction $f_{\mathrm{dep},\gamma}$ decreases to below 1% in model CO138E50. However, about 3.5% of the decay energy of ^{56}Co is carried by positrons (e.g., Axelrod 1980). These are effectively trapped in the ejecta because of the postulated weak magnetic fields (e.g. Colgate & Petschek 1979). Therefore, the energy deposition from positrons makes the dominant contribution to the light curve at $t \gtrsim 400$ days (dotted line in Fig. 10.5).

The luminosity provided by the positrons is given as

$$L_{\mathrm{bol,e^+}} = 2 \times 10^{41} \left(\frac{M_{\mathrm{Ni},\odot}}{0.4} \right) \exp \left(-\frac{t}{111\mathrm{d}} \right) \tag{10.12}$$

where the positrons are assumed to be completely trapped in the ejecta. Therefore, if the observed tail follows the positron-powered, exponentially declining light curve, the ^{56}Co mass can be determined directly. However, the light curve of SN 1998bw after day 400 is still steeper than the positron-dominated light curve. The light curve of SN 1998bw showed a further flattening at around day 800. The latest observed point in Figure 10.5 is the HST observation on June 11, 2000 (day 778) (Fynbo 2000).

The observed magnitude (V = 25.41 $\pm$ 0.25) is consistent with the prediction of CO138E7, but brighter than CO138E50 (Fig.10.5).

One possibility is that, as the density decreases, the recombination time scale becomes longer than the decay time and ionization freeze-out makes the bolometric light curve even flatter (Fransson & Kozma 1993). Another possible source of the excess luminosity is the emission of radiation due to the

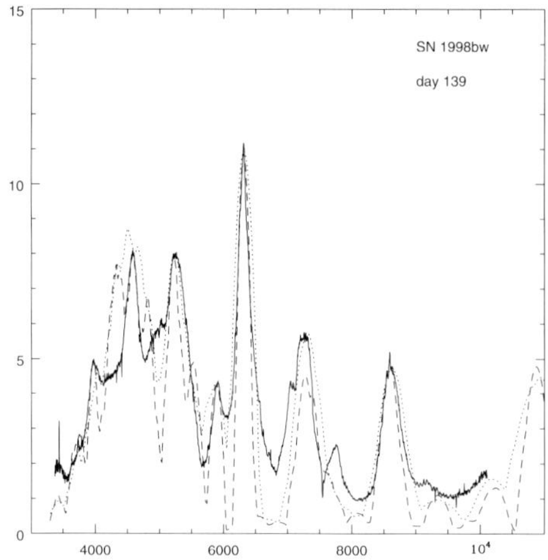

Figure 10.6. A nebular spectrum of SN 1998bw on 12 Sept 1998 (day 139) is compared to synthetic spectra obtained with a NLTE nebular model based on the deposition of gamma-rays from ^{56}Co decay in a nebula of uniform density. Two models were computed. In one model (dotted line) the broad FeII lines near 5300Å is well reproduced. The derived ^{56}Ni mass is 0.65 $M_{\odot}$, and the outer nebular velocity is 11,000km s^{-1}, and the O mass is 3.5$M_{\odot}$. The average electron density in the nebula is log n_e = 7.47 cm^{-3}. In the other model (dashed line), only the narrow OI 6300Å emission line is well reproduced. This model has smaller ^{56}Ni mass (0.35 $M_{\odot}$) and O mass (2.1 $M_{\odot}$), and an outer velocity of 7500km s^{-1}. The density is similar to that of the 'broad-lined' model. The filling factor used is 0.1 for both models (Nomoto et al. 2001ab; Mazzali et al. 2001).

interaction of the ejecta with a circumstellar medium (CSM) (Sollerman et al. 2000). Finally, the flattening may be due to a contribution from an underlying star cluster (Fynbo 2000) rather than to the SN itself.

Late time spectra provide a wealth of information on the elemental abundances and their distributions in velocity space. SN 1998bw seems to have entered the nebular phase between day 65 and 115 (Mazzali et al. 2001; Patat et al. 2001). The spectroscopic features at late times are very similar to those of SNe Ic as shown in Figure 10.6. Dominant emission features include Mg Iλ4571, an Fe II blend around 5,200 Å, the O I$\lambda\lambda$6300,6364 doublet, a feature around 7,200 Å(identified as Ca II and C II by Mazzali et al. 2001), and the Ca II IR triplet. What is different from ordinary SNe Ic, in particular, is the broadness of the line features. Patat et al. (2001) estimated the expansion velocity of the Mg I emitting region and found a value 9,800 $\pm$ 500 km s^{-1}on day 201. Stathakis et al. (2000) also found emission features 45 per cent broader than ordinary SNe Ic for day 94.

Mazzali et al. (2001) showed that the late time spectra of SN 1998bw (Patat et al. 2001) contain both broad and relatively narrow lines. They interpreted the absence of Fe III nebular lines as a sign of clumpiness of the ejecta. Interestingly, some Fe lines are found to be broader than O lines, which is the opposite as to what spherically symmetric models predict. This may imply that the explosion was aspherical (Fig. 10.6; Mazzali et al. 2001).

Aspherical Models

The result that the explosion energy of SN 1998bw was extremely large is based on the analysis of the light curve and spectra and on the assumption that the explosion was spherically symmetric. There is a possible alternative explanation for the large peak luminosity of SN 1998bw. If the explosion was aspherical, the estimate of the explosion energy might be smaller. There is growing evidence that light from SNe Ib/c is weakly polarized ($\sim 1\%$), implying that these explosions might be somewhat aspherical. A similar degree of polarization was observed in SN 1998bw at early photospheric phases (Kay et al. 1998; Iwamoto et al. 1998; Patat et al. 2001).

As discussed earlier, it is difficult to reproduce the entire light curve of SN 1998bw consistently using spherically symmetric models, indicating that there exists some degree of asphericity in the explosion of SN 1998bw. An analysis of the late-time light curve suggested a ^{56}Ni mass $M_{\mathrm{Ni},\odot} > 0.4$ (Sollerman et al. 2000), which is close to the values obtained with spherically symmetric models for early light curves. Thus, the possible effects of asymmetry in SN 1998bw seem to be only moderate.

In an asymmetric explosion, nucleosynthesis should also depart from spherical symmetry. Maeda et al. (2002) calculated nucleosynthesis in aspherical explosion models for SN 1998bw with a 2D hydrodynamical code and a detailed nuclear reaction network. They used the progenitor model CO138 (Iwamoto et al. 1998; Nakamura et al. 2001a) and assumed aspherical initial velocity profiles. Figure 10.7 shows the composition in the ejecta at the homologous expansion phase for $E_{\mathrm{exp}} = 10^{52}$ erg and an initial axial-to-radial velocity ratio $v_z/v_r = 8$. In this model, ^{56}Ni is synthesized preferentially along the polar axis, where the shock is stronger, while a lot of unburned material, dominated by O, is left at low velocity in the equatorial region, where burning is much less efficient.

Maeda et al. (2002) found that the nebular line profiles in SN 1998bw can be reproduced by such an aspherical model if the explosion is observed at an angle of about 15 degrees from the polar axis. At such an angle, one might expect that the GRB is weaker than it would be if observed along the jet axis. The actual aspect ratio of the ejecta is much smaller than 8:1, however, as the jet expands laterally, and this may be consistent with the observed polarization.

Such a highly aspherical explosion could occur in the collapse of a rotating stellar core that forms a system consisting of a rotating black hole and an

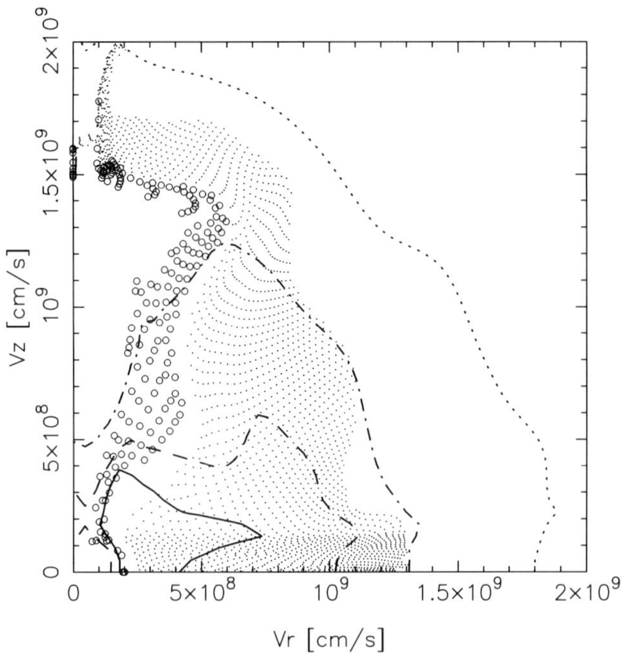

Figure 10.7. Distribution of ^{56}Ni(open circles) and ^{16}O(dots) at the homologous expansion phase in an aspherical explosion model of SN 1998bw (Maeda et al. 2002). Parameters are $E_{\rm exp} = 10^{52}$ erg, $v_z/v_r = 8$. Open circles and dots are test particles of ^{56}Ni and ^{16}O, respectively, indicating local volumes in which mass fractions of these elements are greater than 0.1. Lines are density contours of 0.5(solid), 0.3(dashed), 0.1(dash-dotted), and 0.01(dotted) of the maximum density.

accretion torus around it (MacFadyen & Woosley 1999). Thermal neutrinos from the torus release a large amount of energy as electron-positron pairs. If the black hole is accompanied by a strong magnetic field, rotational energy may be extracted from the black hole via the Blandford-Znajek mechanism (Blandford & Znajek 1977). Jet formation and propagation after energy deposition by the above processes are studied using hydrodynamical simulations (MacFadyen, Woosley, & Heger 2001; Aloy et al. 2000). MacFadyen et al. (2001) suggested that SN 1998bw may be a case in which a black hole was produced by 'fall back' and the resulting jet was less collimated.

The aspherical SN explosion could also be induced in the neutron star formation, e.g., by a strong magnetic field (Nakamura 1998; Wheeler 2001) and convection driven by neutrino heating (e.g., Janka & Muller 1994; Fryer & Warren 2002). Shimizu et al. (2001) pointed out that anisotropy in the neutrino emission would increase the net energy gain by neutrino heating, which leads to a larger explosion energy than in spherically symmetric models.

3. SN 1997ef

This SN was an immediate precursor of SN 1998bw. Its broad-lined spectrum defied interpretation in the context of standard-energy explosion models (Iwamoto et al. 2000). The realization with SN 1998bw that energies much larger than the supposedly standard value 10^{51}erg were possible led to the reinterpretation of SN 1997ef as a hypernova, although of smaller energy than SN 1998bw.

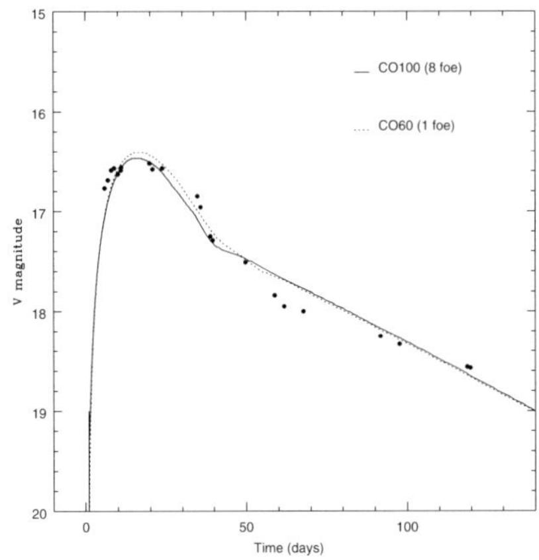

Figure 10.8. Calculated Visual light curves of CO60 and CO100 compared with that of SN 1997ef.

Light Curve

The light curve tail starts only $\sim$ 40 days after maximum, much later than in other SNe Ic. This can be reproduced by different explosion models. In Figure 10.8 we compare the calculated V light curves for the standard energy model CO60 and the energetic model CO100 ($E = 10^{52}$erg) with the observed V light curve of SN1997ef. We adopt a distance of 52.3 Mpc (a distance modulus of $\mu = 33.6$ mag) as estimated from the recession velocity, 3,400 km s^{-1} (Garnavich et al. 1997) and a Hubble constant $H_0 = 65$ km s^{-1} Mpc^{-1}. We assume no color excess, $E(B - V) = 0.00$. The light curve of SN 1997ef has a very broad maximum, which lasts for $\sim$ 25 days.

Since the model parameters of CO100 and CO60 give similar timescales, the light curves of the two models look similar: both have quite a broad peak and reproduce the light curve of SN1997ef reasonably well (Fig. 10.8).

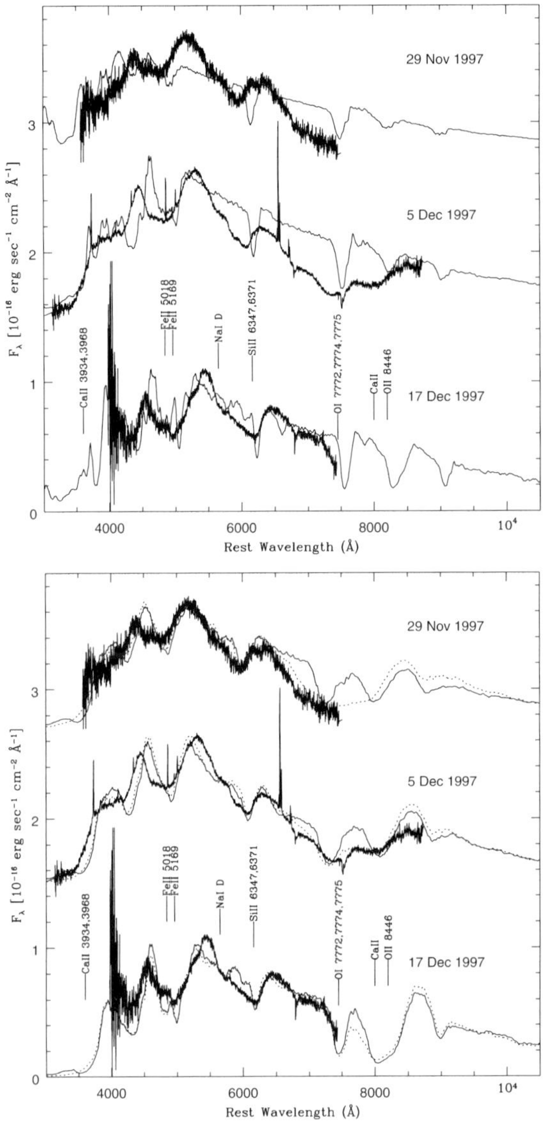

Figure 10.9. Observed spectra of SN 1997ef (bold lines) and synthetic spectra computed using models CO60 (upper) and CO100 (lower). The line features seen in the synthetic spectra for CO60 are much too narrow compared with observations, while the fits are much improved with CO100.

The light curve shape depends also on the distribution of ^{56}Ni, which is produced in the deepest layers of the ejecta. More extensive mixing of ^{56}Ni leads to an earlier rise of the light curve. For SN 1997ef, the best fit is obtained when ^{56}Ni is mixed almost uniformly to the surface for both models. Without such extensive mixing, the rise time to V = 16.5 mag would be $\sim$ 30 d for CO100. However, spectroscopic dating suggests that the peak occurred $\sim$ 18 days after the explosion.

Model CO60 has the same kinetic energy ($E_{\rm K} = 1 \times 10^{51}$ erg) as model CO21, which was used for SN Ic 1994I (see Table 1 for the model parameters). Since the light curve of SN 1997ef is much slower than that of SN 1994I, the ejecta mass of CO60 is $\sim$ 5 times larger than that of CO21.

The ejecta mass of CO100 is a factor of $\sim$ 2 larger than that of CO60, and it is only $\sim$ 20% smaller than that of model CO138, which was used for SN 1998bw (Table 1). Thus the explosion energy of CO100 should be $\sim$ 10 times larger than that of CO60 to reproduce the light curve of SN 1997ef. This explosion is very energetic, but still much weaker than the one in CO138. The smaller $E_{\rm K}$ for a comparable mass allows CO100 to reproduce the light curve of SN 1997ef, which has a much broader peak than that of SN 1998bw.

The light curve of SN 1997ef enters the tail around day 40. Since then, the observed V magnitude declines linearly with time at a rate of $\sim 1.1 \times 10^{-2}$ mag day^{-1}, which is slower than in other SNe Ic and is close to the ^{56}Co decay rate 9.6×10^{-3} mag day^{-1}. Such a slow decline implies much more efficient γ-ray trapping in the ejecta of SN 1997ef than in SN 1994I. The ejecta of both CO100 and CO60 are fairly massive and are able to trap a large fraction of the γ-rays, so that the calculated light curves have slower tails compared with CO21.

However, the light curves of both models decline somewhat faster in the tail than the observations. A similar discrepancy has been noted for the Type Ib supernovae (SNe Ib) 1984L and 1985F (Swartz & Wheeler 1991; Baron, Young, & Branch 1993). The late time light curve decline of these SNe Ib is as slow as the ^{56}Co decay rate, so that the inferred value of M is significantly larger (and/or $E_{\rm K}$ is smaller) than those obtained by fitting the early light curve shape. Baron et al. (1993) suggested that the ejecta of these SNe Ib must be highly energetic and as massive as $\sim 50\ M_{\odot}$. We will suggest that such a discrepancy between the early- and late-time light curves might be an indication of asphericity in the ejecta of SN 1997ef and that it might be the case in those SNe Ib as well.

Spectra

As we have shown, light curve modeling provides direct constraints on $M_{\rm CO}$ and E. However, it is difficult to determine uniquely these values, and hence the characteristics of the explosion, from the light curve shape alone, since models with different values of $M_{\rm ej}$ and E can yield similar light curves. Fortunately, however, models with different values of $M_{\rm ej}$ and E are expected to show different spectral evolution.

Using detailed spectrum synthesis, we can therefore distinguish between different models clearly, because the spectrum contains much more information than a single-band light curve.

Around maximum light, the spectra of SN 1997ef show just a few very broad features, and are quite different from those of ordinary SNe Ib/c, but similar to those of SN 1998bw. However, at later epochs the spectra develop features that are easy to identify, such as the Ca II IR triplet at ~ 8200Å, the O I absorption at 7500 Å, several Fe II features in the blue, and they look very similar to the spectrum of the ordinary SN Ic 1994I.

We computed synthetic spectra with a Monte Carlo spectrum synthesis code using the density structure and composition of the hydrodynamic models CO60 and CO100. We produced synthetic spectra for three epochs near maximum, of SN 1997ef: Nov 29, Dec 5, and Dec 17. These are early enough that the spectra are very sensitive to changes in the kinetic energy. As in the light curve comparison, we adopted a distance modulus of $\mu = 33.6$ mag, and $E(B-V) = 0.0$.

In Figure 10.9 (upper) we show the synthetic spectra computed with the ordinary SN Ic model CO60. The lines in the spectra computed with this model are always much narrower than the observations. This clearly indicates a lack of material at high velocity in model CO60, and suggests that the kinetic energy of this model is much too small.

Synthetic spectra obtained with the hypernova model CO100 for the same 3 epochs are shown in Figure 10.9 (lower). The spectra show much broader lines, and are in good agreement with the observations. In particular, the blending of the Fe lines in the blue, giving rise to broad absorption troughs, is well reproduced. The two 'emission peaks' observed at ~ 4400 and 5200Å correspond to the only two regions in the blue that are relatively line-free.

The spectra are characterized by a low temperature, even near maximum, because the rapid expansion combined with the relatively low luminosity (from the tail of the light curve we deduce that SN 1997ef produced about $0.15 M_\odot$ of ^{56}Ni, compared to about $0.6 M_\odot$ in a typical SN Ia and $0.5 M_\odot$ in SN 1998bw) leads to rapid cooling. Thus the Si II 6355Å line is not very strong.

Model CO100 has $E = 10^{52}$ erg, $M_{\rm ej} = 7.5 M_\odot$, $M(^{56}{\rm Ni}) = 0.15 M_\odot$. From these values, we find $M_{\rm CO} = 10 M_\odot$, $M_{\rm rem} = 2.5 M_\odot$. A $10 M_\odot$ CO core is formed in a $\sim 30 M_\odot$ star. Although model CO100 yields rather good synthetic spectra, it still fails to reproduce the observed large width of the O I - Ca II feature in the only near-maximum spectrum that extends sufficiently far to the red (5 Dec 1997). An improvement can be obtained by introducing an arbitrary flattening of the density profile at the highest velocities (Mazzali et al. 2000; Branch 2001). This leads to higher values of both E and $M_{\rm ej}$.

Possible Aspherical Effects

The light curve, the photospheric velocities, and the spectra of SN 1997ef are better reproduced with the hyper-energetic model CO100 than with the ordinary SN Ic model CO60. However, there remain several features that are still difficult to explain with model CO100.

These discrepancies may be interpreted as a possible sign of asphericity in the ejecta: A part of the ejecta moves faster than average to form the lines at high-velocities at early phases, while the other part of ejecta expands with a lower velocity so that the low-velocity Si II line comes up at later epochs. Having a low-velocity component would also make it easier to reproduce the slow tail and the long duration of the photospheric phase.

Extensive mixing of ^{56}Ni is required to reproduce the short rise time of the light curve. One possibility to induce such mixing in the velocity space is an asymmetric explosion. In the extremely asymmetric cases, material ejection may take place in a jet-like form (e.g., MacFadyen & Woosley 1999; Khokhlov et al. 1999). A jet could easily bring some Ni from the deepest layers out to the high velocity surface. The lack of a strong case of coincidence with a GRB suggests that if a jet was produiced it was either weak or it was not pointing towards us.

Unlike other hypernovae, SN 1997ef does not seem to be a unique case. At least two other SNe, SN 1997dq (Matheson et al. 2001) and SN 1999ey show very similar properties. Unfortunately, these two objects were not very intensively observed. Whether this is just a coincidence, or whether it really indicates that SN 1997ef-like hypernovae are more frequent than others remains an interesting question that must be answered observationally.

4. SN 2002ap

SN Ic 2002ap was discovered in M74 on 2002 January 30 (Hirose 2002). The SN was immediately recognised as a hypernova from its broad spectral features (Kinugasa et al. 2002; Meikle et al. 2002; Gal-Yam, Ofek & Shemmer 2002; Filippenko & Chornock 2002). This indicates high velocities in the ejected material, which is the typical signature of hypernovae. It was therefore followed from several observatories, and the relative proximity also favored observations with small telescopes. Luckily, the SN was discovered very soon after it exploded: the discovery date was January 29, while the SN was not detected on January 25 (Nakano, Kushida, & Li 2002). This is among the earliest any SN has been observed, with the obvious exceptions of SN 1987A and SN 1993J.

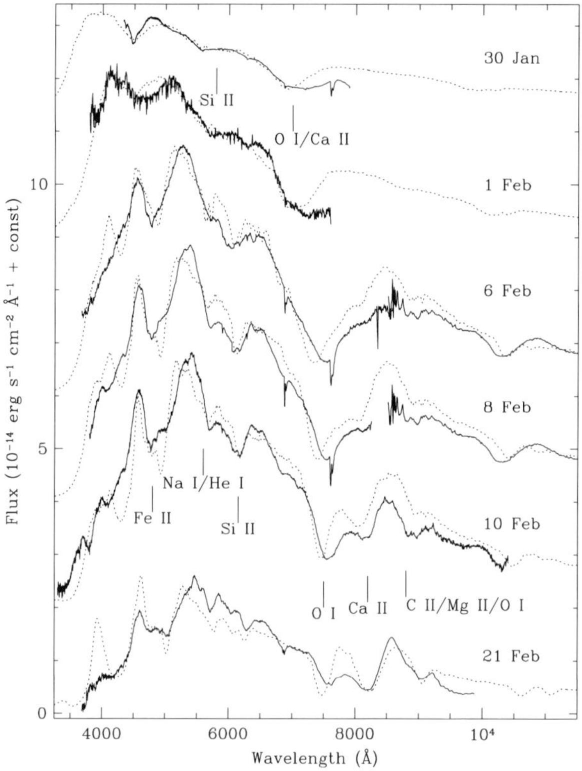

Figure 10.10. A comparison between some observed spectra of SN 2002ap (*thick lines*: January 30 — WHT; February 1 — Gunma Obs.; February 6 — Beijing Obs.; February 8 — Subaru FOCAS; February 7 IR — Subaru CISCO, shown twice; February 10 — Lick Obs.; February 21 — Asiago Obs.) and synthetic spectra computed with model CO100/4 (*dashed lines*) (Mazzali et al. 2002).

Light Curve

Figure 10.2 shows the V-band light curves of the same four SNe as in Figure 10.1. SN 2002ap reached V maximum on about February 8 at $V = 12.3$ mag. SN 2002ap peaks earlier than both hypernovae 1998bw and 1997ef, but later than the normal SN 1994I, suggesting an intermediate value of the ejecta mass $M_{\rm ej}$.

Using a distance to M74 of 8 Mpc ($\mu = 29.5$ mag; Sharina, Karachentsev, & Tikhonov 1996), and a combined Galaxy and M74 reddening of $E(B-V) = 0.09$ mag (estimated from a Subaru HDS spectrum; Takada-Hidai, Aoki, &

Zhao 2002), the absolute magnitude is $M_V = -17.4$. This is comparable to SN 1997ef and fainter than SN 1998bw by almost 2 mag. Since peak brightness depends on the ejected ^{56}Ni mass, SNe 2002ap, 1997ef, and 1994I appear to have synthesized similar amounts of it. Estimates were $\sim$ 0.07 $M_\odot$ for SN 1994I (Nomoto et al. 1994) and 0.13 $M_\odot$ for SN 1997ef (Mazzali et al. 2000). The ^{56}Ni mass for SN 2002ap is estimated to be $\sim$ 0.07 $M_\odot$, which is similar to that of normal core-collapse SNe such as SNe 1987A and 1994I.

Spectra

Figure 10.1 shows the maximum-light spectra of SN 2002ap, of the hypernovae SNe 1998bw and 1997ef, and of the normal SN Ic 1994I. If line width is the distinguishing feature of a hypernova, then clearly SN 2002ap is a hypernova, as its spectrum resembles that of SN 1997ef much more than that of SN 1994I. Line blending in SN 2002ap and SN 1997ef is comparable. However, some individual features that are clearly visible in SN 1994I but completely blended in SN 1997ef can at least be discerned in SN 2002ap (e.g. the Na I–Si II blend near 6000 Å and the Fe II lines near 5000 Å). Therefore, spectroscopically SN 2002ap appears to be located just below SN 1997ef in a "velocity scale," but far above SN 1994I, which appears to confirm the evidence from the light curve.

The spectral evolution of SN 2002ap shown in Figure 10.10 appears to follow closely that of SN 1997ef, at a rate about 1.5 times faster. The spectra and the light curve of SN 2002ap can be well reproduced by a model with ejected heavy-element mass $M_{\rm ej}$= 2.5–5 $M_\odot$ and E_{51} = 4–10. Both $M_{\rm ej}$ and $E_{\rm kin}$ are much smaller than those of SNe 1998bw and 1997ef (but they could be larger if a significant amount of He is present).

Is SN 2002ap a Hypernova or a Supernova?

Although SN 2002ap appears to lie between normal core-collapse SNe and hypernovae, it should be regarded as a hypernova because its kinetic energy is distinctly higher than for normal core-collapse SNe. In other words, the broad spectral features that characterize hypernovae are the results of a high kinetic energy. Also, SN 2002ap was not more luminous than normal core-collapse SNe. Therefore brightness alone should not be used to discriminate hypernovae from normal SNe, while the criterion should be a high kinetic energy, accompanied by broad spectral features. Further examples of hypernovae are necessary in order to establish whether a firm boundary between the two groups exists.

For these values of $E_{\rm kin}$, $M_{\rm ej}$, and $M(^{56}{\rm Ni})$, we can constrain the progenitor's main-sequence mass $M_{\rm ms}$ and the remnant mass $M_{\rm rem}$. Modeling the explosions of C+O stars with various masses, we obtain $M(^{56}{\rm Ni})$ as a function of the parameter set ($E_{\rm kin}$, $M_{\rm CO}$, $M_{\rm rem} = M_{\rm ej} - M_{\rm CO}$). The model which is most consistent with our estimates of (M_{ej}, E) is one with $M_{CO} \approx 5\,M_\odot$, $M_{rem} \approx 2.5\,M_\odot$, and $E_{51} = 4.2$. The 5 $M_\odot$ C+O core forms in a He core of mass $M_\alpha = 7\ M_\odot$, corresponding to a main-sequence mass $M_{ms} \approx$ 20–25 $M_\odot$. The $M_{ms} - M_\alpha$ relation depends on convection and metallicity (e.g., Nomoto & Hashimoto 1988; Umeda & Nomoto 2002).

The estimated progenitor mass and explosion energy are both smaller than those of previous Hypernovae such as SNe 1998bw and 1997ef, but larger than those of normal core-collapse SNe such as SN 1999em. This mass range is consistent with the non-detection of the progenitor in pre-discovery images of M74 (Smartt et al. 2002).

Given the estimated mass of the progenitor, binary interaction including the spiral-in of a companion star (Nomoto et al. 2001a) is probably required in order for it to lose its hydrogen and some (or most) of its helium envelope. This would suggest that the progenitor was in a state of high rotation. It is possible that a high rotation rate and/or envelope ejection are also necessary conditions for the birth of a hypernova.

Possible Aspherical Effects

SN 2002ap was not apparently associated with a GRB. This may actually be not so surprising, since the explosion energy of SN 2002ap is about a factor of 5-10 smaller than that of SN 1998bw, as also indicated by the weak radio signature (Berger, Kulkarni, & Chevalier 2002). The present data show no clear signature of asymmetry, except perhaps for some polarization (Kawabata et al. 2002; Leonard et al. 2002; Wang et al. 2002), which is smaller than that of SN 1998bw. This suggests that the degree of asphericity is smaller in SN 2002ap and that the possible "jet" may have been weaker, which makes GRB generation more difficult.

5. SN 1999as

SN 1999as was discovered on February 18, 1999 by the Supernova Cosmology Project (Knop et al. 1999) in an anonymous galaxy having a redshift of 0.127. The absolute magnitude was exceptionally bright, $M_V < -21.5$, at least nine times brighter than the hypernova SN 1998bw.

SN 1999as is spectroscopically classified as a SN Ic because its photospheric phase spectra show no conspicuous lines of hydrogen, He I, or Si II λ6355. The usual SN Ic spectral lines such as Ca II and O I are very broad, like in other hypernovae. However, some narrow ($\sim$ 2000 km s^{-1}) but highly blueshifted ($\sim$ 11,000 km s^{-1}) lines of Fe II are also present (Hatano et al. 2001).

By fitting the observed light curve, we have obtained the following constraints on the explosion model: the ejected mass $M_{ej} \simeq 10-20\ M_\odot$, the kinetic energy of ejected matter $E \simeq 10^{52} - 10^{53}$ ergs, and the mass of ejected radioactive

^{56}Ni $M_{\rm Ni} > 4\ M_\odot$ (Deng et al. 2001). The progenitor of this may have been as massive as $\sim\ 60 M_\odot$, and the explosion almost certainly resulted in the formation of a Black Hole. Unfortunately, the spectral coverage is not very extensive, and an accurate determination of the properties of this supernova is therefore difficult.

The asymmetric hydrodynamical model of Maeda et al. (2002) could represent this new class of hypernovae. In this model, a clump of freshly synthesized ^{56}Ni exists at high velocity ($\sim$ 15,000 km s^{-1}) near the symmetry axis; such a clump could produce narrow but high–velocity absorption lines if the viewing angle with respect to the symmetry axis is small. On the other hand, the lack of a detected GRB may suggest that the angle was not very small.

6. Type IIn Hypernova: SN 1997cy and SN 1999E

SN 1997cy is a different type of hypernova, as it is of Type IIn. However, the energy deduced from its light curve is extremely large. The energy reveals itself in the strength of the ejecta-CSM interaction. Furthermore, the SN may have a correlated GRB. SN 1999E is very similar to SN 1997cy in the spectra and light curve (Rigon et al. 2003).

SN 1997cy displayed narrow Hα emission on top of broad wings, which lead to its classification as a Type IIn (Germany et al. 2000; Turatto et al. 2000). Assuming $A_V = 0.00$ for the galactic extinction (NED) we get an absolute magnitude at maximum $M_v \leq -20.1$. It is the brightest SN II discovered so far. The light curve of SN 1997cy does not conform to the classical templates of SN II, namely Plateau and Linear, but resembles the slow evolution of Type IIn SN 1988Z. As seen from the *uvoir* bolometric light curve in Figure 10.11, the SN light curve decline is slower than the ^{56}Co decay rate between day 120 to 250, suggesting circumstellar interaction for the energy source. (Here the outburst is taken to be coincident with GRB970514.)

In the interaction model (Turatto et al. 2000), collision of the SN ejecta with the slowly moving circumstellar matter (CSM) converts the kinetic energy of the ejecta into light, thus producing the observed intense light display of the SN. The exploratory model considers the explosion of a massive star of $M = 25M_\odot$ with a parameterized kinetic energy E. The collision is assumed to start near the stellar radius at a distance r_1, where the density of the CSM is ρ_1, and adopt for the CSM a power-law density profile $\rho \propto r^n$. The parameters $E_{\rm K}$, ρ_1, and n, are constrained from comparison with the observations.

The regions excited by the forward and reverse shock emit mostly X-rays. The density in the shocked ejecta is so high that the reverse shock is radiative and a dense cooling shell is formed (e.g., Suzuki & Nomoto 1995; Terlevich et al. 1992). The X-rays are absorbed by the outer layers and the core of the ejecta, and re-emitted as UV-optical photons.

Narrow lines are emitted from the slowly expanding unshocked CSM photoionized by the SN UV outburst or by the radiation from the shocks; intermediate width lines come from the shock-heated CSM; broad lines come from either the cooler region at the interface between ejecta and CSM.

Figure 10.11 shows the model light curve which best fits the observations. The model parameters are: $E = 3 \times 10^{52}$ erg, $\rho_1 = 4 \times 10^{-14}$ g cm^{-3} at $r_1 = 2 \times 10^{14}$ cm (which corresponds to a mass-loss rate of $\dot{M} = 4 \times 10^{-4}$ $M_\odot$ yr^{-1} for a wind velocity of 10 km s^{-1}), and $n = -1.6$. The large mass-loss episode giving rise to the dense CSM is supposed to occur after the progenitor makes a loop in the HR diagram from BSG to RSG. In this model, the mass of the low-velocity CSM is $\sim 5M_\odot$, which implies that the transition from BSG to RSG took place about 10^4 yr before the SN event.

The large CSM mass and density are necessary to have large shocked masses and thus to reproduce the observed high luminosity, and so is the very large explosion energy. For models with low $E_{\rm K}$ and high ρ_1, the reverse shock speed is too low to produce a sufficiently high luminosity. For example, a model with $E = 10^{52}$ erg and ρ_1 as above yields a value of $L_{\rm UVOIR}$ lower than the observed luminosity by a factor of ~ 5. For high E or low ρ_1, the expansion of the SN ejecta is too fast for the cooling shell to absorb enough X-rays to sustain the luminosity. Thus in this model $E_{\rm K}$ and $\dot{M}$ are constrained within a factor of ~ 3 of the reported values.

The shape of the light curve constrains the circumstellar density structure. For $n = -2$, the case of a steady wind, $L_{\rm UVOIR}$ decreases too rapidly around day 200. To reproduce the observed decrease after day ~ 300, the CSM density is assumed to drop sharply at the radius the forward shock reaches at day 300, so that the collision becomes weaker afterwards. (Such a change of the CSM density corresponds to the transition from BSG to RSG of the progenitor $\sim 10^4$ yr before the SN explosion.) This is consistent with the simultaneous decrease in the Hα luminosity.

The observed light curve drops sharply after day 550. The model reproduces such a light curve behavior (Figure 10.11) assuming that when the reverse shock propagates through $\sim 5\ M_\odot$, it encounters exceedingly low density region and thus it dies. In other words, the model for the progenitor of SN 1997cy assumes that most of the core material has fallen into a massive black hole of, say, $\sim 10M_\odot$, while the extended H/He envelope of $\sim 5\ M_\odot$ has not collapsed. Then material is ejected from the massive black hole possibly in a jet-like form, and the envelope is hit by the "jet" and ejected at high velocity.

In this model, the ejecta are basically the H/He layers and thus contain the original (solar abundance) heavy elements plus some heavy elements mixed from the core (before fall back) or jet materials. This might explain the lack of oxygen and magnesium lines in the spectra particularly at nebular phases (Turatto et al. 2000).

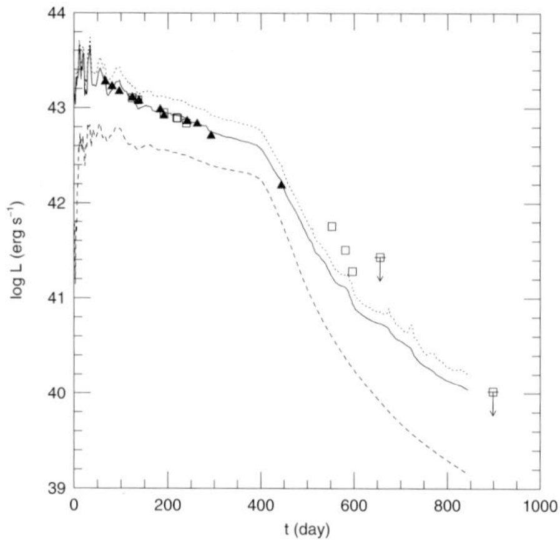

Figure 10.11. The *uvoir* bolometric light curve of SN 1997cy compared with the synthetic light curve obtained with the CSM interaction model (Turatto et al. 2000).

7. Properties of Hypernovae

Based on the observed objects and their interpretation, it is possible to make some generalizations regarding the properties of Hypernovae and their relation to the progenitor stars.

The Explosion Kinetic Energy

In Figure 10.12 we plot E as a function of the main-sequence mass $M_{\rm ms}$ of the progenitor star as derived from fitting the optical light curves and spectra of various hypernovae, of the normal SNe 1987A, 1993J, and 1994I (e.g., Shigeyama & Nomoto 1990; Nomoto et al. 1993; Nomoto et a. 1994; Shigeyama et al. 1994; Iwamoto et al. 1994; Woosley et al. 1994; Young, Baron, & Branch 1995), and of SNe 1997D (Turatto et al. 1998). Properties of Type Ib/c supernovae/hypernovae thus derived are summarized in Table 10.2. It appears that E increases with $M_{\rm ms}$, forming a 'Hypernova Branch', reaching values much larger than the canonical 10^{51} erg. SNe 1997D and 1999br, on the contrary, are located well below that branch, forming a 'Faint SN Branch'.

This trend might be interpreted as follows. Stars with $M_{\rm ms} \lesssim$ 20-25 $M_\odot$ form a neutron star (SN 1987A may be a borderline case between the neutron star and black hole formation). Stars with $M_{\rm ms} \gtrsim$ 20-25 $M_\odot$ form a black hole (e.g., Ergma & van den Heuvel 1998); whether they become hypernovae or faint SNe may depend on the angular momentum in the collapsing core, which in turn depends on the stellar winds, metallicity, magnetic fields, and binarity.

Hypernovae might have rapidly rotating cores owing possibly to the spiraling-in of a companion star in a binary system. The core of faint SNe II might not

	83N	94I	02ap	97ef	98bw
	Ib	Ic	Ic (Hypernovae)		
			Pre-explosion		
$M_{\rm ZAMS}$	15	15	21	34	40
$M_{\rm He}$	4	4	6.6	13	16
$M_{\rm CO}$	2	2	4.5	11	14
$M_{\rm exp}$	4	2.1	4.6	11.1	13.8
			Post-explosion		
$M_{\rm rem}$	1.25	1.2	2.1	1.6	2.9
$M_{\rm ej}$	2.75	0.9	2.5	9.5	10.9
$M_{\rm He}$	2.0	0	0.1	0	0
$M_{\rm CO}$	0.5	0.6	1.8	5.3	8
$M_{\rm IME}$	0.1	0.2	0.5	4	2
$M_{\rm Ni}$	0.15	0.07	0.1	0.13	0.7
E_{51}	1	1	7	19	30

Table 10.2. Properties of supernovae and hypernovae

have a large angular momentum, because the progenitor had a massive H-rich envelope so that the angular momentum of the core might have been transported to the envelope possibly via a magnetic-field effect.

Between these two branches, there may be a variety of SNe. A dispersion in the properties of SNe II-P has been reported (Hamuy 2003).

The Mass of Ejected ^{56}Ni

A similar relation is observed between the mass of ^{56}Ni, $M(^{56}\rm Ni)$, synthesized in core-collapse supernovae and $M_{\rm ms}$ in Figure 10.12, which is important to know for the study of the chemical evolution of galaxies. Stars with $M_{\rm ms} \lesssim$ 20-25 $M_\odot$, forming a neutron star, produce $\sim 0.08 \pm 0.03\ M_\odot$ ^{56}Ni as in SNe 1993J, 1994I, and 1987A.

For stars with $M_{\rm ms} \gtrsim$ 20-25 $M_\odot$, which form black holes, $M(^{56}\rm Ni)$ appears to increase with $M_{\rm ms}$ in the 'Hypernova Branch', while SNe in the ' Faint SN Branch' produced only very little ^{56}Ni. For faint SNe, because of the large gravitational potential, the explosion energy was so small that most of ^{56}Ni fell back onto a compact star remnant.

Asymmetry

All Hypernovae of Type Ic show some signatures of asymmetry, or at least of a departure from purely 1-dimensional spherically symmetric models. This may support the case for their connection with at least some GRB's.

So far only SN 1998bw and 2003dh have the well established connection with GRBs. In the other cases, either a GRB was not generated, or if it was

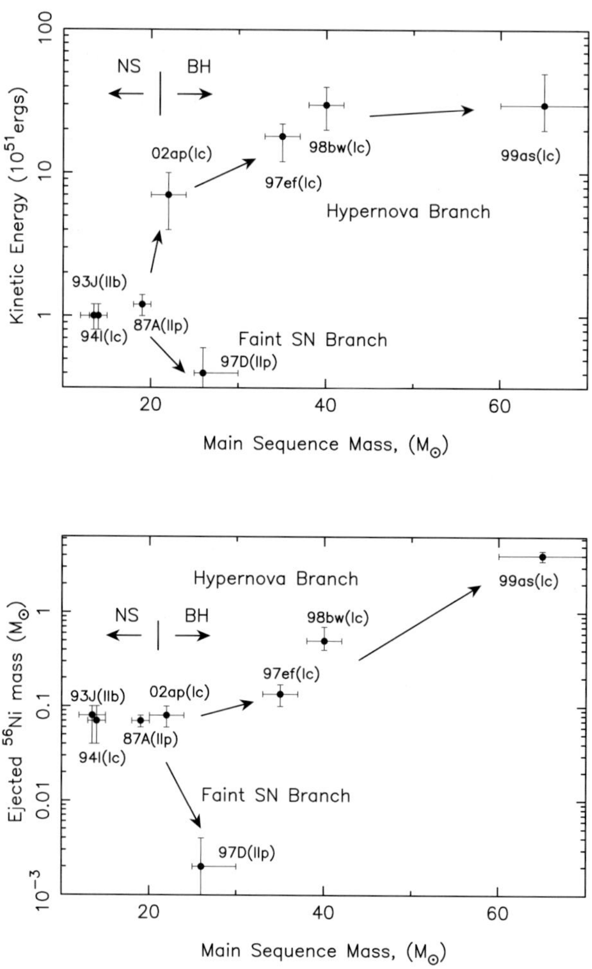

Figure 10.12. Explosion energies and the ejected ^{56}Ni mass against the main sequence mass of the progenitors for several bright supernovae/hypernovae (Nomoto et al. 2002).

it was weak and/or not pointing towards us. The issue of directionality is very important. If hypernovae are aspherical, we expect to find a range of hypernova properties for the same Ni mass. This can be established at late times, independetly of the shape of the ejecta. These objects may be very different at early phases, showing different light curves, velocities, abundances. So far, however, this evidence is missing.

Gamma-Ray Bursts/Supernovae Connection

Candidates for the GRB/SN connection include GRB 980425/SN Ic 1998bw (Galama et al. 1998; Iwamoto et al. 1998), GRB 971115/SN Ic 1997ef (Wang & Wheeler 1998), GRB 970514/SN IIn 1997cy (Germany et al. 2000; Turatto et al. 2000), GRB 980910/SN IIn 1999E (Thorsett & Hogg1999), and GRB 991002/SN IIn 1999eb (Terlevich, Fabian, & Turatto 1999). Recently, GRB 030329 has shown the evidence of a supernova (SN Ic 2003dh) in its optical afterglow spectra, confirming the GRB/SN connection directly (Stanek et al. 2003).

Several GRB's are suggested to be associated with SNe, such as GRB 980326 (Bloom et al. 1999), GRB 970228 (Reichart 1999; Galama et al. 2000), and GRB 011121/SN 2001ke (Bloom et al. 2002; Garnavich et al. 2002). The decline of the light curve of the optical afterglows of these GRBs slowed down at late phases, and this can be reproduced if a red-shifted SN 1998bw-like light curve is superposed on the power-law component.

A question is whether the supernovae associated with GRBs have a uniform maximum luminosity, i.e., whether $\sim 0.5\ M_{\odot}$ ^{56}Ni production as in SN 1998bw is rather common or not. Figure 10.12 shows that the ^{56}Ni mass and thus intrinsic maximum brightness of Hypernovae has a large diversity. We certainly need more examples before we can define the luminosity function and the actual distribution of masses of ^{56}Ni produced in supernovae/hypernovae.

For several hypernovae such as SNe 1998ey and 2002ap, no GRB counterpart has been proposed. These hypernovae were less energetic events than SN 1998bw. It is possible that a weaker explosion is less efficient in collimating the γ-rays to give rise to a detectable GRB (GRB980425 was already quite weak compared to the average GRBs), or that some degree of inclination of the beam axis to the line-of-sight results in a seemingly weaker supernova and in the non-detection of a GRB. Only the accumulation of more data will allow us to address these questions.

Properties of hypernova nucleosynthesis suggest that hypernovae of massive stars may make important contributions to the Galactic (and cosmic) chemical evolution (Nakamura et al. 2001b; Nomoto et al. 2001b). In view of small frequencies of GRBs, this implies that hypernovae are much more frequent than GRBs, i.e., only a special class of hypernovae gives rise to GRBs (Nomoto et al. 2002).

8. Possible Evolutionary Scenarios to Hypernovae

Here we classify possible evolutionary paths leading to C+O star progenitors. In particular, we explore the paths to the progenitors that have rapidly rotating cores with a special emphasis, because the explosion energy of hypernovae may be extracted from rapidly rotating black holes (Blandford & Znajek 1977).

(1) Case of a single star: If the star is as massive as $M_{\rm ms} \gtrsim 40\ M_\odot$, it could lose its H and He envelopes in a strong stellar wind (e.g., Schaller et al. 1992). This would be a Wolf-Rayet star.

(2) Case of a close binary system: Suppose we have a close binary system with a large mass ratio. In this case, the mass transfer from star 1 to star 2 inevitably takes place in a non-conservative way, and the system experiences a common envelope phase where star 2 is spiraling into the envelope of star 1. If the spiral-in releases enough energy to remove the common envelope, we are left with a bare He star (star 1) and a main-sequence star (star 2), with a reduced separation. If the orbital energy is too small to eject the common envelope, the two stars merge to form a single star (e.g., van den Heuvel 1994).

(2-1) For the non-merging case, possible channels from the He stars to the C+O stars are as follows (Nomoto, Iwamoto, & Suzuki 1995).

(a) Small-mass He stars tend to have large radii, so that they can fill their Roche lobes more easily and lose most of their He envelope via Roche lobe overflow.

(b) On the other hand, larger-mass He stars have radii too small to fill their Roche lobes. However, such stars have large enough luminosities to drive strong winds to remove most of the He layer (e.g., Woosley, Langer, & Weaver 1995). Such a mass-losing He star would again be a Wolf-Rayet star.

Thus, from the non-merging scenario, we expect two different kinds of SNe Ic, fast and slow, depending on the mass of the progenitor. SNe Ic from smaller mass progenitors (channel 2-1-a) show faster light-curve and spectral evolutions, because the ejecta become more quickly transparent to both gamma-ray and optical photons. The slow SNe Ic originate from the Wolf-Rayet progenitors (channels 1 and 2-1-b). The presence of both slow and fast SNe Ib/Ic has been noted by Clocchiatti & Wheeler (1997).

(2-2) For the merging case, the merged star has a large angular momentum, so that its collapsing core must be rotating rapidly. This would lead to the formation of a rapidly rotating black hole from which possibly a hyper-energetic jet could emerge. If the merging process is slow enough to eject the H/He envelope, the star would become a rapidly rotating C+O star. Such stars are the candidates for the progenitors of Type Ic hypernovae like SNe 1997ef and 1998bw. If a significant amount of H-rich envelope remains after merging, the rapidly rotating core would lead to a hypernova of Type IIn possibly like SN 1997cy (or Type Ib).

Table 10.3. Models and Results of jet induced explosion models (Maeda & Nomoto 2003). Masses are in solar mass unit ($M_\odot$), and $\theta_{\rm jet}$ is in degree.

Model	$M_{\rm MS}$	$M_{\rm REM0}$	α	$\theta_{\rm jet}$	$E_{\rm total}$	$M_{\rm REM}$	$M(^{56}$Ni)	[S/Si]	[C/O]
40A	40	1.5	0.01	15	10.9	5.9	1.07E-1	-0.46	-1.3
40B	40	1.5	0.01	45	1.2	6.8	8.11E-2	-0.54	-1.2
40C	40	1.5	0.05	15	32.4	2.9	2.40E-1	-0.30	-1.3
40D	40	3.0	0.01	15	8.5	10.5	6.28E-8	-1.1	-1.0
25A	25	1.0	0.01	15	6.7	1.9	7.81E-2	-0.28	-0.80
25B	25	1.0	0.01	45	0.6	1.5	1.51E-1	-0.26	-0.82

equatorial direction. In total, complete Si-burning elements can be enhanced (Maeda & Nomoto 2003; Nomoto et al. 2001ab).

Recently, Maeda & Nomoto (2003) have reported hydrodynamics and nucleosynthesis of jet-induced explosion models. We show some of the results below (See Maeda & Nomoto 2003 for detailes). .

The main ingredient of the above models is a pair of jets propagating through a stellar mantle. At the beginning of each calculation, the central part ($M_r \leq M_{\rm REM0}$) of a progenitor star is displaced by a compact remnant. The jets are injected at the interface with the opening half-angle $\theta_{\rm jet}$. The energy injected by the jets is assumed to be proportional to the accretion rate (i.e., $\dot{E}_{\rm jet} = \alpha \dot{M}_{\rm accretion} c^2$). The accretion at the inner boundary is traced and is used to update the central remnant' mass and the property of the jets. The models are summarized in Table 10.3. The outcome of the explosion depends on the interaction between the jets and the stellar mantle, and on the accretion rate which is affected by the interaction itself. The hydrodynamic evolution and nucleosynthesis in two dimensions are found to be as follows.

Figure 10.15 shows a snapshot of the velocity distribution of Model 40A at 1.5 seconds after the initiation of the jets. The jets propagate through the stellar mantle, depositing their energies into ambient matter at the working surface. The bow shock expands laterally to push the stellar mantle sideways, which reduces the accretion rate. The strong outflow occurs along the z-axis (the jet direction), while matter accretes from the side. As the accretion rate decreases, the jets are turned off. Then the inflow along the r-axis turns to the weak outflow.

The outcome is a highly aspherical explosion. The accretion forms a central dense core. Densities near the center become much higher than those in spherical models. This feature is consistent with the suggestions by the spectroscopic (Mazzali et al. 2000) and the light curve (Nakamura et al. 2001a; Maeda et al. 2003) modeling of hypernovae.

Other hydrodynamic properties are as follows (Tab. 10.3):

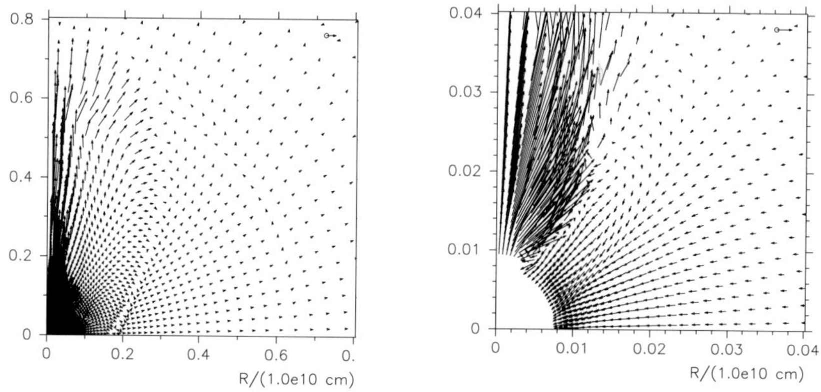

Figure 10.15. Velocity distribution of Model 40A at 1.5 second after the initiation of the jets. The right panel shows that in the central region on an expanded scale. The reference arrow at the upper right represents 2×10^9cm s^{-1}.

(1) A more massive star makes a more energetic explosion. The reason is that a more massive star has a stronger gravity to make the accretion rate higher. This is consistent with the relation seen in the hypernova branch (Fig. 10.12).

(2) A more massive star forms a more massive compact remnant. The remnant's mass increases as the accretion feeds it. The final mass $M_{\rm REM}$ reaches typically $> 5M_\odot$ for a $40M_\odot$ star, and $\sim 2M_\odot$ for a $20M_\odot$ star. The bipolar models provide the way of explosions with black hole formation in a consistent manner. Given the discovery of the evidence of a hypernova explosion that accompanied formation of a black hole of $\sim 5M_\odot$ (X-ray Nova Sco; Israelian et al. 1999; Podsiadlowski et al. 2002), it offers an interesting possibility.

Relatively high temperatures along the z-axis and low temperatures along the r-axis have significant effects on nucleosynthesis. It results in angle-dependent distribution of nucleosynthetic products as shown in Figure 10.16. The distribution of ^{56}Ni (which decays into ^{56}Fe) is elongated along the z-axis, while that of ^{16}O is concentrated in the central region. Though this feature has been already shown in Maeda et al. (2002) (see Fig. 10.7), Maeda & Nomoto (2003) have found that the central concentration of ^{16}O and the enhancement of the density there are much more significant than the previous model of Maeda et al. (2002), because of the continuous accretion from the side. Such a configuration, i.e., high velocity Fe and low velocity O, has been suggested to be responsible for the feature in the late phase spectra of SN1998bw, where the OI] 6300 was narrower than the FeII] 5200 blend (Mazzali et al. 2001; Maeda et al. 2002)

Along the z-axis, heavy isotopes which are produced with high temperatures $T_9 \equiv T/10^9\mathrm{K} \gtrsim 5$ are blown up to the surface. As a result, ^{64}Ge, ^{59}Cu, ^{56}Ni, ^{48}Cr, and ^{44}Ti (which decay into ^{64}Zn, ^{59}Co, ^{56}Fe, ^{48}Ti, and ^{44}Ca, re-

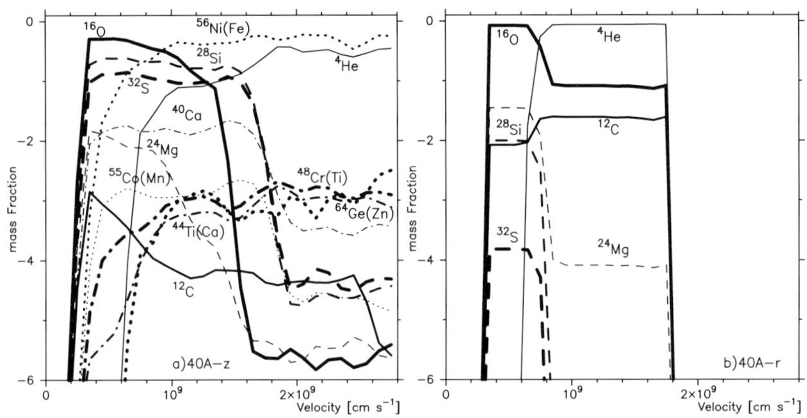

Figure 10.16. Mass fractions of selected isotopes in the velocity space along the z-axis (left panel) and along the r-axis (right panel) of Model 40A (Maeda & Nomoto 2003).

spectively) are ejected at the highest velocities. Isotopes which are synthesized with somewhat lower temperatures ($T_9 = 4 - 5$) are first pushed aside as the jets propagate, then experience circulation to flow into behind the working surface. ^{55}Co, ^{52}Fe (which decay into ^{55}Mn and ^{52}Cr, respectively), ^{40}Ca, ^{32}S, and ^{28}Si are therefore ejected at the intermediate velocities. Isotopes which are not synthesized but are only consumed during the explosion are accreted from the side. ^{24}Mg, ^{16}O, and ^{12}C occupy the innermost region at the lowest velocities.

The distribution of isotopes as a function of the velocity shows inversion as compared with conventional spherical models. This affects the overall abundance patterns in the whole ejecta as shown in Figure 10.17. As noted above, materials which experience higher T_9 are preferentially ejected along the z-axis, while materials with lower T_9 accrete from the side in the bipolar models. Zn and Co are ejected at higher velocities than Mn and Cr, so that the latter accrete onto the central remnant more easily. As a consequence, [Zn/Fe] and [Co/Fe] are enhanced, while [Mn/Fe] and [Cr/Fe] are suppressed.

Interestingly, the abundance pattaerns of [(Zn, Co, Mn, Cr)/Fe] are the same as seen in extremely metal-poor stars. The bipoar models have the effect of "mixing and fallback" which has been suggested by Umeda & Nomoto (2002, 2003).

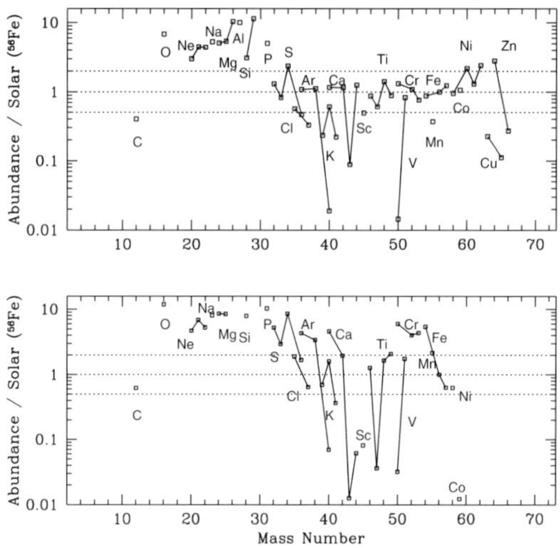

Figure 10.17. Isotropic yields of the bipolar model 40A (upper) and a spherical model (lower) with $E_{51} = 10$, $M(^{56}\mathrm{Ni}) = 0.1 M_\odot$, and $M_{\mathrm{MS}} = 40 M_\odot$ (Maeda & Nomoto 2003).

10. Extremely Metal-Poor (EMP) Stars and Faint Supernovae

Recently the most Fe deficient and C-rich low mass star, HE0107-5240, was discovered (Christlieb et al. 2002). This star has [Fe/H] $= -5.3$ but its mass is as low as $0.8\ M_\odot$. This would challenge the recent theoretical arguments that the formation of low mass stars, which should survive until today, is suppressed below [Fe/H] $= -4$ (Schneider et al. 2002).

The important clue to this problem is the observed abundance pattern of this star. This star is characterized by a very large ratios of [C/Fe] = 4.0 and [N/Fe] = 2.3, while the abundances of elements heavier than Mg are as low as Fe (Christlieb et al. 2002). Interestingly, this is not the only extremely metal poor (EMP) stars that have the large C/Fe and N/Fe ratios, but several other such stars have been discovered (Ryan 2002). Therefore the reasonable explanation of the abundance pattern should explain other EMP stars as well. We show that the abundance pattern of C-rich EMP stars can be reasonably explained by the nucleosynthesis of 20 - 130 $M_\odot$ supernovae with various explosion energies and the degree of mixing and fallback of the ejecta.

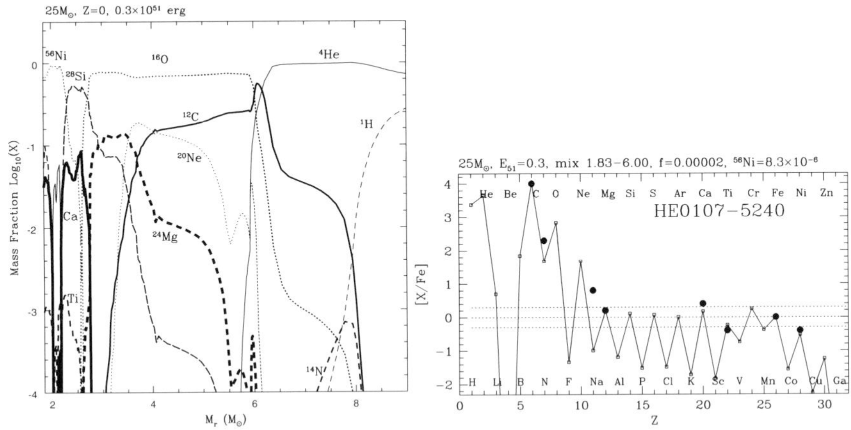

Figure 10.18. (left) The post-explosion abundance distributions for the 25 $M_\odot$ model with the explosion energy $E_{51} = 0.3$ (right). Elemental abundances of the C-rich most Fe deficient star HE0107-5240 (filled circles), compared with a theoretical supernova yield (Umeda & Nomoto 2003).

The Most Metal-Poor Star HE0107-5240

We consider a model that C-rich EMP stars are produced in the ejecta of (almost) metal-free supernova mixed with extremely metal-poor interstellar matter. We use Pop III pre-supernova progenitor models, simulate the supernova explosion and calculate detailed nucleosynthesis (Umeda & Nomoto 2003).

In Figure 10.18 (right) we show that the elemental abundances of one of our models are in good agreement with HE0107-5240, where the progenitor mass is 25 $M_\odot$ and the explosion energy $E_{51} = 0.3$ (Umeda & Nomoto 2003).

In this model, explosive nucleosynthesis takes place behind the shock wave that is generated at $M_r = 1.8\ M_\odot$ and propagates outward. The resultant abundance distribution is seen in Figure 10.18 (left), where M_r denotes the Lagrangian mass coordinate measured from the center of the pre-supernova model (Umeda & Nomoto 2003). The processed material is assumed to mix uniformly in the region from $M_r = 1.8\ M_\odot$ and 6.0 $M_\odot$. Such a large scale mixing was found to take place in SN1987A and various explosion models (e.g., Hachisu et al. 1990; Kifonidis et al. 2000). Almost all materials below $M_r = 6.0\ M_\odot$ fall back to the central remnant and only a small fraction ($f = 2\times 10^{-5}$) is ejected from this region. The ejected Fe mass is $8 \times 10^{-6}\ M_\odot$.

The CNO elements in the ejecta were produced by pre-collapse He shell burning in the He-layer, which contains 0.2 $M_\odot$ ^{12}C. Mixing of H into the He shell-burning region produces $4 \times 10^{-4}\ M_\odot$ ^{14}N. On the other hand, only a small amount of heavier elements (Mg, Ca, and Fe-peak elements) are ejected and their abundance ratios are the average in the region of $M_r = 1.8$ - $6.0\ M_\odot$. The sub-solar ratios of [Ti/Fe] $= -0.4$ and [Ni/Fe] $= -0.4$ are the results of the relatively small explosion energy ($E_{51} = 0.3$). With this "mixing and fallback", the large C/Fe and C/Mg ratios observed in HE0107-5240 are well reproduced (Umeda & Nomoto 2003).

In this model, N/Fe appears to be underproduced. However, N can be produced inside the EMP stars through the C-N cycle, and brought up to the surface during the first dredge up stage while becoming a red-giant star (Boothroyd & Sackmann 1999).

Carbon-rich EMP stars: CS 22949-037 and CS 29498-043

The "mixing and fallback" is commonly required to reproduce the abundance pattern of typical EMP stars. In Figure 10.19 (left) we show a model, which is in good agreement with CS22949-037 (Umeda & Nomoto 2003). This star has [Fe/H] $= -4.0$ and also C, N-rich (Norris et al. 2001; Depagne et al. 2002), though C/Fe and N/Fe ratios are smaller than HE0107-5240. The model is the explosion of a 30 $M_\odot$ star with $E_{51} = 20$. In this model, the mixing region ($M_r = 2.33$ - $8.56\ M_\odot$) is chosen to be smaller than the entire He core ($M_r =$ 13.1 $M_\odot$) in order to reproduce relatively large Mg/Fe and Si/Fe ratios. Similar degree of the mixing would also reproduce the abundances of CS29498-043 (Aoki et al. 2002), which shows similar abundance pattern.

We assume a larger fraction of ejection, 2%, from the mixed region for CS22949-037 than HE0107-5240, because the C/Fe and N/Fe ratios are smaller. The ejected Fe mass is 0.003 $M_\odot$. The larger explosion energy model is favored for explaining the large Zn/Fe, Co/Fe and Ti/Fe ratios (Umeda & Nomoto 2002).

Without mixing, elements produced in the deep explosive burning regions, such as Zn, Co, and Ti, would be underproduced. Without fallback the abundance ratios of heavy elements to lighter elements, such as Fe/C, Fe/O, and Mg/C would be too large. In this model, Ti, Co, Zn and Sc are still underproduced. However, these elements may be enhanced efficiently in the aspherical explosions (Maeda et al. 2002; Maeda & Nomoto 2003). Almost the same effects as the "mixing and fallback mechanism" are realized if the explosion is jet-like, although the total energy can be smaller if the beaming angle of the jet is small (Maeda & Nomoto 2003).

According to Maeda & Nomoto (2003), some bipolar models explain the existence of Fe-poor explosions with very little amount of Fe, which leave a massive central remnant with $M_{\rm REM} \sim 10 M_\odot$. Such explosions would be responsible for the formation of the carbon-rich metal-poor stars, e.g., CS22949-037 (Norris et al. 2001; Aoki et al. 2002; Depagne et al. 2002), as the bipoplar explosions with smaller $M(^{56}\rm Ni)$ lead to larger [C/O] (See also Umeda & Nomoto 2003).

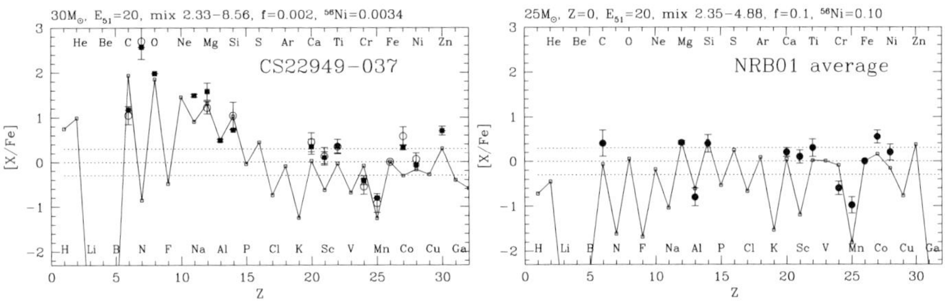

Figure 10.19. (left) Elemental abundances of CS 22949-037 (open circles for Norris et al. (2001), and solid squares for Depagne et al. (2002)), compared with a theoretical supernova yield (Umeda & Nomoto 2003). (right) Averaged elemental abundances of stars with [Fe/H] $= -3.7$ (Norris et al. 2001).

EMP Stars with a Typical Abundance Pattern

Similarly, the "mixing and fall back" process can reproduce the abundance pattern of the typical EMP stars without enhancement of C and N. Figure 10.19 (right) shows that the averaged abundances of [Fe/H] $= -3.7$ stars in Norris et al. (2001) can be fitted well with the model of 25 $M_{\odot}$ and $E_{51} = 20$ but larger fraction ($\sim 10\%$) of the processed materials in the ejecta.

SNe with Small Fe Ejection

In our model, [Fe/H] of several kinds of EMP stars can be understood in the supernova-induced star formation scheme (Audouze & Silk 1995; Shigeyama & Tsujimoto 1998; Nakamura et al. 1999). In this scheme, [Fe/H] of the second-generation stars are determined by the ejected Fe mass divided by the mass of hydrogen swept up by the supernova ejecta. As the swept-up hydrogen mass is roughly proportional to the explosion energy, Fe/H $\propto$ (M(Fe)/0.07$M_{\odot}$) / E_{51}, where M(Fe) is the ejected Fe mass. The average stars of [Fe/H] $\simeq -3.7$ (Norris et al. 2001), CS22949-037, and HE0107-5240 correspond to (M(Fe)/0.07$M_{\odot}$) / E_{51} = 0.07, 0.002, and 0.0004, respectively. This correspondence suggests that [Fe/H] of the EMP stars do not reflect the age of the stars, but rather the properties of the supernovae, such as the degree of mixing and fallback or collimation of a jet.

We have also shown that the most iron-poor star, as well as other C-rich EMP stars, is likely to be enriched by massive supernovae that are characterized by relatively small Fe ejection. Such supernovae are not hypothetical, but have been actually observed, forming a distinct class of type II supernovae ('faint supernovae') (Nomoto et al. 2002). The prototype is SN1997D, which is very underluminous and shows quite narrow spectral lines (Turatto et al. 2000):

these features are well modeled as an explosion of a $25M_\odot$ star ejecting only $2 \times 10^{-3} M_\odot$ ^{56}Ni with small explosion energy $E_{51} \sim 0.4$ (Turatto et al. 2000). SN1999br is a very similar faint supernova (Zampieri et al. 2003). On the other hand, typical EMP stars without enhancement of C and N correspond to the abundance pattern of energetic supernovae ('hypernovae'). For both cases, black holes more massive than$\sim 3 - 10M_\odot$ must be left as results of fallback, suggesting copious formation of the first black holes from the first stars. These black holes may make up some of the dark mass in the Galactic halo.

In our model, HE0107-5240 with [Fe/H]= -5.3 was formed from C- and O-enhanced gases with [(C,O)/H] ~ -1. With such enhanced C and O, the cooling efficiency is large enough to form small-mass stars. In other word, our model predicts that low-mass EMP stars with [Fe/H] < -4 are likely to have enhanced [(C, N, O)/Fe] and [Mg/Fe] in some cases. A consequence of the low-mass EMP stars being carbon-rich is that the population III stars that provided their metals are massive enough to form (the first) black holes.

11. The First Stars

It is of vital importance in current astronomy to identify the first generation stars in the Universe, i.e., totally metal-free, Pop III stars. The impact of the formation of Pop III stars on the evolution of the Universe depends on their typical masses. Recent numerical models have shown that, the first stars are as massive as $\sim 100\ M_\odot$ (Abel, Bryan, & Norman 2002). The formation of long-lived low mass Pop III stars may be inefficient because of slow cooling of metal free gas cloud, which is consistent with the failure of attempts to find Pop III stars.

If the most Fe deficient star, HE0107-5240, is a Pop III low mass star that has gained its metal from a companion star or interstellar matter (Yoshii 1981), would it mean that the above theoretical arguments are incorrect and that such low mass Pop III stars have not been discovered only because of the difficulty in the observations?

Based on the results in the earlier section, we propose that the first generation supernovae were the explosion of $\sim$ 20-130 $M_\odot$ stars and some of them produced C-rich, Fe-poor ejecta. Then the low mass star with even [Fe/H] < -5 can form from the gases of mixture of such a supernova ejecta and the (almost) metal-free interstellar matter, because the gases can be efficiently cooled by enhanced C and O ([C/H] ~ -1).

Pair-Instability Supernovae

We have shown that the ejecta of core-collapse supernova explosions of massive stars can well account for the abundances of EMP stars. We can put further constraint on the typical mass of Pop III. The abundances of all

Table 10.4. The results of the stability analysis for Pop III and Pop I stars. ○ and × represent that the star is stable and unstable, respectively. The e-folding time for the fundamental mode is shown after × in units of 10^4yr (Nomoto et al. 2002).

mass ($M_\odot$)	80	100	120	150	180	300
Pop III	○	○	○	× (9.03)	× (4.83)	× (2.15)
Pop I	○	× (7.02)	× (2.35)	× (1.43)	× (1.21)	× (1.71)

observed EMP stars including the most metal-poor one are inconsistent with the abundance pattern of pair-instability supernovae (PISNe). For typical EMP stars and CS22948-037, enrichment by PISNe cannot be consistent with the observed abundant Zn/Fe and Co/Fe ratios (Umeda & Nomoto 2002; Heger & Woosley 2002). For HE0107-5240 and other C-rich EMP stars, PISNe enrichment is difficult to account for the large C/Mg ratios. Therefore the supernova progenitors that are responsible for the formation of EMP stars are in the range of $M \sim 20 - 130\ M_\odot$, but not more massive than 130 $M_\odot$. This upper limit depends on the stability of massive stars as will be discussed below.

Stability and Mass Loss of Massive Pop III Stars

To determine the upper limit mass of the Zero Age Main Sequence (ZAMS), we analyze a linear non-adiabatic stability of massive ($80M_\odot$ - $300M_\odot$) Pop III stars using a radial pulsation code (Nomoto et al. 2002). Because CNO elements are absent during the early stage of their evolution, the CNO cycle does not operate and the star contracts until temperature rises sufficiently high for the 3α reaction to produce ^{12}C. We calculate that these stars have $X_{\rm CNO} \sim 1.6 - 4.0 \times 10^{-10}$, and the central temperature $T_c \sim 1.4 \times 10^8 K$ on their ZAMS. We also examine the models of Pop I stars for comparison.

Table 10.4 shows the results for our analysis. The critical mass of ZAMS Pop III star is $128M_\odot$ while that of Pop I star is $94M_\odot$. This difference comes from very compact structures (with high T_c) of Pop III stars.

Stars more massive than the critical mass will undergo pulsation and mass loss. We note that the e-folding time of instability is much longer for Pop III stars than Pop I stars with the same mass, and thus the mass loss rate is much lower. These results are consistent with Ibrahim, Boury, & Noels (1981) and Baraffe, Heger, & Woosley (2001). However, the absence of the indication of PISNe may imply that these massive stars above 130 $M_\odot$ undergo significant mass loss, thus evolving into Fe core-collapse rather than PISNe.

12. Concluding Remarks

In this paper, we first describe how the basic parameters of hypernova SN 1998bw are derived from observations and modeling, and discuss the properties of other hypernovae individually. These hypernovae seem to come from rather massive stars, being more massive than $\sim$ 20 - 25 $M_{\odot}$ on the main-sequence, thus forming black holes. On the other hand, there are some examples of massive SNe with only a small kinetic energy. We suggest that stars with non-rotating black holes are likely to collapse "quietly" ejecting a small amount of heavy elements (Faint supernovae). In contrast, stars with rotating black holes are likely to give rise to very energetic supernovae (Hypernovae).

We present distinct nucleosynthesis features of these two types of "black-hole-forming" supernovae. Nucleosynthesis in Hypernovae are characterized by larger abundance ratios (Zn,Co,V,Ti)/Fe and smaller (Mn,Cr)/Fe. Nucleosynthesis in Faint supernovae is characterized by a large amount of fall-back, yielding large [α/Fe]. We show that the abundance pattern of the recently discovered most Fe deficient star, HE0107-5240, and other extremely metal-poor carbon-rich stars are in good accord with those of black-hole-forming supernovae, but not pair-instability supernovae. This suggests that black-hole-forming supernovae made important contributions to the early Galactic (and cosmic) chemical evolution. Finally we discuss the nature of First (Pop III) Stars.

Acknowledgments

This work has been supported in part by the grant-in-Aid for Scientific Research (14047206, 14540223) of the Ministry of Education, Science, Culture, and Sports in Japan.

References

Abel, T., Bryan, G.L., & Norman, M.L. 2002, Science, 295, 93
Aoki, W., Ryan, S.G., Beers, T.C., & Ando, H. 2002, ApJ, 567, 1166
Aloy, M.A., Müller, E., Ibáñez, J.M., Martí, J.M., MacFadyen, A. 2000, ApJ, 531, L119
Arnett, D. 1982, ApJ, 253, 785
Arnett, D. Bahcall, J.N., Kirshner, R.P., Woosley, S.E. 1989, ARA&A, 27, 629
Arnett, D. 2001, in *Supernovae and Gamma-Ray Bursts (Proceedings of the Space Telescope Science Institute Symposium, Baltimore, USA)*, eds. M. Livio, N. Panagia, K. Sahu (Cambridge), 250
Audouze, J., & Silk, J. 1995, ApJ, 451, L49
Axelrod, T.S. 1980, Ph.D. thesis, University of California
Baraffe, I., Heger, A., & Woosley, S.E. 2001, ApJ, 550, 890
Baron, E., Young, T.R., Branch, D. 1993, ApJ, 409, 417
Berger, E., Kulkarni, S.R., Chevalier, R.A. 2002, ApJ, 577, L5
Blake, L.A.J., Ryan, S.G., Norris, J.E., Beers, T.C. 2001, Nucl. Phys. A. 688, 502
Blandford, R.D., & Znajek, R.L. 1977, MNRAS, 179, 433
Bloom, J.S., et al. 1999, Nature, 401, 453
Bloom, J.S., et al. 2002, ApJ, 572, L45
Boothloyd, A.I., & Sackmann, I.-J. 1999, ApJ, 510, 217
Branch, D. 2001, in *Supernovae and Gamma-Ray Bursts (Proceedings of the Space Telescope Science Institute Symposium, Baltimore, USA)*, eds. M. Livio, N. Panagia, K. Sahu (Cambridge), 96
Chevalier, R.A. 1989, ApJ, 346, 847
Christlieb, N., et al. 2002, Nature, 419, 904
Clocchiatti, A., & Wheeler, J.C. 1997, ApJ, 491, 375
Colgate, S.A., & Petschek, A.G. 1979, ApJ, 229, 682
Colgate, S.A., Petschek, A.G., Kriese, J.T. 1980, ApJ, 237, L81
Deng, J., Hatano, K., Nakamura, T., Maeda, K., Nomoto, K., Nugent, P., Aldering, G., & Branch, D. 2001, in *New Century of X-ray Astronomy, ASP Conference Series, 251*, eds. H. Inoue, & H. Kunieda (ASP; San Francisco), 238
Depagne, E., et al. 2002, A&A, 390, 187
Ebisuzaki, T., Shigeyama, T., & Nomoto, K. 1989, ApJ, 344. 65
Ergma, E., & van den Heuvel, E.P.J. 1998, A&A, 331, L29
Filippenko, A.V., & Chornock, R. 2002, IAU Circ., 7825
Fransson, C., & Kozma, C 1993, ApJ, 408, L25
Fryer, C.L., & Warren, M.S. 2002, ApJ, 574, L65

Fynbo, J.U. 2000, ApJ, 542, L89
Galama, T.J., et al. 1998, Nature, 395, 670
Galama, T.J., et al. 2000, ApJ, 536, 185
Gal-Yam, A., Ofek, E.O., & Shemmer, O. 2002, MNRAS, 332, L73
Garnavich, P., et al. 1997, IAU Circ., 6798
Garnavich, P., et al. 2002, ApJ, 582, 924
Germany, L.M., Reiss, D.J., Schmidt, B.P., Stubbs, C.W., Sadler, E.M. 2000, ApJ, 533, 320
Hachisu, I., Matsuda, T., Nomoto, K., Shigeyama, T. 1990, ApJ, 358, L57
Hachisu, I., Matsuda, T., Nomoto, K., Shigeyama, T. 1991, ApJ, 368, L27
Hamuy, M. 2003, ApJ, 582, 905
Hashimoto, M., Nomoto, K., Shigeyama, T. 1989, A&A, 210, L5
Hatano, K., Branch, D., Nomoto, K., Deng, J.S., Maeda, K., Nugent, P., Aldering, G. 2001, BAAS, 198, 3902
Heger, A., & Woosley, S.E. 2002, ApJ, 567, 532
Hirose, Y. 2002, IAU Circ., 7810
Ibrahim, A., Boury, A., & Noels, A. 1981, A&A, 103, 390
Israelian, G., et al. 1999, Nature, 401, 142
Iwamoto, K., Nomoto, K., Höflich, P., Yamaoka, H., Kumagai, S., Shigeyama, T. 1994, ApJ, 437, L115
Iwamoto, K. et al. 1998, Nature, 395, 672
Iwamoto, K. et al. 2000, ApJ, 534, 660
Janka, H.-T., Müller, E. 1994, A&A, 290, 496
Kawabata, K.S., Jeffery, D., Kosugi, G., Sasaki, T., et al. 2002, ApJ, 580, L39
Kay, L.E., Halpern, J.P., Leighly, K.M., Heathcote, S., Magalhaes, A.M., Filippenko, A.V. 1998, IAU Circ., 6969
Khokhlov, A.M., Höflich, P.A., Oran, E.S., Wheeler, J.C., Wang, L., Chtchelkanova, A.Yu. 1999, ApJ, 524, L107
Kifonidis, K., Plewa, T., Janka, H-Th., Müller, E. 2000, ApJ, 531, L123
Kinugasa, K., et al. 2002, ApJ, 577, L97
Kippen, R.M. et al. 1998, GCN Circ., 67
Knop, R., Aldering, G., Deustua, S., et al. 1999, IAU Circ., 7128
Leonardo, D.C., Filippenko, A.V., Chornock, R., Foley, R.J., 2002, PASP, 114, 1333
MacFadyen, A.I., & Woosley, S.E. 1999, ApJ, 524, 262
MacFadyen, A.I., Woosley, S.E., Heger, A. 2001, ApJ, 550, 410
Maeda, K., Nakamura, T., Nomoto, K., Mazzali, P.A., Patat, F., Hachisu, I. 2002, ApJ, 565, 405
Maeda, K., Mazzali, P.A., Deng, J., Nomoto, K., Yoshii, Y., Tomita, H., Kobayashi, Y. 2003, ApJ, 593, in press (astro-ph/0305182)

Maeda, K., & Nomoto, K. 2003, ApJ, submitted (astro-ph/0304172)
Matheson, T., Filippenko, A.V., Li, W., Leonard, D.C., Shields, J.C. 2001, AJ, 121, 1648
Mazzali, P.A., Iwamoto, K., Nomoto, K. 2000, ApJ, 545, 407
Mazzali, P.A., Nomoto, K., Patat, F., Maeda, K. 2001, ApJ, 559, 1047
Mazzali, P.A., et al. 2002, ApJ, 572, L61
McKenzie, E.H., & Schaefer, B.E. 1999, PASP, 111, 964
McWilliam, A., Preston, G.W., Sneden, C., Searle, L. 1995, AJ, 109, 2757
Meikle, P., Lucy, L., Smartt, S., et al. 2002, IAU Circ., 7811
Nagataki, S., Hashimoto, M., Sato, K., & Yamada, S. 1997, ApJ, 486, 1026
Nakamura, T. 1998, Prog. Theor. Phys, 100, 921
Nakamura, T., Umeda, H., Nomoto, K., Thielemann, F.-K., Burrows, A. 1999, ApJ, 517, 193
Nakamura, T., Mazzali, P.A., Nomoto, K., Iwamoto, K. 2001a, ApJ, 550, 991
Nakamura, T., Umeda, H., Iwamoto, K., Nomoto, K., Hashimoto, M., Hix, W.R., Thielemann, F.-K. 2001b, ApJ, 555, 880
Nakano, S., Kushida, R., Li, W. 2002, IAU Circ., 7810
Nomoto, K., & Hashimoto, M. 1988, Phys.Rep., 256, 173
Nomoto, K., Suzuki, T., Shigeyama, T., Kumagai, S., Yamaoka, H., Saio, H. 1993, Nature, 364, 507
Nomoto, K., et al. 1994, Nature, 371, 227
Nomoto, K., Iwamoto, K., Suzuki, T. 1995, Phys. Rep., 256, 173
Nomoto, K., et al. 2001a, in *Supernovae and Gamma-Ray Bursts (Proceedings of the Space Telescope Science Institute Symposium, Baltimore, USA)*, eds. M.Livio, N.Panagia, K.Sahu (Cambridge), 144
Nomoto, K., Maeda, K., Umeda, H., Nakamura, T. 2001b, in *The Influence of Binaries on Stellar Populations Studies*, ed. D. Vanbeveren (Kluwer), 507 (astro-ph/0105127)
Nomoto, K., Maeda, K., Umeda, H., Ohkubo, T., Deng, J., Mazzali, P. 2003, in *IAU Symp 212, A Massive Star Odyssey, from Main Sequence to Supernova*, eds. van der Hucht, A. Herrero, & C. Esteban (ASP, San Francisco 2003), 395 (astro-ph/0209064)
Norris, J.E., Ryan, S.G., & Beers, T.C. 2001, ApJ, 561, 1034
Ohkubo, T., Umeda, H., & Nomoto, K., 2003, Nuc.Phys. A. 718, 632c
Patat, F. et al. 2001, ApJ, 555, 917
Pinto, P., & Eastman, R. 2000, ApJ, 530, 744
Podsiadlowski, Ph., Nomoto, K., Maeda, K., Nakamura, T., Mazzali, P.A., Schmidt, B. 2002, ApJ, 567, 491
Primas, F., Reimers, D., Wisotzki, L., Reetz, J., Gehren, T., Beers, T.C. 2000, in *The First Stars*, ed. A. Weiss, et al. (Springer), 51

Reichart, D.E. 1999, ApJ, 521, L111
Rigon, L., Turatto, M., Benetti, S., et al. 2003, MNRAS, 340, 191
Ryan, S.G., Norris, J.E., Beers, T.C. 1996, ApJ, 471, 254
Ryan, S.G. 2002, in *CNO in the Universe*, eds. C. Charbonnel, D. Schaerer, & G. Meynet, in press (astro-ph/0211608)
Schaller, G., Schaerer, D., Meynet, G., Maeder, A. 1992, A&AS, 96, 269
Schneider, R., Ferrara, A., Natarajan, P., & Omukai, K. 2002, ApJ, 571, 30
Sharina, M.E., Karachentsev, I.D., Tikhonov, N.A. 1996, A&AS, 119, 499
Shigeyama, T., & Nomoto, K. 1990, ApJ, 360, 242
Shigeyama, T., Suzuki, T., Kumagai, S., Nomoto, K., Saio, H., Yamaoka, H. 1994, ApJ, 420, 341
Shigeyama, T., & Tsujimoto, T. 1998, ApJ, 507, L135
Shimizu, T.M., Ebisuzaki, T., Sato, K., Yamada, S. 2001, ApJ, 552, 756
Smartt, S.J., Vreeswijk, P., Ramirez-Ruiz, E., et al. 2002, ApJ, 572, L147
Sneden, C., Gratton, R.G., Crocker, D.A. 1991, A&A, 246, 354
Sollerman, J., Kozma, C., Fransson, C., Leibundgut, B., Lundqvist, P., Ryde, F., Woudt, P. 2000, ApJ, 537, L127
Stanek, K.Z., et al. 2003, ApJ, in press (astro-ph/0304173)
Stathakis, R.A., et al. 2000, MNRAS, 314, 807
Suzuki, T., & Nomoto, K. 1995, ApJ, 455, 658
Swartz, D.A., & Wheeler, J.C. 1991, ApJ, 379, L13
Takada-Hidai, M., Aoki, W., Zhao, G. 2002, PASJ, 54, 899
Terlevich, R., Tenorio-Tagle, G., Franco, J., Melnick, J. 1992, MNRAS, 255, 713
Terlevich, R., Fabian, A., Turatto, M. 1999, IAU Circ., 7269
Thielemann, F.-K., Nomoto, K., Hashimoto, M. 1996, ApJ, 460, 408
Thorsett, S.E., & Hogg, D.W. 1999, GCN Cir., 197
Turatto, M., et al. 1998, ApJ, 498, L129
Turatto, M., Suzuki, T., et al. 2000, ApJ, 534, L57
Umeda, H., & Nomoto, K. 2002, ApJ, 565, 385
Umeda, H., Nomoto, K., Tsuru, T.G., Matsumoto, H. 2002, 578, 855
Umeda, H., & Nomoto, K. 2003, Nature, 422, 871
van den Heuvel, E.P.J. 1994, in *Interacting Binaries*, ed. H. Nussbaumer & A.Orr (Berlin:Springer Verlag), 263
Wang, L., & Wheeler, J.C. 1998, ApJ, 504, L87
Wang, L., Baade, D., Fransson, C., et al. 2002, ApJ, submitted (astro-ph/0206386)
Wheeler, J.C. 2001, ApJ, 504, 87
Woosley, S.E., Eastman, R.G., Weaver, T.A., Pinto, P.A. 1994, ApJ, 429, 300
Woosley, S.E., Langer, N., & Weaver, T.A. 1995, ApJ, 448, 315
Woosley, S.E., & Weaver, T.A. 1995, ApJS, 101, 181

Woosley, S.E., Eastman, R., Schmidt, B. 1999, ApJ, 516, 788
Yoshii, Y. 1981, A&A, 97, 280
Young, T., Baron, E., Branch, D. 1995, ApJ, 449, L51
Zampieri et al. 2003, MNRAS, 338, 711

Chapter 11

THE COLLAPSAR ENGINE FOR GRBS AND HYPERNOVAE

Weiqun Zhang and
Astronomy Department, UC Santa Cruz, Santa Cruz, CA 95064

Chris L. Fryer
Theoretical Astrophysics, Los Alamos National Laboratories, Los Alamos, NM, 87545

Abstract

For most collapsing stars, neutrinos from the proto-neutron star are able to drive a strong supernova explosion with a neutron star remnant. But very massive stars ($> 20M_\odot$) will ultimately form black holes. These stars have the potential to drive enormous explosions by tapping the energy in a black hole accretion disk. The engine behind this black hole accretion disk explosion has been termed "collapsar". In this chapter, we review the collapsar mechanism for gamma-ray bursts, hypernovae, and X-ray flashes from progenitor formation through jet propogation.

1. Introduction

A star that collapses down to a black hole need not end its life in a whimper. Woosley (1993) proposed that when a rotating massive star collapsed to a black hole, the accretion energy of the high angular momentum material around this black hole could easily power a robust explosion. Woosley (1993) proposed this energy source to explain gamma-ray bursts (GRBs). Little did he know that this model would not only become the favored model for long-duration gamma-ray bursts, but it would also become the model for the related, but undiscovered at that time, hypernovae discussed by Nomoto et al. (Ch. 10). This miniature version of a an active galactic nuclei relied on the enormous accretion rates (up to $1M_\odot s^{-1}$) from the imploding star to drive a relativistic explosion out through the star. Studies by MacFadyen & Woosley (1999) found that the star

C.L. Fryer (ed.), Stellar Collapse, 327–355.

collimated the explosion into a narrow relativistic jet which they argued met the energy and Lorentz-factor requirements of the observed sample of GRBs. Winds from the disk and shocks from the outflowing jet ultimately eject the stellar envelope, producing a supernova-like explosion along with the GRB.

This model led to a number of predictions: GRBs are formed from massive stars and hence should be associated with star-forming regions, GRBs should be beamed, and GRBs should have supernova-like features in their lightcurves. Even as MacFadyen & Woosley were writing up their results, supernova 1998bw, also known as GRB 980425, occured, providing the first evidence that GRBs have supernova-like light curves (Galama et al. 1998). This quickly skyrocketed the collapsar model as one of the leading models for GRBs and stellar collapse has remained the primary model for long duration GRBs to this date. Since these observations, a number of additional observations of GRBs with supernova-like outbursts have been observed: GRB 980326 (Bloom et al. 1999), GRB011121/SN2001ke (Garnavich et al. 2003, Bloom et al. 2002), and now GRB030329/SN2003dh (Stanek et al. 2003). With the association with supernovae, there has been an insurgence in the study of extremely energetic supernovae (or hypernovae) and it is likely that hypernovae and GRBs are related (see Nomoto et al. Ch. 10).

On top of these GRB/supernova associations, a great deal of indirect data also supports a massive star origin for long-duration gamma-ray bursts. For instance, it appears that long-duration GRBs occur in star-bursting galaxies (Hogg & Fruchter 1999) and in star forming regions (e.g. Holland & Hjorth 1999) of these galaxies. Some of the afterglow light-curves for GRBs imply that the GRB is moving through wind-swept media (e.g. Li & Chevalier 2001). Finally, evidence is growing suggesting that GRBs are indeed beamed (Frail et al. 2001) at the same level (within observation errors) predicted by MacFadyen & Woosley (1999).

As this observational evidence continues to amass, the GRB and supernova community have begun to combine forces to investigate the nature of GRBs and Hypernovae and, with these investigations, have begun to the better determine the relationship between these objects and normal supernovae. It is our belief that GRBs and Hypernovae are produced by those stars that collapse to form black holes (either through direct collapse or fallback) that retain sufficient angular momentum to form an accretion disk from which energy can be extracted to drive these energetic explosions. In this chapter, we discuss collapsar progenitors, the engine driving these explosions, and the resultant collimated jets that arise from these objects.

2. Collapsar Progenitor Models

The collapsar model carries with it several constraints on the collapsing progenitor star. First and foremost, the star must not explode in a normal strong supernova explosion. Such stars do not have enough fallback to collapse to black holes. The collapsing star must either weakly explode and endure rapid fallback (MacFadyen, Woosley, & Heger 2001) or fail to explode (MacFadyen & Woosley 1999) under the neutrino-driven supernova mechanism. Core-collapse simulations by Fryer (1999) predicted that above $\sim 20 M_\odot$, stars would only explode weakly and form black holes through fallback and above 40-45$M_\odot$, stars would collapse directly to black holes. Note that this $20 M_\odot$ limit agrees with the expectations by Nomoto et al. (Ch. 10) for the beginning of the hypernova branch of stellar explosions. The fact that the minimum stellar mass for black hole formation coincides with the beginning of the hypernova branch suggests that a black hole engine (e.g. collapsar) may be responsible for these explosions. But if we are limited to stars which collapse to form black holes, hypernovae and GRBs can make up, at most, 10-20% of all stellar explosions. Indeed, as we shall see, further constraints will lower this limit even further.

For GRBs, an additional constraint is the burst duration: $<$100-1000 s. Any engine behind these bursts must not last much longer than this timescale. However, the engine must power a jet long enough for it to plow through the star which is collapsing around it (otherwise the jet is not able to clear its path and produce baryon-free, relativistic outflows). To satisfy both these constraints requires that the jet take only 100-1000 s to break through the star. This is too short a time for even a relativistic jet to plow through the hydrogen envelope of a red supergiant star. Instead, it has been argued (MacFadyen & Woosley 1999) that the star must have lost its hydrogen envelope ($\equiv$ a bare helium star). So any progenitor scenario must be able to eject the bulk of the hydrogen envelope of these stars.

Lastly, the star must be rotating sufficiently rapidly at collapse that there is enough angular momentum to form a disk in the collapsed material (see §3 for more details). This is one of the most stringent constraints and is currently the most difficult aspect of making collapsar progenitors. These three constraints:

- **I:** Star must form Black Hole: Initial Stellar Mass Above $\sim 20 M_\odot$,
- **II:** No Hydrogen Envelope, and
- **III:** Rapid Rotation

all must be satisfied by any prospective GRB progenitor (only I and III apply for hypernovae).

The first complete study of the progenitors for the collapsar models was done by Fryer, Woosley, & Hartmann (1999). They devised 3 mechanisms to

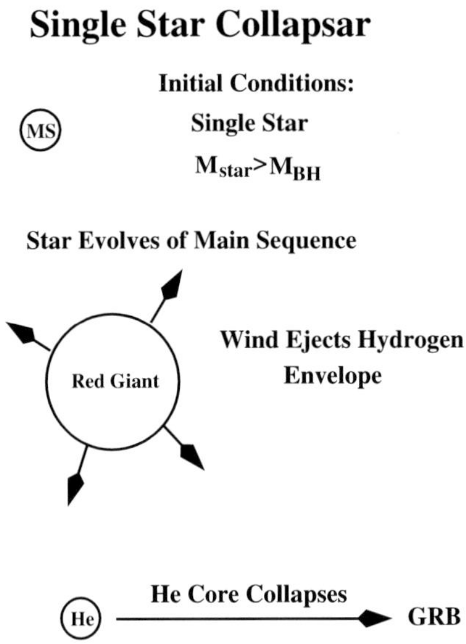

Figure 11.1. Evolutionary phase for a single star GRB progenitor. This star must have a combination of high-metallicity and high mass to lose its hydrogen envelope before collapse. It must then retain its core angular momentum to then produce a disk at collapse. The largest weakness of this model is due to the angular momentum loss during evolution.

form collapsar GRBs. The first (Fig. 11.1) assumed that the progenitor was a massive, rapidly rotating, single star. Even at moderate metallicities, stars with initial masses above 30-40 $M_\odot$ will eject their hydrogen envelope through wind mass-loss. Hence, single mass stars easily satisfy conditions I and II. However, mass loss from winds tends to also reduce the angular momentum of the star and it is now generally agreed that this scenario will not have enough angular momentum to make suitable black hole accretion disks.

To produce sufficient angular momentum in the collapsing star, Fryer et al. (1999) turned to binary systems to form collapsars. The first of such scenarios invokes a progenitor consisting of a close binary system. In this binary, when the primary component (most massive star) moves off the main sequence to become a giant, it envelopes its companion (Fig. 11.2). During the ensuing common envelope phase, the companion ejects all/most of the primary's hydro-

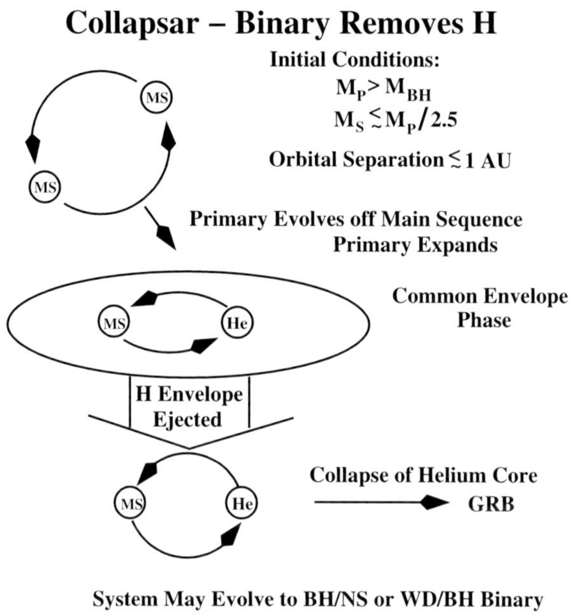

Figure 11.2. The most common binary progenitor for GRBs utilizes a binary companion to remove the massive star's hydrogen envelope. This may alleviate the angular momentum loss, as there will be less angular momentum loss in winds.

gen envelope, leaving it as a bare helium core. For such systems, the primary does not have to have strong winds to remove the hydrogen envelope, and the mass constraint drops down to $20\,M_\odot$. If the uncovered helium star does not lose too much mass through winds, it will ultimately form a black hole. Since the star lost its mass through binary orbital energy instead of a wind, it will not lose angular momentum as it sheds its hydrogen envelope. Such a system could end up with a rapidly spinning core. However, recent stellar evolution calculations with recipes for magnetic fields produced by shear flows suggest that even these stars would not have high enough rotation rates to produce collapsar disks (Heger et al. 2003)

Fryer et al. (1999) proposed another binary scenario which may overcome the magnetic field barriers set by Heger et al. (2003). In this scenario, the two stars evolve roughly at the same time and both stars merge (Constraint I requires that the merged helium core needs to be at least $\sim 7\ M_\odot$), ejecting

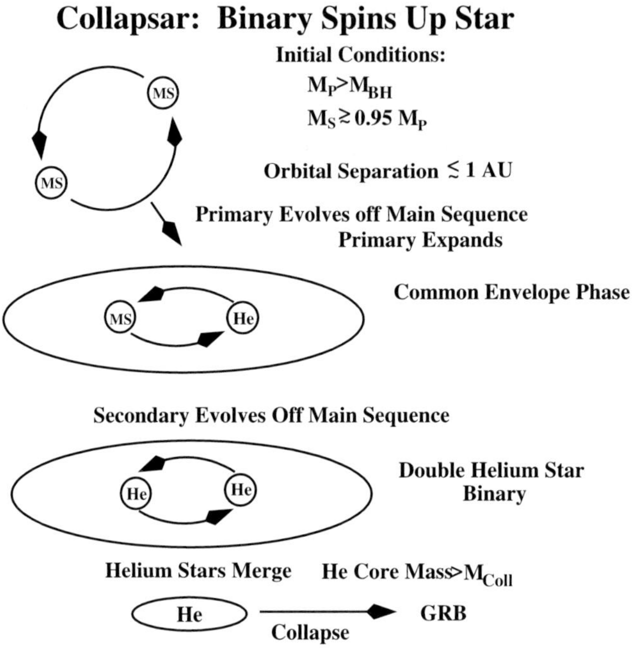

Figure 11.3. In rare cases, the two binary stars may be so close in initial mass that they evolve off the main sequence roughly at the same time. In this scenario, the primary star evolves off the main sequence. But before it collapses, its companion also evolves off the main sequence and the two cores coalesce in their combined hydrogen envelopes. If, in the ensuing common envelope phase, these two stars merge, they produce a rapidly spinning core. Although rare, this binary merger scenario provides the most natural solution to the angular momentum problem.

their combined hydrogen envelopes (Fig. 11.3). The hope is that the rapid spin-up of the star during the the merger of the two helium cores will produce a rapidly spinning core that would retain enough of its angular momentum to produce a disk upon collapse. However, much more detailed models must be performed to verify this scenario.

All of the above progenitors rely upon a massive star which eventually will collapse to a black hole. However, a similar GRB progenitor, dubbed the "He Merger" progenitor (Fryer & Woosley 1998) can also produce collapsar-like explosions and is often included in collapsar progenitors. In this model, Fryer & Woosley (1998) argued that a compact object (neutron star or black hole) in a binary with a massive companion will merge with its companion when the companion leaves the main sequence and expanded into a giant (Fig. 11.4). The compact object would spiral into the center of its companion, accreting rapidly until it becomes a black hole (if it was not a black hole at the onset of inspiral). The inspiral spins up the companion star, easily providing the rotation necessary for the formation of a collapsar disk. Zhang & Fryer (2001) have modeled this inspiral process and found that the inspiral did spin up the star (Fig. 11.5), perhaps even too much (Narayan, Piran, & Kumar 2001). Whether the spin rate is too much will depend upon the compact object's binary companion and initial spin, the compact objects magnetic field, and the field of the companion star. Again, this progenitor begs further study.

Notice that these progenitors have increasingly complicated evolutionary paths. One might wonder if any of these scenarios occur frequently enough to explain the observed GRBs and hypernovae. But remember, these events are fairly rare, occurring roughly once per every 1000 supernovae, or 10^{-5} per year in a Milky Way-massed galaxy. Even though the scenarios are becoming quite specific, within the errors of population synthesis studies, the rates of formation of all of them can exceed this value (Fryer et al. 1999).

To summarize, the current list of possible progenitors of collapsars are:

- **Single Star:** Single star progenitor with strong winds. Satisfies constraints I and II, but not III. It is very unlikely that the cores of these stars will rotate fast enough.

- **Common Envelope:** Binary system where the Hydrogen envelope is lost due to common envelope phase. Satisfies I and II and may satisfy III, depending on stellar model. Most common progenitor.

- **Core Merger** The helium cores of 2 massive stars merge during the common envelope phase. Satisfies constraints I, II, and probably III (but more studies are needed). Highly specified criteria make it difficult, but not impossible, to produce enough collapsars.

- **He Star-Compact Object Merger** A compact object (neutron star or black hole) merges with its companion. Satisfies constraints I, II, and III (but may make too much angular momentum). Highly specified criteria make it difficult, but not impossible, to produce enough collapsars.

3. Disk Formation and Accretion Energy

In §2, we listed a number of paths in stellar evolution that may produce stars that will collapse down to black holes and with sufficient angular momentum to form a disk when the star collapses. In this section, we discuss the conditions required to form an accretion disk and extract the disk accretion energy to drive a jet through the star.

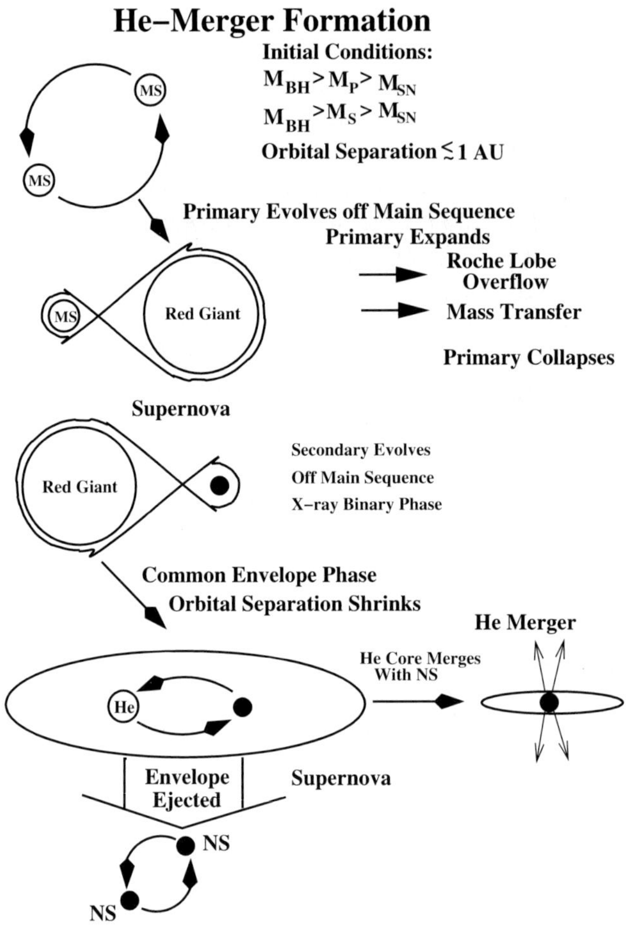

Figure 11.4. Most collapsar progenitors assume the collapse of a massive star. However, if a neutron star or black hole merges with a companion star, it will have many of the same features to the true collapsar progenitors and thes "He-mergers" have been lumped into the list of collapsar progenitors. In this progenitor scenario, the primary star evolves in a close binary system, possibly going through mass transfer or a common envelope phase, but not merging with its companion. It collapses to form a black hole or neutron star. When the companion then evolves off the stellar main-sequence, the binary can go through a second mass transfer stage. If the primary's compact remnant merges with its companion, it will spin up the companion's helium core which will then accrete onto the compact remnant through a disk. This model may actually have too much angular momentum.

The formation of an accretion disk/black hole system which will accrete fast enough to produce gamma-ray bursts requires considerable fine-tuning. There must be enough angular momentum to hang up material in a disk:

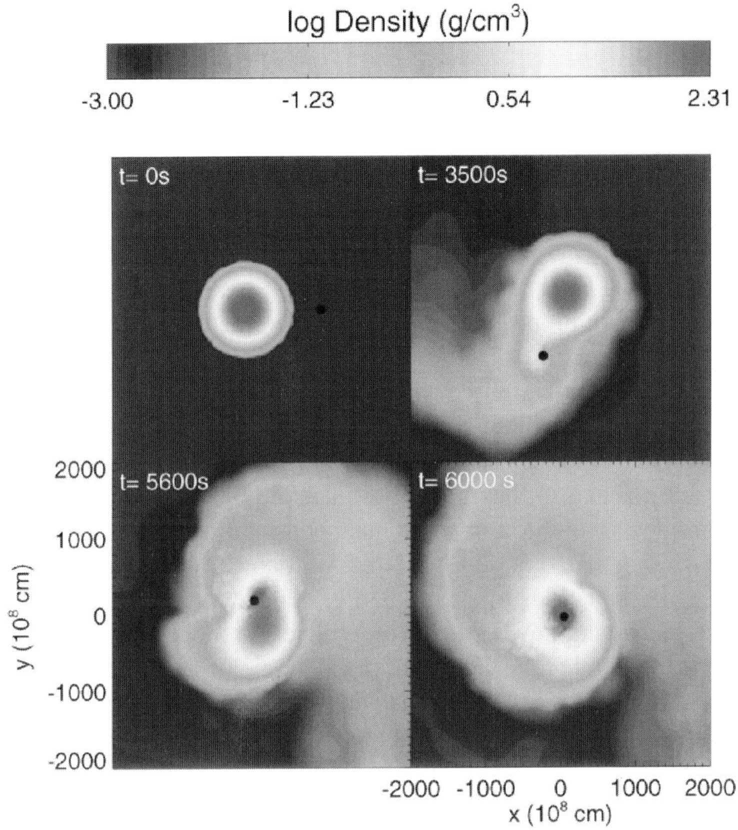

Figure 11.5. Simulation of the merger of a helium star with a compact remnant. Within a few orbit times, the compact remnant spirals into the center of the companion helium star, spinning up that star. The resultant angular momentum may be too high to produce a fast-accreting disk around the black hole (Narayan, Piran, & Kumar 2001).

$$j > GM_{\rm BH}/c \gtrsim 10^{16}{\rm cm}^2{\rm s}^{-1} \tag{11.1}$$

where G is the gravitational constant, $M_{\rm BH} \approx 3M_{\odot}$ is the black hole mass, and c is the speed of light. But the disk must not have too much angular momentum either. Popham et al. (1999) found that the accretion rate must be very high for the neutrino annihilation engine to work (these constraints might be more relaxed for a magnetic field mechanism) and hence the disk must be fairly compact. In addition, if the disk has a lot of angular momentum, it will lose much of its mass in a wind that may prematurely disrupt the star (MacFadyen, Woosley, & Heger 2001; Narayan et al. 2001). The accretion time is roughly set to the orbital period where the material hangs up divided by the disk viscosity ($\alpha \approx 0.1$), and by limiting the accretion time to $< 100\,$s, one can get a rough upper limit on the angular momentum:

$$j < (100G^2 M_{\rm BH}\alpha/2\pi)^{1/3} \approx 3 \times 10^{17} {\rm cm}^2 {\rm s}^{-1}. \quad (11.2)$$

Such a narrow range in angular momentum insures that collapsars will be rare events, but even if 1% of black hole forming stars form collapsar outbursts, we would have more than enough explosions to explain the GRB, and probably the hypernova, observations. More details on the disk evolution can be found in Popham et al. (1999) and Narayan et al. (2001).

Two methods have been proposed to extract the thermal and kinetic energy of these disks: neutrino annihilation and magnetic dynamos. A number of magnetic field dynamos have been proposed which will extract kinetic energy of the disk or rotational energy from the black hole to drive a jet (see Popham et al. 1999 for a summary). Most of these mechanisms rely upon the poorly understood jet generation mechanism behind active galactic nuclei (AGN). Almost certainly, AGN produce jets through some magnetic-field driven mechanism (e.g. Blandford & Znajek 1977), but how this mechanism works physically is still unknown. Extrapolations of such a poorly understood mechanism are dangerous at best, and quantitative estimates for the jets produced by magnetic fields can not be made at this time. Even so, magnetic fields are very likely to be the medium through which disk energy is converted into a jet for GRBs.

However, neutrino annihilation above the disk can be calculated and these calculations suggest that neutrino annihilation alone may well explain the observed bursts. As an example of how an accretion disk explosion occurs, let's look at the neutrino case in more detail.

Formation of Jets: A Neutrino-Driven Example

Just as in core-collapse supernovae, a collapsar explosion is launched when the pressure in the energy gain region overcomes the pressure of the infalling stellar material. The physics behind the neutrino driven mechanism is fairly straightforward and a reasonably accurate picture can be obtained semi-analytically (Fryer & Meszaros 2003).

The accretion disk that forms around the black hole can reach extremely high densities ($> 10^{11} {\rm g\,cm}^{-3}$) and temperatures ($> 10$MeV). The neutrino flux emitted under these conditions can exceed that in core-collapse supernovae ($> 10^{53} {\rm ergs\,s}^{-1}$). Some of this neutrino energy is absorbed above the disk by the infalling stellar material, slowing down the infall. More importantly, neutrino-annihilation above the disk can inject considerable energy and momentum and can halt the infalling material and cause it to turn around in a collapsar explosion. Neutrino-annihilation is not very efficient in core-collapse supernovae, but in collapsars, where neutrino trajectories are more nearly head-on and the neutrino fluxes are extremely high, the neutrino/anti-neutrino cross-section can become large enough to dominate the energy deposition above the disk and drive an explosion.

As the star collapses, the high angular momentum material along the rotation plane slows down and builds up in a disk around the black hole. The low-angular momentum material in the rotation axis falls directly into the black hole. As the low angular momentum material accretes, the region along the rotation axis is cleared. As the density decreases, it becomes easier and easier to drive an explosion in this evacuated region. As a shell of matter falls towards the black hole, it is supported by the momentum of the scattered and absorbed neutrinos:

$$a_\nu \approx \frac{\kappa_t dr}{m_{\text{shell}}} \frac{L_\nu}{c} \approx \frac{1.5 \times 10^{-17} (kT_{\nu_e}/4\text{MeV})^2 L_\nu}{4\pi r^2 c}, \tag{11.3}$$

where L_ν is the neutrino luminosity, $m_{\text{shell}} = 4\rho\pi r^2 dr$ is the mass, dr the thickness, r the radius of the shell, c is neutrino velocity $\approx$ the speed of light.

The corresponding acceleration from neutrino annihilation requires detailed models of the black hole accretion disk system. The disk models and annihilation calculations from Popham et al. (1999) led to a value for the energy deposited along a surface per unit path length ($[\dot{e}] = \text{ergs}\,\text{s}^{-1}\,\text{cm}^{-1}$). With this deposition rate $\dot{e}(r)$, which is a function of height above the disk, we can calculate the acceleration due to neutrino annihilation inside a $30°$ cone along the rotation axis:

$$a_{\nu\bar{\nu}} \approx \frac{\dot{e}(r) dr}{c} \frac{1}{m_{\text{shell}}} \approx \frac{\dot{e}(r)}{\pi c r^2 \rho}. \tag{11.4}$$

The onset of the explosion occurs when $a_g + a_\nu + a_{\nu\bar{\nu}} > 0$ where $a_g = -GM_{\text{BH}}/r^2$ is the gravitational acceleration with gravitational constant G, and black hole mass M_{BH}[1]. Using equations 4 and 5, the acceleration condition for the onset of the explosion can be translated into a threshold condition on the density:

$$\rho_{\text{crit}} < \frac{4\dot{e}(r)}{4\pi G M_{\text{BH}} c - 1.5 \times 10^{-17} (kT_{\nu_e}/4\text{MeV})^2 L_\nu}. \tag{11.5}$$

For our estimates, we take the radii where the energy deposition from neutrino annihilation peaks (~20 km). Figure 6 gives critical densities for set values of $\dot{e}$ versus an effective luminosity $[(kT_{\nu_e}/4\text{MeV})^2 L_\nu]$. The critical density rises sharply as the "Eddington" luminosity for neutrinos is reached. The data ($\dot{e}$ at 20 km and L_ν) for a series of disks from Popham et al. (1999) are also shown for comparison.

These critical densities correspond to an accretion rate inside the $30°$ cones along the rotation axes of

$$\dot{M}_{\text{crit}} = 0.536 \pi r^2 \rho_{\text{crit}} v_{\text{ff}} \tag{11.6}$$

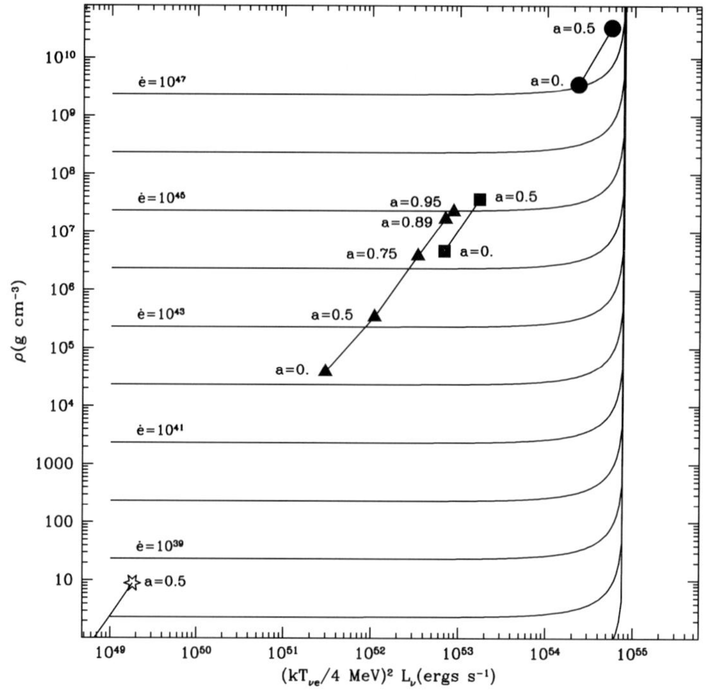

Figure 11.6. Critical densities along the pole versus effective luminosity $[(kT_{\nu_e}/4\text{MeV})^2 L_\nu]$ for set values of the annihilation deposition energy ($\dot{e}$). At high neutrino luminosities, the critical density rises sharply as the "Eddington" luminosity for neutrinos is reached. The data ($\dot{e}$ at 20 km and L_ν) for a series of disks from Popham et al. (1999) are placed on for comparison (tessalated hexagons: $\dot{M}_D = 0.01 M_\odot s^{-1}$, triangles: $\dot{M}_D = 0.1 M_\odot s^{-1}$, squares: $\dot{M}_D = 1 M_\odot s^{-1}$, circles: $\dot{M}_D = 10 M_\odot s^{-1}$).

where $r = 20\,\text{km}$ and $v_{\text{ff}} = \sqrt{2GM_{\text{BH}}/r}$. The accretion rates for massive stars are initially much higher than these critical values, but the accretion rate decreases as material from increasingly outward layers of the star falls onto the black hole. In figure 7 we plot the critical accretion rates for disks accreting 0.1 and 10 $M_\odot s^{-1}$. Note that the accretion rate through the disk need not be related to (and at late times, is much higher than) the accretion along the rotation axis. For comparison, we plot the accretion rates of stars with masses ranging from 20-60 $M_\odot$ (Rauscher et al. 2002) versus mass coordinate of the accreting mass layer.

When the accretion rate of stellar material falling along the polar axis drops below the critical accretion rate for a given disk, the neutrinos from the disk will drive an explosion and, very likely, disrupt the entire star. For a given disk

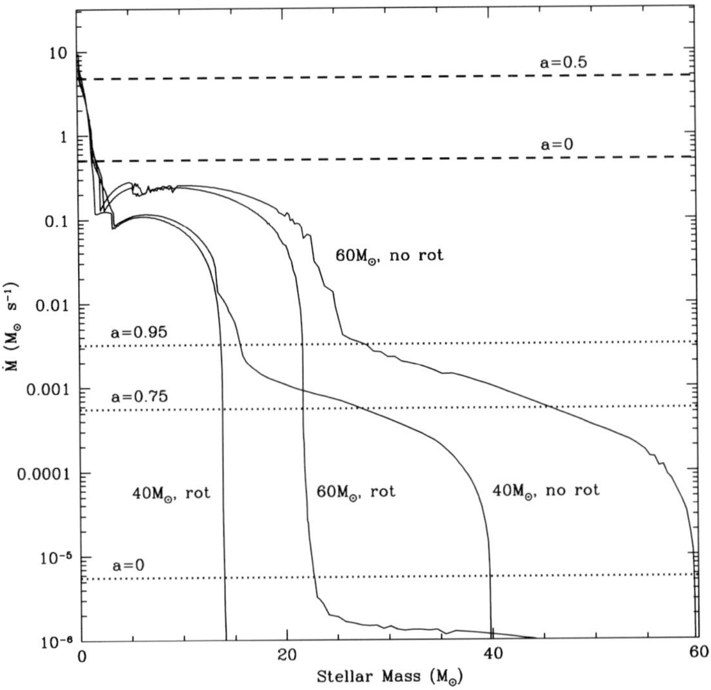

Figure 11.7. Accretion rates versus mass for rotating and non-rotating 40,60 $M_{\odot}$ stars (Heger 2002), along with the critical disk accretion rates from Fig.11.6.

structure, then, this crossing point gives a rough estimate of the remnant mass (e.g. from Fig.11.7 we see that for a disk accretion rate of $0.1 M_{\odot} s^{-1}$ onto a black hole rotating near breakup: a=0.95, the remnant black hole mass for a $30 M_{\odot}$ star is roughly $10 M_{\odot}$.) Note also, that if the neutrino emission was as high as that given by the disks accreting at $10 M_{\odot} s^{-1}$ from Popham et al. (1999), then the explosion would occur immediately. However, Di Matteo, Perna, & Narayan (2002) have found that these disks actually produce a much lower neutrino flux. Even if such high neutrino luminosities could be constructed, they would disrupt their star immediately.

4. Jets

In §1, we outlined the growing observational evidence that supports the proposition a collapsar model for GRBs. In §2 we found that it is possible to make fast rotating stars which will collapse to form black hole accretion disk systems and can drive explosions where most of the energy is injected in a narrow cone alone the rotation axis (§3). Whether the energy is deposited in a neutrino-annihilation cone or in a magnetic field driven jet, the star quickly collimates the explosion into a relativistic jet with very small baryon loading. Before this jet can make gamma-rays, it must break out of the star and then propagate in the stellar wind environment above the star. In this section, we will discuss the propagation of collapsar jets inside the stars and in the nearby stellar environment. We will discuss how the jet is changed by the interaction between the jet and the star. We will also discuss the role of jet cocoons and the sideways expansion of the jet. Finally, we will discussed a unified model of cosmological high-energy transients, including gamma-ray bursts and x-ray flashes .

Physics of Relativistic Jet Propagation

Before we discuss collapsar jets, we start with a short discussion on the propagation of relativistic jets in this section. Supersonic jets are associated with many objects in astrophysics, including young stellar objects, active galactic nuclei and gamma-ray bursts. A comprehensive review of the theories of astrophysical jets is far beyond the scope of this book. More details of this subject may be found in Begelman, Blandford & Rees (1984), Ferrari (1998) or Müller (1998).

Relativistic jets were initially studied in the early seventies in the framework of modeling double radio sources (e.g. Rees 1971; Blandford & Rees 1974; Scheuer 1974). The jet consists of a jet beam, shocked jet material and medium gas, a Mach disk shock, a leading bow shock, and a contact discontinuity. The jet beam is decelerated by the Mach disk shock, and its bulk kinetic energy is randomized by the shock. The medium is shocked by the leading bow shock. The shocked jet beam and the shocked medium meet at the contact discontinuity and then they move sideways and form a cocoon surrounding the jet beam. See Müller (1998) for a cartoon of overall morphology. The jet head is propagating at the speed of the contact discontinuity, which is lower than the actual fluid velocity of the original jet beam. We can estimate the velocity of the jet head by equating the momentum flux of the beam gas and the external medium in the frame of the contact discontinuity (see e.g. Martí et al. 1997):

$$v_j = \frac{v_b}{1 + \sqrt{(\rho_m h_m)/(\rho_b h_b)}\sqrt{1 - v_b^2}} \tag{11.7}$$

where, v_b is the beam flow velocity measured in unit of c, ρ_b and ρ_m are the density of the beam and external medium, and h_b and h_m are the specific enthalpy.

Numerical simulation is a very important tool of studying astrophysics including astrophysical relativistic jets. If the collisional mean-free-paths are small, relativistic jets can been treated as a fluid. Though the mean-free-paths in astrophysical jets are often bigger than the radius of the jets, collective plasma effects will decrease the effective mean-free-paths. Astrophysical jets are also often magnetized. Even a very weak magnetic field which has no effects on dynamics can usually make the Larmor radius of a proton much smaller to the radius of jets (Blandford & Rees 1974). Thus, hydrodynamic simulations of astrophysical relativistic jets are justified.

In their pioneering work, Norman et al. (1982) did a series of Newtonian hydrodynamic simulations of axisymmetric jets injected into a homogeneous medium to study the structure and stabilities of supersonic jets. Their results confirmed the basic jet components proposed by Blandford & Rees (1974). Magnetic fields can be carried by jets. The influence of the magnetic fields might be important for not only the synchrotron radiation such as radio emission of AGNs, but also the dynamics of jets in some cases. Numerical results from MHD simulations have shown some interesting features, such as a "nose cone" (Clarke, Norman, & Burns 1986) and knot-like structures (Thiele & Camenzind 2002).

Astrophysical jets in many cases, such as some AGNs and GRBs, are believed to be highly relativistic. There has been much progress in relativistic numerical simulations during the past decade due to the increasing power of computers and the improvement of numerical methods. Two-dimensional simulations of relativistic jets have been preformed to study the morphology and dynamics by many authors (e.g., van Putten 1993; Martí, Müller & Ibáñez 1994; Duncan & Hughes 1994; Martí et al. 1997; Komissarov & Falle 1998). These results confirmed the formation of the basic features of a jet, i.e., beam, cocoon, Mach disk and bow shock, which have been observed in Newtonian hydrodynamic simulations. It was also found that relativistic jets are more stable and propagate more efficiently than Newtonian ones (e.g., Martí et al. 1997). Three-dimensional relativistic hydrodynamic simulations of jets have also been performed to study three-dimensional effects such as precession of jets (Hughes, Miller & Duncan 2002; Aloy et al. 2003).

Multi-dimensional simulations of relativistic magnetized jets are very important to explore the interaction between magnetic fields and jets and their observational consequences. Knotted structures in extragalactic jets can been explained in terms of MHD (van Putten 1996). Parallel magnetic fields besides relativistic effects can collimate jets (Koide, Nishikawa & Mutel 1996; Nishikawa et al. 1997). An oblique magnetic field can cause a relativistic jet to bend at the jet head (Nishikawa et al. 1998). These multi-dimensional numerical simulations have shown that various observations of extragalactic radio sources can been explained in the frame of relativistic jets.

Numerical Techniques

Given the importance of numerical studies of jets, we will give a brief review of common computational techniques. The governing equations of relativistic hydrodynamics can be written as conservation laws (see, e.g., Aloy et al. 1999)

$$\frac{\partial \mathbf{U}}{\partial t} + \sum_{j=1}^{3} \frac{\partial \mathbf{F}^j}{\partial x^j} = 0, \tag{11.8}$$

where the conserved variable $\mathbf{U}$ is given by

$$\mathbf{U} = (D, S^1, S^2, S^3, \tau)^T, \tag{11.9}$$

and the fluxes are given by

$$\mathbf{F}^i = (Dv^i, S^1v^i + p\delta^{1i}, S^2v^i + p\delta^{2i}, S^3v^i + p\delta^{3i}, S^i - Dv^i)^T, \tag{11.10}$$

here δ is the Kronecker symbol, v is the velocity, and p is the pressure. The conserved variables $\mathbf{U}$ include rest mass density, D, the momentum density, S^j, and energy density, τ. They are measured in the lab frame, and are given by (assuming the speed of light $c = 1$)

$$D = \rho W \tag{11.11}$$

$$S^j = \rho h W^2 v^j \tag{11.12}$$

$$\tau = \rho h W^2 - p - \rho W, \tag{11.13}$$

where $j = 1, 2, 3$, W is the Lorentz factor and h is the specific enthalpy given by $h = 1 + \epsilon + p/\rho$, here ϵ is the specific internal energy.

There are many numerical methods for special relativistic hydrodynamics. We refer the reader to a comprehensive "living review" by Martí & Müller (1999). Here we only mention the so-called high-resolution shock capturing (HRSC) methods, which are based on the fact that special relativistic hydrodynamics with causal equation of state is a hyperbolic system of conservation laws (Anile 1989). HRSC methods are particularly promising in modern numerical simulations of astrophysical jets. They are very stable. They can achieve very high accuracy and handle ultrarelativistic flows, strong shocks and contact discontinuities extremely well. Numerical simulations of relativistic jets are very challenging because of the ultrarelativistic speed involved in the problem. The simulations have been done successfully by several authors (Aloy et al. 2000; Zhang, Woosley, & MacFadyen 2003) using the GENESIS method. GENESIS is a very efficient scheme based on HRSC techniques. We give a brief overview of the method and refer the reader to Aloy et al. (1999) for details.

To numerically solve Eq. 11.8, each spatial dimension is discretized into cells. Using the method lines, the time dependent evolution of Eq. 11.8 can be expressed in the form

$$\frac{d\mathbf{U}_{i,j,k}}{dt} = -\frac{\mathbf{F}^x_{i+1/2,j,k} - \mathbf{F}^x_{i-1/2,j,k}}{\Delta x} - \frac{\mathbf{F}^y_{i,j+1/2,k} - \mathbf{F}^y_{i,j-1/2,k}}{\Delta y} - \frac{\mathbf{F}^z_{i,j,k+1/2} - \mathbf{F}^z_{i,j,k-1/2}}{\Delta z}, \quad (11.14)$$

where $\mathbf{F}^x_{i\pm1/2,j,k}$, $\mathbf{F}^x_{i\pm1/2,j,k}$ and $\mathbf{F}^x_{i\pm1/2,j,k}$ are the fluxes at the cell interface.

In order to achieve high order accuracy in time, the time integration is done using a high order Runge-Kutta scheme (Shu & Osher 1988), which involves prediction and correction. For example, the third order accuracy can be achieved via

$$\mathbf{U}^{(1)} = \mathbf{U}^n + \Delta t L(\mathbf{U}^n) \quad (11.15)$$

$$\mathbf{U}^{(2)} = \frac{3}{4}\mathbf{U}^n + \frac{1}{4}\mathbf{U}^{(1)} + \frac{1}{4}\Delta t L(\mathbf{U}^{(1)}) \quad (11.16)$$

$$\mathbf{U}^{n+1} = \frac{1}{3}\mathbf{U}^n + \frac{2}{3}\mathbf{U}^{(2)} + \frac{2}{3}\Delta t L(\mathbf{U}^{(2)}), \quad (11.17)$$

where $L(\mathbf{U})$ is the right hand side of Eq. 11.14, $\mathbf{U}^{n+1}$ is the final value after advancing one time step from $\mathbf{U}^n$.

The Marquina's algorithm is used to compute the numerical fluxes $\mathbf{F}$ from the physical variables at the cell interface. The values of the physical fluid variables: pressure, rest mass density and velocity at the cell interface are interpolated using reconstruction schemes (e.g., piecewise parabolic method, Colella & Woodward 1984). The reconstruction procedure ensures high order accuracy in spatial dimensions.

The conserved variable $\mathbf{U}$ is evolved directly by the scheme. In each time step, physical variables, such as pressure, rest mass density and velocity, which are necessary in calculating the numerical fluxes, can be recovered from conserved variables by means of a root solver such as Newton-Raphson iteration.

We assume Cartesian coordinates and no gravity in the above discussion. However, coordinate geometry and approximate gravity can be easily implemented by slightly modifying the right hand side of Eq. 11.14, It is also very straightforward to use non-uniform grids and a general equation of state.

Simulations of Collapsar of Jets

Most numerical simulations of relativistic jets have concentrated on AGN jets. Here we discuss the relatively new field of collapsar jets, concentrating on the specific aspects of relativistic jets in the collapsar GRB model. Relativistic jets from AGNs have lower Lorentz factor ($\sim$ 10) compared with those from collapsars ($\sim$ 100). In addition, in collapsars, the density of the jet can be much lower than its surroundings and the density in the surrounding medium changes by over 10 orders of magnitude during the jet progression. In contrast, the density of AGN jets are comparable to the medium through which they propogate and the density of the medium does not change dramatically over the jet progression. These differences mean that we can not rely entirely upon our understanding of AGN jets and merely extrapolate this knowledge to understand the propogation of collapsar jets. Instead, we must calculate relativistic jets for the specific case of collapsar jets. By calculating these details, we produce observational predictions that can ultimately lead to tests of the collapsar GRB model.

Numerical simulations of the collapsar model were first performed by using a two-dimensional Newtonian piecewise parabolic method (PPM) code (MacFadyen & Woosley 1999; MacFadyen, Woosley, & Heger 2001). They explored the jet proprogation through rotating, massive stars and found the energy released at the polar region by either neutrinos or MHD processes can power energetic jetlike outflows. Their simulations produced highly relativistic jets emerge out of the star with a very small beaming angle. But these results were limited by the fact that the simulations were Newtonian in nature. Since collapsars are highly relativistic, it is very important to perform the study in fully relativistic simulations.

Aloy et al. (2000) did a series of two-dimensional special relativistic simulations of collapsar jets using the GENESIS code. In their simulations, energy is continuously deposited into the cone region around the rotational axis. They found that after a small latency of less than one second, a relativistic jet starts to form and propagates along the axis. The jet breaks out the star after a few seconds. The jet is very narrowly beamed to less than 10 degrees. Its Lorentz factor increases up to 33 before it breaks out. After it breaks out, the Lorentz factor of the jet continues to increase. Meanwhile, the jet starts to expand sideways, but the jet beam stays collimated. Their simulations demonstrated that a very well collimated relativistic jet can be born from a collapsing star due to sufficient energy being released at the inner polar region. As one expects, the jet is spatially inhomogeneous, and time dependent.

Using the same method of the GENESIS code, Zhang, Woosley & MacFadyen (2003) examined the propogation of two-dimensional special relativistic jets in collapsar in more detail and very high resolution. Zhang et al. (2003) began with the jet already 2000 km from the black hole. The jet, which presumably forms in the very inner region ($<$ 100 km), is initiated in a parametric way based on its power, opening angle, Lorentz factor, and internal energy. The details of the formation of the jet and what has gone deep inside the star are still unclear. However, values of jets at, say, 200 km, can be estimated from observational evidence and theoretical considerations.

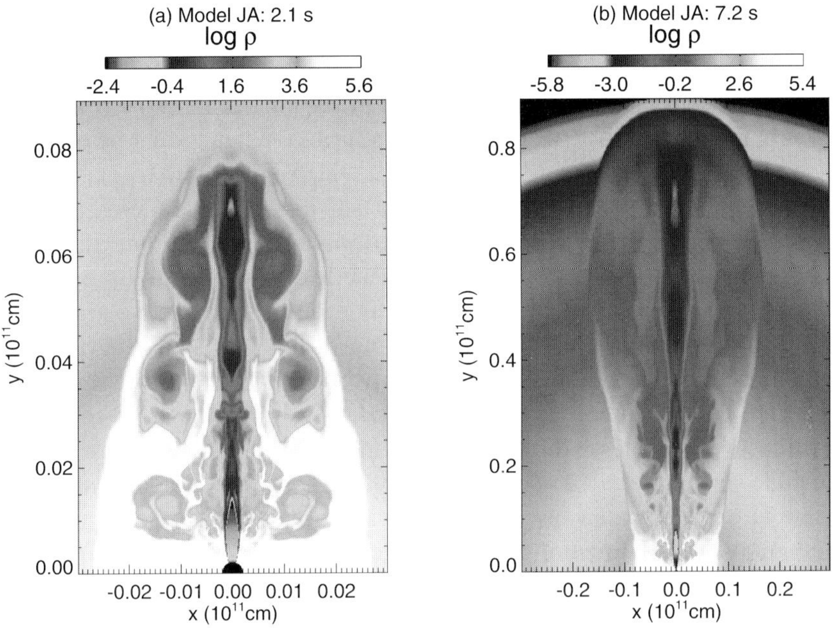

Figure 11.8. Density structure in the local rest frame for Model JA of Zhang, Woosley, & MacFadyen (2003) at (a) $t = 2.1\,\mathrm{s}$ and (b) $7.2\,\mathrm{s}$. In (a), only the central region of the star is shown. The radius of the star is 8×10^{10} cm.

In all models of Zhang, Woosley & MacFadyen (2003), the jet propagates along the rotational axis of the progenitor star. The jet is narrowly collimated. In agreement with previous studies by Aloy et al., it was found that a relativistic jet is collimated by its passage through the stellar mantle. Starting with an initial half-angle of up to 20 degrees, it emerges with half-angles that, though variable with time, are around 5 degrees. The jet consists of a supersonic beam, a cocoon consisting of shocked jet material and shocked medium gas, a terminal bow shock, a working surface, and backflows, which are components of a jet in the model of Blandford & Rees (1974). Figure 8 shows two snapshots of Model JA of Zhang, Woosley, & MacFadyen (2003).

Along the polar axis, the beam in each model is divided into two regions by a reverse shock: an unshocked region and a shocked region. The injected jet material is initially expanding and its internal energy was converted into kinetic energy until it reaches the reverse shock. The reverse shock, which carries the message of the need for deceleration forms because the jet head runs into stationary stellar material. As this reverse shock propogates back through the star, it converts most of the jet's kinetic energy back into internal energy, and the jet beam becomes moderately relativistic. The jet beam meets the shocked stellar matter at the contact discontinuity.

Due to the negative density gradient of the stellar matter along the rotation axis, the jet accelerates. Eventually, the jet breaks out of the star a few seconds after it is introduced into the computational grids. When the jet breaks out of the star, its Lorentz factor is about 10-50, which is too low to make GRBs. However, the jet is still very hot and most of its energy remains in internal energy, not kinetic energy. When the jet continues to expand outside the star, the internal energy will be converted into kinetic energy and the Lorentz factor will reach more than 100 before GRBs are produced. The above argument is confirmed by simulations of jet propagation in the near stellar environment. When it reaches 10^{12} cm, the jet has already obtained a peak Lorentz factor of about 150.

As the jet propagates in the near stellar winds, sideways expansion of the jet is inevitable since it is only moderately relativistic and has a large internal energy. As the jet expands sideways, the velocity along its edge does not increase as rapidly in the radial direction as at the center. A small amount of material ends up at larger angles, greater than the commonly expected $\theta_0 + 1/\Gamma_0$, where θ_0 and Γ_0 are the opening angle and the Lorentz factor at the breakout, respectively. An interesting consequence of the sideways expansion of a jet in the stellar winds is that the jet will gain a structure across the opening angle with higher energy and Lorentz factor near the core even if it is initially uniformly distributed across the opening angle. So what we see depends upon the viewing angle. The structured jet model were used by many to explain observational results of GRB afterglow lightcurves (e.g., Salmonson 2000; Rossi, Lazzati, & Rees 2002; Zhang & Meszaros 2002).

According to the currently favored fireball shock model of GRBs, GRBs are produced by internal shocks in relativistic outflows. The GRB outflows are expected to have very rich time structures so that internal shocks can occur. An encouraging aspect of the collapsar model is that it *does* produce relativistic jets with rich time structures. It was found by numerical simulations that the instabilities during the jet propagation inside the helium star can imprint a lot of structure even if a constant jet is initially introduced. Interaction of the the jet beam with the cocoon causes mixing. The entrainment of the cocoon into the jet beam brings the variabilities needed in the fireball shock scenario.

Since the jet head moves much slower than the jet itself, some energy is "wasted" inside the star before the jet breaks out of the star. Most of the wasted energy is in the nonrelativistic hot material near the jet which can power hypernovae (a.k.a. jet-driven supernovae) (Zhang, Woosley, & MacFadyen 2003).

The simulations of Zhang, Woosley, & MacFadyen (2003) were performed in two separate stages because of the difficulty of carrying too large a range of

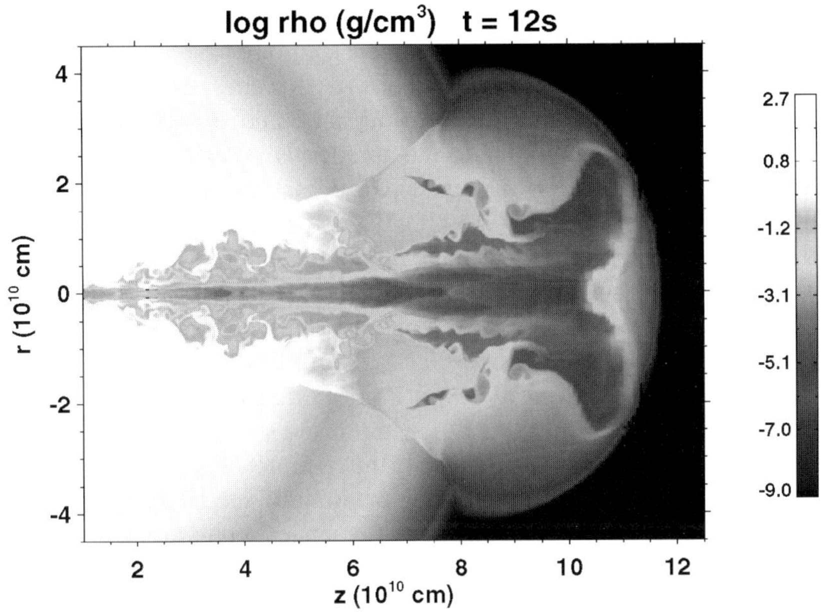

Figure 11.9. Density structure in the local rest frame for one of the models of Zhang, Woosley, & Heger (2003). The jet is breaking out the star. At the head of the jet, the cocoon is expanding sideways very quickly. Also note the plug at the head of the jet.

radii on the computational grid. The first stage is for propagation inside the star and the second stage for the stellar wind. So the emergence of the jet, which will definitely lead to some sort of "precursor" activity, is not accurately simulated. To address more accurately the breakout of the jet, two-dimensional calculations covering a large range of radii with very high resolution at the stellar surface were simulated (Zhang, Woosley, & Heger 2003). As seen in previous studies, after it breaks out the star, both the jet and its cocoon are loaded with a lot of internal energy. Although its current Lorentz factor is only about 10, the final Lorentz factor of the jet can be ~ 200 when it expands adiabatically and gains higher Lorentz factor.

After the jet breaks out the star, its cocoon expands both outwards and sideways (Figure 9). An interesting effect of the cocoon is that it prevents the jet beam from expanding sideways. Without the cocoon, a bare "hot" jet with a Lorentz of about 10 will inevitably expand sideways. The cocoon emerges out the star with an energy close to the total energy injected at the base of the jet before the jet emerges from the star (Ramirez-Ruiz, Celotti, & Rees 2002). The conversion of its internal energy into kinetic will give the cocoon a Lorentz factor of $\sim 5 - 10$ and the cocoon can reach angular sizes of up to 30°. For instance, the isotropic equivalent energy of relativistic material at 30° is about 10^{50} erg for one of their models. The deceleration of the cocoon by external shocks is likely to produce γ-rays and x-rays (Ramirez-Ruiz, Celotti, & Rees 2002). The "anomalous" GRB 980425 might be a cocoon viewed from a large angle. It might also contribute to the afterglows and precursors of a normal GRB and may be the origin of the new type of fast transient: hard x-ray flashes (Heise et al. 2001; Kippen et al 2001). These flashes have many properties of GRBs (energy, isotropy, approximate duration, distribution with redshift), but have a much softer spectrum (and so far no optical afterglow). The association of this kind of softer transient with gamma-ray bursts seen off-axis has been predicted for many years (Woosley & MacFadyen 1999; Woosley 2000, 2001).

The velocity of the head of the jet in the star is subrelativistic, and the jet beam is highly relativistic. Shocked medium and shocked jet material piles up near the contact discontinuity and feeds the cocoon. This forms a "plug" at the head of the jet (Fig. 9). After breakout, the initially subrelativistic plug will be accelerated to Lorentz factors of $\sim 10-20$ since it continues to be pushed by the relativistic jet below it and the stellar wind cannot hinder its acceleration. Meanwhile, the jet beam is slowed down as its energy is transfered to the plug and part of its kinetic energy is converted into its internal energy. The fast moving jet beam pushing a "slowly" moving plug leads to the production of a reverse shock which moves backward and slows down the jet beam, above which is a contact discontinuity and a forward shock. Since the plug has a limited length, the forward shock will reflect at the end of the plug. This further complicates the dynamics (e.g., Waxman & Mészáros 2003). The interaction of the jet and the plug will affect both of them substantially, and may lead to short-hard GRBs (Zhang, Woosley, & MacFadyen 2003).

So far, the numerical simulations of collapsar jets we discussed above were perform in two-dimension. Increasing computer power has allowed recent three-dimensional simulations. There is no need to emphasize the importance three-dimensional simulations (Zhang, Woosley, & Alex in preparation). Will the results of three-dimensional simulations be different from those of two-dimensional simulations? Will the jet survive from instabilities arisen from three-dimensional effects? Will the plug near the head of the jet still be prominent?

The method of the GENESIS code, which was used in the three-dimensional simulation we are discussing, has an excellent ability of keeping symmetry (Aloy et al. 1999). If starting initially with a perfectly axisymmetric jet, it was found that the symmetry was not broken by the numerical scheme and there were negligible differences between results of three-dimensional simulations and those of two-dimensional simulations. But it is reasonable to assume that some precesses would break the symmetry. In order to break the perfect symmetry of the cylindrical initial conditions, a 1% imbalance in power was given to one side

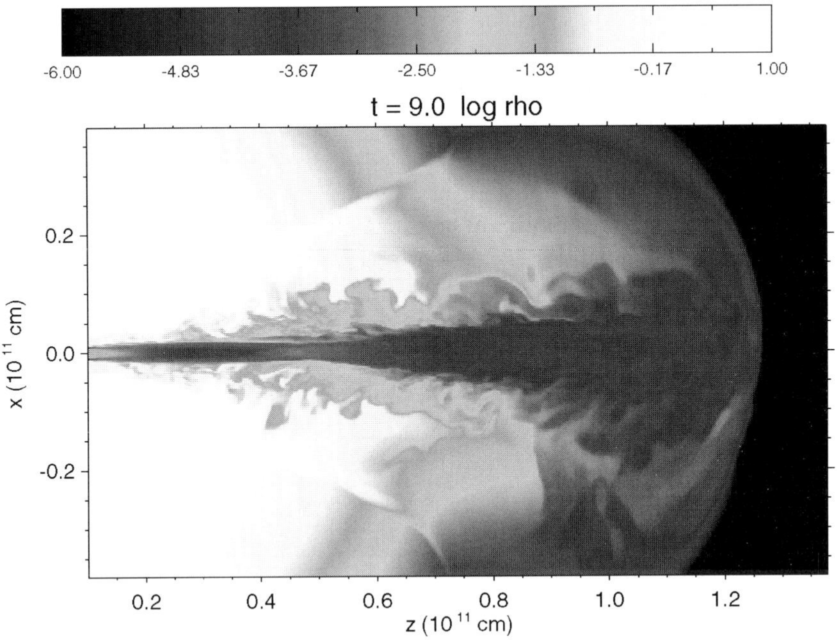

Figure 11.10. Density structure in the local rest frame for one of the three-dimensional simulations of asymmetric jets (Zhang, Woosley, & Heger 2003).

of the base of the jet in Model 3C. Fig. 10 shows a snapshot of that simulation. It is about 1 s after leading edge of the jet reaches the edge of the star. At that moment, the Lorentz factors in the central jet are $\sim$10 and the Lorentz factors in the cocoon are $\sim$ 2. But both will end up moving much faster as the jet accelerates in the wind. It was also found that the existence of a plug will not change due to three-dimensional effects in case of unbalanced initial jets.

Jets produced by systems of a black hole and an accretion disk are likely to precess according to both theoretical models (e.g., Pringle 1996; Lai 2003) and observational evidence (e.g., Cowley et al. 1998; Ogilvie & Dubus 2001). In order to explore effects of jet precession, a series of simulation were performed with precessing injected jets. In their simulations, the axis of injected jet was kept a fixed angle to the rotational axis of the star, which is a parameter in the simulation. Meanwhile, the jet axis is processing around the rotational axis with a period of 2 s. Fig. 11 shows the results of one model with an angle of 3 degrees between the jet axis and the rotational axis. As one expects, that model shows very rich structures. It was also found that there is a critical angle, which depends on the initial conditions of jets. If the inclining angle of the jet is less

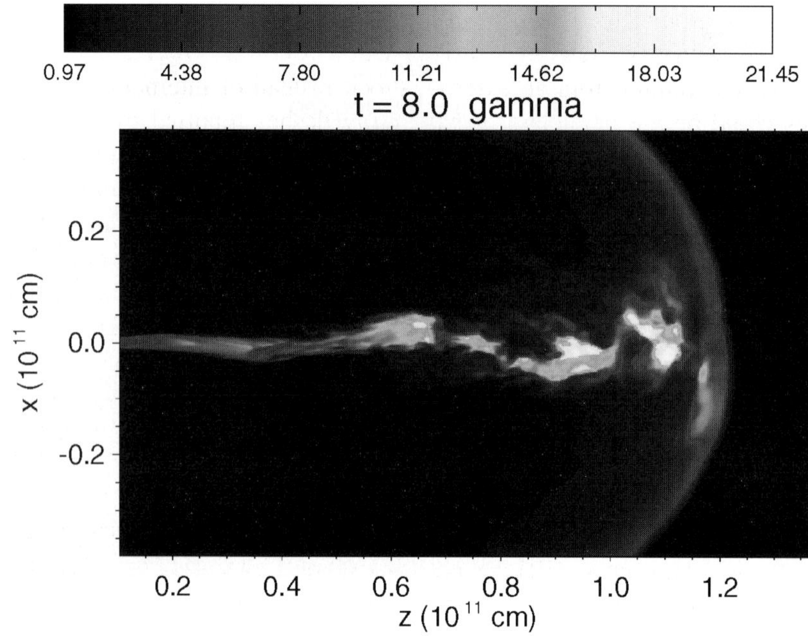

Figure 11.11. Lorenz factor 8 s after initiation in a three-dimensional simulation of precessing jets. In this model, the angle between the jet axis and the rotational axis is three degrees.

than that value, ~ 5 degrees, the jet will be able to emerge from the star with high Lorentz factor. If the inclining angle is too large, no material with a high Lorentz factor can emerge from the star even if the jet can actually penetrate the star. This is because entrainment of stellar material into the jet beam is too much to maintain high Lorentz factor. In summary, small-angle precession can help build time structures, which are needed for making GRBs, whereas a jet with a large precession angle cannot make GRBs. But it remains a question and is worth exploring whether that can make x-ray flashes.

We conclude this section with discussing a possible unified model for high energy cosmological transients based upon a collapsar whose properties would vary depending upon the angle at which it was observed (Woosley, Zhang, & Heger 2003; Zhang, Woosley, & MacFadyen 2003).

In the equatorial plane of a collapsar - the common case - probably little more is seen than an extraordinary supernova. Along the axis (< 5 degrees), one sees an ordinary GRB. Moving off axis, one expects a smooth decline in the Lorentz factor and energy of relativistic ejecta according to numerical simulations. One may see a low energy wing of the jet up to ~ 15 degrees, and then a hot mildly relativistic cocoon up to ~ 30 degrees, as moving off axis. So a large region of the sky, much larger than that which sees the main GRB, will see a soft γ or hard x-ray transient with less power, lower Lorentz factor, and perhaps coming from an external shock instead of internal ones. These events could be the new class of hard x-ray flashes reported by Heise et al. (2001). These flashes would have many properties of GRBs: energy, isotropy, approximate duration, distribution with redshift. However, because they are less luminous, their log N-log S distribution may not exhibit the same roll over. Their median redshift should be smaller. XRFs may, most frequently be seen in isolation and will be characterized by softer spectra, but there should also be an underlying XRF in every GRB since the emission of the mildly relativistic cocoon material is beamed to a larger angle that includes the poles. In some cases these XRFs should be seen as precursors or extended hard X-ray emission following a common GRB. They should be associated with supernovae. Indeed XRFs may more frequently serve as guideposts to jet powered supernovae than GRBs, especially the nearby ones.

GRB 980425 also has a special place in this unified model for high energy cosmological transients. GRB 980425 was a GRB viewed off axis, but not in the manner some envision. While an observer along the axis would have seen a much more powerful burst, what we saw in GRB 980425 was not just the weak emission of a highly relativistic jet seen from the side, but the mildly relativistic, lower energy outflow from the explosion of the cocoon with $\Gamma \sim 10$ interacting with the pre-explosive wind of the star. This suggests that GRB 980425 was a transition object between common GRBs and XRFs (see Fig 12). It may also be that the collapsar itself was somewhat subenergetic.

Although the unified model can explain a variety of observations, bear in mind that not all collapsars are the same. It is equally possible that different phenomena are from different kind of collapsars and/or some other objects, instead of viewing the same object from different angles. Observations, detailed numerical simulations and analytic calculations might ultimately test these speculations.

Future

So far, the present studies on the collapsar jets have shown that the collapsar model, the currently favored model for long-soft GRBs, is able to explain many of the observed characteristics of GRBs. In the future, many issues require further examination. Hopefully, they will further strengthen our knowledge on relativistic jets from collapsars. Here, we discuss some of them.

First, the long term evolution of the jet is interesting and critical in linking the numerical collapsar model to observations. The interaction of the jet and the plug, which should have important implications for observations, needs

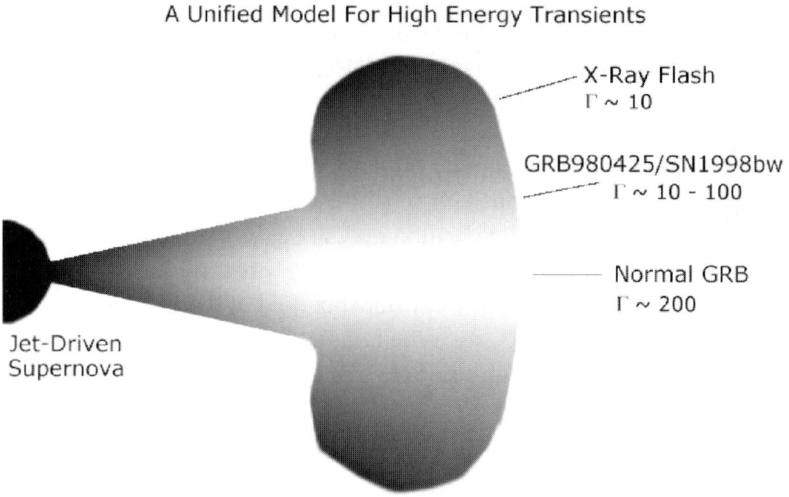

Figure 11.12. Diagram of the unified model. Observers looking directly along the head of the jet will see a GRB. As they move off-axis, the GRB will become weaker and ultimately the observer will only see an X-ray flash. Viewing angle is critical in GRB observations.

more detailed study. As it sweeps the external medium, the jet will eventually be slowing down and experiencing sideways expansion. The transition from spherical to jet-like expansion has been studied under some simple assumptions (e.g., Rhoads 1997; Sari, Piran, & Halpern 1999), but it needs further examinations under more realistic assumptions. It is also important to include physics of radiation mechanism and transport for the late-time evolution of the jet when it is optically thin.

Second, a jet-driven supernova is expected from both the collapsar model (MacFadyen, Woosley, & Heger 2001) and other GRB models involving jets and collapse of massive stars, such the magnetar model (Khokhlov et al. 1999). It is expected that more GRB-SN associations, such as, GRB 980425/SN 1998bw and GRB 030329/SN2003dh will be discovered. The future observational data on these should be able to shed more light on the progenitor and rule in or out those GRB progenitor models. Studying supernovae driven by collapsar jets will be very crucial in understanding the collapsar model.

Third, future study will hopefully provide more tests for the unified model we discussed in the last section. It is also interesting to study what we expect to see from different types of collapsars if there are.

Fourth, it is still unclear whether a GRB can be made from a supergiant star. It was shown by Newtonian numerical simulations that even a very powerful collapsar jet can emerge from a red supergiant star and make a GRB (Mac-

Fadyen, Woosley, & Heger 2001). Whereas Meszaros & Rees (2001) found that a jet can emerge from the hydrogen envelope of a blue supergiant star and in principle make a GRB if the inner engine can last long enough for the jet to break through the helium core. High-resolution relativistic simulations should be able to tell us the answer about whether a GRB progenitor could have an extended hydrogen envelope. If not, what is the consequence of a relativistic jet choked inside a supergiant star?

Last but not least, magnetic field is possibly very important to collapsar jet propagation. The recent discovery of large linear polarization in the prompt γ-ray emission from GRB 021206 (Coburn & Boggs 2003) seems to imply a dynamically-dominant magnetic field in the outflow of that burst (Lyutikov, Pariev, & Blandford 2003). It is important to study electromagnetic explosions of collapsars and magnetized collapsar jets.

Acknowledgments

Notes

1. Note that we have neglected the disk mass which, in Popham et al. (1999) can be as high as 0.5 $M_\odot$. Even for this high disk mass, the change in the gravitational acceleration is less than 20%.

References

Aloy, M. A., Ibáñez, J. Mª., Martí, J. Mª., & Müller, E. 1999, ApJS, 122, 151
Aloy, M. A., Müller, E., Ibáñez, J. Mª., Martí, J. Mª., & MacFadyen, A. I. 2000, ApJ, 531, L119
Aloy, M. A., Martí, J. Mª., Gómez, J.L., Agudo, I., Müller, E., & Ibáñez, J. Mª. 2003, ApJ, 585, L109
Anile, A.M. 1989, Relativistic Fluids and Magneto-Fluids (Cambridge: Cambridge Univ. Press)
Begelman, M.C., Blandford, R.D., & Rees, M.J. 1984, Reviews of Modern Physics, 56, 255
Blandford, R.D. ,& Rees, M.J. 1974, MNRAS, 169, 395
Blandford, R.D. ,& Znajek, R.L. 1977, MNRAS, 179, 433
Clarke, D.A., Norman, M.L., & Burns, J.O. 1986, ApJ, 311, L63
Coburn, W., & Boggs, S.E. 2003, Nature, 423, 415
Colella, P., & Woodward, P.R. 1984, J. Comput. Phys, 54, 174
Cowley, A.P., et al. 1998, ApJ, 504, 854
Duncan, G.C., & Hugnes, P.A. 1994, ApJ, 436, L119
Ferrari, A. 1998, Annu. Rev. Astron. Astrophys., 36, 539
Frail, D., et al. 2001, ApJ, L55
Fryer, C. L., & Woosley, S.E. 1998, ApJ, 502, L9
Fryer, C. L. 1999, ApJ, 522, 413
Fryer, C. L., Woosley, S. E., & Hartmann, D.H. 1999, ApJ, 526, 152

Fryer, C. L., & Kalogera, V. 2001, ApJ, 554, 548
Fryer, C. L., & Meszaros, P. 2003, ApJ, 588, L25
Heger, A., Woosley, S.E., & Spruit, H. in preparation for ApJ
Heise, J., in't Zand, J., Kippen, R. M., Woods, P. M. 2001, GRBs in the Afterglow Era, eds. Costa, Frontera, & Hjorh, ESO Astrophysics Symposia, (Springer), 16
Garnavich, P. et al. 2003, ApJ, 582, 924
Hogg, D. W., & Fruchter, A. S. 1999, ApJ, 520, 54
Holland, S. T., & Hjorth, J. 1999, A&A, 344, L67
Hughes, P.A., Miller, M.A., & Duncan, G.C. 2002, ApJ, 572, 713
Khokhlov, A. M., Höflich, P. A., Oran, E. S., Wheeler, J. C., Wang, L., & Chtchelkanova, A. Yu. 1999, ApJ, 524, L107
Kippen, R. M., Woods, P. M, Heise, J., in 't Zand, J. J. M., Briggs, M. S., & Preece, R. D. 2001, in proceedings of the Woods Hole GRB Workshop, ed. G.R. Ricker & R.K. Van der Speck, 244
Koide, S., Nishikawa, K.-I., Mutel, R.L. 1996, ApJ, 463, L71
Komissarov, S.S., & Falle, S.A.E.G. 1998, MNRAS, 297, 1087
Lai, D. 2003, ApJ, submitted, astro-ph/0306012
Li, Z.-Y., & Chevalier, R. A. 2001, ApJ, 551, 940
Lyutikov, M., Pariev, V.I., & Blandford, R. 2003, ApJ, submitted, astro-ph/0305410
MacFadyen, A. I., & Woosley, S. E. 1999, ApJ, 524, 262
MacFadyen, A. I., Woosley, S. E., & Heger, A. 2001, ApJ, 550, 410
Martí, J. Mª., Müller, E. 1999, Living Reviews in Relativity, Vol. 2, http://www.livingreviews.org/Articles/Volume2/1999-3marti
Martí, J. Mª., Müller, E., & Ibáñez, J. Mª. 1994, A&A, 282, L9
Martí, J. Mª., Müller, E., Font, J. A., Ibáñez, J. Mª., & Marquina, A. 1997, ApJ, 479, 151
Meszaros P., & Rees, M.J. 2001, ApJ, 556, L37
Müller, E. 1998, in Saas-Fee Advanced Couse 27, Computational Methods for Astrophysical Fluid Flow, LeVeque, R.J., Mihalas, D., Dorfi, E.A., & Müller, E.
Narayan, R. Piran, T., Kumar, P. 2001, ApJ, 557, 949
Nishikawa, K., Koide, S., Sakai, J., Christodoulou, D.M., Sol, H. & Mutel, R.L., 1997, ApJ, 483, L45
Nishikawa, K., Koide, S., Sakai, J., Christodoulou, D.M., Sol, H. & Mutel, R.L., 1998, ApJ, 498, 166
Norman, M.L., Smarr, L., Winkler, K.-H.A., & Smith, M.D. 1982, A&A, 113, 285
Ogilvie, G.I., & Dubus, G. 2001, MNRAS, 320, 485
Popham, R., Woosley, S. E., & Fryer, C. 1999, ApJ, 518, 356
Pringle, J.E. 1996, MNRAS, 281, 857
Ramirez-Ruiz, E., Celotti, A., & Rees, M.J. 2002, MNRAS, 337, 1349

Rees, M.J. 1971, Nature, 229, 312
Rhoads, J.E. 1997, ApJ, 487, L1
Rossi, E., Lazzati, D., & Rees, M.J. 2002, MNRAS, 332, 945
Salmonson, J. 2000, ApJ, 544, L115
Sari, R., Piran, T., & Halpern, J. 1999, ApJ, 519, L17
Scheuer, P.A.G. 1974, MNRAS, 166, 513
Shu, C.W., & Osher, S.J. 1988, J. Comput. Phys, 77, 439
Stanek, K. Z. et al. 2003, astro-ph/0304173
Thiele, M., & Camenzind, M. 2002, A&A, 381, 53
van Putten, M.H.P.M. 1993, ApJ, 408, L21
van Putten, M.H.P.M. 1996, ApJ, 467, 157
Waxman, E., & Mészáros, P. 2003, ApJ, 584, 390 submitted, astro-ph/0206392
Woosley, S.E., & MacFadyen, A. I. 1999, A&AS, 138, 499
Woosley, S.E. 2000, GRBs, 5th Huntsville Symposium, eds. Kippen, Mallozzi, & Fishman, AIP, Vol 526, 555
Woosley, S.E. 2001, GRBs in the Afterglow Era, eds. Costa, Frontera, & Hjorh, ESO Astrophysics Symposia, (Springer), 555
Woosley, S.E., Zhang, W., & Heger, A. 2002, Gamma-Ray Bursts in the Afterglow Era, the 3rd Rome Workshop
Zhang, B., & Meszaros, P. 2002, ApJ, 571, 876
Zhang, W. & Fryer, C. L. 2001, ApJ, 550, 357
Zhang, W., Woosley, S.E, & MacFadyen, A.I. 2003, ApJ, 586, 356
Zhang, W., Woosley, S.E, & Heger, A. 2003, in preparation

Chapter 12

GAMMA-RAY BURSTS FROM SUPERNOVAE

Solutions to the Baryon Loading Problem

Christopher D. Matzner*

Deptartment of Astronomy and Astrophysics, University of Toronto

matzner@cita.utoronto.ca

Abstract The cosmological gamma-ray bursts (GRBs) are thought to emit their energy in gas moving toward Earth at nearly light speed. A popular class of models places the source of this outflow in the collapsing core of a massive star. Massive stars also have massive envelopes, which pose a barrier to the development of relativistic motion: a baryon-loading problem. I review several models for segregating energy from mass in the context of supernova-GRB models.

Keywords: Supernovae; Gamma-Ray Bursts; Baryon Loading Problem

1. Introduction: Brief review of GRB properties

The properties of cosmic gamma-ray bursts are thoroughly discussed in several detailed reviews, notably the observational review by Fishman & Meegan (1995) and the theoretical ones by Piran (1999) and Mészáros (2002). I mention here only a few pertinent facts to introduce the subject.

Since their chance discovery in 1969, GRBs have challenged theorists to account for several remarkable properties. They are brief (10^{-1} – 10^3 s), incredibly brilliant (~ 0.1 erg per square meter) flashes of γ-ray (keV to MeV and higher) photons. GRBs have a broad range of spectral and timing properties; many are seen to flicker on timescales down to the photon-counting limits of the detectors, i.e., tens of milliseconds.

To this list of attributes have been added many others, especially from the Compton Gamma-Ray Observatory (CGRO) and Beppo-Sax sattelites: for in-

*Canada Research Chair

C.L. Fryer (ed.), Stellar Collapse, 357–372.

stance, a bimodal distribution of durations (short-hard and long-soft bursts; Kouveliotou et al 1993), a seemingly isotropic sky distribution, and the appearance of a turnover in the luminosity histogram (at least for long bursts) indicating an outer edge to their spatial distribution. The combination of isostropy and an outer edge is most naturally explained in a cosmological model where the population extends to redshifts $z > 1$ (Meegan et al. 1992). This was dramatically verified in 1997, once Beppo-Sax provided X-ray positions fine enough to allow optical follow-up: an optical afterglow was discovered with absorption lines indicating $z = 1.8$ (Metzger et al. 1997).

At cosmological redshifts, the flux at Earth implied total energies of order $10^{52} - 10^{54}$ ergs $= 0.01 - 1 M_\odot c^2$, assuming our view was not a priveliged one (i.e., for unbeamed emission: energies derived this way are isotropic-equivalent or just "isotropic" values). The vast energies and brief timescales of GRBs both point to the release of energy by compact objects – i.e., newly formed, solar-mass black holes or neutron stars emitting a large fraction of their rest energy.

Beaming would alleviate the energy requirements at the expense of requiring many more bursts. Some of the bursts with optical, infrared, and radio afterglows have shown a steepening of their decay rates. This is interpreted (Rhoads 1999) as a viewing effect from a beamed jet: as it decelerates, more and more of a relativistically-expanding shell becomes visible to the observer. A shell moving at Lorentz factor Γ beams its emission into an angle of roughly $1/\Gamma$. Jets of opening angle θ dim more quickly when their edges become visible, roughly when $\Gamma < 1/\theta$. Working backwards with a model for Γ gives an estimate of θ. This, in turn, allows one to estimate the true (rather than isotropic) gamma-ray energy E_γ. Using this technique, Frail et al. (2001) found that all GRBs with redshifts and derived opening angles were consistent with $E_\gamma = 10^{50.5\pm0.5}$ erg – a remarkable result. E_γ is, of course, only a lower limit for the explosion that caused the GRB: γ-rays may account for only a fraction (ϵ_γ) of the total, and one must have $\epsilon_\gamma < 1$ for there to be energy remaining in the afterglow. Also, one expects a counterjet for each observed jet.

The γ-rays themselves are argued to arise from within a relativistic outflow. GRB spectra often extend smoothly across (and well above) the electron rest energy, a regime where photons can produce $e^\pm$ pairs. If the GRB photons were confined to a region small enough that they could cross it in the shortest observed timescales ($c\delta t$: roughly white dwarf sizes), the produced pairs would become optically thick and their nonthermal spectra would be ruined. Relativistic expansion evades this, both by increasing the emission radius corresponding to δt to $\sim \Gamma^2 c\delta t$, and by reducing the photon energies by a factor Γ in the emitting frame. Lithwick & Sari (2000) show that lower limits of $\Gamma \gtrsim 10^2$ can be placed on GRBs with cosmological redshifts.

It is important to bear in mind that GRBs are anything but uniform in their time profiles and spectral properties and that our information on their distance, beaming, and afterglow properties are for now restricted to the long-duration bursts because of the limitations of Beppo-Sax (which are being addressed by HETE-II and, soon, by Swift). Bursts of γ-rays are also emitted by solar flares and by soft gamma repeaters (neutron stars). Among cosmic GRBs, there are the long and short classes discussed by Kouveliotou et al. (1993), and various subclasses (e.g., no-high-energy bursts) have been identified.

A case in point is GRB 980425, the only GRB with a reasonably firm supernova identification (in SN 1998bw). It was somewhat unusual in several of its γ-ray properties, consisting of a smooth, single pulse that was devoid of photons above about 300 keV. What really sets it apart is the proximity of the associated SN: $z = 0.0085$ (rather than $z > 0.45$ for all other bursts with redshifts), placing the isotropic energy at $\sim 10^{48}$ erg (rather than 10^{51-54} erg). This combination suggests 980425 was either a typical long-duration GRB viewed from the side (Höflich et al 1996) or a different phenomenon altogether, as argued by Matzner & McKee (1999) and Tan et al. (2001). The interpretation of this burst is important in the context of energetic phenomena from supernovae, and might have a bearing on the cosmological bursts. A critical question is the minimum Γ for the emission, which is lessened by the dimness and smoothness of the pulse. As Matzner et al. (2001) argue, the lack of photons with energies approaching $m_e c^2$ means that the emitting region could have been only semi-relativistic ($\Gamma \sim 2$).

Supernovae and Baryon Loading

Supernovae, being the most energetic explosions familiar to astrophysicists, are natural candidates for the sources of GRBs. Typical core-collapse supernovae emit $\sim 0.1 M_\odot c^2$ in 10 seconds, albeit in neutrinos rather than γ-rays. Supernova-based models are now the popular explanation for long-duration GRBs, thanks in part to the projection of afterglows against the star-forming regions of galaxies, in part to a series of computational demonstrations of principle, and in part to the association of SN 1998bw with GRB 980425. Further indications, though circumstantial, include SN-like bumps in several optical afterglows, the possible detection of heavy elements and large hydrogen absorbing columns in X-ray afterglows, and the indication of a wind-like $\rho \propto r^{-2}$ external medium in the models for some afterglows.

To achieve a Lorentz factor Γ, a flow's kinetic energy must exceed its rest-mass energy (and internal energy, if any) by the same factor Γ. If the total energy of a GRB jet is of order 10^{52} erg, then, the swept-up baryons can weigh no more than $\sim 10^{-4}(10^2/\Gamma)\, M_\odot$. This is an exceptionally small amount given that core-collapse supernovae occur in which possess tens of stellar masses at formation. The discrepancy amounts to a severe baryon-loading problem which must be addressed by any viable SN-GRB model.

Two broad classes of solution have been proposed. In the first class, energy responsible for the GRB is transmitted *through* the star's envelope – odd though this sounds – to appear in a relatively baryon-free state at its surface. In the second class, the stellar envelope is moved away, clearing a channel for a clean, relativistic outflow all the way from the star's core to its surface. The following sections examine these proposals in detail.

2. Hydrodynamic shock acceleration and weak GRBs

The very first theoretical model for GRBs involved SN explosions, and solved the baryon loading problem by pushing the GRB energy through the stellar envelope. Colgate (1974) proposed that GRBs result from a hydrodynamical effect in the outer envelopes and subsurface layers of supernova stars. This proposal was revived by Matzner & McKee (1999) and Tan et al (2001) to explain smooth, soft, and relatively weak GRBs like 980425. Although there is reason to believe this class of model cannot fit the properties of most cosmological GRBs, it is nevertheless an important element in the SN-GRB connection.

Stars, being gas balls of a definite mass, have density distributions that fall from very high values in the center to essentially zero at the stellar surface (or sonic point, for wind-emitting stars). Near the surface, the density can be reasonably approximated as a power-law of the depth (distance to the surface). The pressure is the weight of the overlying material, which (for constant gravity) is proportional to its mass: i.e., if $p \propto \rho^{\gamma_p}$ for some polytropic index γ_p, then the overlying mass satisfies $m_{\rm ext} \propto \rho_p^{\gamma}$. But since $m_{\rm ext} \propto \rho \times$depth, $\rho \propto ({\rm depth})^{1/(\gamma_p - 1)}$.

In an early phase of a supernova, this density profile will be swept up by the shock front that bounds the explosion's blastwave (outwardly-expanding fireball). As a blastwave sweeps up mass it must generally decelerate due to its energy being shared around (according to $Mv^2 \propto E$). However, the blastwave behaves starkly differently in the outer regions of stars, which contain relatively little mass. As it travels down the steeply declining "density ramp," the shock experiences a phase much like the crack of a whip. A declining fraction of the SN's energy is contained in a more rapidly-declining fraction of its mass – causing the shock to accelerate. Conditions permitting, this acceleration continues to the point that the flow becomes relativistic and then ultrarelativistic. The segregation of energy relative to mass then amounts to an effective, if rather inefficient, solution to the baryon loading problem.

A familiar analogy

The efficiency of this shock-acceleration process is determined by the details of the hydrodynamic evolution. Before discussing this, I wish to review the basic physics underlying the phenomenon. These are best illustrated by the introductory mechanics demonstration in which a tennis ball is stacked on top of a basketball, then the pair dropped together onto a hard floor. Anyone who has seen the demo will recall the tennis ball bouncing far above its initial height (often at the peril of the instructor!). If extended with larger and larger balls at the bottom and smaller and smaller balls at the top, and if idealized with perfectly elastic collisions, this experiment becomes a reasonable analogue to the outer dynamics of supernovae – with collisions between balls standing in for the shock front.

In the experiment the balls begin by falling toward the floor, but the analysis is easiest in a frame where the floor rises up to meet the balls (which are at rest). For the purpose of studying shock propagation, we can consider only the first two collisions of each ball: the one where it is struck by its lower neighbor, and the one where it strikes its upper neighbor. Suppose $\alpha < 1$ is the ratio of masses between from one ball to the next. If the previous ball hit is now moving at velocity v, then the center of the mass (CM) for the next collision moves at $v/(1+\alpha)$. In the CM frame the ball to be hit approaches the CM at this speed before the collision and then recedes at this speed afterwards, so in the lab frame it goes from 0 to $2v/(1+\alpha)$. This is also the shock velocity, since it is the speed of the ball between collisions that transmits information about the shock front. Of course the second collision will reduce the ball's velocity by a factor $(1-\alpha)/(1+\alpha)$ (unless it is last in the chain).

Will there only be these two collisions? The fact that each ball moves forward after its second collision faster than the last did, and slower than the next will, do not point to further collisions. However, note that the speed of a ball after its second collision limits to 0 as $\alpha \to 1$. If the boundary condition that started the motion (the approaching floor) requires net motion, further bounces must occur. For instance, if the first ball to be hit recoils from the floor at $2v_{\rm floor}$ and then slows to $2v_{\rm floor}(1-\alpha)/(1+\alpha)$ after its second collision. If $\alpha > 1/3$, then the latter velocity is less than $v_{\rm floor}$ and it is due to strike the floor again. In the limit $\alpha \to 1$, the many internal bounces of balls constitute a gas pressure; the balls become a gas with adiabatic index $\gamma_{\rm ad} = 3$. (In general, $\gamma_{\rm ad} = 1 + 2/D$ for a nonrelativistic monatomic gas moving in D dimensions.)

To complete the analogy, we must choose how to divide the stellar mass profile into balls. Suppose each e-folding of stellar density (density scale height) is divided evenly into N balls. Then, since $m_{\rm ext} \propto \rho^{\gamma_p}$, one e-fold of ρ is γ_p e-folds of $m_{\rm ext}$; dividing into N balls requires a mass ratio $\alpha = e^{-\gamma_p/N}$ from one to the next. The shock velocity increases by $2/(1+\alpha)$, N times per scale height: $\Delta \ln v = N \ln[2/(1+\alpha)]$, in the space where $\Delta \ln \rho = -1$. Using the expression for α,

$$\frac{d \ln v}{d \ln \rho} = -N \ln \frac{2}{1 + e^{-\gamma_p/N}} \overset{N\to\infty}{\longrightarrow} -\frac{\gamma_p}{2}.$$

In the continuum limit, then, $v \propto \rho^{-\gamma_p/2}$.

In the outer density ramp of a supernova (Sakurai 1960, Matzner & McKee 1999), the shock velocity obeys $v_s \propto \rho^{-0.18}$ to $v_s \propto \rho^{-0.2}$ for various γ_p. Hydrodynamic shock acceleration is thus several times slower than the laboratory analogue, due primarily to the fact that gas atoms move in all three dimensions, rather than only one, which implies $\gamma_{\rm ad} = 5/3$ rather than 3: gas has a softer equation of state. (In fact, the pressure in SN explosions is typically dominated by photons, which, being relativistic, are even softer: $\gamma_{\rm ad} = 4/3$.)

Shock dynamics and breakout transients

Given the shock acceleration law $v_s \propto \rho^{-0.19}$ appropriate for shocks at the outer layers of stars, several other questions must be answered before a transient can be predicted.

First, what is the proper coefficient for this power law? In the laboratory example, there is a floor velocity $v_{\rm floor}$ that depends on the height from which the balls were dropped. The lab analogue is closest in this sense to the accretion-induced collapse of white dwarfs (Tan et al. 2001), which drop inward and bounce. For ordinary core-collapse supernovae, on the other hand, the velocity scale has something to do with the relation $Mv^2 \sim E$ mentioned above, and the only density scale at hand is a typical stellar envelope density $\rho_\star$. Matzner & McKee (1999) showed that simply multiplying the deceleration law $v_s(r) \simeq \sqrt{E/M(r)}$ with the acceleration law $v_s(r) \propto [\rho(r)/\rho_\star]^{-0.19}$ and adjusting the coefficient to match a well-known solution, gives a remarkably accurate prediction of the shock velocity. (A slightly more accurate coefficient was developed by Tan et al. 2001.)

Second, suppose we wish to know the final velocity of a gas element long after it has been hit by the shock. The planar, nonrelativistic solution (Sakurai 1960, Matzner & McKee 1999) shows an asymptotic speed of just over twice v_s. However stars are spherical rather than planar, which causes the postshock acceleration to be truncated sooner for shells deeper within the star (Kazhdan & Murzina 1992; Matzner & McKee 1999), leading to final velocities less than twice v_s.

Third, will the acceleration become relativistic prior to shock breakout? Since the postshock pressure is (usually) dominated by photons, acceleration stops if the photons can leak out of the star ahead of the shock front. Photons diffuse upstream a distance L and optical depth τ in a time $\sim \tau L/c$; setting this equal to the time L/v_s for the shock to cross L, we expect the shock front to be spread over an optical depth $\tau \simeq c/v_s$. If this is less than the optical depth to the stellar surface, photons diffuse away. The end of shock acceleration thus gives rise to a flash of photons at the postshock temperature (Klein & Chevalier 1978, Imshennik & Nadëzhin 1989, Matzner& McKee 1999). The energy of the flash increases as the optical depth of the envelope is decreased (all other factors being equal), so it is larger for extend'ed red supergiant stars ($R \sim 500 R_\odot$) than for more compact stars like blue supergiants ($R \sim 50 R_\odot$) or Wolf-Rayet stars ($R \lesssim R_\odot$). At the same time, shock acceleration ends sooner in the more diffuse stars.

One criterion for relativistic shock acceleration is therefore to have a sufficiently compact (i.e., optically thick) stellar envelope. If this criterion is satisfied, the most important factor is the velocity scale $\sqrt{E/M}$: energetic explosions in low-mass envelopes are faster overall. Also somewhat important is the distribution of envelope mass: more centrally concentrated envelopes (denser centers relative to their atmospheres) have more fast ejecta. Tan et al. (2001) discuss these criteria and demonstrate that stars close to their Eddington luminosity are more centrally concentrated. For instance, the model developed for SN 1998bw has $E = 3 \times 10^{52}$ erg and $M = 4M_\odot$ in the envelope, as well as a near-Eddington luminosity. It is also compact enough for the shock to achieve relativistic speeds. Matzner & McKee (1999) and Tan et al. (2001) show that several 10^{48} erg of relativistic ejecta should develop – just enough to power GRB 980425.

Fourth, what changes as the shock becomes relativistic? Johnson & McKee (1971) address the ultrarelativistic limit, and Tan et al. (2001) investigate the transition. The shock Lorentz factor obeys $\Gamma_s \propto \rho^{-0.232}$ down the density ramp. And, the postshock acceleration leads to a final Lorentz factor for each shell that scales as $\Gamma \propto \Gamma_s^{2.73} \propto \rho^{-0.633} \propto m_{\rm ex}^{-0.633/\gamma_p}$. This means that the ejecta kinetic energy above some 4-velocity Γv varies as $v^{-(5.35\gamma_p-2)}$ and $\Gamma^{-0.63\gamma_p/(\gamma_p-0.63)}$ in the nonrelativistic and relativistic limits. (For $\gamma_p = 4/3$, for instance, $E_k \propto v^{-5.0}$ and $E_k \propto \Gamma_p^{-1.2}$ in the two regimes.) The energy distribution thus flattens out at the transition, so that a large fraction of the total energy in relativistic ejecta appears at significant Γ's.

How does the production of relativistic ejecta produce a GRB? The most obvious source is that considered by Colgate (1974): the flash from breakout itself. However, recall that this flash signals the loss of postshock pressure; the faster the shock goes, the less energy should exist in the flash. One thus expects weak flashes for shocks that become relativistic – clearly much less than the energy in relativistic motion. For example, for the SN 1998bw progenitor, I estimate $\sim 10^{44}$ ergs emerging in ~ 30 keV photons (and a maximum ejecta Lorentz factor of ~ 60). This flash has much less than the 10^{48} ergs required for GRB 980425.

The last question, then, is how to convert the kinetic energy of relativistic ejecta into soft γ-ray photons that arrive at the observer in a flash of about 15 seconds duration. For this we can turn to several helpful features of relativistic dynamics (Piran 1999): 1. Ejecta with kinetic energy E will convert it into heat (available to be radiated) if they strike only $\Gamma^{-2}E/c^2$ of ambient mass; and 2. the observer's view of the emission is sped up by a factor $2\Gamma^2$ relative to the lab frame. Somewhat surprisingly, these facts make it possible for GRB 980425 to be the result of SN 98bw's ejecta running into its presupernova stellar wind (Tan et al 2001). The production of soft γ rays, however, is unexplained.

The shock-acceleration mechanism has drawbacks if applied to GRBs like those observed at cosmological distances. The relativistic ejecta constitute only the tail of an ejecta distribution, so much more energy must be supplied (to re-emerge in an afterglow). Also, the energy is extracted by means of an external shock: a process argued to produce only smooth, single-pulse light curves like GRB 980425's.

3. Baryon removal

The alternative is to get the baryons out of the way of the energy that makes the GRB. All SN models for the cosmological GRBs do this in one way or another, as I discuss in the following sections. In keeping with the theme of this review I shall continue to concentrate on the evasion of baryon loading, rather than the specifics of the energy injection. See Matzner (2002) for more detailed discussions.

Infall

The first possibility to be considered is that the entire stellar envelope falls inward, partly directly onto the forming compact object, and partly into a rotation-supported accretion disk (as determined by its specific angular momentum j). The path along the axis would then be relatively clear, as only low-j material can exist there and it too should drain away. Since GRB models often appeal to the rapid accretion of the stellar core, it is certainly reasonable to expect the core material to have cleared out in this manner. But what about the envelope? It takes about twice the envelope's free-fall time for it to accrete, because of its initial hydrostatic pressure (MacFadyen, Woosley & Heger 2001). It would be unreasonable to expect to see a GRB much shorter than the time it takes to clear the baryons. But, to get a free-fall time of 10 s, for instance, requires mean stellar density of 4×10^4 g/cm^3. Only very compact stars with radii less than $R_\odot$ are this dense, the carbon-oxygen Wolf-Rayets being one possibility (Matzner 2003). Of course, all pre-supernova stars have such dense regions in their cores and mantle regions; it is only the overlying, more diffuse envelopes that necessarily pose a baryon loading problem.

Jets

Rather than falling away, the stellar envelope can be shoved aside by the same jet that gives rise to γ-ray emission. Once again, the relevant criterion is that the time it takes to shove aside the star (as seen by the observer) should not be long compared to the GRB itself. To apply this criterion we must model the progress of a jet on its way through the stellar envelope.

Head motion. Similar models have long existed for the expansion of radio lobes from active galaxies (Begelman & Cioffi 1989, figure 12.1): one pictures a jet travelling forward, then striking the surrounding (ambient) material in a splash region called the jet head. The jet and ambient material, now heated and pressurized, flow sideways into a bubble, the jet cocoon, that surrounds the jet. The forward motion of the jet head is determined by momentum conservation in its own frame: the momentum flux from the jet catching up behind the head must balance that of the ambient material arriving in front of it. For a relalativistic fluid, the energy and momentum flux are given by the stress tensor

$$T^{\mu\nu} = p\eta^{\mu\nu} + hU^{\mu}U^{\nu}$$

where $\eta^{\mu\nu}$ =diag$(-1,1,1,1)$ is the metric of flat space, $U^{\mu} = \gamma(1,0,0,\beta)$ is the 4-velocity for motion in the $\hat{z}$-direction, and $h = (p + u_{\rm int} + \rho c^2)$ is the (comoving) enthalpy. The equation of state gives $u_{\rm int} = p/(\gamma_{\rm ad} - 1)$ for an ideal gas, e.g., $u_{\rm int} = 3p$ if the pressure comes from photons or other relativistic particles. The energy density in the lab frame is thus $T^{0\,0} = \gamma^2 h - p$; the $\hat{z}$-momentum density and the (energy flux)$_z/c$ are both $T^{0\,3} = T^{3\,0} = \gamma^2\beta h$; and the momentum flux is $T^{3\,3} = \gamma^2\beta^2 h + p$. The form of the stress tensor is discussed in detail by Pauli (1958).

The momentum flux of ambient material arriving at the jet head from its front, then, is $p_a + \Gamma_h\beta_h h_a$, which $\simeq \Gamma_h\beta_h\rho_a c^2$ since stars cannot have relativistic internal energies ($\rho_a c^2 > u_{{\rm int},a}$).

In the jet head's frame, the momentum flux of jet material coming from behind is $p_j + (\gamma\beta)^2_{jh} h_j$, where $(\gamma\beta)_{jh} = \Gamma_j\Gamma_h(\beta_j - \beta_h)$ is the relative 4-velocity between the jet and its head. The leading p_j can be dropped if either $p_j \ll \rho_j$ (a cold jet), or $\Gamma_j^2 \gg 1$ (a relativistic jet); this simplifies the result. Making this approximation and balancing the two fluxes, one finds after a little algebra

$$\beta_h = \frac{\beta_j}{1 + \tilde{L}^{-1/2}} \quad \text{where} \quad \tilde{L} \equiv \frac{\Gamma_j^2 h_j}{\rho_a c^2}.$$

The total fluence of energy through the jet $\dot{E}_j$ is the energy flux $\Gamma_j^2\beta_j h_j c$ times the jet's cross-sectional area $A_j = \pi R_j^2$; therefore $\tilde{L} = \dot{E}_j/(\beta_j A_j \rho_a c^3)$. Now, $\dot{E}_a$ includes the rest energy carried by baryons ($\dot{M}_j c^2$) and the luminosity

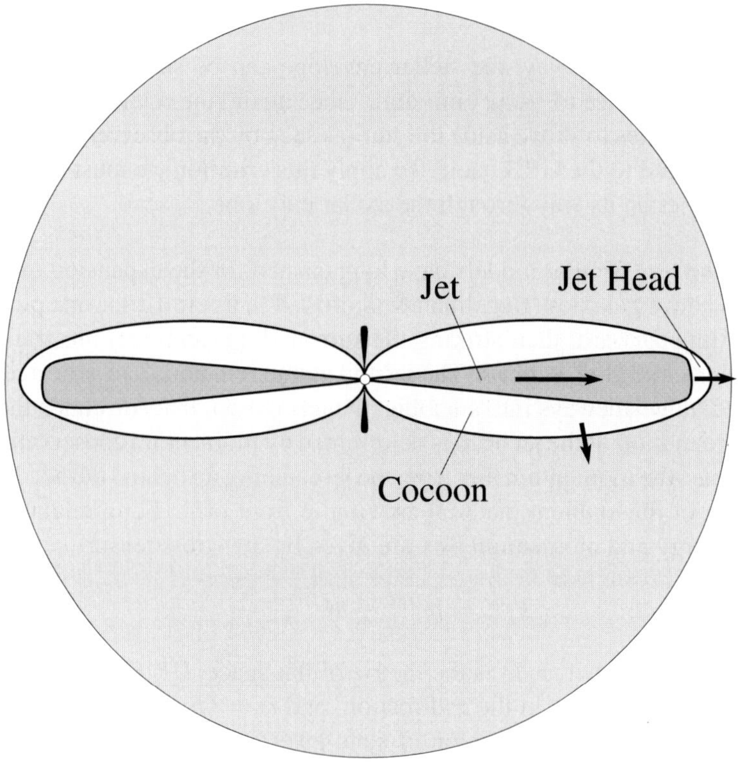

Figure 12.1. Schematic of a jet traversing a stellar envelope. The two collide at the jet head, from which material flows sideways into the cocoon.

of free energy by the central engine (L_j):

$$\tilde{L} = \frac{L_j + \dot{M}_j c^2}{\beta_j A_j \rho_a c^3} \simeq \frac{L_j/c^3}{A_j \rho_a}, \quad \text{or} \quad \frac{L_{j,\text{iso}}/c^3}{dM_a/dr}. \tag{12.1}$$

The approximation assumes the jet is relativistic, and the last expression employs the jet's isotropic luminosity $L_{j,\text{iso}} = L_j 4\pi r^2/A_j$ as well as the relation $dM_a/dr = 4\pi r^2 \rho_a$ for the mass of spherically-distributed ambient material.

What does this imply for the observed delay δt_{obs} between the onset of jet activity and the GRB? Ignoring the cosmological redshift z (which stretches all observed times by $1+z$), the observed time is related to the lab-frame time and radius $t_{\text{obs}} = t - r/c$, so

$$\frac{c\, dt_{\text{obs}}}{dr} = \frac{c\, dt}{dr} - 1 = \frac{1}{\beta} - 1$$

for an object moving directly toward the observer at velocity βc. In the case of the jet head, we have $1/\beta_h - 1 = (\tilde{L}^{-1/2} + 1)/\beta_j - 1 \simeq \tilde{L}^{-1/2}$ to high accuracy. Thus

$$c\frac{dt_{\rm obs}}{dr} = \tilde{L}^{-1/2}, \tag{12.2}$$

a remarkably simple relation that holds for an arbitrarily relativistic or nonrelativisitc jet head.

Cocoon inflation. Before deriving the time delay $\delta t_{\rm obs}$ from these results, let's pause to consider the effect of the spent jet energy on the surrounding envelope. As mentioned above, one expects the shocked jet and ambient material to flow out of the pressurized head region and into a cocoon around the jet. As long as the cocoon's pressure exceeds that of the surrounding envelope (and its infall ram pressure, if it is moving), it will expand sideways at a speed $\beta_c \simeq p_c/(\rho_a c^2)$ – where p_c is the cocoon pressure. Assume a relation $p_c = E_c/(3V_c)$ between the cocoon's energy, volume and pressure. Its energy E_c is that part of the emitted energy that has caught up to the head; if $\Gamma_j \gg \Gamma_h$, this reduces to $L_j t_{\rm obs}$. The volume V_c must be the product of its length (r, the distance the jet head has traveled) and the square of its sideways extent [$\sim (c\beta_c t)^2$], up to a constant. Assuming a nonrelativistic jet head, algebra gives

$$\beta_c \simeq \tilde{L}^{3/8}\theta_j^{1/2}$$

where θ_j is the jet's opening angle. For a given $\tilde{L}$, wider jets carry more energy but do not move faster, so they blow fatter cocoons. It is possible for the cocoon to get ahead of the jet [if $\tilde{L} < (\theta_j/90^\circ)^4$, it turns out], in which case the jet powers a spherical explosion rather than clearing the envelope out of its path. (This places a rather weak constraint on the type of star that can launch a GRB.) It is also possible for the cocoon to be crushed by ambient or ram pressure, but only in the dense stellar core; see Matzner (2002) for more details.

Are jets pressure-confined?. One would like to know what values to give the jet luminosity L_j and area A_j when evaluating $\tilde{L}$. The observation of an isotropic-equivalent γ-ray luminosity $L_{{\rm iso},\gamma}$ and the inference, at least, of a jet opening angle θ_γ give a sense of how to do this. If γ-rays were produced from the jet at an efficiency ϵ_γ, then $L_j = (\theta_\gamma^2/4)L_{{\rm iso},\gamma}/\epsilon_\gamma$. If the opening angle has been preserved since the jet was inside the star, then in addition $\theta_j = \theta_\gamma$ and $A_j = \pi r^2\theta_\gamma^2$.

But how can we be sure that the jet did not spread in angle after breaking out of the star? For this to be possible, the jet would have to have a relativistic internal energy ($p_j \gg \rho_j c^2$) at the time of breakout. This would lead both to spreading and acceleration as the internal is converted into kinetic energy. However, lateral spreading is limited by relativistic effects to an angle $1/\Gamma_j$ (for the same reason that isotropic emission of photons gets boosted into a $1/\Gamma$

cone by relativistic motion):

$$\theta_\gamma \simeq \max(\theta_j, 1/\Gamma_j),$$

where the two cases refer to cold jets (incapable of spreading) and to hot, spreading jets, respectively. If the latter is true, θ_γ still yields useful information: it gives the jet's Lorentz factor at the time of breakout. Moreover, if the jet is pressure-confined, its own cocoon provides the confining pressure. This suggests a picture in which the jet expands laterally until the cocoon pressure matches its internal pressure (which decreases sharply with increasing θ_j).

Adiabatic confined jets: Suppose, for example, the jet expands adiabatically from an initial radius $r_0 \sim 10$ km, the scale of the launching region. Conserving both the jet's luminosity $L_j = (4\Gamma_j^2 p_j)\pi(\theta_j r)^2$ (since $u_{\mathrm{int},j} = 3p_j \gg \rho_j$) and its mass flux $\dot{M}_j = \Gamma_j \rho_j A_j$, and taking $p_j \propto \rho_j^{4/3}$,

$$\Gamma_j(r) = \frac{R_j}{r_0} \quad \text{and} \quad p_j(r) = \frac{L}{4\pi r_0^2 c}\Gamma_j^{-4}. \qquad \text{(adiabatic jet)}$$

The Lorentz factor goes up and the pressure drops precipitously as the jet radius R_j increases. The cocoon pressure $\rho_a \beta_c^2 c^2$, by contrast, varies as $R_j^{-1/2}$; matching the two,

$$\Gamma_j = \left(\frac{R_{j1}^{1/2} r}{4 r_0^{3/2}}\right)^{2/7}, \quad \text{where} \quad R_{j1}(r) \equiv \left[\frac{L}{\pi c^3 \rho_a(r)}\right]^{1/2}$$

is the jet radius at which $\tilde{L} = 1$ [given $\rho_a(r)$ locally]. The jet radius is $R_j = \Gamma_j r_0$ and the final beaming angle is given by $\theta_\gamma = \Gamma_j^{-1}$.

Evaluating in terms of solar masses and radii, and taking $r \to R_\star$, $L \to 10^{52} L_{52}$ erg/s and $\rho_a \to M_{\mathrm{env}}/(4\pi R_\star^3)$ appropriate for a star with an r^{-2} density profile, one finds

$$\theta_\gamma = 0.88^\circ \left(\frac{R_\star}{R_\odot}\right)^{-1/2} \left(\frac{r_0}{10\ \mathrm{km}}\right)^{3/7} \left(\frac{M_{\mathrm{env}}}{M_\odot}\right)^{1/14} L_{52}^{-1/14}.$$

At first glance this looks reasonable, but note that $\theta_j \sim 3 - 10^\circ$ is considered typical, and that $p_j/p_c \propto \theta_j^{-3.5}$. For this reason it seems most likely that pressure confinement ceases in the core regions where $R \ll R_\odot$, and the jet continued as a cold, ballistic flow with fixed θ_γ from there on out (so $\theta_j = \theta_\gamma$ in the outer envelope). The same conclusion was reached by Mészáros & Rees (2001).

Nonadiabatic confined jets: Now suppose instead the jet's expansion is *not* adiabatic: for instance, it may have mixed with some of the surrounding material on its travels toward the surface. Intuitively, this would erase the jet's recollection of its small launching scale r_0. If a much larger effective r_0 were used in the above expression, it might be consistent with pressure confinement all the way to the stellar surface.

To test this idea, we need a theory for the mixing of a jet with its surroundings – i.e., with the cocoon material (density $\rho_c < \rho_a$). This is most easily done by imagining a jet of fluid, instantaneously uniform (in the lab frame) with area A_j, that mixes with an area δA of the cocoon. The total energy, momentum, and rest mass are each conserved through the mixing process. Since the length of the jet is not changing (by construction), one can write

$$\delta\left(\left\{\begin{array}{c}\Gamma_j^2 h_j - p_j\\ \Gamma_j^2\beta_j h_j\\ \Gamma_j\rho_j\end{array}\right\}A_j\right) = \left\{\begin{array}{c}\rho_c c^2\\ 0\\ \rho_c\end{array}\right\}\delta A_j.$$

This gives expressions for δp_j, $\delta\rho_j$ and $\delta\gamma_j$ given δA_j. As suspected, the effective launching scale $r_0 \equiv R_j/\Gamma_j$ increases during this type of mixing, because R_j increases and Γ_j either stays constant (if $\rho_c = 0$) or decreases. Values of $r_0 = \{175, 2900\}\mathrm{km} \times (R_\star/R_\odot)^{7/6}(M_{\mathrm{ej}}/M_\odot)^{-1/6}$ would be sufficient to make $\theta_\gamma = \{3^\circ, 10^\circ\}$.

Thus, non-adiabatic mixing between jet and cocoon *might* make the observed GRBs consistent with a jet that was pressure-confined within the outer envelopes of compact stars. However, note the following caveats:

- The final Lorentz factor must be on the order of $10^2 - 10^3$; this constrains how much mass can mix into the jet.

- Numerical simulations cannot hope to predict the degree of mixing faithfully, primarily because $r_0 \ll R_\star$ and both scales must be resolved in the same simulation.

- Most seriously: The natural internal timescale of the jet is R_j/c. Since the jet is hot, one expects it to filter fluctuations on shorter scales. The observer sees these fluctuations on a timescale $\Gamma_j R_j/c = r_0/(\theta_\gamma^2 c) \sim 0.2(R_\star/R_\odot)^{7/6}$ s. Since this is much longer than the shortest fluctuations seen (milliseconds), a hot jet would have to recreate short-time fluctuations by some mechanism. (Perhaps, for instance, from entropy fluctuations in the jet at breakout.)

Ballistic jets: constraints on progenitors. Keeping in mind that GRB jets might be confined by cocoon pressure throughout the entirety of their star's envelope, it is worthwhile to calculate what one expects for *ballistic* jets – jets whose opening angle is set deep below the outer envelope. In this case, as discussed at the beginning of §12.1, $\theta_j = \theta_\gamma$ and $L_j = (\theta_\gamma^2/4)L_{\mathrm{iso},\gamma}/\epsilon_\gamma$.

In that case the formulae for the motion of the jet head derived in §3 apply directly.

Why should the jet shut off just as it finishes burrowing through the envelope? The requirement that removing baryons should not take long compared to GRB durations is, as in §3, the most important constraint. Putting numbers into equations (12.1) and (12.2), and adding the redshift-dependent time dilation,

$$\left[\frac{t_{\rm obs}}{20(1+z)\ {\rm s}}\right]^2 \gtrsim \frac{\varepsilon_\gamma}{L_{\gamma,{\rm iso}52}} \frac{M_{\rm env}}{M_\odot} \frac{R}{R_\odot}. \tag{12.3}$$

This seems a rather stringent contraint when one considers that core-collapse supernova progenitors usually have $M_{\rm env}R \gg M_\odot R_\odot$. An exception are the carbon-oxygen post-Wolf-Rayet progenitors of type Ic supernovae, like the SN 1998bw progenitor.

Delayed GRBs

In some theories, the stellar envelope is stripped or blown away in a precursor event that then triggers the GRB. One likely scenario involves the coalescence of a helium star and its compact companion (Fryer & Woosley 1998). The outer envelope would be stripped away in the early phases of the event, leaving behind dense stellar core material to accrete onto a pre-existing or newly-formed black hole. This sequence of events can affect baryon loading different ways. If the stripped material flies off entirely in the orbital plane and the GRB is beamed along the orbital axis, then the barrier posed by the envelope is avoided geometrically. However: if some of the envelope is cast in the jet direction, it will smother the jet *more* effectively as it expands, because the product $M_{\rm env}R$ increases with time in equation (12.3). It is only after the ejecta have expanded long enough to be optically thin to γ-rays that they no longer matter.

The ‘supranova’ model of Vietri & Stella (1998) invokes a supernova explosion to remove the stellar envelope and to produce a neutron star too massive to exist without spinning. This then spins down electromagnetically, eventually collapsing into a black hole and producing the γ-ray burst. An attractive aspect of this model is the vast spindown energy ($\sim 10^{53}$ erg) available to clear away or aside the stellar envelope – a process likely to be accelerated by the Rayleigh-Taylor instability. However, viscous accretion of neutron star material onto a black hole cannot match the duration of long GRBs (Janka et al. 1999).

4. Conclusions

The conclusion most easily drawn from these considerations is that stars are perilous environments for highly relativistic flow, and that baryon loading is a serious problem even in models that successfully predict a GRB central engine. Every viable model discussed here to evade this problem rests on the progenitor star being quite compact – e.g., a post-Wolf-Rayet star, with no hydrogen envelope. It must be compact enough to develop a relativistic outer shock (§2), to collapse in time (§3), to provide the pressure to confine a relativistic jet (§12.1), or to be blown aside in time by a ballistic jet (§12.1). Helium star coalescence models also appeal to quite compact progenitors. Only the 'supranova' does not; but, it does not seem relevant to long-duration GRBs. Only post-Wolf-Rayet stars (of helium or carbon-oxygen composition), and perhaps the cores of very massive stars (Bond, Arnett, & Carr 1984), are compact enough to be SN progenitors for γ-ray bursts.

References

Begelman M. C., Cioffi D. F., 1989, *Ap. J. (Letters)*, 345, L21

Bond J. R., Arnett W. D., Carr B. J., 1984, *Ap. J.*, 280, 825

Colgate, S. A., 1974, *Ap. J.*, 187, 333

Fishman, G. J. & Meegan, C. A. 1995, Ann. Rev. Astr. Ap, 33, 415

Frail, D. A. et al. 2001, *Ap. J. (Letters)*, 562, L55

Fryer, C. L. & Woosley, S. E. 1998, *Ap. J. (Letters)*, 502, L9

Kouveliotou, C., Meegan, C. A., Fishman, G. J., Bhat, N. P., Briggs, M. S., Koshut, T. M., Paciesas, W. S., & Pendleton, G. N. 1993, *Ap. J. (Letters)*, 413, L101

Lithwick Y., Sari R., 2000, in ApJ submitted, astro-ph/0011508

Höflich P., Wheeler J. C., Hines, D. C., & Trammell, S. R. 1996, *Ap. J.*, 439, 307556L

Imshennik, V. S. & Nadëzhin, D. K. 1989, Ast. Space Phys., 8, 1

Janka, H.-T., Eberl, T., Ruffert, M., & Fryer, C. L. 1999, *Ap. J. (Letters)*, 527, L39

Kazhdan, Y. M. & Murzina, M. 1992, *Ap. J.*, 400, 192

Klein, R. I. & Chevalier, R. A. 1978 *Ap. J. (Letters)*223, 109

MacFadyen, A. I., Woosley, S. E. & Heger, A.,

Matzner, C. D. 2002, MNRAS, submitted, astro-ph/0203085

Matzner C. D. & McKee C. F., 1999, *Ap. J.*, 510, 379

Matzner C. D., Tan J. C., McKee C. F., 2001, in Gamma 2001, eds. S. Ritz, N. Gehrels, and C. Shrader, astro-ph/0106338

Johnson, M. H. & McKee, C. F. 1971, Phys.Rev.D., 3, 858

Meegan, C. A., Fishman, G. J., Wilson, R. B., Horack, J. M., Brock, M. N., Paciesas, W. S., Pendleton, G. N., & Kouveliotou, C. 1992, Nature, 355, 143

Mészáros, P. 2002, Ann. Rev. Astr. Ap, 40, 137

Mészáros P., Rees M. J., 2001, *Ap. J. (Letters)*, 556, L37

Metzger, M. R., Djorgovski, S. G., Kulkarni, S. R., Steidel, C. C., Adelberger, K. L., Frail, D. A., Costa, E., & Frontera, F. 1997, Nature, 387, 878

Pauli, W. *Theory of Relativity*, 1958, Pergamon Press (NY)

Piran, T. 1999, Physics Reports, 314, 575
Rhoads, J. E. 1999, *Ap. J.*, 525, 737
Sakurai A., 1960, Comm. Pure Appl. Math., 13, 353
Tan J. C., Matzner C. D., McKee C. F., 2001, *Ap. J.*, 551, 946
Vietri, M. & Stella, L. 1998, *Ap. J. (Letters)*, 507, L45

Chapter 13

STELLAR COLLAPSE AND GRAVITATIONAL WAVES

Chris L. Fryer
Theoretical Astrophysics, Los Alamos National Laboratories, Los Alamos, NM, 87545

Daniel E. Holz, Scott A. Hughes and
Kavli Institute for Theoretical Physics, UC Santa Barbara, Santa Barbara, CA 93106

Michael S. Warren
Theoretical Astrophysics, Los Alamos National Laboratories, Los Alamos, NM, 87545

Abstract

The new generation of gravitational wave (GW) detectors have the potential to open a new window onto the violent dynamics of core collapse. Although it is certain that core collapse events generate gravitational radiation, understanding the characteristics of the radiation—whether it can be measured with the new detectors, and the best way to go about doing so—is a challenging problem. In this chapter we review the promise of GWs as observational probes, including a discussion of the current state of new detectors, and discuss the status of work to understand the waves generated by stellar core collapse.

1. Introduction

For quite some time the collapse of massive stellar cores in supernova explosions has been regarded as likely to be an interesting and important source of gravitational radiation (see, for example, Eardley 1983 for an early review of the subject). Core collapse events certainly produce GWs: large amounts of mass ($\sim$ 1–100 $M_{\odot}$) flow in a compact region ($\sim 10^8$–10^9 cm) at relativistic velocities ($v/c \sim 1/5$). These characteristics are *necessary* conditions for a source to be an interesting GW source; however, they are *not* sufficient con-

C.L. Fryer (ed.), Stellar Collapse, 373–402.

ditions. Detailed analysis is needed to understand whether these potentially interesting sources are in fact sufficiently asymmetric to be strong radiators, and whether we can understand their robust features well enough to search for them with GW observatories.

Theoretical modeling of stellar collapse has become increasingly sophisticated in recent years: collapse theorists can now model collapses in three dimensions (albeit requiring enormous amounts of CPU time), and can include important physics such as neutrino transport and the equation of state of very dense matter. Previously, most estimates of GW emission from stellar collapse were (of necessity) rather idealized, typically studying the evolution of instabilities in rotating fluid distributions (Chandrasekhar 1969, Centrella et al. 2000, New et al. 2000). Although extremely valuable in establishing the conditions under which strongly radiating instabilities can occur, these analyses do not incorporate important physical effects which occur during stellar collapse. Using modern collapse models, estimates of GW emission can now be constructed that more realistically reflect the phenomenology of stellar collapse. Coupling these modern models with the wisdom that has been gleaned from decades of study from the idealized cases, we should now be able to assemble a robust picture of the GWs arising from core collapse.

The maturity of modern collapse simulations is quite timely: they arrive just as broad-band GW detectors based on laser interferometry are coming into existence. Science runs with the initial generations of these detectors have just begun, although these instruments have not yet reached their design sensitivities. Furthermore, the design sensitivities are not really at the level where stellar core collapse can be considered a realistically interesting signal (Fryer, Holz, & Hughes 2002, hereafter FHH). However, the detectors' sensitivites have improved quite rapidly[1], and a vigorous R&D program promises to push the sensitivites to better levels very rapidly (Gustafson, Shoemaker, Strain, & Weiss 1999). The possibility of seeing core collapse GWs is stronger than ever[2].

In this chapter we review the state of our present knowledge of GW events. We first present an overview of GW physics, focusing on the properties of the waves themselves and how one detects them, plus a brief review of the current state of the various GW interferometers. (We note that the discussion of the detectors' state is quite likely to be out of date by the time this article appears in print.) This material is largely taken from a recent review article (Hughes et al. 2001) with some updating to reflect the current status of the detectors. We then turn to an in-depth discussion of GWs from core collapse events, comparing the likely properties of these waves to the detection thresholds of the new interferometric detectors. This discussion is based to a large extent on the review of FHH, with some updating based on recent work on r-modes and 3-D simulations.

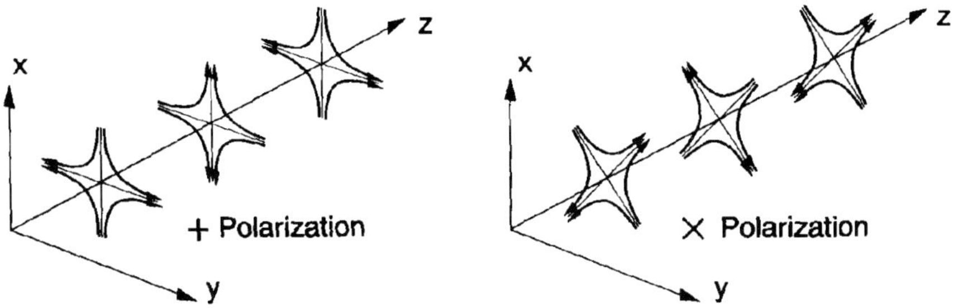

Figure 13.1. The lines of force associated with the two polarizations of a GW (from Abramovici et al. 1992).

2. GWs and detectors: overview

Gravitational radiation is a natural consequence of general relativity, first described more-or-less correctly by Albert Einstein (1918). GWs are tensor perturbations to the metric of spacetime, propagating at the speed of light, with two independent polarizations. As electromagnetic radiation is generated by the acceleration of charges, gravitational radiation arises from the acceleration of masses. Electromagnetic waves are created (at lowest order) by the time changing charge dipole moment, and are thus dipole waves. Monopole EM radiation would violate charge conservation. At lowest order, GWs come from the time changing quadrupolar distribution of mass and energy; monopole GWs would violate mass-energy conservation, and dipole waves violate momentum conservation.

GWs act tidally, stretching and squeezing objects as they pass through. Because the waves arise from quadrupolar oscillations, they are themselves quadrupolar in character, squeezing along one axis while stretching along the other. When the size of the object that the wave acts upon is small compared to the wavelength (which is the case for all ground-based detectors), forces arising from the two GW polarizations act as in Fig. 13.1. The polarizations are named "+" (plus) and "×" (cross), as a result of the orientation of the axes associated with their force lines.

Interferometric GW detectors measure this tidal field via their action upon a widely-separated set of test masses, arranged as in Fig. 13.2. Each mass is suspended with a sophisticated pendular isolation system to eliminate the effect of local ground noise. Above the resonant frequency of the pendulum (typically of order 1 Hz), the mass moves freely. In the absence of a GW, the sides L_1 and L_2 shown in Fig. 13.2 are taken to have the same length, L.

Suppose the interferometer in Fig. 13.2 is arranged so that its arms lie along the x and y axes of Fig. 13.1. Suppose further that a wave impinges on the detector down the z axis, and the "+" polarization axes are aligned with the detector's arms. The tidal force of this wave stretches one arm while squeezing the other; each arm oscillates between stretch and squeeze as the wave itself oscillates. The wave is detectable by measuring the separation of the test masses in each arm. In particular, since one arm is always stretched while the other is squeezed, we can monitor the length difference of the two arms:

$$\delta L(t) \equiv L_1(t) - L_2(t) \; . \tag{13.1}$$

For the case discussed above, this change in length turns out to be the armlength times the $+$ polarization amplitude:

$$\delta L(t) = h_+(t) L \; . \tag{13.2}$$

The GW acts as a dimensionless strain in the detector; h is often referred to as the "wavestrain". Equation (13.2) is derived by applying the equation of geodesic deviation to the separation of the test masses, using a GW tensor on a flat background spacetime to develop the curvature tensor; see Thorne (1987), Sec. 9.2.2 for details. We obviously do not expect astrophysical GW sources to align themselves in as convenient a manner as described above. Generally, both polarizations of the wave influence the test masses:

$$\frac{\delta L(t)}{L} = F^+ h_+(t) + F^\times h_\times(t) \equiv h(t) \; . \tag{13.3}$$

The antenna response functions F^+ and $F^\times$ weight the two polarizations in a quadrupolar manner as a function of a source's position and orientation relative to the detector; see Thorne (1987), Eqs. (104a,b) and associated text.

The test masses at the ends of each arm are made of a highly transparent material (fused silica in present designs; perhaps sapphire in future upgrades). All of the test masses are faced with dielectric coatings such that they are extremely good reflectors of the 1.064 micron infrared laser light. The mirrors at the far end of each arm have amplitude reflectivities approaching unity. The mirrors at the corner joining the arms are less reflective, since they must couple the light into the Fabry-Perot cavity arms. The corner mirrors' multilayer dielectric coatings have power reflectivities $T \sim 3\%$. A very stable laser beam is divided at the beamsplitter, directing light into the two arm cavities. If the finesse of the cavity is $\mathcal{F}$ and the amplitude reflectivity of the corner mirrors is $r_{\rm corner}$, then each photon makes on average $\mathcal{F}/\pi \simeq \sqrt{r_{\rm corner}}/(1-r_{\rm corner}) \sim 65$ bounces. The light from the two arms then recombines at the beamsplitter. The mirrors are positioned so that, in the absence of a GW, all of the light goes back to the laser and the photodiode reads no signal. If a signal is present, the relative

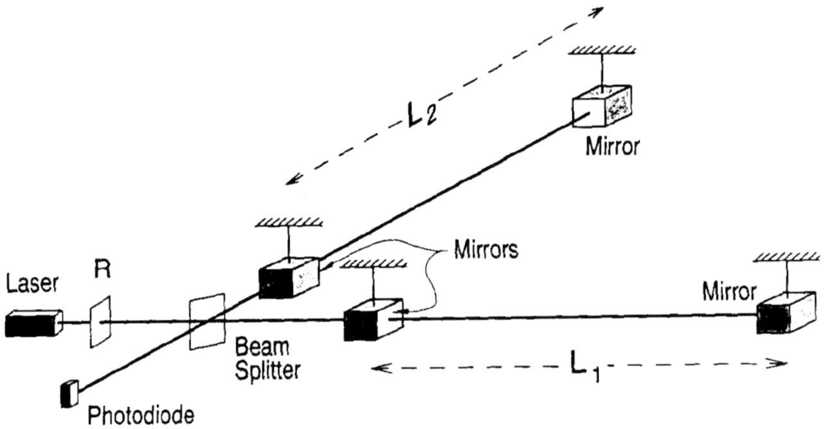

Figure 13.2. Layout of an interferometer for detecting GWs (from Abramovici et al. 1992).

phase Φ of the two beams must have changed by an amount proportional to h, changing the light's interference pattern. With no intervention, this would cause light to leak into the photodiode. In principle, the wavestrain h could be read from the intensity of this light. In practice, a system of servo loops controls the system such that destructive interference is guaranteed — the photodiode is kept dark, and is thus called the "dark port". The wavestrain h is encoded in the servo signals used to keep the dark port dark.

The energy flux carried by GWs scales with $\dot{h}^2$ (where the overdot represents a time derivative). To conserve energy flowing through large spheres, the wavestrain falls off with distance as $1/r$. We have already argued that the lowest order contribution to the waves is due to the changing quadrupole moment of the source. To order of magnitude, this moment is given by $Q \sim$ (source mass)(source size)^2. Since the wavestrain is dimensionless, the scaling must take the form

$$h \sim \frac{G}{c^4}\frac{\ddot{Q}}{r}\,. \tag{13.4}$$

The second time derivative of the quadrupole moment is given approximately by $\ddot{Q} \simeq 2Mv^2 \simeq 4E^{\rm ns}_{\rm kin}$, where v is the source's internal velocity, and $E^{\rm ns}_{\rm kin}$ is the nonspherical part of its internal kinetic energy. Strong sources of gravitational radiation are sources that have strong non-spherical dynamics; hence, core collapse events must be quite asymmetric in order to give off considerable energy in GWs.

For an interesting rate of observable events, detectors must be sensitive to sources at rather large distances. For example, to detect several stellar collapse event in a year, our detectors must reach $r \sim 10$ Mpc. For stellar collapse, a reasonable estimate of the non-spherical kinetic energy is $E_{\rm kin}^{\rm ns}/c^2 \sim 0.1\,M_\odot$. Plugging these numbers into Eq. (13.4) yields the estimate

$$h \sim 10^{-21}\text{–}10^{-22}\ . \tag{13.5}$$

These tiny numbers set the sensitivity required to measure GWs. Combining this scale with Eq. (13.3) says that for every kilometer of baseline L, we must measure a distance shift between the arm lengths, δL, of better than 10^{-16} centimeters.

The prospect of achieving such stringent displacement sensitivities often strikes people as insane. How can light, whose wavelength $\lambda \sim 10^{-4}$ cm is 10^{12} times larger than the typical displacement, be used to measure that displacement? For that matter, how is it possible that thermal motions don't wash out these tiny effects?

That such measurement is possible with laser interferometry was first analyzed thoroughly by Weiss (1972). [3] We examine first how a 1 micron laser can measure a 10^{-16} cm effect. As mentioned above, the light bounces back and forth roughly 100 times before leaving the arm cavity (corresponding to about half a cycle of a 100 Hz GW). The light's accumulated phase shift during those 100 round trips is

$$\Delta\Phi_{\rm GW} \sim 100 \times 2 \times \Delta L \times 2\pi/\lambda \sim 10^{-9}\ . \tag{13.6}$$

This phase shift is measurable provided that the shot noise at the photodiode, $\Delta\Phi_{\rm shot} \sim 1/\sqrt{N}$, is less than $\Delta\Phi_{\rm GW}$. N is the number of photons accumulated over the measurement; $1/\sqrt{N}$ is the magnitude of phase fluctuation in a coherent state, appropriate for describing a laser. We therefore must accumulate 10^{18} photons over the roughly 0.01 second measurement, which translates to a laser power of about 100 watts. In fact, as was pointed out by Drever (1983), one can use a much less powerful laser: even in the presence of a GW, only a tiny portion of the light that comes out of the interferometer's arms goes to the photodiode. The vast majority of the light is sent back to the laser. An appropriately placed mirror bounces this light back into the arms, *recycling* it. The recycling mirror is shown in Fig. 13.2, labeled "R". With that mirror, a laser of $\sim$ 10 watts drives several hundred watts to circulate in the "recycling cavity" (the optical cavity between the recycling mirror and the arms), and $\sim$ 10 kilowatts to circulate in the arms.

Thermal excitations are overcome by averaging over many many vibrations. For example, the atoms on the surface of the interferometers' test mass mirrors oscillate with an amplitude

$$\delta l_{\rm atom} = \sqrt{\frac{kT}{m\omega^2}} \sim 10^{-10}\,{\rm cm} \tag{13.7}$$

at room temperature T, with m the atomic mass, and with a vibrational frequency $\omega \sim 10^{14}\,\mathrm{s}^{-1}$. This amplitude is huge relative to the effect of the GW—why doesn't it wash out the wave? GWs are detectable because the atomic vibrations are random and incoherent. The $\sim$ 7 cm wide laser beam averages over about 10^{17} atoms and at least 10^{11} vibrations in a typical measurement. These atomic vibrations cancel out, becoming irrelevant compared to the coherent effect of a GW. Other thermal vibrations, however, end up dominating the detectors' noise spectra in certain frequency bands. For example, the test masses' normal modes are thermally excited. The typical frequency of these modes is $\omega \sim 10^5\,\mathrm{s}^{-1}$, and they have mass $m \sim 10\,\mathrm{kg}$, so $\delta l_{\mathrm{mass}} \sim 10^{-14}\,\mathrm{cm}$. This, again, is much larger than the effect we wish to observe. However, the modes are very high frequency, and so can be averaged away provided the test mass is made from material with a very high quality factor Q (so that the mode's energy is confined to a very narrow band near ω, and thus doesn't leak into the band we want to use for measurements). Understanding the physical nature of noise in GW detectors is an active field of current research; see Levin (1998), Liu & Thorne (2000), Santamore & Levin (2001), Buonanno & Chen (2001a,b), Hughes & Thorne (1998), Creighton (2000), and references therein for a glimpse of recent work. In all cases, the fundamental fact to keep in mind is that a GW acts *coherently*, whereas noise acts *incoherently*, and thus can be beaten down provided one is able to average away the incoherent noise sources.

Current detectors

The first generation of long baseline, kilometer-scale interferometric GW detectors are in operation or are being constructed and commissioned at several sites around the world. Briefly, the major ground-based interferometric GW projects are as follows:

- **LIGO**. Three LIGO (Laser Interferometer Gravitational-wave Observatory) interferometers are currently operating: two in Hanford, Washington (with 2 and 4 km arms, sharing the same vacuum system), and one in Livingston, Louisiana (4 km arms). An aerial view of the Hanford site is included in Fig. 13.3. The LIGO detectors are designed to operate as power recycled Michelson interferometers with arms acting as Fabry-Perot cavities. The large distance between sites (about 3000 km) and differing arm lengths are designed to support coincidence analysis. Much current research and development is focused on advanced LIGO detector design. The goal of these planned improvements is to provide a broader frequency band and a $\sim$ 10-fold increase in range for sources via a lowered noise floor.

- **Virgo**. Virgo, the Italian/French long baseline GW detector, is under construction near Pisa, Italy (Marion 2000). It has 3 km arms and advanced passive seismic isolation systems. In most respects, Virgo is similar to LIGO; a major difference is that it should achieve better low frequency sensitivity in its first generation due to its advanced seismic isolation. Virgo will usefully complement the LIGO detectors, strengthening coincidence analysis and significantly improving source position determination.

- **GEO600**. GEO600 is a 600 meter interferometer constructed by a British-German collaboration near Hannover, Germany (Lück et al. 2000). It uses advanced interferometry and advanced low noise multiple pendulum suspensions, serving as a testbed for advanced detector technology, and allowing it to achieve sensitivities comparable to the multi-kilometer instruments.

- **TAMA300**. The TAMA detector near Tokyo, Japan can already claim significant observation time, with more than 1000 hours of operation (Ando et al. 2001). It has achieved a peak strain sensitivity of $h \sim 3 \times 10^{-19}\,\mathrm{Hz}^{-1/2}$ at frequencies near 1000 Hz. TAMA has 300 meter arms and is operated in the recombined Michelson configuration with Fabry-Perot arms. A much improved 3 km detector is currently under design (Kuroda et al. 2000).

- **ACIGA**. The Australian Consortium for Interferometric Gravitational Astronomy plans to build an observatory near Perth, Australia (McCleland et al. 2000). They are presently engaged in the construction of an 80 meter research interferometer, which can be extended to kilometer scale. They are studying advanced detection methods and technologies which could lead to much decreased noise floors in advanced interferometers.

Since interferometric GW detectors have nearly equal sensitivity in all directions, it is essentially impossible to deduce pointing information from the output of a single detector. To get accurate information about the source direction it is necessary to make use of time-of-flight differences between detectors. To do this the detectors must be widely spaced, and cannot be collinear. At the very minimum, three sites are needed for acceptable pointing. A fourth detector, widely removed from the plane of the other three, is particularly valuable for improving directional information. Thus a detector in Australia would greatly add to the science output of the GW observatory network, which is otherwise confined entirely to the northern hemisphere.

LIGO is representative of the design and operation principles of these interferometers, so we shall focus on it for the remainder of our discussion.

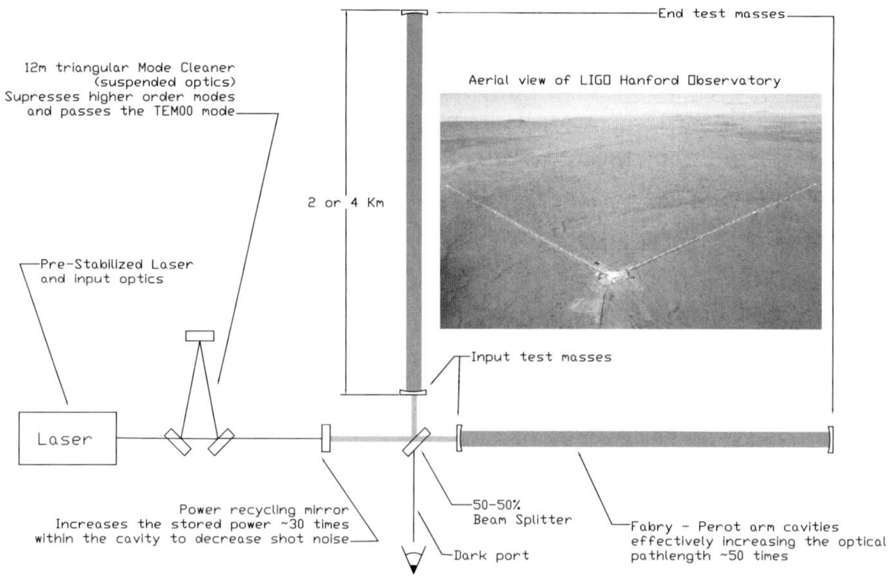

Figure 13.3. Simplified optical layout of a LIGO interferometer. Shown here are the prestabilized laser, the input mode cleaner, the recycling mirror, and the test mass mirrors. As discussed in Sec. 2, servo loops ensure that the recombined light destructively interferes so that the dark port is kept dark. The GW signal is read out from the forces needed to keep the recombined light in destructive interference.

LIGO overview

Construction of both LIGO facilities and vacuum systems was completed in early 2000. Both observatories have mostly concentrated to date on detector commissioning and a series of "Engineering Runs", leading to the first "Science Run", which began in August 2002. This will be followed by more frequent and longer Science Runs until 2006/7, when major detector upgrades are scheduled. In parallel with the commissioning effort and Engineering/Science Runs, the LIGO collaboration is also focused on the research and development of advanced detectors, promising a wider detector band and greatly improved sensitivity.

The LIGO detectors operate as power recycled Michelson interferometers with Fabry-Perot arms; see Fig. 13.3. Very high duty cycle is needed for each interferometer in order to effectively use the full network for coincidence analysis, which is necessary for achieving a low false detection rate and high confidence observations. The wide (3000 km) separation between the LIGO sites is large enough that the chance of environmentally induced coincidence events is small. Both sites are equipped with environmental sensors that cover a wide range of possible disturbances that otherwise could cause false detections. For example, LIGO monitors the local seismic background, electromagnetic fluctuations, acoustic noise, cosmic radiation, dust, vacuum status, weather, power line transients, and magnetic fields (using ultra-sensitive magnetometers at several locations at each observatory).

We now briefly describe the operating principles of LIGO, the major sources of noise that limit sensitivity, and plans for future upgrades.

Laser, optics, and configuration. The basic optical layout of the LIGO detectors is shown in Figure 13.3. LIGO uses a Nd:YAG near infrared laser (wavelength 1064 nm) with peak power $\sim$ 10W as the light source. Various electro-optical components and servo loops are used to stabilize both the frequency and power of the laser.

The light from the pre-stabilized laser passes through the input optics and is coupled into the 12 meter, triangular mode cleaner cavity. The mode cleaner passes only the TEM_{00} mode, eliminating higher order modes. Starting with the mode cleaner, every major optical component is within a large vacuum system, operating at 10^{-9} Torr. After conditioning by the mode cleaner, the light enters the interferometer. All major optics in the interferometer are suspended on a single steel wire loop, mechanically isolated from the ground by vibration isolators and controlled by multiple servo loops. The mirrors are made of fused silica with extremely high mechanical Q and polished to within $\sim$ 1 nanometer RMS. They have high homogeneity, low bulk loss, and multi-layer coatings with less than 50 ppm scattering loss. Each mirror is actuated by four precision coils, each positioned around a permanent magnet glued to the back side of the test masses. The coil assembly also features a sensitive shadow sensor for local control. Additional optical levers and wavefront sensors provide more precise sensing. The laser beam is coupled into the arms by a beam splitter. Each arm is a Fabry-Perot optical cavity, increasing the effective length of the arm to magnify the phase shift (proportional to cavity finesse) caused by the wave. The stored power within the interferometer is built up by the partially transmitting recycling mirror.

An operating interferometer tries to keep the dark port perfectly dark, adjusting the positions of the various optical components such that light coming out of the arms destructively interferes and no light goes to the photodetector. When this is achieved, the interferometer is on *resonance*, with maximum power circulating in the arms. Several interconnected control loops are used to achieve and then maintain resonance. An interferometer on resonance is usually described as *locked*. Keeping lock must be highly automated, requiring minimal operator interaction for high uptime. The GW signal is extracted from the servo signals used to maintain the lock and correct for the changing length difference between the arms.

Noise sources. The sensitivity of GW interferometers is limited by a large number of noise sources. We list here some of the most important and interesting fundamental noise sources; many of these were originally recognized, and had their magnitudes estimated, by Weiss (1972).

- **Seismic noise**. Ambient seismic waves (e.g., loading of the continental shelf by surf striking the coast) or culturally induced seismic waves (e.g., passing trucks, logging, cattle guards) continuously pass under the test masses of the detector. The natural motion of the surface peaks around 150 mHz; this is called the "microseismic peak". Cultural noises tend to be at higher frequencies, near several Hz. The test masses must be carefully isolated from the ground to effectively mitigate seismic noise. Seismic noise will limit the low frequency sensitivity of first generation ground-based GW detectors; the only way to get good performance below ~ 1 Hz is to put one's detector space. This is a major motivator for the LISA[4] gravitational-wave detector.

- **Thermal noise**. Thermally excited vibrational modes of the test mass or the suspension system will couple with the resonances of the system. By improving the Q of the components one can isolate the thermally induced noise to the resonant frequency.

- **Shot noise**. The number of photons in the input laser beam fluctuates; this surfaces as noise at the dark port. This noise is proportional to $1/\sqrt{\mathrm{recycling\ gain} \times \mathrm{input\ laser\ power}}$. Increasing the recycling gain and/or increasing the laser power lowers the shot noise. Unfortunately, high power in the cavities induces other unwanted effects, such as radiation pressure noise (discussed in the next item) or thermal lensing (local deformation of the optical surfaces of the cavity). The right choice of laser power and recycling gain involves a compromise among many sources of noise.

- **Radiation pressure noise**. Fluctuations in the number of photons reflecting from the mirrors induces a fluctuating force on the mirrors. This effect scales as $\sqrt{\mathrm{recycling\ gain} \times \mathrm{input\ laser\ power}}$—the inverse of the proportionality entering the shot noise: there is a penalty to increasing the laser power. Reducing shot noise and radiation pressure noise in tandem is a topic of advanced detector R&D; see Buonanno & Chen (2001a,b) and references therein for further discussion.

- **Gravity gradient noise**. When seismic waves, atmospheric pressure fluctuations, cars, animals, tumbleweeds, etc., pass near a GW detector, they act as density perturbations on the neighboring region. This in turn can produce significant fluctuating gravitational forces on the interferometer's test masses (Hughes & Thorne 1998; Creighton 2000), which is expected to become the limiting noise source at low frequencies for advanced ground-based detectors with high quality seismic isolation systems.

- **Laser intensity and frequency noise**. All lasers have some inherent noise, which causes fluctuations in both the laser's intensity and frequency. This noise does not cancel out perfectly when the signal from the two arms destructively recombines, and so some noise leaks into the dark port.

- **Scattered light**. Some laser light can scatter out of the main beam, and then be scattered back, coupling into the interferometer's signal. This light will carry information about its scattering surface, and will generally be out of phase with the beam, contaminating the desired signal. A dense baffling system has been installed to greatly reduce this source of noise.

- **Residual gas**. Any vacuum system contains some trace amount of gas that is extremely difficult to reduce; in LIGO, these traces (mostly hydrogen) are at roughly 10^{-9} Torr. Density fluctuations from these traces in the beam path will induce index of refraction fluctuations in the arms. Residual gas particles bouncing off the mirrors can also increase the displacement noise.

- **Beam jitter**. Jitter in the optics will cause the beam position and angle to fluctuate slightly, causing noise at the dark port.

- **Electric fields**. Fluctuations in the electric field around the test masses can couple into the interferometric signal via interaction between the field and the induced or parasitic surface charge on the mirror surface.

- **Magnetic fields**. Fluctuations in the local magnetic field can affect the test masses when interacting with the actuator magnets bonded to the surface of the mirrors.

- **Cosmic showers**. High energy penetrating muons can be stopped by the test masses and induce a random transient due to recoil.

The initial LIGO detectors will be limited by seismic noise at low frequencies ($\lesssim 50\,\mathrm{Hz}$), by thermal noise in the test mass suspensions at intermediate frequencies (~ 50–$200\,\mathrm{Hz}$), and by shot noise at high frequencies ($\gtrsim 200\,\mathrm{Hz}$). Present detector noise is above the target level, particularly at low and intermediate frequencies, though there has been much progress recently in approaching the target noise curve.

Detector upgrades. Much research within the experimental GW community focuses on developing technologies for improving the sensitivity of LIGO and other ground-based detectors. The first stage detectors are somewhat conservatively designed, ensuring that they can be operated without the excessive introduction of new technology. The price for this conservatism is limited astrophysical reach: current LIGO sensitivity is such that detection of sources is plausible, but not particularly probable, based on our current understanding of sources.

To broaden the astrophysical reach of these instruments, major upgrades are planned for 2006/7. The goal of these upgrades is to push the lower frequency "wall" to lower frequencies, and push the noise level down by a factor ~ 10 across the band. This will increase the distance to which sources can be detected by a factor of 10 – 15, and the volume of the universe which LIGO samples by a factor of 1000–3000. This will dramatically boost any measured event rates.

Discussion of these plans is given in Gustafson et al. (1999). Major changes include a redesigned seismic isolation system (pushing the wall down to about 10 Hz), a more powerful laser (pushing the shot noise down by about a factor of 10), and replacement of the optical and suspension components with improved materials to reduce the impact of thermal noise. In addition, the system will allow "tunable" noise curves—experimenters will be able to shape the noise curve to "chase" particularly interesting sources.

3. GW emission mechanisms

Several distinct physical mechanisms could drive GW emission in stellar core collapse. Following our earlier discussion (cf. Sec. 2 of FHH), we begin with a quick overview of GW generation; further detail and references can be found in Thorne (1987). We then discuss numerical methods of calculating wavestrains given a distribution of masses and mass currents, and the GWs produced by certain instabilities—bar modes, fragmentations, and r-modes.

Unlike FHH, we do *not* discuss the ringing of newly born black holes. A black hole that is distorted from its stationary Kerr configuration will radiate GWs that drive it back to the stationary state. A black hole formed in core collapse will certainly be distorted; if large amounts of material accrete onto it, it will be continually driven into new states of distortion. This in principle could form an interesting GW source. In practice, however, FHH show that it is not likely to be interesting as a LIGO source: the waves are emitted at high frequencies where detector sensitivities are poor, and the associated strain is unlikely to be interestingly large. The interested reader should consult FHH for further details.

Formal overview

The conventional approach to calculating the GW emission of a given mass distribution is via a multipole expansion of the perturbation $h_{\mu\nu}$ to a background spacetime $g^{\mathrm{B}}_{\mu\nu}$. The transverse-traceless projection of this metric, evaluated in the radiation zone, is the metric of the radiation field. The lowest (quadrupole) order piece of this field is (Thorne 1980)

$$h^{\mathrm{TT}}_{jk} = \frac{1}{d}\left[2\frac{G}{c^4}\frac{d^2}{dt^2}\mathcal{I}_{jk}(t-r) + \frac{8}{3}\frac{G}{c^5}\epsilon_{pq(j}\frac{d^2}{dt^2}\mathcal{S}_{k)p}(t-r)n_q\right]^{\mathrm{TT}} . \quad (13.8)$$

$\mathcal{I}_{jk}$ and $\mathcal{S}_{jk}$ are the mass and current quadrupole moments of the source, d is the distance from the source to the point of measurement, ϵ_{ijk} is the antisymmetric tensor, and n_q is the unit vector pointing in the propagation direction. Parentheses in the subscripts indicate symmetrization over the enclosed indices, and the superscript TT indicates that one is to take the transverse-traceless projection. Most GW estimates are based on Eq. (13.8). When bulk mass motions dominate a source's dynamics, the first term describes the radiation that is generated; for example, it produces the well-known "chirp" associated with binary inspiral. The second term dominates for a system whose dynamics are dominated by mass currents, as is the case for radiation from the r-mode instability.

When the background spacetime is flat (or nearly so) the mass and current moments have particularly simple forms. For example, in Cartesian coordinates the mass quadrupole is given by

$$\mathcal{I}_{jk} = \int d^3x\,\rho\left[x^j x^k - \frac{1}{3}r^2\delta_{jk}\right], \quad (13.9)$$

where ρ is the mass density, and $\delta_{jk} = 1$ for $j = k$ and 0 otherwise. The δ_{jk} term ensures that the resulting tensor is trace free.

GWs carry energy and angular momentum from the source (Isaacson 1968). The lowest order contribution to the power, P, emitted in GWs is due to variations in the quadrupole moment:

$$P = \frac{dE}{dt} = \frac{1}{5}\frac{c^5}{G}\left\langle\frac{d^3\mathcal{I}_{jk}}{dt^3}\frac{d^3\mathcal{I}_{jk}}{dt^3}\right\rangle . \quad (13.10)$$

Although radiated power is useful for understanding the effect of GW emission on a source's dynamical evolution, P is not useful for detectability estimates. Instead, one needs an estimate of the wavestrain h, which is directly measured by the detectors. When averaged over all source orientations and sky positions, the power and strain are related by

$$P = \frac{\pi^2 c^3}{G} f^2 d^2 h^2, \quad (13.11)$$

where f is the GW frequency and d is the luminosity distance to the source. For a given strain, higher frequency waves radiate more energy. Because of detector noise, however, higher frequency waves are not necessarily more detectable.

Bar modes

Bar modes are an instability in which the material in the stellar core forms a rapidly rotating bar-like structure. Such a structure has a rapidly varying quadrupole moment, and as such is potentially a copious emitter of GWs. Bar mode instabilities occur in objects whose rotational kinetic energy exceeds some fraction of their potential energy, with the ratio generally written as $\beta \equiv T/|W|$. Standard lore (Chandrasekhar 1969) tells us that an object is unstable on a secular time scale if $\beta \gtrsim 0.14$, and is dynamically unstable if $\beta \gtrsim 0.27$. This lore may be violated if the density profile is not centrally peaked: if centrifugal forces produce a peak in the density off the source's rotational center, dynamical instabilities can set in at much lower values of β (Centrella et al. 2000). Most core-collapse simulations with up-to-date progenitor models, however, find that the mass distribution does not suffer a centrifugal hang-up. This follows from the fact that, in these modern progenitor models, the rotation speeds of the collapsing stellar cores is relatively small (Heger 1998), particularly when compared to earlier models (Rampp, Müller, & Ruffert 1998). In these simulations, the density is centrally concentrated, and thus the low critical values of Centrella et al. (2000) do not apply. Nonetheless, the rotational energy of these models can be quite large (cf. Fig. 13.4), so β is likely to be large enough that bar-mode instabilities will occur.

Heartened by this possibility, we review here expressions for bar mode GW emission. Consider a bar of mass m and length $2r$, rotating with angular frequency ω. The GW energy radiated is given, in the quadrupole approximation, by

$$P_{\rm bar} = \frac{32}{45}\frac{G}{c^5} m^2 r^4 \omega^6. \tag{13.12}$$

A detector at a distance d from the source would measure an rms strain

$$h_{\rm bar} = \sqrt{\frac{32}{45}}\frac{G}{c^4}\frac{mr^2\omega^2}{d}. \tag{13.13}$$

Note that, due to symmetry, the frequency of the emitted GWs is twice the bar's rotation frequency.

Fragmentation instability

To set a physically motivated upper limit to the GW emission that might be produced in stellar collapse, imagine that the collapse material fragments into clumps, which then orbit for some number of cycles as the collapse proceeds. For concreteness, consider the material fragmenting into a binary system

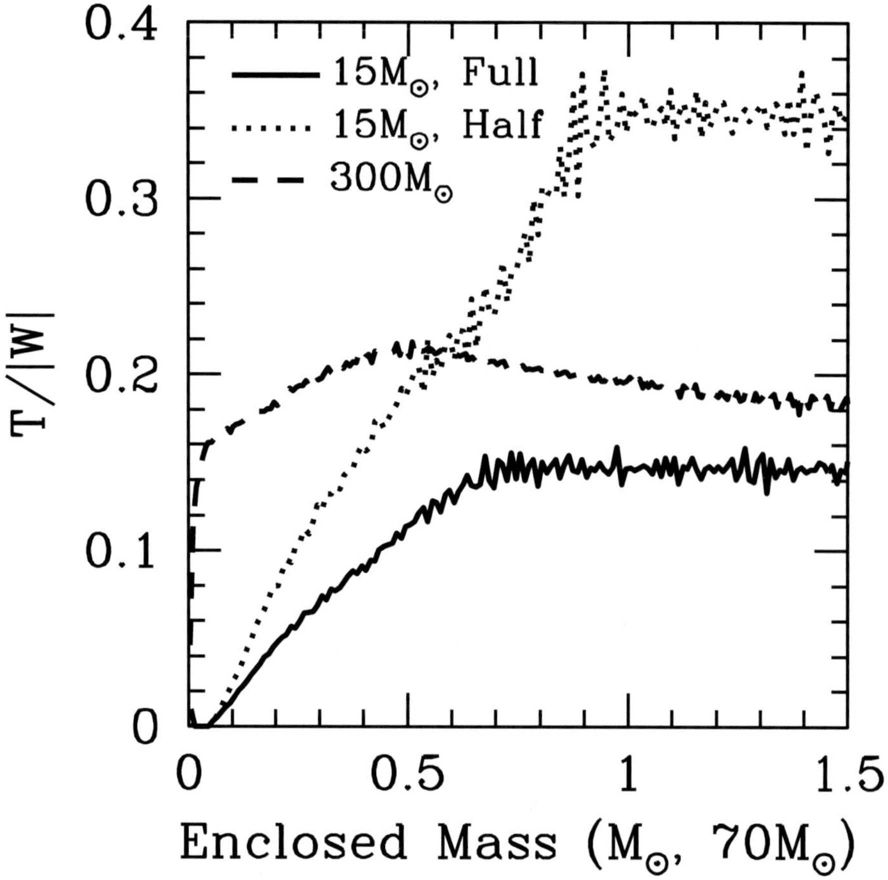

Figure 13.4. Rotational energy divided by gravitational energy ($\beta \equiv T/|W|$) vs. mass for collapsing stars. The horizontal axis is in solar masses for the 15 $M_\odot$ stellar models, in fractions of 70 $M_\odot$ for the 300 $M_\odot$ models. Shown are rotating core collapse (full rotation), 1.6 seconds after bounce; core collapse with half rotation, 1.4 seconds after bounce; and direct collapse of a 300 $M_\odot$ star, 1.9 seconds after bounce. Note that $T/|W|$ for the half rotating progenitor is actually larger than for the fully rotating progenitor. This is because that star is more compact.

(though it could very well fragment into more objects). Such a matter distribution is plausible if the density distribution during collapse peaks off center, as is indicated by some simulations of collapsing Population III stars (Fryer, Woosley, & Heger 2001).

Two bodies, each of mass m, in circular orbit about one another at a frequency ω and with separation $2r$, radiate GWs with power and mean strain

$$P_{\rm bin} = \frac{128}{5}\frac{G}{c^5}m^2r^4\omega^6 \tag{13.14}$$

$$h_{\rm bin} = \sqrt{\frac{128}{5}\frac{G}{c^4}}\frac{mr^2\omega^2}{d}. \tag{13.15}$$

These formulae make no assumption about orbital frequency, and thus apply to, for example, pressure supported as well as Keplerian orbits. For Keplerian orbits, $4\omega^2r^3 = Gm$, so the above expressions become

$$P_{\rm bin} = \frac{2}{5}\frac{G^4}{c^5}\frac{m^5}{r^5} \tag{13.16}$$

$$h_{\rm bin} = \sqrt{\frac{8}{5}}\frac{G^2}{c^4}\frac{m^2}{r\,d}. \tag{13.17}$$

Note that if the "horizons" of the two bodies touch ($r = 2m\,G/c^2$), the power radiated reaches a maximum of $P = c^5/80G \sim 10^{57}$ ergs s^{-1}, independent of the system's mass. The length of time such emission can be sustained scales with the total mass—supermassive black hole binaries do radiate more than microscopic ones.

R-modes

R-modes are peculiar instabilities that may occur in neutron stars. These modes are large scale oscillations in the *current* distribution of the fluid—they drive very little change in the star's density, and as such, their GW generation is described using the second term of Eq. (13.8). R-modes have been of particular interest to GW studies (e.g., Lindblom, Owen, & Morsink 1998; Andersson, Kokkotas, & Schutz 1999; Owen et al. 1998) because they are unstable to GW emission: gravitational radiation tends to *increase* the amplitude of the mode. Lindblom, Owen, & Morsink (1998) first mapped out the range of stellar spin and temperature for which the viscosity was unlikely to damp away this runaway growth, and concluded that many hot, young neutron stars were likely to be important sources of GWs. Current thinking is that the r-modes of young hot neutron stars are actually unlikely to be important sources of GWs. Most analyses assume that the growth of the mode will eventually be limited by nonlinear hydrodynamic effects. Arras et al. (2002) find, after a careful analysis of mode-mode coupling effects, that the saturation amplitude of the modes is far smaller—by four orders of magnitude—than had been assumed previously. Their rather forceful conclusion is that r-modes are completely undetectable.

In FHH we examined the r-mode waves that could be produced by various collapse scenarios that lead to neutron star formation. Our conclusions were broadly in line with those of Owen et al. (1998). In fact, we reinforced those conclusions somewhat by pointing out that material falling back onto the newly-born neutron star could spin it up and reawaken an r-mode instability that had previously decayed away. Re-examining the conclusions in light of the wisdom of Arras et al. (2002), we are forced to a rather different outcome—the r-mode waves are unlikely to ever be detected.

As a consequence of this new understanding, we will end our r-modes discussion here. Interested readers can re-examine Sec. 2.4 of FHH, noting that the amplitude factor α should be around 10^{-5}–10^{-4}, rather than the 0.1–1 considered there. With this in mind, the results for r-modes from young remnants of stellar collapse (Secs. 3.2.2 and 4.2.2 of FHH) are *much* more pessimistic as far as direct detection is concerned.

Direct numerical calculation

If one has a computational model of a core-collapse scenario, one can calculate the GW emission predicted by that model by numerically applying Eq. (13.8) to the mass and current distributions predicted by that model. Naive implementation of Eq. (13.8) works poorly: computing the required derivatives numerically (by evaluating the quadrupole moments on multiple time slices and differencing), introduces spurious numerical noise. However, if the potentials which drive the motion of the matter are known, the time derivatives can be rewritten as spatial derivatives, and the GW emission can be calculated using data from a single time slice. Calculations demonstrating this technique are given in Finn & Evans (1990), Blanchet, Damour, & Evans (1990), and Centrella & McMillan (1993). The Centrella & McMillan paper is specialized to smooth particle hydrodynamics (SPH).

Fryer & Warren (2002) have recently modeled core collapse with a 3-dimensional code using SPH. One of the goals of this work will be to test whether the instabilities discussed above, particularly bar modes and fragmentation, actually occur in a computational stellar collapse model. At present, the only GWs produced by these models come from the large scale, convective motions of matter near the protoneutron star (cf. Fig. 13.5). The GWs from these motions are computed using the formulae given in Centrella & McMillan (1993). We discuss these results at length below.

The work of Fryer & Warren (2002) coupled an equation of state for dense matter and a flux-limited diffusion neutrino transport scheme into a parallel 3-dimensional SPH code. This code includes all of the physics used in 2-dimensional models less than a decade ago (e.g. Herant et al. 1994) and shows the remarkable progress in simulations of core-collapse. Fig. 13.5 shows the convective upwells of the first such 3-dimensional simulations of the collapse of a symmetric star. These simulations assumed spherically symmetric im-

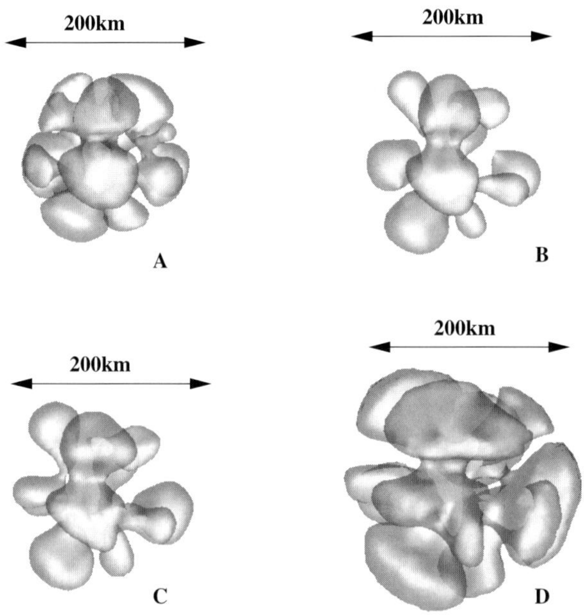

Figure 13.5. Isosurfaces of upward moving bubbles (radial velocities moving outward at $1000\,\mathrm{km\,s^{-1}}$) as a function of time ($t_A < t_B < t_C < t_D$) from the 3-dimensional simulations of Fryer & Warren (2002). Note that the number of modes decreases slightly as the flows merge over time (compare panel A with later panels).

plementation of gravity for better comparison with past 2-dimensional work. However, the code developed by Fryer & Warren (2002) has a tree-based algorithm for calculating gravity and can be used to model the collapse of rotating or asymmetric cores. The first rotating models are running and, as predicted by the 2-dimensional simulations (Fryer & Heger 2000), show convection that is primarily limited to the polar regions (Fig. 13.6).

Improvements still must be made: a more sophisticated neutrino algorithm is needed to accurately model neutrino heating, general relativistic and equation of state effects also contribute to uncertainties. However, progress along all these fronts is being made.

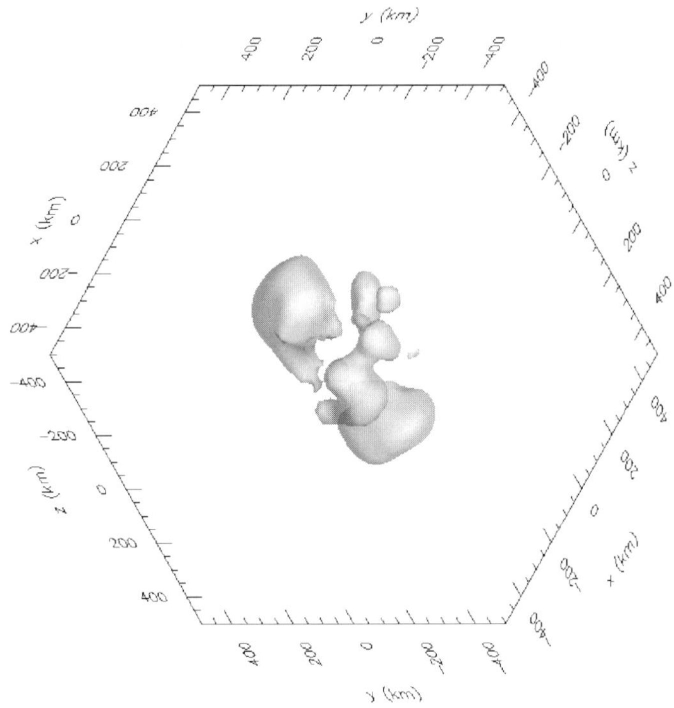

Figure 13.6. Isosurfaces of upward moving bubbles (radial velocities moving outward at $1000\,\mathrm{km\,s^{-1}}$) from the rotating 3-dimensional simulations of Fryer & Warren (2003). Due to the angular momentum gradient, which prevents convection along the equator, most of the convection occurs along the rotation (z) axis.

4. Results for GW emission

Applying the various mechanisms discussed in the previous section, we now estimate GW strengths for two important stellar collapse scenarios—supernovae, and the death of Population III stars. In each case, we discuss what is known about the rates of these events and their likely angular momentum distribution (both of which strongly impact their importance as GW sources), and then estimate their GW strengths. In both cases, we sketch the wavestrains likely for waves arising from bar mode instabilities, and a possible fragmentation instability, comparing these waves to the noise level that is the goal of the second generation LIGO-II detectors (in particular, the broad-band configuration of those detectors; cf. Gustafson, Shoemaker, Strain, & Weiss

1999). For supernovae, we also discuss the recent results of Fryer & Warren (2002), for gravitational waves generated in a 3-D numerical calculation of convective motions in a supernova explosion.

Supernovae

Formation rate and angular momentum. The supernova rate is fairly well known, lying somewhere between 1 per 50–140 years in the Galaxy (Cappellaro et al. 1997). It is not clear, however, what fraction of these core-collapse events (if any) are rotating rapidly enough to develop these instabilities, and thereby emit detectable GWs.

Insight into the angular momentum distribution can be gained by studying pulsars , the compact remnants of core-collapse supernovae. From measurements of young pulsars , we know that at least some neutron stars are born with periods faster than 20 ms. Whether or not any neutron stars are born with millisecond periods is harder to ascertain—pulsars spin down as they emit radiation, and the spindown rate is not particularly well determined. A recent analysis of Chernoff & Cordes (private communication) found that the initial spin periods could be fit with a Gaussian distribution peaking at 7 ms, with sub-ms pulsars lying beyond the 2-sigma tail. Does this mean that less than 10% of pulsars are born spinning with millisecond periods, or does it mean that many pulsars are born spinning rapidly, and GW emission removes a considerable amount of their angular momentum? It is to be noted that the analysis of Chernoff & Cordes is very sensitive to their choice of spindown rates and other uncertainties in their population study; they stress that such results should be taken with a great deal of caution. Though such large and important uncertainties complicate efforts to estimate the likely magnitudes of GW emission following supernovae, they also illustrate the impact that GW observations could have. Even a solid null result (no GWs seen from core-collapse with a high degree of confidence) would have important impact, telling us that rapidly rotating newly born neutron stars are very rare.

Stellar theorists have now produced models of core-collapse progenitors which include angular momentum (Heger 1998). Though these simulations build in a number of assumptions about the angular momentum transport in the massive star, they provide some handle on the angular momentum distribution in the collapsing core. We base our analysis on the angular momentum profiles from the core-collapse simulations of $15\,M_\odot$ stars by Fryer & Heger (2000), which use these latest progenitors and model the core-collapse through supernova explosion (Models 1,5; cf. Fig. 2 of FHH).

GWs from bar modes and fragmentation. Because the angular momentum distributions used by Fryer & Heger (2000) have peak values significantly lower than those used in the past, there is no centrifugal hang-up. The collapse proceeds nearly identically to a non-rotating star, with a density distribution peaked at the center of the star. This makes it harder for bar-mode instabilities to develop, and produces weaker GW emission. During bounce, the neutron star is not compact enough to quickly drive bar-mode instabilities.

However, the explosion produced by these rotating core-collapse supernovae is much stronger along the poles than along the equator (Fryer & Heger 2000), causing much of the low angular-momentum material to be ejected. Hence, after the explosion—~ 1 s after collapse—β can increase to high enough values that bar-mode instabilities are likely to develop (cf. Fig. 13.4). The amount of matter enclosed by the proto-neutron star extends in all cases beyond $\sim 1\,M_\odot$, corresponding to values of β which are certainly above the secular instability limit ($\beta \sim 0.14$), and probably also above the dynamic instability limit ($\beta \sim 0.27$). Notice in Fig. 13.4 that β is actually larger for the model which has less initial angular momentum. This is because this model has contracted more, and is spinning more rapidly. Notice that we have to push to these conditions merely to produce bar-mode instabilities. It is even less likely that fragmentation will occur in core-collapse supernovae.

The Fryer & Heger (2000) simulations are axisymmetric (2D) and so, by construction, cannot produce a bar mode instability. However, we can get a handle on the *potential* of these models to radiate by a bar mode instability by taking the mass and angular momentum distributions of the Fryer & Heger models and imagining that some fraction of the mass in these models participates in such an instability. We do this by assuming that all of the matter up to some enclosed mass becomes unstable and forms a bar (conserving angular momentum), and then calculate the GW emission as a function of total unstable mass. One should bear in mind that we assume *all* of the enclosed mass ends up in the bar. These estimates are thus relatively strong upper limits (although the strain could increase if we allowed the bar to contract and spin up). We illustrate the detectability of waves from this bar model in Figure 13.7. Each point on this plot illustrates a different possible bar, varying the amount of mass that participates in the instability. Open circles illustrate the wavestrain for a single GW cycle; filled circles give the characteristic strain obtainable if the bar radiates coherently for 100 cycles. This plot demonstrates that bar modes are potentially promising sources of waves if bars remain coherent for at least a moderate ($\sim$ 50–100) number of GW cycles.

Since the density is centrally peaked, a fragmentation instability is unlikely to occur in core-collapse supernovae. However, if it did occur, even the single-cycle strain would be quite large ($\sim 10^{-22}$), and fall in an interesting band (from a few hundred Hz up to a kilohertz or so).

Numerically calculated GW estimates. As computational models improve we can expect to reach the point where a 3-dimensional numerical calcu-

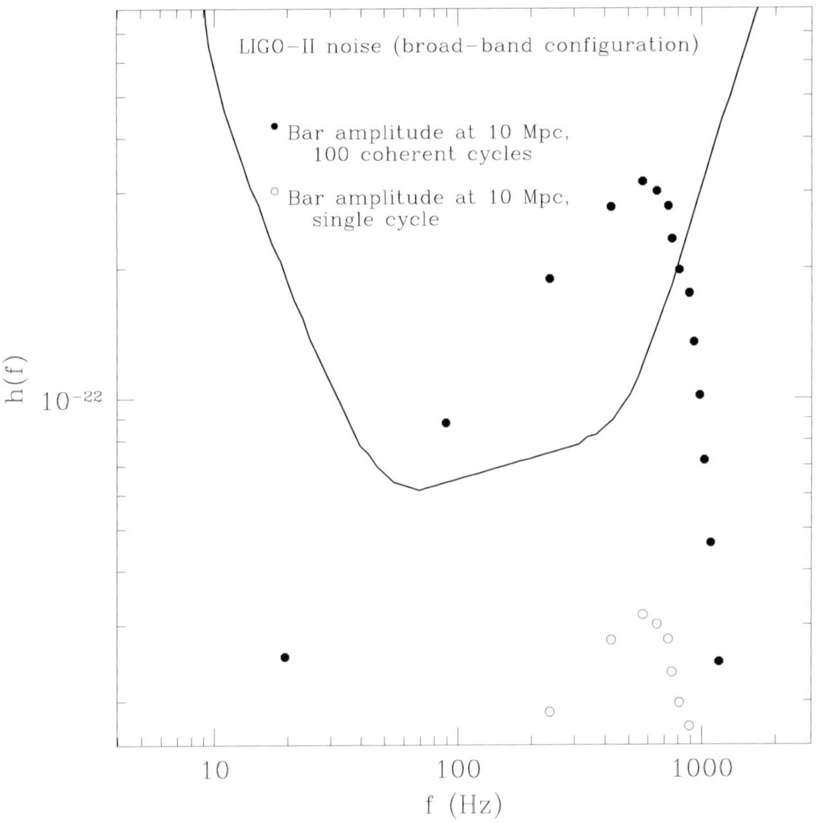

Figure 13.7. Bar mode gravitational waves compared to broad-band LIGO-II noise. We use the Fryer & Heger (2000) models for core collapse, and then estimate GW emission as follows: we assume that all mass inside a given radius participates in a bar mode instability and forms a bar, conserving angular momentum as it does so. Each point represents the waves produced by a particular choice of radius, moving to larger radii from right to left. An open circle is the strain from a single wave cycle of the bar; a closed circle is the integrated strain that would be measured if the bar were to remain coherent for 100 GW cycles. The range between the open and closed circles suggests that bar-mode waves could be of interesting strength provided that they remain coherent for a minimum of $\sim$ 50–100 cycles.

lation can adequately model an asymmetric supernova explosion, and diagnose whether the various instabilities discussed above are likely to actually occur. We're not quite there yet, but recent progress has been impressive. As our physical understanding of supernovae improves, and as more powerful computational resources become available, we can look forward to serious numerical studies that model realistic supernova explosions.

At present most relevant models that can be studied to estimate GW emission in a 3-D calculation are those produced by Fryer & Warren (2002). The only

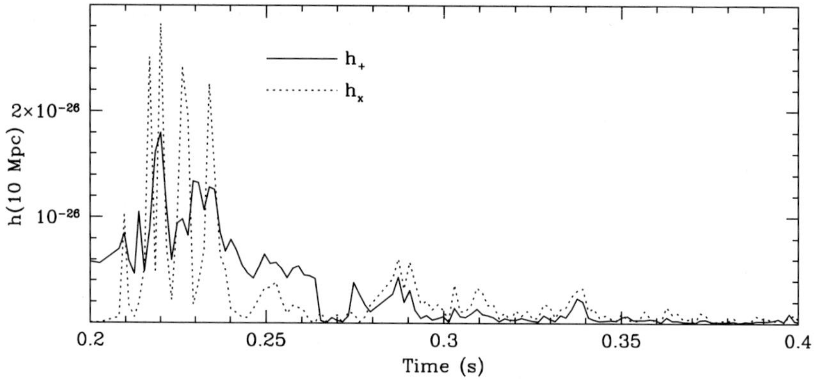

Figure 13.8. Magnitude of the wavestrain for convective motions in a 3-D stellar collapse model (cf. Fig. 13.5), for a supernova at 10 Mpc.

dynamical aspect of this model with sufficient asymmetry to produce GWs are the convective cells bubbling through the stellar core (cf. Fig. 13.5). The GW strengths in these bubbles is simply estimated by postprocessing the model's data using the formulae of Centrella & McMillan (1993). The magnitude of the GW strain expected from this model's convective behavior is shown in Fig. 13.8 for an explosion at 10 Mpc.

The first rotating simulations are now being run (Fryer & Warren 2003), but with the rotation speeds expected from stellar evolution models (Heger 1998), it does not appear that bar-mode instabilities develop both because of the low, but realistic angular momentum in the Heger models and because the convection and viscosity (dominated by numerical viscosity) transport out this angular momentum. If bar-modes do not develop, the dominate gravitational wave signal occurs at bounce where models by Dimmelmeier, Font, & Müller (2002a,2002b) are among the most physically accurate.

These convective GWs are extremely weak—even though they arise from the rapid overturning motions of dense matter in the star, the multipolar distribution of this matter is not ideally distributed to be a strong sources of GWs. Even in a galactic explosion (distance to source $\sim$ 10 kpc), these waves would be weak ($h \sim 10^{-23}$). Preliminary results from rotating simulations from Fryer & Warren (2003) also do not develop strong bar modes and hence do not have strong GW signals (only a $\sim$ 2 order of magnitude increase). Even higher angular momenta are required to cause fragmentation. If the results of Fryer & Warren (2003) are correct, fragmentation can be ruled out in core-collapse supernovae. Future 3-D models will study the effects of angular momentum transport and will also include the effects of asymmetries in collapse to find the maximum gravitational wave signals that may develop from core-collapse supernovae.

Collapse of very massive stars

At solar metallicity, stellar winds severely limit the pre-collapse mass of massive stars—very few massive stars will remain massive up to the time of collapse. As these winds are driven by the opacity of metals in the stellar envelope, it is likely that mass loss from winds will decrease as the fraction of metals in the envelope is reduced. Population III stars are the first generation of stars formed in the early universe, before stars formed the metals that abound today. Here we review the death of very massive (100–500 $M_\odot$) Population III stars. Like Chandrasekhar-massed white dwarfs, these stars must suffer one of two fates: either they explode in a giant thermonuclear explosion ("hypernova"), or they collapse to form black holes. The fate is determined by the stellar mass. If the star's mass exceeds $\sim 260\,M_\odot$, it will collapse to a black hole (Fryer, Woosley, & Heger 2001; Baraffe, Heger, & Woosley 2001). However, if the star is rotating, rotational (plus thermal) support prevents the star from undergoing immediate collapse to a black hole (Fryer, Woosley, & Heger 2001). Rotating, very massive stars collapse and bounce, forming a much larger compact core than those produced by core-collapse supernovae: a 50–70 $M_\odot$, 1000–2000 km proto-black hole instead of the 1 $M_\odot$, 100 km proto-neutron star. This rotating proto-black hole is susceptible to bar instabilities and may produce a strong GW signal (see also Madau & Rees 2001).

Formation rate and angular momentum. Estimating the rate of core-collapse for very massive stars depends on two rather uncertain quantities: the amount of matter found in Population III stars, and the number of these stars which actually collapse to form black holes. The mass distribution of stars at birth is known as the initial mass function (IMF). Today the IMF is peaked toward low mass stars, such that 90% of stellar core-collapse occurs in stars between 8 and $\sim 20\,M_\odot$, and only 1% of core-collapse occurs in stars more massive than 40 $M_\odot$. However, it has long been believed that the first generation of stars after the Big Bang tended to be more massive than stars formed today (e.g., Silk 1983; Carr & Rees 1984). Recent simulations by Abel, Bryan, & Norman (2000) suggest that the typical mass of first generation stars is $\sim 100\,M_\odot$, and it is possible that a majority of Population III stars had masses in excess of 100 $M_\odot$.

The light from these very massive stars re-ionizes the early universe; from this we can derive a constraint on the formation rate of these stars. Although we expect that their photons ionized a significant fraction of the early universe, there should not be so many stars that they ionize the universe several times over. Using our best estimates of the re-ionization fraction, the amount of ultraviolet photons produced by these massive stars, and the ionization efficiency of massive stars, one finds that 0.01%–1% of the baryonic matter in the universe was incorporated into very massive stars (Abel, Bryan, & Norman 2000). This corresponds to about 10^4–10^7 very massive stars produced in a $10^{11}\,M_\odot$ galaxy, or a rate of massive stellar collapse as high as one every few thousand years.

We should temper these optimistic statements with two caveats. First, these are Population III stars, and so are born at high redshift ($z \gtrsim 5$). As they evolve to collapse in less than a few million years (Baraffe, Heger, & Woosley 2001), they will only be observed near to the high redshifts of their birth. In addition, although we may believe our formation rate of very massive stars (within a few orders of magnitude), it is currently impossible to determine what fraction of very massive stars are produced with masses beyond the $\sim 260\,M_\odot$ mass limit necessary for black hole formation. The Galaxy could produce a million of these objects, or perhaps just a few hundred. Assuming 1–10 million very massive stars per galaxy beyond $z = 5$ gives us a secure upper limit.

The rotation of these stars has again been calculated using the stellar evolution code developed by Heger (1998); for this analysis we use the Fryer et al. (2001) rotation profiles.

GWs from bar modes and fragmentation. We expect the proto-black hole formed in the collapse of a massive star to become secularly unstable (Fig. 13.4), and these secular instabilities are likely to develop before the proto-black hole collapses to a black hole (Fryer, Woosley, & Heger 2001). Given the large amount of mass ($\sim 70\,M_\odot$) and angular momentum these objects possess, it is not surprising that these objects can produce strong GW signals. However, the cosmological redshift moves the peak of the source waves out of the band of LIGO detectors: even at the relatively low value $z = 5$ (corresponding to a luminosity distance ~ 48 Gpc in the currently popular "concordance cosmology" [Wang, Tegmark, & Zaldarriaga 2002]), the strain from bar modes peaks at frequencies less than 10 Hz, with a strain 8×10^{-23}. This is well below the LIGO II threshold. Even coherent integration over ~ 100 cycles is unlikely to produce a detectable signal; see Fig. 13.9.

The waves from massive star collapse may be detectable, however, if a fragmentation instability occurs. With our crude model of fragmentation, we find that both strain and frequency are boosted if the core splits into two pieces which then fall into a Keplerian orbit, conserving angular momentum. If this instability occurs and the pieces orbit coherently for ~ 10 cycles, these waves may be detectable at redshifts $z \sim 5$; see Figure 13.9.

5. Summary & concluding thoughts

One clear conclusion can be drawn from this survey: despite the impressive recent progress in our understanding of core collapse, we *still* have a relatively

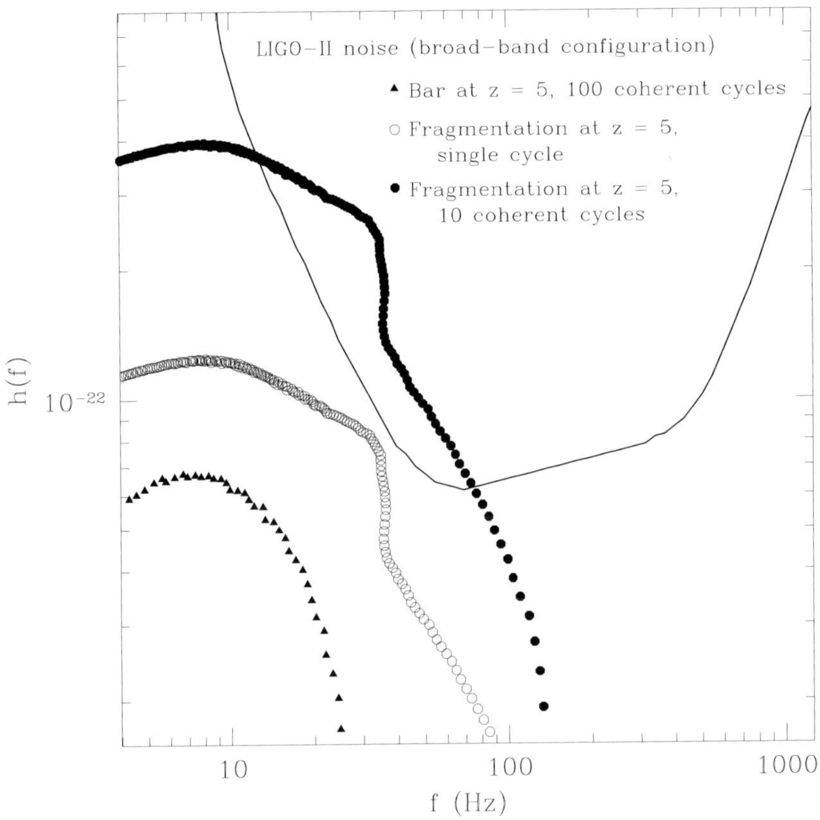

Figure 13.9. Possible bar mode and fragmentation instability GW signals for collapse of $300\,M_\odot$ Population III star. As in Fig. 13.7, each point corresponds to all mass inside a particular radius participating in the instability, conserving angular momentum. The prospects for detecting bar modes from these collapse events are extremely bad: the cosmological redshift pushes this signal far out of the LIGO band. Waves generated by a fragmentation instability are potentially interesting: both their strains and frequencies are significantly higher, and might be accessible to LIGO, particularly if the signal remains coherent for some number of cycles.

poor grasp of the processes that drive GW emission in these events. Past work on instabilities, when coupled with 2-D models of stellar collapse, provide a decent estimate of the range of wave strengths and frequencies that are possible, but do not provide specifics for what is to be expected from collapse.

Progress and future understanding will only come as these models are further generalized, building in the rotation or core oscillations that seem needed (Fryer & Heger 2000) to explain asymmetries in supernova (Wang et al. 2001). Fully 3-D simulations, such as those from Fryer & Warren 2002, hold great promise for extending our knowledge of the collapse process. Better stellar models, better neutrino transport algorithms and the inclusion of general relativistic effects will allow modelers to unambiguously extract the effects of of GWs. If the GW detector sensitivities continue to dramatically improve, one can hope that GW data will meaningfully impact core-collapse work later in the decade. As noted in Sec. 4, even a solid null result would provide useful information. An unambiguous *non*-null result (perhaps in coordination with other astronomical instruments) would be most exciting of all, and may open our clearest window onto the processes occuring deep inside collapsing stellar cores.

Acknowledgments

We thank Peter Bender, Craig Hogan, and Szabolcs Marká for allowing us to use material from Hughes et al. (2001), and we thank Kip Thorne for permission to reproduce Figs. 13.1 and 13.2. The 3-D simulations were funded by a *Scientific Discovery through Advanced Computing* grant, and were performed on the IBM SP at the National Energy Research Scientific Computing Center. The work of DEH and SAH is supported at the KITP by NSF Grant PHY–9907949.

Notes

1. The LIGO detectors have improved their sensitivities by several orders of magnitude across a wide frequency band in the first nine months of 2002.
2. It's worth noting that there is always a chance—albeit a small one—that a relatively nearby star could go supernova and produce an anomalously strong signal. The current configuration of LIGO would likely have seen SN 1987a, were it to have been operating back then. The last thing that the GW detection community wants is to miss another such event!
3. It should be noted that the possibility of detecting GWs with laser interferometers has an even longer history, reaching back to Pirani in 1956, and was independently proposed by several workers: Gertsenshtein and Pustovoit in 1962, Weber in the 1960s, and Weiss c. 1970. See Sec. 9.5.3 of Thorne (1987) for further discussion and references.
4. The Laser Interferometer Space Antenna (Danzmann et al. 1998; Bender 2001).

References

Abel, T., Bryan, G. L., Norman, M. L. 2000, ApJ, 540, 39.
Abramovici, A. et al. 1992, Science, 256, 325
Ando, M. et al. 2001, Phys. Rev. Lett. 86, 3950.
Andersson, N., Kokkotas, K., & Schutz, B. F. 1999, ApJ, 510, 846.
Arras, P., Flanagan, E. E., Morsink, S. M., Schenk, A. K., Teukolsky, S. A., and Wasserman, I. 2002, ApJ, submitted; astro-ph/0202345.
Baraffe, I., Heger, A., & Woosley, S. E. 2001, ApJ, 550, 890.
Bender, P.L. 2001, in Gravitational Waves, eds. I. Ciufolini, V. Gorini, V. Moschella, and P. Fre, (Institute of Physics Publishing, Bristol, UK), p.115.
Blanchet, L., Damour, T., & Schäffer, G. 1990, MNRAS, 242, 289.
Buonanno, A. & Chen, Y. 2001, Phys. Rev. D, 64, 042006.
Buonanno, A. & Chen, Y. 2001, Class. Quantum Grav. 18, L95.

Cappellaro, E., Turatto, M., Tsvetkov, D. Tu., Bartunov, O. S., Pollas, C., Evans, R., & Hamuy, M. 1997, A&A, 322, 431.

Carr, B. J. & Rees, M. J. 1984, MNRAS, 206, 315.

Centrella, J. M., New, K. C. B., Lowe, L. L., & Brown, J. D. 2000, ApJ, 550, 193.

Chandrasekhar, S. 1969, Ellipsoidal Figures of Equilibrium (New Haven: Yale University Press).

Creighton, T. 2000, Phys. Rev. D, submitted; gr-qc/0007050.

Danzmann, K. et al. 1998, LISA Pre-Phase A Report, 2nd Edition (Report MPQ-233, Max-Planck Institut für Quantenoptik, Garching, Germany), p. 1

Dimmelmeier, H., Font, J., & Müller, E. 2002, A&A, 388, 917

Dimmelmeier, H., Font, J., & Müller, E. 2002, A&A, 393, 523

Drever, R. W. P. 1983, in Gravitational Radiation, eds. N. Deruelle and T. Piran (North Holland: Amsterdam), p. 321.

Eardley, D. M. 1983, in Gravitational Radiation, eds. N. Deruelle and T. Piran (North Holland: Amsterdam), p. 257.

Einstein, A. 1918, Königlich Preußische Akademie der Wissenschaften Berlin, Sitzungsberichte, p. 154.

Finn, L. S. & Evans, C. 1990, ApJ, 351, 588.

Fryer, C. L. & Heger, A. 2000, ApJ, 541, 1033.

Fryer, C. L., Holz, D. E., & Hughes, S. A. 2002, ApJ, 565, 430; referred to in the text as FHH.

Fryer, C. L. & Warren, M. S. 2002, ApJ, 574, L65.

Fryer, C. L. & Warren, M. S. 2003, in preparation

Fryer, C. L., Woosley, S. E., & Heger, A. 2001, ApJ, 550, 372.

Gustafson, E., Shoemaker, D., Strain, K., & Weiss, R. 1999, LSC White Paper on Detector Research and Development, LIGO Document T990080-00-D.

Heger, A. 1998, Ph.D. thesis, Technische Univ. München.

Hughes, S. A., Marká, S., Bender, P. L., & Hogan, C. J. 2001, to appear in the Proceedings of the 2001 Snowmass Meeting; astro-ph/0110349.

Hughes, S. A. & Thorne, K. S. 1998, Phys. Rev. D, 58, 122002.

Isaacson, R. A. 1968, Phys. Rev., 166, 1272.

Kuroda, K. et al. 2000, Int. J. Mod. Phys. D, 8, 557.

Levin, Yu. 1998, Phys. Rev. D, 57, 659.

Lindblom, L., Owen, B. J., & Morsink, S. M. 1998, Phys. Rev. Lett., 80, 4843.

Liu, Y. T. & Thorne, K. S. 2001, Phys. Rev. D, 62, 122002.

Lück et al. 2000, in AIP Conf. Proc. 523, Procedings of the 3rd Edoardo Amaldi COnference, ed. S. Meshkov (AIP: Melville), p. 119.

Madau, P. & Rees, M. J. 2001, ApJ, 551, L27.

Marion, F. 2000, in AIP Conf. Proc. 523, Procedings of the 3rd Edoardo Amaldi Conference, ed. S. Meshkov (AIP: Melville), p. 110.

McCleland, D. E. et al. 2001, in AIP Conf. Proc. 523, Procedings of the 3rd Edoardo Amaldi COnference, ed. S. Meshkov (AIP: Melville), p. 140.
New, K.C.B., Centrella, J.M., & Tohline, J.E. 2000, Phys. Rev. D, 62, 064019
Owen, B. J., Lindblom, L., Cutler, C., Schutz, B. F., Vecchio, A., & Andersson, N. 1998, Phys. Rev. D, 58, 084020.
Rampp, M., Müller, E., & Ruffert, M. 1998, A&A 332, 969.
Santamore, D. H. & Levin, Yu. 2001, Phys. Rev. D, 64, 042006.
Silk, J. 1983, MNRAS, 205, 705.
Thorne, K. S. 1980, Rev. Mod. Phys., 52, 299.
Thorne, K. S. 1987, in Three Hundred Years of Gravitation, eds. S. W. Hawking and W. Israel (Cambridge: Cambridge University Press), p. 330.
Wang, L., Howell, D. A., Höflich, P., Wheeler, J. C. 2001, ApJ, 550, 1030.
Wang, X., Tegmark, M., & Zaldarriaga, M. 2002, Phys. Rev. D, 65, 123001.
Weiss, R. 1972, Quarterly Progress Report of the Research Laboratory of Electronics of MIT, 105, 54.

Index

Volume 299: *Open Issues in Local Star Fomation,* edited by Jacques Lépine, Jane Gregorio-Hetem
Hardbound, ISBN 1-4020-1755-3, December 2003

Volume 298: *Stellar Astrophysics - A Tribute to Helmut A. Abt,* edited by K.S. Cheng, Kam Ching Leung, T.P. Li
Hardbound, ISBN 1-4020-1683-2, November 2003

Volume 297: *Radiation Hazard in Space,* by Leonty I. Miroshnichenko
Hardbound, ISBN 1-4020-1538-0, September 2003

Volume 296: *Organizations and Strategies in Astronomy, volume 4,* edited by André Heck
Hardbound, ISBN 1-4020-1526-7, October 2003

Volume 295: *Integrable Problems of Celestial Mechanics in Spaces of Constant Curvature,* by T.G. Vozmischeva
Hardbound, ISBN 1-4020-1521-6, October 2003

Volume 294: *An Introduction to Plasma Astrophysics and Magnetohydrodynamics,* by Marcel Goossens
Hardbound, ISBN 1-4020-1429-5, August 2003
Paperback, ISBN 1-4020-1433-3, August 2003

Volume 293: *Physics of the Solar System,* by Bruno Bertotti, Paolo Farinella, David Vokrouhlický
Hardbound, ISBN 1-4020-1428-7, August 2003
Paperback, ISBN 1-4020-1509-7, August 2003

Volume 292: *Whatever Shines Should Be Observed,* by Susan M.P. McKenna-Lawlor
Hardbound, ISBN 1-4020-1424-4, September 2003

Volume 291: *Dynamical Systems and Cosmology,* by Alan Coley
Hardbound, ISBN 1-4020-1403-1, November 2003

Volume 290: ***Astronomy Communication,*** edited by André Heck, Claus Madsen
Hardbound, ISBN 1-4020-1345-0, July 2003

Volume 287/8/9: ***The Future of Small Telescopes in the New Millennium,*** edited by Terry D. Oswalt
Hardbound Set only of 3 volumes, ISBN 1-4020-0951-8, July 2003

Volume 286: ***Searching the Heavens and the Earth: The History of Jesuit Observatories,*** by Agustín Udías
Hardbound, ISBN 1-4020-1189-X, October 2003

Volume 285: ***Information Handling in Astronomy - Historical Vistas,*** edited by André Heck
Hardbound, ISBN 1-4020-1178-4, March 2003

Volume 284: ***Light Pollution: The Global View,*** edited by Hugo E. Schwarz
Hardbound, ISBN 1-4020-1174-1, April 2003

Volume 283: ***Mass-Losing Pulsating Stars and Their Circumstellar Matter,*** edited by Y. Nakada, M. Honma, M. Seki
Hardbound, ISBN 1-4020-1162-8, March 2003

Volume 282: ***Radio Recombination Lines,*** by M.A. Gordon, R.L. Sorochenko
Hardbound, ISBN 1-4020-1016-8, November 2002

Volume 281: ***The IGM/Galaxy Connection,*** edited by Jessica L. Rosenberg, Mary E. Putman
Hardbound, ISBN 1-4020-1289-6, April 2003

Volume 280: ***Organizations and Strategies in Astronomy III,*** edited by André Heck
Hardbound, ISBN 1-4020-0812-0, September 2002

Volume 279: ***Plasma Astrophysics , Second Edition,*** by Arnold O. Benz
Hardbound, ISBN 1-4020-0695-0, July 2002

Volume 278: ***Exploring the Secrets of the Aurora,*** by Syun-Ichi Akasofu
Hardbound, ISBN 1-4020-0685-3, August 2002

Volume 277: ***The Sun and Space Weather,*** by Arnold Hanslmeier
Hardbound, ISBN 1-4020-0684-5, July 2002

Volume 276: ***Modern Theoretical and Observational Cosmology,*** edited by Manolis Plionis, Spiros Cotsakis
Hardbound, ISBN 1-4020-0808-2, September 2002

Volume 275: ***History of Oriental Astronomy,*** edited by S.M. Razaullah Ansari
Hardbound, ISBN 1-4020-0657-8, December 2002

Volume 274: ***New Quests in Stellar Astrophysics: The Link Between Stars and Cosmology,*** edited by Miguel Chávez, Alessandro Bressan, Alberto Buzzoni,Divakara Mayya
Hardbound, ISBN 1-4020-0644-6, June 2002

Volume 273: ***Lunar Gravimetry,*** by Rune Floberghagen
Hardbound, ISBN 1-4020-0544-X, May 2002

Volume 272:***Merging Processes in Galaxy Clusters,*** edited by L. Feretti, I.M. Gioia, G. Giovannini
Hardbound, ISBN 1-4020-0531-8, May 2002

Volume 271: ***Astronomy-inspired Atomic and Molecular Physics,*** by A.R.P. Rau
Hardbound, ISBN 1-4020-0467-2, March 2002

Volume 270: ***Dayside and Polar Cap Aurora,*** by Per Even Sandholt, Herbert C. Carlson, Alv Egeland
Hardbound, ISBN 1-4020-0447-8, July 2002

Volume 269: ***Mechanics of Turbulence of Multicomponent Gases,*** by Mikhail Ya. Marov, Aleksander V. Kolesnichenko
Hardbound, ISBN 1-4020-0103-7, December 2001

Volume 268: ***Multielement System Design in Astronomy and Radio Science,*** by Lazarus E. Kopilovich, Leonid G. Sodin
Hardbound, ISBN 1-4020-0069-3, November 2001

Volume 267: ***The Nature of Unidentified Galactic High-Energy Gamma-Ray Sources,*** edited by Alberto Carramiñana, Olaf Reimer, David J. Thompson
Hardbound, ISBN 1-4020-0010-3, October 2001

Volume 266: ***Organizations and Strategies in Astronomy II,*** edited by André Heck
Hardbound, ISBN 0-7923-7172-0, October 2001

Volume 265: ***Post-AGB Objects as a Phase of Stellar Evolution,*** edited by R. Szczerba, S.K. Górny
Hardbound, ISBN 0-7923-7145-3, July 2001

Volume 264: ***The Influence of Binaries on Stellar Population Studies,*** edited by Dany Vanbeveren
Hardbound, ISBN 0-7923-7104-6, July 2001

Volume 262: ***Whistler Phenomena - Short Impulse Propagation,*** by Csaba Ferencz, Orsolya E. Ferencz, Dániel Hamar, János Lichtenberger
Hardbound, ISBN 0-7923-6995-5, June 2001

Volume 261: ***Collisional Processes in the Solar System,*** edited by Mikhail Ya. Marov, Hans Rickman
Hardbound, ISBN 0-7923-6946-7, May 2001

Volume 260: ***Solar Cosmic Rays,*** by Leonty I. Miroshnichenko
Hardbound, ISBN 0-7923-6928-9, May 2001

Volume 259: ***The Dynamic Sun,*** edited by Arnold Hanslmeier, Mauro Messerotti, Astrid Veronig
Hardbound, ISBN 0-7923-6915-7, May 2001

Volume 258: ***Electrohydrodynamics in Dusty and Dirty Plasmas-Gravito-Electrodynamics and EHD,*** by Hiroshi Kikuchi
Hardbound, ISBN 0-7923-6822-3, June 2001

Volume 257: ***Stellar Pulsation - Nonlinear Studies,*** edited by Mine Takeuti, Dimitar D. Sasselov
Hardbound, ISBN 0-7923-6818-5, March 2001

Volume 256: ***Organizations and Strategies in Astronomy,*** edited by André Heck
Hardbound, ISBN 0-7923-6671-9, November 2000

Volume 255: ***The Evolution of the Milky Way- Stars versus Clusters,*** edited by Francesca Matteucci, Franco Giovannelli
Hardbound, ISBN 0-7923-6679-4, January 2001

Volume 254: ***Stellar Astrophysics***, edited by K.S. Cheng, Hoi Fung Chau, Kwing Lam Chan, Kam Ching Leung
Hardbound, ISBN 0-7923-6659-X, November 2000

Volume 253: ***The Chemical Evolution of the Galaxy***, by Francesca Matteucci
Paperback, ISBN 1-4020-1652-2, October 2003

Hardbound, ISBN 0-7923-6552-6, June 2001

Volume 252: *Optical Detectors for Astronomy II*, edited by Paola Amico, James W. Beletic
Hardbound, ISBN 0-7923-6536-4, December 2000

Volume 251: *Cosmic Plasma Physics*, by Boris V. Somov
Hardbound, ISBN 0-7923-6512-7, September 2000

Volume 250: *Information Handling in Astronomy*, edited by André Heck
Hardbound, ISBN 0-7923-6494-5, October 2000

Volume 249: *The Neutral Upper Atmosphere*, by S.N. Ghosh
Hardbound, ISBN 0-7923-6434-1, July 2002

Volume 247: *Large Scale Structure Formation*, edited by Reza Mansouri, Robert Brandenberger
Hardbound, ISBN 0-7923-6411-2, August 2000

Volume 246: *The Legacy of J.C. Kapteyn*, edited by Piet C. van der Kruit, Klaas van Berkel
Paperback, ISBN 1-4020-0374-9, November 2001
Hardbound, ISBN 0-7923-6393-0, August 2000

Volume 245: *Waves in Dusty Space Plasmas,* by Frank Verheest
Paperback, ISBN 1-4020-0373-0, November 2001
Hardbound, ISBN 0-7923-6232-2, April 2000

Volume 244: *The Universe*, edited by Naresh Dadhich, Ajit Kembhavi
Hardbound, ISBN 0-7923-6210-1, August 2000

Volume 243: *Solar Polarization*, edited by K.N. Nagendra, Jan Olof Stenflo
Hardbound, ISBN 0-7923-5814-7, July 1999

Volume 242: *Cosmic Perspectives in Space Physics*, by Sukumar Biswas
Hardbound, ISBN 0-7923-5813-9, June 2000

Missing volume numbers have not yet been published.
For further information about this book series we refer you to the following web site: http://www.wkap.nl/prod/s/ASSL

To contact the Publishing Editor for new book proposals:
Dr. Harry (J.J.) Blom: harry.blom@wkap.nl